Handbook of Genetics and Genomics

Handbook of Genetics and Genomics

Editor: Rosanna Mann

R CALLISTO
REFERENCE

www.callistoreference.com

Callisto Reference,
118-35 Queens Blvd., Suite 400,
Forest Hills, NY 11375, USA

Visit us on the World Wide Web at:
www.callistoreference.com

ISBN: 978-1-64116-110-7 (Hardback)

Cataloging-in-Publication Data

Handbook of genetics and genomics / edited by Rosanna Mann.
    p. cm.
Includes bibliographical references and index.
ISBN 978-1-64116-110-7
1. Genetics. 2. Genomics. 3. Genomes. 4. Molecular genetics. I. Mann, Rosanna.
QH430 .H36 2019
576.5--dc23

Table of Contents

Preface

Genetics is a field of biology that is concerned with the study of genes, variation and heredity in living organisms. Trait inheritance and molecular inheritance are the pillars of genetic studies. However, modern genetics has now extended into structural, functional and behavioral investigations of genes. Genetic processes in combination with the environmental factors influence behavior and traits. Genomics is an interdisciplinary field of science that studies the structure, behavior and evolution of the entire set of DNA of an organism. This book traces the progress of these fields and highlights some of its key concepts and applications. It strives to provide a fair idea about these disciplines and to help develop a better understanding of the latest advances. Students, researchers, experts and all associated with the fields of genetics and genomics will benefit alike from this book.

This book is the end result of constructive efforts and intensive research done by experts in this field. The aim of this book is to enlighten the readers with recent information in this area of research. The information provided in this profound book would serve as a valuable reference to students and researchers in this field.

At the end, I would like to thank all the authors for devoting their precious time and providing their valuable contribution to this book. I would also like to express my gratitude to my fellow colleagues who encouraged me throughout the process.

Editor

Epistasis, physical capacity-related genes and exceptional longevity: *FNDC5* gene interactions with candidate genes *FOXOA3* and *APOE*

Noriyuki Fuku[1][*][†], Roberto Díaz-Peña[2,3][†], Yasumichi Arai[4], Yukiko Abe[4], Hirofumi Zempo[1], Hisashi Naito[1], Haruka Murakami[5], Motohiko Miyachi[5], Carlos Spuch[6], José A. Serra-Rexach[8], Enzo Emanuele[7], Nobuyoshi Hirose[1] and Alejandro Lucia[9]

Abstract

Background: Forkhead box O3A (*FOXOA3*) and apolipoprotein E (*APOE*) are arguably the strongest gene candidates to influence human exceptional longevity (EL, i.e., being a centenarian), but inconsistency exists among cohorts. Epistasis, defined as the effect of one locus being dependent on the presence of 'modifier genes', may contribute to explain the missing heritability of complex phenotypes such as EL. We assessed the potential association of epistasis among candidate polymorphisms related to physical capacity, as well as antioxidant defense and cardiometabolic traits, and EL in the Japanese population. A total of 1565 individuals were studied, subdivided into 822 middle-aged controls and 743 centenarians.

Results: We found a *FOXOA3* rs2802292 T-allele-dependent association of fibronectin type III domain-containing 5 (*FDNC5*) rs16835198 with EL: the frequency of carriers of the *FOXOA3* rs2802292 T-allele among individuals with the rs16835198 GG genotype was significantly higher in cases than in controls ($P < 0.05$). On the other hand, among non-carriers of the *APOE* 'risk' ε4-allele, the frequency of the *FDNC5* rs16835198 G-allele was higher in cases than in controls (48.4% vs. 43.6%, $P < 0.05$). Among carriers of the 'non-risk' *APOE* ε2-allele, the frequency of the rs16835198 G-allele was higher in cases than in controls (49% vs. 37.3%, $P < 0.05$).

Conclusions: The association of *FDNC5* rs16835198 with EL seems to depend on the presence of the *FOXOA3* rs2802292 T-allele and we report a novel association between *FNDC5* rs16835198 stratified by the presence of the *APOE* ε2/ε4-allele and EL. More research on 'gene*gene' and 'gene*environment' effects is needed in the field of EL.

Keywords: Exceptional longevity, Centenarians, FOXO3A, FNDC5, APOE, Ageing

* Correspondence: noriyuki.fuku@nifty.com
†Equal contributors
[1]Graduate School of Health and Sports Science, Juntendo University, Chiba, Japan
Full list of author information is available at the end of the article

Background

Exceptional longevity (EL), defined as reaching 100+ years of age, is a generally accepted model of healthy aging with major age-related diseases usually delayed [1], and sometimes even avoided, in centenarians [2]. Age at death during adulthood has a heritability of ~ 25% [3] and consequently EL is likely to be, at least in part, a heritable phenotype. However, genome wide association studies (GWAS) of individuals with EL have mostly yielded poor results, with single nucleotide polymorphisms (SNPs) in the apolipoprotein E (*APOE*) and/or forkhead box O3A (*FOXOA3*) genes being usually the only variants achieving genome-wide significance [3], although there is no unanimity for *FOXOA3* [4]. The *APOE* ε4-allele, which is linked to a higher risk of cardiovascular and Alzheimer's disease [5, 6], is associated with a lower likelihood of reaching EL, whereas the ε2-allele seems to show a positive association with EL and is more frequently found in centenarians than in younger controls [7]. Nonetheless, the aforementioned associations can vary by population [7]. Similarly, there is evidence that genetic polymorphisms in the *FOXO3A* gene can be associated with EL, especially the rs2802292 SNP [8–12], but such an association has not been corroborated in some cohorts [13, 14]. On the other hand, aging is inevitably associated with a decline physical function, especially in muscle mass and function (i.e., sarcopenia), with an acceleration of this process increasing the risk of mortality [15]. As such, there is a rationale for postulating the potential influence on EL of several variations in genes that are candidates to influence physical capacity and muscle function; however, up to date such candidate genes [e.g., α-actinin-3 (*ACTN3*), thyrotropin-releasing hormone (*TRHR*) or acyl-CoA synthetase long-chain family member 1 (*ACSL1*)], have not shown individual associations with EL [16, 17].

Epistasis, defined as the effect of one locus on a given phenotype being dependent on the presence of one or more 'modifier genes', may contribute to explain the missing heritability of complex phenotypes [18]. Though research on the role of epistasis in EL has been rather limited in the past [19], evidence for an impact of genetic interactions in determining human longevity is quickly growing. For instance, longevity may be partly determined by epistatic interactions involving mitochondrial DNA (mtDNA) loci [20]. Also, a combination of functional SNPs within adenosine deaminase (*ADA*) and tumor necrosis factor alpha (*TNF-α*) genes can influence life-expectancy in a gender-specific manner [21]. More recently, Tan et al. reported that interactions of SNPs in the *FOXO* gene family influences longevity [22]. To our knowledge, only two studies have included the *APOE* gene for evaluating gene x gene effects on longevity [23, 24]. Jazwinski et al. reported that the GTPase HRAS (also

known as transforming protein p21, *HRAS1*) and ceramide synthase 1 (also known as LAG1, longevity assurance homolog, *LASS1*) genes interact with the *APOE* gene to reduce age-related increases in lipotoxic events, and haplotypes conformed by the two genes are associated with EL [23]. In the other study, epistasis analysis was performed between *APOE* haplotypes and haptoglobulin gene (*HP*) functional polymorphisms, with *HP* *1/*1 genotype protecting *APOE* ε4-carriers from age-related negative selection; in light of these results, the authors called for further research on *APOE/HP* interactions in age-related diseases such as Alzheimer's and Parkinson's disease [24].

Japan has the longest life expectancy in the world, as well as the highest number of individuals reaching EL [25]. Thus, Japanese long-lived people represent an attractive model to study potential genetic contributors to EL. The purpose of the present study was to assess whether the interaction between variants in genes that are candidates to influence physical capacity and energy metabolism [*ACTN3*, *TRHR*, *ACSL1*], interact with *APOE* and *FOXO3A* genes, and potentially associate with EL in the Japanese population. We also studied the potential interaction with candidate genes involved in antioxidant defense [glutathione peroxidase 1 (*GPX1*), superoxide dismutase 2 (*SOD2*)] and cardiometabolic traits [fibronectin type III domain-containing 5 (*FNDC5*), cyclin-dependent kinase inhibitor 2B antisense noncoding RNA (*CDKN2B-AS1*)],

Results

The observed genotypic distributions of SNPs were consistent with Hardy–Weinberg Equilibrium (HWE) expectations both in controls and centenarians ($P > 0.05$). Table 1 shows the single marker allelic association results of the 12 SNPs included in the study. The frequency of the *APOE* ε2-allele was higher in cases than in controls, whereas ε4-allele frequency was lower in the former (all $P < 0.01$), as shown previously by us using part of the present cohort [26]. None of the other SNPs, including *FOXOA3* rs2802292, was associated with EL, and genotypes did not differ significantly between cases and controls in the recessive, dominant or additive models (all $P > 0.05$) (data not shown). No significant differences were observed in sex-based analyses (data not shown).

With regards to epistasis, we found a *FOXOA3* rs2802292 T-allele-dependent association of fibronectin type III domain-containing 5 (*FDNC5*) rs16835198 with EL, that is, among individuals with the rs16835198 GG genotype, the frequency of carriers of the *FOXOA3* rs2802292 T-allele was significantly higher in cases than in controls ($P < 0.05$; Table 2). On the other hand, among non-carriers of the *APOE* 'risk' ε4-allele, the frequency of the *FDNC5* rs16835198 G-allele was higher in

Table 1 Single marker allelic association results of the 12 genetic variants included in the study

Chromosome	SNP	Gene	Position	Minor allele nucleotide	Minor allele frequency		P_{BONF}	OR	lower 95%CI	Upper 95%CI
					Centenarians	Controls				
1	rs16835198	FNDC5	32,861,080	G	0.48	0.45	0.488	1.16	1.00	1.33
2	rs1050450	GPX1	49,357,401	T	0.08	0.08	1	0.95	0.73	1.25
4	rs6552828	ASCL1	1.85E + 08	G	0.40	0.41	1	0.95	0.82	1.10
6	rs4880	SOD2	1.6E + 08	G	0.13	0.14	1	0.94	0.77	1.16
8	rs7832552	TRHR	1.09E + 08	T	0.50	0.50	1	0.97	0.85	1.12
8	rs7460	PTK2	1.41E + 08	T	0.36	0.36	1	1.00	0.860	1.15
8	rs7843014	PTK2	1.41E + 08	A	0.25	0.27	1	0.88	0.75	1.04
9	rs1333049	CDKN2B-AS1	22,125,504	C	0.47	0.46	1	1.05	0.91	1.21
9	rs2802292	FOXO3A	108,587,315	G	0.29	0.28	1	1.05	0.90	1.23
11	rs1815739	ACTN3	66,560,624	C	0.46	0.47	1	0.96	0.84	1.11
19	rs429358	APOE	44,908,684	C	0.05	0.11	6.19×10^{-10}	0.39	0.29	0.52
19	rs7412	APOE	44,908,882	T	0.07	0.04	0.002	1.80	1.32	2.46

Allele frequencies were compared using chi-square statistics; ORs and 95% confidence interval (95% CI) were calculated using Plink software

Abbreviations: 95%CI 95% confidence interval, *OR* odds ratio, *P_{BONF}* P-value after Bonferroni correction, *SNP* single nucleotide polymorphism

cases than in controls (48.4% vs. 43.6%, $P < 0.05$; Table 3). In turn, among carriers of the 'non-risk' *APOE* ε2-allele, the frequency of the rs16835198 G-allele was higher in cases than in controls (49% vs. 37.3%, $P < 0.05$). No other significant gene interaction was found to be associated with EL.

Discussion

Excluding *APOE*, the present findings do not show replicable association of individual candidate genes, included those related to physical capacity and muscle function, with EL. However, we also assessed the effects of interactions among candidate polymorphisms on the one hand, and EL, on the other hand. In this regard, we report an association of *FDNC5* rs16835198 with EL that seems to depend on the presence of the *FOXOA3* rs2802292 T-allele. Moreover, we found a novel association between *FNDC5* rs16835198 stratified by the presence of the *APOE* ε2/ε4-allele, and EL. We believe this is an interesting finding because *FNDC5* gene encodes the precursor of irisin which, although there is controversy [27], was recently identified a myokine -that is, a molecule released by muscles [28]. Irisin induces expression

of uncoupling protein 1 and other brown adipose tissue-associated genes [partly via increased peroxisome proliferators-activated receptor α (PPAR-α)] in white adipocytes, and thus increases thermogenesis and switching of these cells towards a brown fat-like phenotype [28]. In fact irisin has been postulated as a therapeutic agent against cardiometabolic disorders and a major component of the 'exercise polypill' [29].

Several mechanisms may link genetic variations in *FOXO3A* and *FNDC5* genes with EL. For example, dysregulation of the nutrient-sensing somatotropic axis [comprising growth hormone and its secondary mediator, insulin-like growth factor-1 (IGF-1)] is a major hallmark of human aging [30]. IGF-1 and insulin signaling collectively represent the "insulin and IGF-1 signaling" pathway. Among the multiple targets of this pathway is the FOXO family of transcription factors, which are also involved in aging and show striking evolutionary conservation [31]. High serum irisin levels may contribute to successful aging, with circulating levels of this biomarker being in fact significantly higher in disease-free centenarians than in young healthy controls and being particularly higher than in young patients with acute

Table 2 Genotypic frequency distribution of *FNDC5* rs16835198 according to the *FDNC5* rs16835198 T-allele presence

FDNC5 (rs16835198) Genotype	FOXO3A Presence of rs2802292 T-allele		P-value	OR (95% CI)
	Controls (n = 754)	Centenarians (n = 668)		
GG	147 (19.5%)	162 (24.3%)	0.035	1.32 (1.03–1.70)
GT	378 (50.1%)	323 (48.3%)	NS	–
TT	229 (30.4%)	183 (27.4%)	NS	–

Abbreviations: 95%CI 95% confidence interval, *NS* not significant, *OR* odds ratio

Table 3 Allele frequency distribution of *FNDC5* rs16835198 alleles according to *APOE* ε2/ε4 status in the study population

FNDC5 (rs16835198) allele		Absence of *APOE* ε4-allele[a]		Presence of *APOE* ε4-allele[a]	
		Controls (2n = 1306)	Centenarians (2n = 1340)	Controls (2n = 336)	Centenarians (2n = 128)
G	N	570	648	165	62
	%	43.6%	48.4%	49.1%	48.4%
	P-value	0.0167		NS	
	OR (95% CI)	1.21 (1.04, 1.41)		–	
T	N	736	692	171	66
	%	56.4%	51.6%	50.9%	51.6%
	P-value	0.0167		NS	
	OR (95% CI)	0.83 (0.71–0.96)		–	
FNDC5 (rs16835198) allele		Absence of *APOE* ε2-allele[a]		Presence of *APOE* ε2-allele[a]	
		Controls (2n = 1508)	Centenarians (2n = 1258)	Controls (2n = 134)	Centenarians (2n = 210)
G	n	685	607	50	103
	%	45.4%	48.3%	37.3%	49%
	P-value	NS		0.035	
	OR (95% CI)	–		1.62 (1.04–2.52)	
T	n	823	651	84	107
	%	54.6%	51.7%	62.7%	51%
	P-value	NS		0.035	
	OR (95% CI)	–		0.62 (0.40–0.96)	

Abbreviations: 95%CI 95% confidence interval, *NS* not significant, *OR* odds ratio
[a]All individuals underwent genotyping for rs429358 and rs7412. The 3 major isoforms of human APOE gene (E2, E3, and E4) coded by 3 alleles (ε2, ε3, and ε4) differ in amino acid sequence at 2 sites, residue 112 (rs429358) and residue 158 (rs7412)

myocardial infarction [32]. Moreover, genetic variants in the *FNDC5* gene have been associated with in vivo insulin sensitivity [33], and aging is associated with alterations in insulin sensitivity/signaling [34]. Irisin might also play a significant role in reducing the risk of obesity or several related diseases [35]. In a previous report, we found no significant association between genotype and allele frequencies of polymorphisms in *FNDC5* and EL [36]. However, the rs16835198 G-allele was associated with a trend towards lower luciferase gene repoter activity in vitro [36].

Insulin sensitivity could be a link explaining the association of the combination of the two loci in *FOXO3A* and *FNDC5*, as well as in *APOE* and *FNDC5*, with EL. Insulin metabolism as well as insulin-altering therapies in Alzheimer's disease are modulated by *APOE* status [37, 38], while irisin reduces diet-induced obesity and insulin resistance in vivo [28]. The *FNDC5* rs16835198 G-allele could favor longevity in combination with the *APOE* ε2-variant, contributing to situations in which rs16835198 SNP distribution is found among long-lived subjects in proportions similar to those found in younger controls. Thus, the conditional effects of genes on the phenotype of aging could play a role in longevity. Most studies show a lower frequency of the *APOE* ε4-allele in centenarians than in younger controls, while the

APOE ε2-allele might be more frequent in centenarians than in younger controls [7]. A "gene x gene" interaction would introduce a modifying allele in another locus, changing the impact of the risk/protective allele from deleterious to beneficial, or vice versa. Here, we report a significant effect of the rs16835198-G allele in the *FNDC5* gene on the association between *APOE* ε2 carriage/non-carriage and EL. There are some limitations in our study. First, our results should be ideally replicated in one or more ethnically and geographically-independent cohorts to account for potential differences in 'gene x environment' interactions. Another drawback of case: control designs as the present one is the choice of an appropriate control group, with demographic factors, notably differences in year of birth between centenarians and controls, being potential confounders (e.g., the centenarians and controls of our study were born in the early 1900s and after 1930, respectively) [39]. Differences in gender distribution between cases and controls are also to be accounted for because genetic factors influence survival at advanced age in a sex-specific manner [40]. In this regard, the potential demographic biases of cross-sectional comparisons of genotype/allele frequencies between controls and long lived individuals could be overcome by adding more complete demographic information to genetic data, allowing for estimation of survival rates related to candidate genes [40–42]. Because we have only analyzed one SNP within the *FOXO3A* and *FNDC5* genes, further research with different variants of this or other genes involved in insulin-IGF-1, and insulin signaling collectively representing the "insulin and IGF-1 signaling" pathway, must be undertaken. On the other hand, the need for using independent populations for replication may generate some controversy. Some variants may have opposing consequences on the phenotype of interest, and the effect of gene x gene interactions on human health and lifespan might depend on several factors (e.g., internal and external exposure, including medication or different genetic backgrounds of individuals comprising the populations).

Conclusion

We found an association between *FDNC5* rs16835198 stratified by the presence of the *FOXOA3* rs2802292 T-allele, and EL, as well as between *FNDC5* rs16835198 stratified by the presence of the *APOE* alleles, and EL. Our results suggest the need for further investigation on the possible influence of *FOXO3A/FNDC5* interactions and *APOE/FNDC5* interactions with EL, but also with other age-related diseases such as atherosclerosis, Alzheimer's, or Parkinson's disease. Moreover, identification of gene x gene and gene x environmental factors effects on lifespan may significantly improve our understanding of the association between genetic and non-genetic regulators of aging.

Methods

Study population

A total of 1565 individuals were recruited: 822 controls (aged 23–65 years; 604 women, 218 men) and 743 cases (centenarians aged 100–115 years; 623 women, 1120 men). The controls were enrolled from the Nutrition and EXercise Intervention Study (NEXIS) registered on ClinicalTrials.gov (Identifier: NCT 00926744). Inclusion criteria for the control group were being a man or woman aged 23–65 years with no history of stroke, cardiovascular disease, chronic renal failure, or walking difficulties related to knee or back pain [43]. Cases were collected from two cohorts that have been described in detail elsewhere [44]: the Tokyo Centenarians Study (TCS) and the Semi-Supercentenarians Study in Japan (SSC-J). The prevalence rates of hypertension, coronary artery disease and dementia in the Japanese centenarians were 63.6, 28.8 and 59.4%, respectively [26].

Genotyping

Total DNA was extracted from venous blood with the QIAamp DNA Blood Maxi Kit (Qiagen, Hilden, Germany). SNP genotyping was performed using the following TaqMan® genotyping assays (Applied Biosystems,

Foster City, CA): C__34204885_10 for rs16835198, Custom TaMan® Assay for rs1050450, C__30469648_10 for rs6552828, C___8709053_10 for rs4880, C__29085798_10 for rs7832552, C____243385_10 for rs7460, C__11605645_10 for rs7843014, C___1754666_10 for rs1333049, C___1841568_10 for rs2802292, C____590 093_1 for rs1815739, C____904973_10 for rs7412 and C___3084793_20 for rs429358. Several quality control procedures, such as repeating the genotyping on a random 10% of the samples were carried out to assure no discrepant results in samples.

Statistical analysis

Statistical analysis of high-density SNP genotyping data was carried out as follows: allele frequencies in cases and controls were compared using chi-square statistics, and odds ratios (OR) and 95% confidence intervals (95% CI) were calculated using Plink software [45]. Bonferroni correction was applied by dividing the *P*-value by the number of SNPs tested to give the corrected *P*-value. Allele frequencies obtained for each SNP were tested for deviations from HWE expectations. The combined effects of *APOE* and *FOXO3A* alleles and SNPs studied on the risk of EL were analyzed by binary logistic regression with SPSS v.22 statistical software (IBM, Somers, NY, USA) and OR and *P* values were calculated.

Abbreviations

ACSL1: Acyl-CoA synthetase long-chain family member 1; ACTN3: Actinin-3; ADA: Adenosine deaminase; APOE: Apolipoprotein E; CDKN2B-AS1: Cyclin-dependent kinase inhibitor 2B antisense noncoding RNA; EL: Exceptional longevity; FNDC5: Fibronectin type III domain-containing 5; FOXOA3: Forkhead box O3A; GPX1: Glutathione peroxidase 1; GWAS: Genome wide association studies; HP: Haptoglobulin gene; HWE: Hardy–Weinberg Equilibrium; IGF-1: Insulin-like growth factor-1; mtDNA: Mitochondrial DNA; NEXIS: Nutrition and EXercise Intervention Study; OR: Odds ratios; SNPs: Single nucleotide polymorphisms; SOD2: Superoxide dismutase 2; SSC-J: Semi-Supercentenarians Study in Japan; TCS: Tokyo Centenarians Study; TNF-α: Tumor necrosis factor alpha; TRHR: Thyrotropin-releasing hormone

Acknowledgements

The authors acknowledge Dr. Kenneth McCreath for his editorial work.

Funding

Publication of this article is supported by grants from the Grant-in-Aid for Scientific Research (B) (15H03081 to N.F.) and Challenging Exploratory Research (16 K13052 to N.F.) programs of the Ministry of Education, Culture, Sports, Science and by a Grant-in-Aid for Scientific Research from the Ministry of Health, Labor, and Welfare of Japan (to M. M). Research by A. Lucia is supported by grants from the Spanish Ministry of Economy and Competitiveness and [*Fondo de Investigaciones Sanitarias* (FIS)] and *Fondos Feder* (grant number PI15/00558).

About this supplement

This article has been published as part of *BMC Genomics* Volume 18 Supplement 8, 2017: Proceedings of the 34th FIMS World Sports Medicine Congress. The full contents of the supplement are available online at https://bmcgenomics.biomedcentral.com/articles/supplements/volume-18-supplement-8.

Authors' contributions

All authors were involved in drafting the article or revising it critically for important intellectual content, and all authors approved the final version to be published. Study conception and design. DPR, FN and LA. Acquisition of data. Arai Y, Abe Y, ZH, Naito H, MH, MM and Hirose N. Analysis and interpretation of data. DPR, SRJA, SC and EE.

Competing interests

The authors declare that they have no competing interests.

Author details

[1]Graduate School of Health and Sports Science, Juntendo University, Chiba, Japan. [2]Hospital Universitari Institut Pere Mata, IISPV, URV. CIBERSAM, Reus, Spain. [3]Facultad de Ciencias de la Salud, Universidad Autónoma de Chile, Talca, Chile. [4]Center for Supercentenarian Medical Research, Keio University School of Medicine, Tokyo, Japan. [5]Department of Physical Activity Research; National Institutes of Biomedical Innovation, Health and Nutrition, Tokyo, Japan. [6]Neurology Group, Galicia Sur Health Research Institute (IIS Galicia Sur), Centro de investigación biomédica en red del área de salud mental (CIBERSAM), Vigo, Spain. [7]2E Science, Robbio, (PV), Italy. [8]Centro de investigación biomédica en Envejecimiento y Fragilidad (CIBERFES), Madrid, Spain. [9]European University and Research Institute i+12, Madrid, Spain.

References

1. Soerensen M, Nygaard M, Debrabant B, Mengel-From J, Dato S, Thinggaard M, et al. No association between variation in longevity candidate genes and aging-related phenotypes in oldest-old Danes. Exp Gerontol. 2016;78:57–61. [cited 2016 Nov 5] Available from: http://www.ncbi.nlm.nih.gov/pubmed/26946122

2. Terry DF, Sebastiani P, Andersen SL, Perls TT. Disentangling the roles of disability and morbidity in survival to exceptional old age. Arch Intern Med. 2008;168:277–83. Available from: http://www.pubmedcentral.nih.gov/articlerender.fcgi?artid=2895331&tool=pmcentrez&rendertype=abstract

3. Brooks-Wilson AR. Genetics of healthy aging and longevity. Hum Genet. 2013;132:1323–38.

4. Sebastiani P, Gurinovich A, Bae H, Andersen S, Malovini A, Atzmon G, et al. Four Genome-Wide Association Studies Identify New Extreme Longevity Variants. J Gerontol A. 2017;72:1453-64. Available from: https://academic.oup.com/biomedgerontology/article/3072309/Four.

5. Kumar NT, Liestol K, Loberg EM, Reims HM, Brorson SH, Maehlen J. The apolipoprotein E polymorphism and cardiovascular diseases–an autopsy study. Cardiovasc Pathol. 2012;21:461–9. Available from: http://www.ncbi.nlm.nih.gov/pubmed/22440829

6. Liu C-C, Liu C-C, Kanekiyo T, Xu H, Bu G. Apolipoprotein E and Alzheimer disease: risk, mechanisms and therapy. Nat Rev Neurol. 2013;9:106–18. Available from: https://doi.org/10.1038/nrneurol.2012.263

7. Garatachea N, Marín PJ, Santos-Lozano A, Sanchis-Gomar F, Emanuele E, Lucia A. The ApoE gene is related with exceptional longevity: a systematic review and meta-analysis. Rejuvenation Res. 2015;18:3–13. Available from: http://online.liebertpub.com/doi/abs/10.1089/rej.2014.1605, http://www.ncbi.nlm.nih.gov/pubmed/25385258.

8. Willcox BJ, Donlon T a, He Q, Chen R, Grove JS, Yano K, et al. FOXO3A genotype is strongly associated with human longevity. Proc Natl Acad Sci U S A. 2008;105:13987–92.

9. Anselmi CV, Malovini A, Roncarati R, Novelli V, Villa F, Condorelli G, et al. Association of the FOXO3A locus with extreme longevity in a southern Italian centenarian study. Rejuvenation Res. 2009;12:95–104. Available from: http://online.liebertpub.com/doi/abs/10.1089/rej.2008.0827, http://www.ncbi.nlm.nih.gov/pubmed/19415983.

10. Li Y, Wang WJ, Cao H, Lu J, Wu C, Hu FY, et al. Genetic association of FOXO1A and FOXO3A with longevity trait in Han Chinese populations. Hum Mol Genet. 2009;18:4897–904.

11. Sun L, Hu C, Zheng C, Qian Y, Liang Q, Lv Z, et al. FOXO3 variants are beneficial for longevity in southern Chinese living in the Red River basin: a case–control study and meta-analysis. Sci Rep. 2015;5:9852. Available from: http://www.nature.com/srep/2015/150408/srep09852/full/srep09852.html

12. Soerensen M, Dato S, Christensen K, McGue M, Stevnsner T, Bohr VA, et al. Replication of an association of variation in the FOXO3A gene with human longevity using both case–control and longitudinal data. Aging Cell. 2010;9:1010–7.

13. Flachsbart F, Caliebe A, Kleindorp R, Blanché H, von Eller-Eberstein H, Nikolaus S, et al. Association of FOXO3A variation with human longevity confirmed in German centenarians. Proc Natl Acad Sci U S A. 2009;106:2700–5.

14. Fuku N, Díaz-Peña R, Arai Y, Abe Y, Pareja-Galeano H, Sanchis-Gomar F, et al. rs2802292 polymorphism in the FOXO3A gene and exceptional longevity in two ethnically distinct cohorts. Maturitas. 2016;92:110–4.

15. Fuku N, Alis R, Yvert T, Zempo H, Naito H, Abe Y, et al. Muscle-related polymorphisms (mstn rs1805086 and actn3 rs1815739) are not associated with exceptional longevity in Japanese centenarians. PLoS One. 2016;11:e0166605.

16. Fuku N, He ZH, Sanchis-Gomar F, Pareja-Galeano H, Tian Y, Arai Y, et al. Exceptional longevity and muscle and fitness related genotypes: a functional in vitro analysis and case–control association replication study with SNPs THRH rs7832552, IL6 rs1800795 and ACSL1 rs6552828. Front Aging Neurosci. 2015;7:59.

17. Garatachea N, Pareja-Galeano H, Sanchis-Gomar F, Santos-Lozano A, Fiuza-Luces C, Morán M, et al. Exercise attenuates the major hallmarks of aging. Rejuvenation Res. 2015;18:57–89. Available from: http://online.liebertpub.com/doi/abs/10.1089/rej.2014.1623

18. Cordell HJ. Epistasis: what it means, what it doesn't mean, and statistical methods to detect it in humans. Hum Mol Genet. 2002;11:2463–8. Available from: http://www.ncbi.nlm.nih.gov/pubmed/12351582

19. Ukraintseva S, Yashin A, Arbeev K, Kulminski A, Akushevich I, Wu D, et al. Puzzling role of genetic risk factors in human longevity: "risk alleles" as pro-longevity variants. Biogerontology. 2016;17:109–27.

20. Niemi A-K, Moilanen JS, Tanaka M, Hervonen A, Hurme M, Lehtimäki T, et al. A combination of three common inherited mitochondrial DNA polymorphisms promotes longevity in Finnish and Japanese subjects. Eur J Hum Genet. 2005;13:166–70.

21. Napolioni V, Carpi FM, Giannì P, Sacco R, Di Blasio L, Mignini F, et al. Age- and gender-specific epistasis between ADA and TNF-α influences human life-expectancy. Cytokine. 2011;56:481–8.

22. Tan Q, Soerensen M, Kruse TA, Christensen K, Christiansen L. A novel permutation test for case-only analysis identifies epistatic effects on human longevity in the FOXO gene family. Aging Cell. 2013;12:690–4.

23. Jazwinski SM, Kim S, Dai J, Li L, Bi X, Jiang JC, et al. HRAS1 and LASS1 with APOE are associated with human longevity and healthy aging. Aging Cell. 2010;9:698–708.

24. Napolioni V, Giannì P, Carpi FM, Predazzi IM, Lucarini N. APOE haplotypes are associated with human longevity in a Central Italy population: evidence for epistasis with HP 1/2 polymorphism. Clin Chim Acta. 2011;412:1821–4.

25. Santos-Lozano A, Sanchis-Gomar F, Pareja-Galeano H, Fiuza-Luces C, Emanuele E, Lucia A, et al. Where are supercentenarians located? A worldwide demographic study. Rejuvenation Res. 2015;18:14–9. Available from: http://www.ncbi.nlm.nih.gov/pubmed/25386976

26. Garatachea N, Emanuele E, Calero M, Fuku N, Arai Y, Abe Y, et al. ApoE gene and exceptional longevity: insights from three independent cohorts. Exp Gerontol. 2014;53:16–23.

27. Timmons JA, Baar K, Davidsen PK, Atherton PJ. Is irisin a human exercise gene? Nature. 2012;488:E9–10.

28. Boström P, Wu J, Jedrychowski MP, Korde A, Ye L, Lo JC, et al. A PGC1-α-dependent myokine that drives brown-fat-like development of white fat and thermogenesis. Nature. 2012;481:463–8. [cited 2016 Nov 9] Available from: http://www.pubmedcentral.nih.gov/articlerender.fcgi?artid=3522098&tool=pmcentrez&rendertype=abstract

29. Fiuza-Luces C, Garatachea N, Berger NA, Lucia A. Exercise is the real Polypill. Physiology. 2013;28:330–58. Available from: http://physiologyonline.physiology.org/cgi/doi/10.1152/physiol.00019.2013

30. López-Otín C, Galluzzi L, Freije JMP, Madeo F, Kroemer G. Metabolic control of longevity. Cell. 2016;166:802–21.

31. Kenyon CJ. The genetics of ageing. Nature. 2010;464:504–12.

32. Emanuele E, Minoretti P, Pareja-Galeano H, Sanchis-Gomar F, Garatachea N, Lucia A. Serum irisin levels, precocious myocardial infarction, and healthy exceptional longevity. Am J Med. 2014;127:888–90. Available from: http://www.ncbi.nlm.nih.gov/pubmed/24813865.

33. Staiger H, Böhm A, Scheler M, Berti L, Machann J, Schick F, et al. Common genetic variation in the human FNDC5 locus, encoding the novel muscle-derived "Browning" factor Irisin, determines insulin sensitivity. PLoS One. 2013;8:e61903.

34. Barbieri M, Gambardella A, Paolisso G, Varricchio M. Metabolic aspects of the extreme longevity. Exp Gerontol. 2008;43:74–8.

35. Spiegelman BM. Banting lecture 2012: regulation of adipogenesis: toward new therapeutics for metabolic disease. Diabetes. 2013;62:1774–82.

36. Sanchis-Gomar F, Garatachea N, He ZH, Pareja-Galeano H, Fuku N, Tian Y, et al. FNDC5 (irisin) gene and exceptional longevity: a functional replication study with rs16835198 and rs726344 SNPs. Age (Dordr). 2014;36:9733.

37. Craft S, Asthana S, Schellenberg G, Baker L, Cherrier M, Boyt AA, et al. Insulin effects on glucose metabolism, memory, and plasma amyloid precursor protein in Alzheimer's disease differ according to apolipoprotein-E genotype. Ann N Y Acad Sci. 2000;903:222–8. Available from: http://www.ncbi.nlm.nih.gov/entrez/query.fcgi?db=pubmed&cmd=Retrieve&dopt=AbstractPlus&list_uids=10818510

38. Aisen PS, Berg JD, Craft S, Peskind ER, Sano M, Teri L, et al. Steroid-induced elevation of glucose in Alzheimer's disease: relationship to gender, apolipoprotein E genotype and cognition. Psychoneuroendocrinology. 2003;28:113–20.

39. Lewis SJ. Methodological problems in genetic association studies of longevity–the apolipoprotein E gene as an example. Int J Epidemiol. 2004;33:962–70. [cited 2016 Nov 9]Available from: http://www.ncbi.nlm.nih.gov/pubmed/15319409

40. Passarino G, Montesanto A, Dato S, Giordano S, Domma F, Mari V, et al. Sex and age specificity of susceptibility genes modulating survival at old age. Hum Hered. 2006;62:213–20.

41. Yashin AI, De Benedictis G, Vaupel JW, Tan Q, Andreev KF, Iachine IA, et al. Genes, demography, and life span: the contribution of demographic data in genetic studies on aging and longevity. Am J Hum Genet. 1999;65:1178–93. Available from: http://www.cell.com/article/S0002929707626214/fulltext

42. Dato S, Carotenuto L, Benedictis G. Genes and longevity: a genetic-demographic approach reveals sex- and age-specific gene effects not shown by the case–control approach (APOE and HSP70.1 loci). Biogerontology. 2007;8:31–41.

43. Murakami H, Iemitsu M, Fuku N, Sanada K, Gando Y, Kawakami R, et al. The Q223R polymorphism in the leptin receptor associates with objectively measured light physical activity in free-living Japanese. Physiol Behav. 2014;129:199–204.

44. Gondo Y, Hirose N, Arai Y, Inagaki H, Masui Y, Yamamura K, et al. Functional status of centenarians in Tokyo, Japan: developing better phenotypes of exceptional longevity. J Gerontol A Biol Sci Med Sci. 2006;61:305–10. [cited 2016 Nov 4] Available from: http://www.ncbi.nlm.nih.gov/pubmed/16567382.

45. Purcell S, Neale B, Todd-Brown K, Thomas L, Ferreira MAR, Bender D, et al. PLINK: a tool set for whole-genome association and population-based linkage analyses. Am J Hum Genet. 2007;81:559–75.

2

Identification of recent cases of hepatitis C virus infection using physical-chemical properties of hypervariable region 1 and a radial basis function neural network classifier

James Lara[*], Mahder Teka and Yury Khudyakov

Abstract

Background: Identification of acute or recent hepatitis C virus (HCV) infections is important for detecting outbreaks and devising timely public health interventions for interruption of transmission. Epidemiological investigations and chemistry-based laboratory tests are 2 main approaches that are available for identification of acute HCV infection. However, owing to complexity, both approaches are not efficient. Here, we describe a new sequence alignment-free method to discriminate between recent (R) and chronic (C) HCV infection using next-generation sequencing (NGS) data derived from the HCV hypervariable region 1 (HVR1).

Results: Using dinucleotide auto correlation (DAC), we identified physical-chemical (PhyChem) features of HVR1 variants. Significant ($p < 9.58 \times 10^{-4}$) differences in the means and frequency distributions of PhyChem features were found between HVR1 variants sampled from patients with recent vs chronic (R/C) infection. Moreover, the R-associated variants were found to occupy distinct and discrete PhyChem spaces. A radial basis function neural network classifier trained on the PhyChem features of intra-host HVR1 variants accurately classified R/C-HVR1 variants (classification accuracy (CA) = 94.85%; area under the ROC curve, AUROC = 0.979), in 10-fold cross-validation). The classifier was accurate in assigning individual HVR1 variants to R/C-classes in the testing set (CA = 84.15%; AUROC = 0.912) and in detection of infection duration (R/C-class) in patients (CA = 88.45%). Statistical tests and evaluation of the classifier on randomly-labeled datasets indicate that classifiers' CA is robust ($p < 0.001$) and unlikely due to random correlations (CA = 59.04% and AUROC = 0.50).

Conclusions: The PhyChem features of intra-host HVR1 variants are strongly associated with the duration of HCV infection. Application of the PhyChem biomarkers to models for detection of the R/C-state of HCV infection in patients offers a new opportunity for detection of outbreaks and for molecular surveillance. The method will be available at https://webappx.cdc.gov/GHOST/ to the authenticated users of Global Hepatitis Outbreak and Surveillance Technology (GHOST) for further testing and validation.

* Correspondence: xzl5@cdc.gov
Division of Viral Hepatitis, National Center for HIV, Hepatitis, TB and STD
Prevention, Centers for Disease Control and Prevention, Atlanta, GA 30333,
USA

Background

Hepatitis C is a liver inflammation caused by HCV. Approximately 80% of HCV-infected individuals develop a life-long (chronic) infection, while the other experience a short-term infection and clear the virus [1]. Accurate identification of acute or recent hepatitis C infection is essential for identification of outbreaks and for devising timely public health interventions to interrupt transmissions. In outbreak settings, epidemiological investigation allows for the detection of recent infection. In surveillance settings, however, epidemiological support may be limited, and information on duration of HCV infection may not be available. Recent infection can be also identified by detection of HCV seroconversion and/or by gauging anti-HCV IgG avidity [2, 3]. However, detection of seroconversion is time-consuming, and avidity tests are not broadly available, thus rendering both approaches of impractical for surveillance. To date, there are not cost-effective and reliable methods suitable for large-scale identification of recently acquired HCV infection.

We have recently shown that genetic diversity of intra-host HVR1 variants is associated with duration of HCV infection and can be applied for the detection of recent (R) or chronic (C) infections [4]. The study showed that the R/C state of infection correlated with position-specific amino-acid PhyChem properties in HVR1. However, methods that utilize sequence-specific features require multiple sequence alignment (MSA), which can be an NP-complete problem [5] or computationally expensive, especially when applied to the next-generation sequencing (NGS) data. In addition, extraction and identification of high-quality biomarkers from nucleotide sequences, beyond sequence patterns and population diversity, are not trivial and remain largely unexplored.

There are myriads of ways for transforming DNA/RNA sequence data into numerical representations. One of the most informative representations is based on using scads of PhyChem properties for individual nucleotides or various combinations of nucleotides [6]. The aim of this study was two-fold: firstly, to investigate DNA data transformation techniques for identifying the PhyChem features of HVR1 variants from unaligned sequences; and, secondly, to evaluate the identified features for the accurate detection of the R/C states of HCV infection. Here, we investigated applicability of the HVR1 NGS data for the differential assessment of duration of HCV infection. We describe the application of the DNA dinucleotide-based auto-covariance (DAC) method to effectively identify relevant PhyChem features of HVR1 variants, and the implementation of a radial basis function neural network (RBFNN) classifier to discriminate between R- and C-associated intra-host HVR1 variants without

need of MSA prior to the classification test. We also discuss the use of this approach in the domain of cyber-molecular technology for rapid detection of the R/C state of HCV infection in surveillance settings.

Methods

HVR1 sequence data

Sequences of the intra-host HVR1 variants ($n = 15,041$) sampled from 301 HCV-infected patients diagnosed with chronic ($n = 123$) or recent ($n = 178$) infection – patients infected for more than 1 year or less than a year, respectively– were described in our previous study [4]. The four nucleotide (nt) bases (A, G, U and C) present in the HVR1 of HCV RNA genomes were converted to the corresponding DNA format (A, G, T and C) because of the greater availability of PhyChem properties for the DNA-specific base T than for the RNA-specific U.

For statistical and classification tests, the data were divided into two datasets (training/testing). Sequences of intra-host HVR1 variants ($n = 5681$) derived from 222 persons (R, $n = 124$; C, $n = 98$) were used for training of the classifier, while remainder of the data ($n = 9360$) from 79 persons (R, $n = 54$; C, $n = 25$) were used for testing of the classifier. HVR1 variants comprising the training and test datasets were represented as feature vectors of 148 PhyChem indexes and assigned to the R or C class based on the R/C infection status of the corresponding patient.

To examine effects of data randomization on performance of RBFNN classifier, five training datasets were generated from the HVR1 sequence data, where instances in each dataset were randomly shuffled using different randomization seeds. In addition, to account for the possibility of random correlations in data, four random datasets were generated from the training dataset, each generated by randomly class-labeling the instances using different randomization seeds.

PhyChem features

The PhyChem indices of DNA nt dimers used to generate feature vectors representing the PhyChem features of HVR1 variants were derived from [6, 7]. Correlation measures for the same PhyChem index between two nt dimers separated by a distance (*Lag*) along the sequence were calculated using the following equation (described in [8]):

$$DAC(u, Lag) = \sum_{i=1}^{L-Lag-1} \left(P_u\left(R_iR_{i+1}\right)-\bar{P}_u\right)\left(P_u\left(R_{i+Lag}R_{i+Lag+1}\right)-\bar{P}_u\right)/(L-Lag-1)$$

where u is a PhyChem index, L is the length of the HVR1 sequence, (R_iR_{i+1}) term is the numerical value of PhyChem index u for the Nt dimer R_iR_{i+1} at position i, and $\bar{P}_u$ is the average value of the PhyChem index u along the HVR1 sequence, which is calculated as follows:

$$\bar{P}_u = \sum_{j=1}^{L-1} P_u \left(R_j R_{j+1}\right)/(L-1)$$

Calculations were performed as implemented in the Pse-in-One software (v1.0.3, 2015–08-21 dev) [8], and done in a manner so that length of the PhyChem feature vector is N^*Lag, where N is the number of DNA PhyChem indices ($N = 148$) and $Lag = 1$.

Comparative analysis of the HVR1 PhyChem variants

The HVR1 PhyChem variants derived from sequences of intra-host HVR1 variants from chronically infected patients were compared with PhyChem profiles of variants derived from recently infected patients. We examined the differences between the population means for a given PhyChem index of HVR1 variants sampled from acute and chronic patients. To illustrate differences in binned plots, values for the same PhyChem index between two contiguous nt dimers were binned into equal-width bins (threshold = 0.006). Statistical analysis of differences in means of nt frequencies between the R/C patient-derived HVR1 variants were also conducted. In addition, differences in the PhyChem properties between HVR1 PhyChem variants were examined by the multi-dimensional scaling (MDS) technique as implement in [9]. Briefly, the MDS algorithm iteratively moves the points around in a kind of simulation of a physical model, where there is a force pushing them apart or together. A Euclidean distance matrix was computed to represent the spacing of the HVR1 PhyChem variants comprising the training dataset in Euclidean space. The two-dimensional MDS projection was initialized by randomizing the positions of the instances (or points). Sammon stress [10] was used as the stress function to define how the difference between the desired and the actual distance between points translates into the forces acting on the points.

RBFNN classifier and classification schemes
RBFNN classifier model

A machine-learning approach based on feed-forward neural networks (FFNNs) was used to examine the practical significance of DAC-based PhyChem features generated from sequences of HVR1 variants for developing computer applications for the R/C assessment. We implemented the Gaussian RBFNN classifier technique as described in [11]. Briefly, the RBFNN is a type of FFNN that uses a Gaussian radial basis function and consists of units divided into three layers: an input layer, a hidden (or radial basis) layer and an output layer (the linear

model). The hidden layer of such types of networks are commonly trained using unsupervised learning by k-means clustering and the output layer using supervised learning by logistic regression (for classification tasks) or by linear regression (for regression tasks). For either task, penalized squared error, using a quadratic penalty on the non-bias weights in the output layer, is used as the loss function to find the model's parameters.

The constructed RBFNN classifier had 2 output units (one output unit per class of infection durations), and the learned model for the lth output unit (i.e., class value) is described by the follow formula:

$$f_l(x_1, x_2, ..., x_m) = g\left(w_{l,0} + \sum_{i=1}^{b} w_{l,i} \exp\left(-\sum_{j=1}^{m} \frac{a_j^2 \left(x_j - c_{i,j}\right)^2}{2\sigma_{i,j}^2} \right) \right)$$

where $x_1, x_2, ..., x_m$ is the feature vector for the HVR1 PhyChem variant concerned, the activation function $g(.)$ is the logistic function, b is the number of basis functions, w_i is the weight for each basis function, a_j^2 is the weight of the jth feature, and $c_{i,j}$ and $\sigma_{i,j}^2$ are the basis function centers and variances, respectively.

Settings for the parameters $w_{(l,)i}$, a_j^2, $c_{i,j}$ and $\sigma_{i,j}^2$ were established by finding a local minimum of the penalized squared error on the training dataset using the following error function:

$$L_{SSE} = \left(\frac{1}{2} \sum_{i=1}^{n} \sum_{l=1}^{k} \left(y_{i,l} - f_l\left(\vec{x}_i\right)\right)^2 \right) + \lambda \sum_{l=1}^{k} \sum_{i=1}^{b} w_{l,i}^2$$

where k classes = 2, y_i is the class value for training instance $\vec{x}_i$, the first sum ranges over all n instances in the training dataset and λ is the ridge parameter establishing the size of the penalty on the weights to control overfitting.

A value setting of 39 that was used for the b parameter, which was determined empirically based on the well-known strategy of grid search with cross-validation (GridSearchCV). The hidden unit centers and variances were initialized as follows: the k-means implementation in [12] was used to initialize the $c_{i,j}$, where the number of k clusters was set at 39 and the minimum standard deviation for the clusters set at 1×10^{-3}; and the initial value of all variance parameters $\sigma_{i,j}^2$ in the network was set to the maximum squared Euclidean distance between any pair of cluster centers to prevent initial value of the variance parameters from being too small [11]. The parameter λ for the logistic regression was set at 1×10^{-8}.

Tuning of the b parameter

The number of basis functions (i.e., number of hidden units) that are employed in RBF networks is a relevant parameter that requires particular attention as it directly impacts complexity of the model. The GridSearchCV method was used to search through the hyper-parameter space for the best value for parameter b. Briefly, GridSearchCV implements a fit and a score method to optimize parameters of a model by cross-validated grid-search over a parameter grid (i.e., a range of values). The lower boundary of the grid was set at 2 and the upper boundary limit was set at 66, which was inferred by clustering the training dataset using an expectation-maximization (EM) algorithm (discussed in [12]). The GridSearchCV implementation used here is as follows: the initial grid is worked on with 2-fold cross-validation (2× CV) to determine the values of parameter b based on an evaluation metric(s) (hereafter, classification accuracy). The best point in the grid is then taken and 10× CV is performed with the adjacent point. If a better point is found, then this will act as new center and another 10× CV is performed. This process is repeated until no better point is found or the best (optimal) point is on the border of the grid.

Classification schemes
The RBFNN classifier was trained and evaluated on the training dataset comprising PhyChem variants of HVR1 labeled according to the actual R/C class associations, and with the randomly-labeled datasets where class-labels were randomly assigned to the variants. Classification performances of the RBFNN classifier derived from each scheme was also evaluated on the other dataset (i.e., unseen data).

Applied statistical tests
The Welch two sample t-test was used to examine the statistical significance of differences between the population means in HVR1 variants sampled from acute ($n = 124$) and chronic ($n = 98$) patients. The null hypothesis is that the difference between the means is 0 (making the difference between these two groups not statistically significant) and the alternative hypothesis is that their difference is not zero. The variance parameter was set to 'false' to account for the difference in sample size.

The strength of the association of DNA nt's and PhyChem variables to the R/C durations of infection was measured using the Pearson product-moment correlation coefficient (r). Additionally, the heuristic Merit metric [13] was used to measure the importance of different subsets of DNA PhyChem features for establishing the association between HVR1 and R/C states. Merit scores for various subsets of PhyChem variables were computed using the following formula,

$$Merit_S = \frac{k \times \overline{r_{ca}}}{\sqrt{k + (k-1) \times \overline{r_{aa}}}}$$

where $\overline{r_{ca}}$ is the average feature-class correlation and $\overline{r_{aa}}$

is the average feature-feature inter-correlation in a feature subset S containing k features.

Measures of statistical significance of the pairwise comparisons of classifying schemes performed in this study was done using the corrected resampled two-tailed T-test [14]. The Welch two sample t-test and Pearson product-moment correlation coefficient (r) were implemented in R (v3.0.1). Computations of the corrected two-tailed T-test and of the Merit scores were implemented as discussed in [12].

Classifier performance evaluation
Four metrics used to evaluate the RBFNN classifier(s) are reported herein: classification accuracy (CA), F_1 measure, the Mathews correlation coefficient (MCC) and the Receiver Operating Characteristic (ROC) curve, which was summarized as a single value by computing the area of the convex shape below the ROC curve (AUROC). These metrics were computed as follows:

$$CA = \frac{TP + TN}{TP + TN + FP + FN} \times 100\%$$

$$F_1 = 2 \cdot \left(\frac{\left(\frac{TP}{TP+FP} \right) \times \left(\frac{TP}{TP+FN} \right)}{\left(\frac{TP}{TP+FP} \right) + \left(\frac{TP}{TP+FN} \right)} \right)$$

$$MCC = \frac{TP \times TN - FP \times FN}{\sqrt{(TP + FP)(TP + FN)(TN + FP)(TN + FN)}}$$

where TP is the number of true positives; TN, the number of true negatives; FP, the number of false positive and FN, the number of false negatives. The interpolated curve (TPR vs FPR), made of points whose coordinates are functions of the *threshold* $= \theta \in [0, 1]$, was generated using the following equations:

$$ROC_x(\theta) = FPR(\theta) = \frac{FP(\theta)}{(FP(\theta) + TN(\theta))}$$

$$ROC_y(\theta) = TPR(\theta) = \frac{TP(\theta)}{(FN(\theta) + TP(\theta))}$$

where FPR is the false positive rate and TPR is the true positive rate. Computation of AUROC values was done by computing the probability that the RBFNN classifier ranks a randomly chosen positive instance above a randomly chosen negative instance, which was accomplished by calculating the ρ statistic from the U statistic. Equations and description of the method can be found in [12].

Results
HVR1 PhyChem features specifically associated with R/C states
The Welch's t-test was used to examine variances in nt and DAC-based PhyChem features in HVR1 sequence

data obtained from R ($n = 124$) and C ($n = 98$) patients. Small (range of mean differences: 0.003–0.019) but significant ($p < 2.20X10^{-16}$) variance in nt frequencies was observed between the R- and C-associated HVR1 sequence variants (Table 1). In addition, the four DNA nt bases were found to have small but significant ($p < 2.2 \times 10^{-16}$) correlation to R/C classes by Pearson's product-moment correlation tests (Table 1). Differences in frequency distributions of DNA nt's in HVR1 sequence variants from R/C patients are shown in Fig. 1.

With exception of three PhyChem features, Welch's t-test produced values that fell inside the 95% confidence interval (C.I.) and t-values > 3.30 for the remaining 145 PhyChem indexes of DNA dimers used to represent PhyChem variants of HVR1. Differences in the means of such indexes between the R- and C-associated HVR1 PhyChem variants (range of mean differences: 0.003 to 0.068) were found statistically significant (p-values ranging from $< 9.58 \times 10^{-4}$ to $< 2.2 \times 10^{-16}$). Differential distribution of the R/C-associated HVR1 PhyChem variants was observed in equal-width binning plots (Fig. 2) and pairwise scatter plots (Fig. 3).

Among all tested, 145 DNA PhyChem features of HVR1 were found to have small-to-medium correlation with the R/C classes at the statistical significance level of $p \leq 0.001$, of which 104 features performed similar to nt bases in terms of the degree of correlation (range of R-values: 0.137–0.539) and statistical significance ($p < 2.2 \times 10^{-16}$). The HVR1 DNA PhyChem features ($n = 15$) with statistically significant ($p < 2.2 \times 10^{-16}$) medium correlation (R-values ≥ 0.5) to R/C classes are shown in Table 1. Evaluation of the feature-class relationship of several feature subsets ($n = 10,927$) by a merit scoring method [13] showed that a relevant association ($Merit \leq 0.416$) to the R/C classes could be observed for feature subsets comprised of only 22 DNA PhyChem features of HVR1. Moreover, such feature-class associations were not found in the randomly-labeled datasets ($Merit = 0$ in 37,000 evaluated feature subsets/per random dataset).

Similar analyses on the HVR1 QS data from 25 C- and 54 R-patients indicated no major differences between the training/test data in terms of the minimum/maximum range of values for the 148 DNA PhyChem

Table 1 Differences in the population means of DNA nt and PhyChem features of HVR1 and correlation to the R/C classes[§]

Features[a]	t-value (p-value)[b]	Means in R/C	Difference in means (95% C.I.)	R-value (95% C.I.)
Nt A	47.86 (<2.20X10⁻¹⁶)	0.162/0.181	0.019 (0.019, 0.020)	0.497 (0.477, 0.516)
Nt G	28.64 (<2.20X10⁻¹⁶)	0.322/0.313	0.009 (0.008, 0.010)	0.346 (0.323, 0.367)
Nt C	24.26 (<2.20X10⁻¹⁶)	0.294/0.286	0.008 (0.007, 0.008)	0.332 (0.309, 0.355)
Nt T	9.61 (<2.20X10⁻¹⁶)	0.218/0.215	0.003 (0.002, 0.003)	0.138 (0.112, 0.163)
Twist-tilt	43.39 (<2.20X10⁻¹⁶)	0.006/−0.010	0.016 (0.015, 0.017)	0.539 (0.520, 0.557)
Slide-rise	42.22 (<2.20X10⁻¹⁶)	−0.037/−0.058	0.021 (0.020, 0.022)	0.500 (0.480, 0.519)
Enthalpy	41.01 (<2.20X10⁻¹⁶)	−0.206/−0.250	0.044 (0.041, 0.045)	0.497 (0.477, 0.516)
Breslauer-dH	41.01 (<2.20X10⁻¹⁶)	−0.184/−0.231	0.047 (0.044, 0.048)	0.494 (0.474, 0.513)
Breslauer-dG	40.17 (<2.20X10⁻¹⁶)	−0.298/−0.273	0.025 (0.024, 0.026)	0.477 (0.457, 0.497)
Protein-DNA twist	37.96 (<2.20X10⁻¹⁶)	−0.326/−0.381	0.055 (0.051, 0.057)	0.472 (0.451, 0.492)
Slide-2	36.90 (<2.20X10⁻¹⁶)	−0.380/−0.448	0.068 (0.064, 0.072)	0.471 (0.450, 0.490)
SE-ZDNA[c]	36.46 (<2.20X10⁻¹⁶)	−0.264/−0.313	0.049 (0.045, 0.050)	0.468 (0.447, 0.488)
Twist-1	36.91 (<2.20X10⁻¹⁶)	−0.302/−0.358	0.056 (0.052, 0.058)	0.462 (0.442, 0.483)
G-content	37.79 (<2.20X10⁻¹⁶)	−0.375/−0.434	0.059 (0.056, 0.062)	0.457 (0.436, 0.477)
Helix coil transition	34.05 (<2.20X10⁻¹⁶)	−0.285/−0.350	0.065 (0.062, 0.070)	0.455 (0.434, 0.475)
MGD[d]	35.72 (<2.20X10⁻¹⁶)	0.321/0.353	0.032 (0.030, 0.033)	0.454 (0.433, 0.475)
Sugimoto_dG	37.43 (<2.20X10⁻¹⁶)	0.502/0.462	0.040 (0.037, 0.042)	0.450 (0.429, 0.470)
Sugimoto_dS	38.17 (<2.20X10⁻¹⁶)	0.520/0.475	0.045 (0.043, 0.048)	0.450 (0.429, 0.471)
Propeller twist	34.58 (<2.20X10⁻¹⁶)	0.196/0.148	0.048 (0.045, 0.050)	0.448 (0.427, 0.469)

[a]the four DNA-specific nt's and the 15 DNA-specific PhyChem properties of HVR1 sequences with R-values ≥0.5 are shown. Detailed description of the DNA PhyChem features used herein is available in [6, 7]
[b]p-value is the same for the Welch two sample t-test and Pearson's product-moment correlation test
[c]abbreviation for: Stabilizing Energy of Z DNA
[d]abbreviation for: Minor Groove Distance
[§]R-values, t-values and differences in means are reported as absolute values

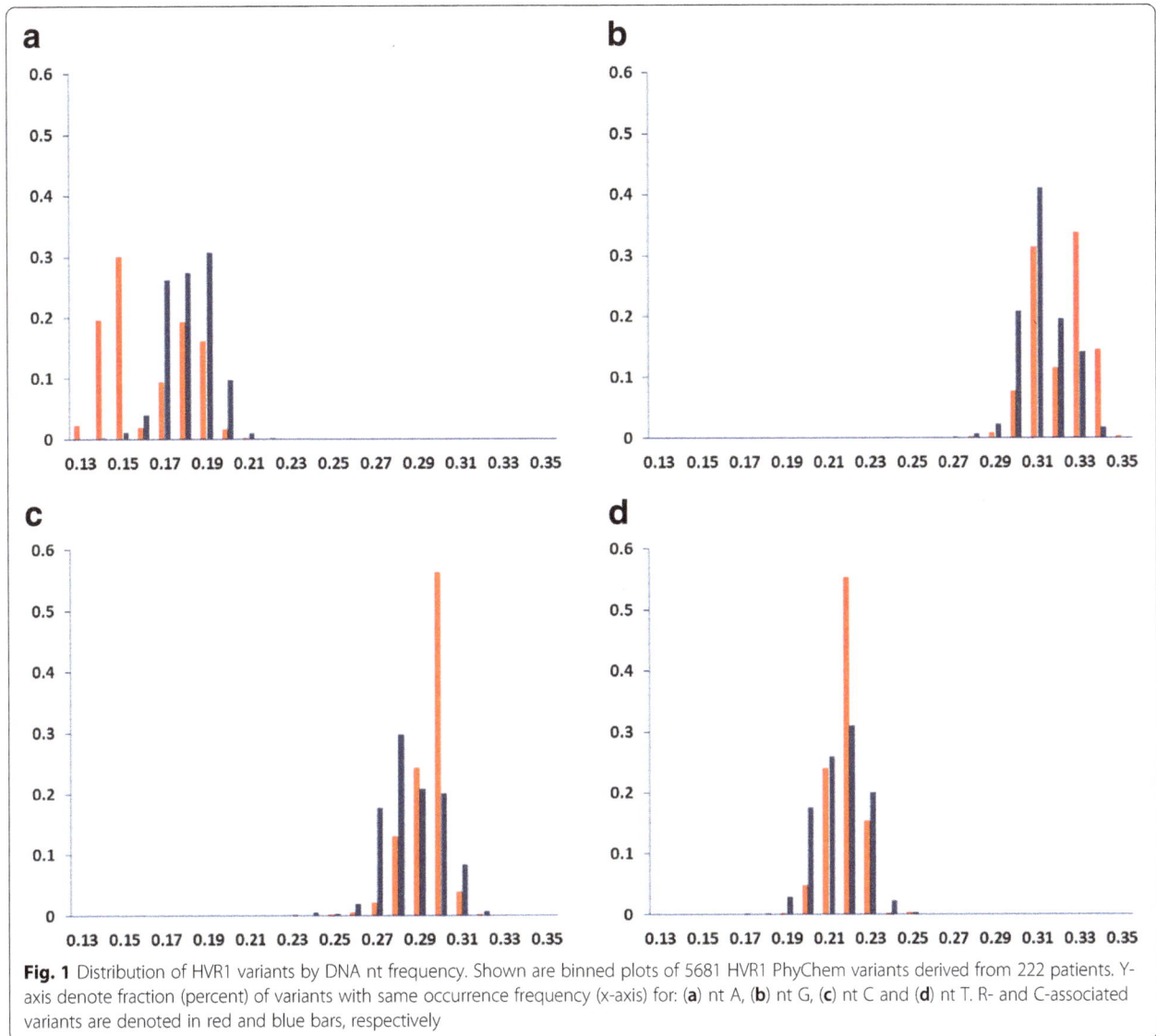

Fig. 1 Distribution of HVR1 variants by DNA nt frequency. Shown are binned plots of 5681 HVR1 PhyChem variants derived from 222 patients. Y-axis denote fraction (percent) of variants with same occurrence frequency (x-axis) for: (**a**) nt A, (**b**) nt G, (**c**) nt C and (**d**) nt T. R- and C-associated variants are denoted in red and blue bars, respectively

features investigated here, as well as in terms of the differential R/C-association in the distributions of the PhyChem variants in binned plots (data not shown).

Spatial distribution of HVR1 PhyChem variants from R/C patients

Differential distribution of various PhyChem properties for the R- and C-HVR1 variants (Figs. 2 and 3) suggests association between the HVR1 PhyChem structure and duration of HCV infection. The R-variants have a less uniform distribution of properties, indicating the existence of preferred PhyChem states for HVR1 variants detected during recent infection. A non-linear unsupervised mapping method was used to examine the PhyChem structure of HVR1 data sampled from R ($n = 124$) and C ($n = 98$) patients. In a MDS plot, the R-associated PhyChem variants of HVR1 were observed to occupy a more central

and restricted PhyChem space than the C-associated variants, which displayed a much broader distribution (Fig. 4). Such differences in spatial distribution between the R/C-HVR1 PhyChem variants suggest applicability of the properties for developing computational models to discriminate between the R/C-states of infection.

Classification tests

The RBFNN machine-learning technique was applied to the data representation of DNA PhyChem variants to generate a classifier for identification of R- and C-associated HVR1 variants. Classification performance evaluation of the RBFNN classifier indicates a high accuracy in identification of R- and C-associated HVR1 PhyChem variants in 10xCV tests (Table 2). The individual R/C-variants in the testing dataset were classified with ~84% accuracy whereas the randomly-labeled

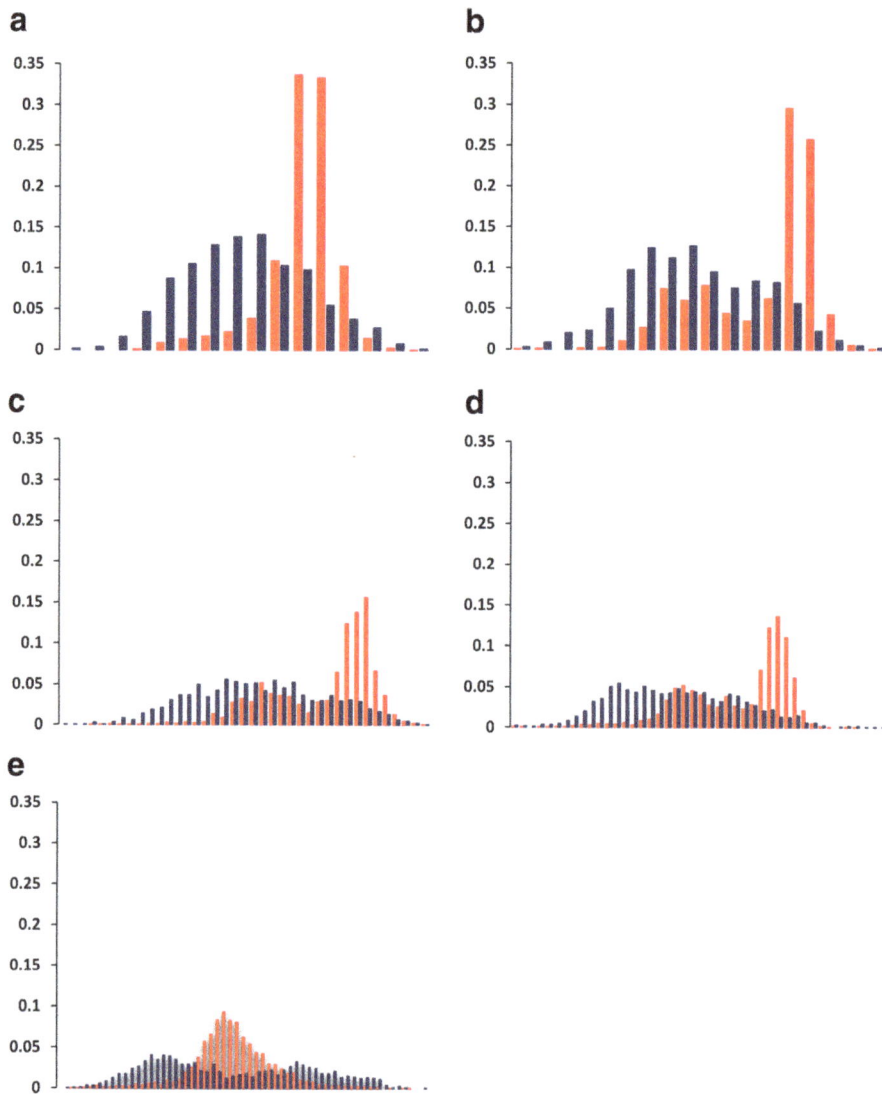

Fig. 2 Distribution of HVR1 variants by DNA PhyChem property. Shown are binned plots of 5681 HVR1 PhyChem variants derived from 222 patients. Y-axis denotes fraction (percent) of variants with same range of values (x-axis) for PhyChem indexes: (**a**) Twist_tilt, (**b**) Slide_rise, (**c**) Enthalpy, (**d**) Breslauer_dH and (**e**) Sugimoto_dH. The Sugimoto_dH index illustrates an example of a DNA PhyChem property found to have small but significant correlation ($r = 0.102$; $p < 1.38 \times 10^{-14}$) to the R/C classes. R- and C-associated variants are denoted in red and blue, respectively

testing and training datasets were classified at a significantly lower accuracy level (AUROC = 0.5) of ~40% and 60%, respectively. The model was applied to classification of patients using a majority vote rule when the duration of infection is defined by the R/C-class comprising > 50% of all intra-host HVR1 variants sampled from the patient. Duration of infection was classified with accuracy of 88.0% for C-patients and 88.89% for R-patients, intra-host HVR1 variants of which were used in the testing dataset, with the overall classification accuracy being 88.45%.

In addition, the RBFNN classifier exhibited a near identical classification performance on three randomized training datasets and showed no significant ($p < 0.001$) differences with the training set used to initialize the RBFNN classifier prior to validation with the test dataset (Table 3). Such observation, taken together with small variations in performance between the full training and the 10xCV training (Table 2), indicate robustness of the RBFNN classifier.

Discussion

Here, we explored a data transformation approach based on DAC of 148 DNA PhyChem properties for identification of association between intra-host HVR1 variants and duration of HCV infection. HCV is an RNA virus. Considering a limited availability of RNA PhyChem properties, we used a DNA-specific representation, which may not be entirely accurate when applied to

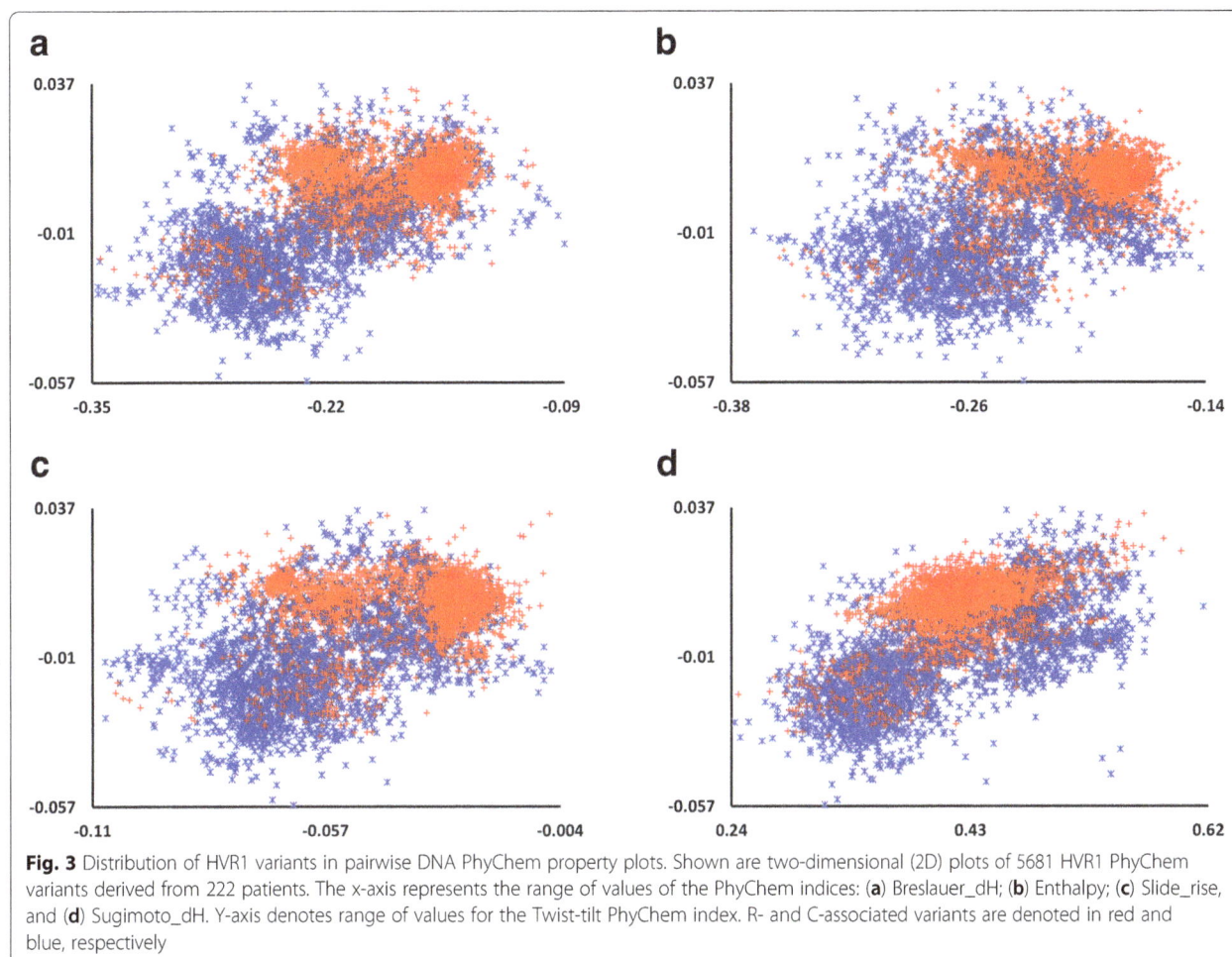

Fig. 3 Distribution of HVR1 variants in pairwise DNA PhyChem property plots. Shown are two-dimensional (2D) plots of 5681 HVR1 PhyChem variants derived from 222 patients. The x-axis represents the range of values of the PhyChem indices: (**a**) Breslauer_dH; (**b**) Enthalpy; (**c**) Slide_rise, and (**d**) Sugimoto_dH. Y-axis denotes range of values for the Twist-tilt PhyChem index. R- and C-associated variants are denoted in red and blue, respectively

RNA. However, the classification accuracy achieved here testifies to applicability of DNA PhyChem properties, at least those selected in this study, to the detection of R/C-state of HCV infection. The framework applied here is readily extendable to using RNA DAC features. It can only be expected that performance of the model will improve when the corresponding RNA PhyChem property data become available.

Welch's t-test indicated that the nt frequency and PhyChem property distributions along the HVR1 genomic region significantly ($p < 0.001$) differ between the R- and C-associated HCV strains examined in this study. However, differences between the R/C HVR1 variants became markedly appreciable after applying the DNA DAC transformation method to the NGS data (Table 1 and Figs. 1 and 2). Furthermore, the finding that 70.3% of the PhyChem features had a significant ($p < 2.20 \times 10^{-16}$) correlation to R/C, with R-values ranging between 0.137–0.539, represents a 96.2% increase of correlative features over the original 4 nt-based information (range of R-values 0.138–0.497). In addition, in binned plots (Figs. 1 and 2), the difference between the R/C-

associated variants was notably greater for PhyChem representation. Taken together with the Merit scores observed for feature subsets, findings suggest that: (*i*) the DNA DAC-based features provide a better discrimination for differentiation between the R/C classes than the nt diversity alone; and (*ii*) more importantly, there are substantial differences in the PhyChem structure between HVR1 variants from the R and C classes.

Association between the PhyChem structure of HVR1 variants and the R/C classes is complex. The data indicate that, although HVR1 variants from both classes are intermixed in all plots (Figs. 2, 3 and 4), majority of the R-variants appear to cluster. This observation indicates that the R-HVR1 variants have preferred PhyChem properties and majority of them constitute only a fraction of the entire PhyChem space occupied by C-HVR1 variants. Thus, the dominant HCV population established during the early stage of infection has HVR1 variants with certain PhyChem properties and evolves during infection into a population containing HVR1 with a wide range of the properties. Frequent establishment of dominant populations early during infection in

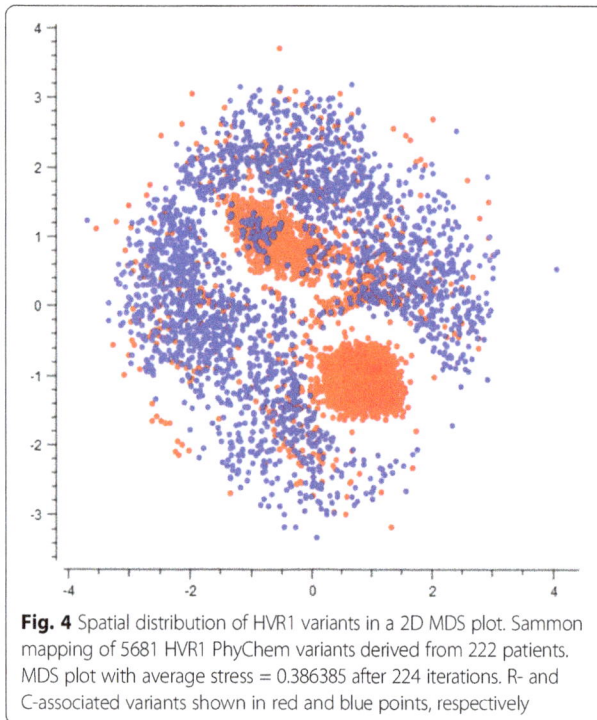

Fig. 4 Spatial distribution of HVR1 variants in a 2D MDS plot. Sammon mapping of 5681 HVR1 PhyChem variants derived from 222 patients. MDS plot with average stress = 0.386385 after 224 iterations. R- and C-associated variants shown in red and blue points, respectively

Table 3 Comparison of RBFNN performance on randomized datasets in 100 10xCV tests[§]

Dataset	No. CV runs	CA	F_1 measure	MCC	AUROC
Train set 1[a]	1000	94.943% (±1.067)	0.960 (±0.009)	0.892 (±0.023)	0.981 (±0.005)
Train set 2	1000	95.958% (±0.717)	0.974 (±0.005)	0.887 (±0.020)	0.986 (±0.003)
Train set 3	1000	96.014% (±0.719)	0.974 (±0.005)	0.889 (±0.020)	0.987 (±0.003)
Train set 4	1000	95.981% (±0.699)	0.974 (±0.005)	0.889 (±0.019)	0.986 (±0.003)

[§]Comparisons are based on the corrected two-tailed T-test at a significance level of $p < 0.001$
[a]Dataset used to train (fit) the RBFNN classifier (1st and 2nd rows in Table 2)

performance on the test dataset (Table 1), suggest that association to R/C is likely due to specific HVR1 traits rather than to the biased sample selection or existence of random statistical correlations in the data. This conclusion is in concordance with prior observations. Previously, we showed that the intra-host HVR1 evolution is associated with the R/C-states of HCV infection [4] as well as with age, gender and ethnicity of hosts and response to interferon treatment [16, 17]. The product of HVR1 expression belongs to a class of proteins known as intrinsically disordered proteins (IDPs) or regions (IDPR) [18, 19]. In general, IDPs/IDPRs have been strongly associated with a multitude of biological functions [20] and play a significant role in evolution [21]. Thus, it seems reasonable to suggest that HVR1, as IDPR, actively participates in the intra-host HCV adaptation and plays specific roles at different stages of HCV infections. The HVR1 functions are likely reflected in changing genetic composition, which is detected using the model developed in this study.

The classification accuracy of the RBFNN classifier (Tables 1 & 2) indicates that the features representing the PhyChem structure of HVR1 can serve as reliable biomarkers of the R/C-association. Based on our findings, we propose that the DNA-specific formulation used herein for the PhyChem representation provides general, information-rich features for detection of trait-specific HVR1 associations beyond the R/C-states of HCV infection, and is potentially applicable to any genomic region. Continued research of such types of features may contribute further to improvement of computational models for the detection of various biological and epidemiological traits from genetic data.

Conclusions and future work

The HVR1 NGS data contain genetic information, which is pertinent for the identification of the R/C-state of HCV infection. Clustering of the R-HVR1 variants in the PhyChem space suggests a particular way of the intra-host HCV evolution in the space during infection and offers a

recipients from minority HCV variants transmitted from the source cases during outbreaks [15] is in concert with this supposition.

Differences in PhyChem properties between HVR1 from R/C classes are substantial. Although the identified here associations may be affected by variation in sampling of intra-host HVR1 variants, the data indicate that the duration of HCV infection is reflected in evolution of HVR1 through the PhyChem space in each infected host. Performance of the RBFNN classifier on the randomized training datasets (Table 2) and on the randomly-labeled dataset (Table 1), in conjunction with

Table 2 RBFNN performance in R/C classification of Intra-host HVR1 PhyChem variants[a]

Dataset	CA	F_1 measure	MCC	AUROC
Full train set[b]	95.795%	0.958	0.910	0.986
Train set	94.847%[c]	0.948[c]	0.890[c]	0.979[c]
Test set	84.145%[d]	0.842[d]	0.670[d]	0.912[d]
Random-labeled train set	59.038%[e] (±1.28)	0.521[e] (±0.007)	−0.007[e] (±0.022)	0.501[e] (±0.012)
Test set	39.965%[f] (±1.948)	0.280[f] (±0.070)	0.003[f] (±0.145)	0.385[f] (±0.144)

[a]For description of train/test data, see Methods Section
[b]Values obtained from RBFNN classifier trained on entire training dataset without CV
[c]Overall value represents averaged values of 10xCV data
[d]Value obtained from RBFNN classifier trained on training dataset by 10xCV
[e]Overall value represents averaged values of 10xCV data obtained from 4 datasets. Standard deviation (SD), in parenthesis
[f]Overall value represents averaged values obtained from 4 RBFNN classifiers trained on randomly-labeled data by 10xCV (SD)

new approach to the detection of R/C-infections. Identification of new features, which can be extracted from NGS data directly and without using MSA, and development of the model, which accurately detects duration of HCV infection, paves a way for designing cyber-molecular diagnostics for the identification of traits of clinical and epidemiological relevance using genetic data.

Unlike the laboratory diagnostic methods for identification of acute HCV infection, our approach is based on extracting PhyChem features from NGS data and using an RBFNN classifier for identification of the R/C-infections, and, thus, suitable for being hosted by Global Hepatitis Outbreak and Surveillance Technology (GHOST) – a web-based virtual diagnostic system for extraction of public health information from sequence data (see paper in this issue). In addition, our study highlights the importance of considering genomic regions that encode IDPs or IDPRs as potential sources of predictive biomarkers, as well as relevance of the examination of HVR1 in biomarker discovery projects for detection of HCV-related traits. We are currently expanding investigation into DNA PhyChem features expressing higher tiers of interaction between nt-dimers (i.e., Lag > 1) and finalizing a python-based script, which will be made available to authenticated users of GHOST (https://webappx.cdc.gov/GHOST/) for further testing and validation.

Acknowledgments
We thank anonymous reviewers for their constructive comments and suggestions to improve the manuscript.

Funding
This study was supported by CDC intramural funding, and by APHL postdoctoral fellowship funding (2016–2017) to MT. Publication costs are funded by an internal program of CDC.

About this supplement
This article has been published as part of *BMC Genomics* Volume 18 Supplement 10, 2017: Selected articles from the 6th IEEE International Conference on Computational Advances in Bio and Medical Sciences (ICCABS): genomics. The full contents of the supplement are available online at https://bmcgenomics.biomedcentral.com/articles/supplements/volume-18-supplement-10.

Authors' contributions
JL and YK conceived and designed experiments. JL and MT implemented and conducted all bioinformatics analyses. JL and YK wrote paper with contributions from MT. All authors read and approved the final version of the manuscript.

Competing interests
Authors declare no competing interests. CDC Disclaimer: The findings and conclusions of this manuscript are those of the authors and do not necessarily represent the official views of the Centers for Disease Control and Prevention.

References
1. Alberti A, Chemello L, Benvegnu L. Natural history of hepatitis C. J Hepatol. 1999;31(Suppl 1):17–24.
2. Araujo AC, Astrakhantseva IV, Fields HA, Kamili S. Distinguishing acute from chronic hepatitis C virus (HCV) infection based on antibody reactivities to specific HCV structural and nonstructural proteins. J Clin Microbiol. 2011; 49(1):54–7.
3. Klimashevskaya S, Obriadina A, Ulanova T, Bochkova G, Burkov A, Araujo A, Stramer SL, Tobler LH, Busch MP, Fields HA. Distinguishing acute from chronic and resolved hepatitis C virus (HCV) infections by measurement of anti-HCV immunoglobulin G avidity index. J Clin Microbiol. 2007;45(10):3400–3.
4. Astrakhantseva IV, Campo DS, Araujo A, Teo CG, Khudyakov Y, Kamili S. Differences in variability of hypervariable region 1 of hepatitis C virus (HCV) between acute and chronic stages of HCV infection. In Silico Biol. 2011; 11(5–6):163–73.
5. Bin MA, Wang Z, Zhang K. Alignment between two multiple alignments. Lect Notes Comput Sci. 2003;2676:254–65.
6. Chen W, Lei TY, Jin DC, Lin H, Chou KC. PseKNC: a flexible web server for generating pseudo K-tuple nucleotide composition. Anal Biochem. 2014; 456:53–60.
7. Friedel M, Nikolajewa S, Suhnel J, Wilhelm T. DiProDB: a database for dinucleotide properties. Nucleic Acids Res. 2009;37(Database issue):D37–40.
8. Liu B, Liu F, Wang X, Chen J, Fang L, Chou KC. Pse-in-one: a web server for generating various modes of pseudo components of DNA, RNA, and protein sequences. Nucleic Acids Res. 2015;43(W1):W65–71.
9. Demšar J, Curk T, Erjavec A, Gorup C, Hočevar T, Milutinovič M, Možina M, Polajnar M, Toplak M, Starič A, Štajdohar M, Umek L, Žagar L, Žbontar J, Žitnik M, Zupan B. Orange: data mining toolbox in python. J Mach Learn Res. 2013;14(Aug):2349–53.
10. Sammon JW. A nonlinear mapping for data structure analysis. IEEE Trans Comput. 1969;18:401–9.
11. Frank E. Fully supervised training of Gaussian radial basis function networks in WEKA. In: Computer science working papers. Hamilton, New Zeland: Department of Computer Science, The University of Waikato; 2014.
12. Witten I, Frank E, Hall MA. Data mining: practical machine learning tools and techniques. Third ed. San Francisco, USA: Morgan Kaufmann; 2011.
13. Hall MA. Correlation-based feature selection for machine learning. Hamilton, New Zeland: Waikato University; 1999.
14. Nadeau C, Bengio Y. Inference for the generalization error. Mach Learn. 2003;52(3):239–81.
15. Sagar M. HIV-1 transmission biology: selection and characteristics of infecting viruses. J Infect Dis. 2010;202(Suppl 2):S289–96.
16. Lara J, Khudyakov Y. Epistatic connectivity among HCV genomic sites as a genetic marker of interferon resistance. Antivir Ther. 2012;17(7 Pt B):1471–5.
17. Lara J, Tavis JE, Donlin MJ, Lee WM, Yuan HJ, Pearlman BL, Vaughan G, Forbi JC, Xia GL, Khudyakov YE. Coordinated evolution among hepatitis C virus genomic sites is coupled to host factors and resistance to interferon. In Silico Biol. 2011;11(5–6):213–24.
18. Khan AG, Whidby J, Miller MT, Scarborough H, Zatorski AV, Cygan A, Price AA, Yost SA, Bohannon CD, Jacob J, et al. Structure of the core ectodomain of the hepatitis C virus envelope glycoprotein 2. Nature. 2014;509(7500):381–4.
19. Kong L, Giang E, Nieusma T, Kadam RU, Cogburn KE, Hua Y, Dai X, Stanfield RL, Burton DR, Ward AB, et al. Hepatitis C virus E2 envelope glycoprotein core structure. Science. 2013;342(6162):1090–4.
20. Oldfield CJ, Dunker AK. Intrinsically disordered proteins and intrinsically disordered protein regions. Annu Rev Biochem. 2014;83:553–84.
21. Chakrabortee S, Byers JS, Jones S, Garcia DM, Bhullar B, Chang A, She R, Lee L, Fremin B, Lindquist S, et al. Intrinsically disordered proteins drive emergence and inheritance of biological traits. Cell. 2016;167(2):369–81. e312

3

Robust transcriptional signatures for low-input RNA samples based on relative expression orderings

Huaping Liu[1,3], Yawei Li[1], Jun He[1], Qingzhou Guan[1], Rou Chen[1], Haidan Yan[1], Weicheng Zheng[1], Kai Song[3], Hao Cai[1], You Guo[1], Xianlong Wang[1*] and Zheng Guo[1,2,3,4*]

Abstract

Background: It is often difficult to obtain sufficient quantity of RNA molecules for gene expression profiling under many practical situations. Amplification from low-input samples may induce artificial signals.

Results: We compared the expression measurements of low-input mRNA samples, from 25 pg to 1000 pg mRNA, which were amplified and profiled by Smart-seq, DP-seq and CEL-seq techniques using the Illumina HiSeq 2000 platform, with those of the paired high-input (50 ng) mRNA samples. Even with 1000 pg mRNA input, we found that thousands of genes had at least 2 folds-change of expression levels in the low-input samples compared with the corresponding paired high-input samples. Consequently, a transcriptional signature based on quantitative expression values and determined from high-input RNA samples cannot be applied to low-input samples, and vice versa. In contrast, the within-sample relative expression orderings (REOs) of approximately 90% of all the gene pairs in the high-input samples were maintained in the paired low-input samples with 1000 pg input mRNA molecules. Similar results were observed in the low-input total RNA samples amplified and profiled by the Whole-Genome DASL technique using the Illumina HumanRef-8 v3.0 platform. As a proof of principle, we developed REOs-based signatures from high-input RNA samples for discriminating cancer tissues and showed that they can be robustly applied to low-input RNA samples.

Conclusions: REOs-based signatures determined from the high-input RNA samples can be robustly applied to samples profiled with the low-input RNA samples, as low as the 1000 pg and 250 pg input samples but no longer stable in samples with less than 250 pg RNA input to a certain degree.

Keywords: Low-input RNA samples - amplification artificial signals - relative expression orderings - transcriptional signatures

Background

Gene expression profiling based on microarray and RNA sequencing techniques allows us to comprehensively characterize RNA transcripts present in a biological sample. However, it is often difficult to obtain sufficient quantity of RNA molecules for gene expression profiling under many practical situations. For example, minimally invasive tissue biopsy techniques, such as fine needle aspiration cytology, core needle biopsy and gastrointestinal endoscopy, are widely used clinically but minimum samples are extracted [1–3]. For another example, in formalin-fixed paraffin-embedded tissue samples with abundant clinical information, the amount of RNA is often limited due to partial RNA degradation [4, 5]. In the studies of rare cell population, single cell [6, 7] or the samples taken with the laser capture microdissection [8] technique, the amount of RNA molecules is also extremely low.

It is critical to overcome this challenge to leverage the power of low-input sampling techniques for biomedical applications. For this type of samples, multiple rounds of pre-amplification are necessary prior to the measurements of gene expression levels. Thus, a number of low-

* Correspondence: wang.xianlong@139.com; guoz@ems.hrbmu.edu.cn
[1]Department of Bioinformatics, Key Laboratory of Ministry of Education for Gastrointestinal Cancer, School of Basic Medical Sciences, Fujian Medical University, Fuzhou 350122, China
Full list of author information is available at the end of the article

input RNA amplification techniques prior sequencing have been developed using PCR or in vitro transcription (IVT) to synthesize enough cDNA or cRNA, such as Smart-seq (switching mechanism at 5′-end of the RNA transcript) [9], DP-seq (primer-based RNA-sequencing strategy) [10] and CEL-seq (cell expression by linear amplification and sequencing) [11]. However, current low-input amplification techniques usually bring a large bias due to the inherent defects in the amplification principles [12]. For example, CEL-seq incorporating IVT can result in 3′ biases due to two rounds of reverse transcription before the linear amplification [11, 13]. Smart-Seq, using PCR to synthesize cDNA, is a nonlinear amplification process, and its efficiency is sequence-dependent [9, 13]; a long transcript may be truncated due to inefficient cDNA synthesis during the amplification process [14, 15]. It has been reported that the amplification bias always exists in lowly expressed genes and genes with abundant CG and long length [16–18]. As a result, it is uncertain whether the expression values measured after the amplification can represent the real gene expression levels or not.

Several studies attempted to prove that gene expression profiling can be performed on low-input RNA samples like high-input RNA samples by showing that the gene expression profiles of low-input RNA samples are significantly correlated with those of the matched high-input RNA samples [19–21]. However, a high correlation between two measurements does not guarantee that the two measurements are congruent, which brings uncertainty to the application of most current disease signatures based on risk scores which are calculated using the measurement values of the signature genes. Therefore, for a transcriptional signature based on the quantitative expression levels, the risk score thresholds determined from high-input RNA samples may be not applicable to low-input RNA samples, and vice versa. It has been reported that quantitative transcriptional signatures lack robustness for clinical applications due to measurement batch effects [22], variations of the tumor epithelial cell proportions in tissues sampled from different sites of a tumor [23, 24] and partial RNA degradation during sample preparation [25, 26]. Another type of disease signature is based on the within-sample relative expression orderings (REOs) of gene pairs [27, 28], which have been identified for predicting the prognosis of colorectal cancer [29], non-small cell lung cancer [30], ER+ breast cancer [31] and other cancers [32, 33]. These REOs-based signatures are robust against various measurement biases introduced by experimental batch effects and platform differences [34], partial RNA degradation [26] and uncertain sampling sites within the same cancer tissue [24]. Thus, we hypothesized that the REOs of gene pairs within individual samples, especially those with large

rank differences, might also be robust against the biases introduced by the RNA amplification procedures.

Through comparing gene expression profiles between the samples with low-input mRNA, ranging from 25 pg to 1000 pg mRNA profiled by the Illumina HiSeq 2000 platform, and their paired high-input 50 ng mRNA samples, we found that there were thousands of genes with at least 2 folds-change (FC) in their expression values even when the input mRNA was 1000 pg. We evaluated the proportions of REOs of gene pairs in the high-input RNA samples maintained in the low-input RNA samples, and found that the proportions were approximately 90% even when the input mRNA was as low as 1000 pg and the input total RNA samples was as low as 250 pg, which suggests that REOs measured in the low-input samples were robust against amplification. Similar results were also found in the low input total RNA ranging from 10 pg to 1000 pg profiled by the Illumina HumanRef-8 v3.0 platform compared with the 100 ng input total RNA samples. As a case study to demonstrate the robustness of REOs-based signatures, we developed REOs-based signatures from high-input RNA samples for discriminating cancer tissues and showed that they can be robustly applied to low-input RNA samples.

Results
Large amplification bias of low-input RNA samples
Based on two datasets (GSE50856 and GSE17565, see Fig. 1) measured by Illumina HiSeq 2000 and Illumina HumanRef-8 v3.0 platforms, respectively, we evaluated the amplification fidelity of low-input RNA samples amplified by several techniques through comparison with the corresponding high-input RNA samples using the FC values.

In the SFM-Smart group of dataset GSE50856, there were respectively 60.56, 64.00, 65.76 and 66.95% of genes with a FC value larger than or equal to 2 in the expression values between 1000 pg, 100 pg, 50 pg and 25 pg mRNA samples compared with the paired high-input samples. As the amount of RNA in the diluted low input samples decreased, the percentage of genes with at least 2 FC increased. Similar results were also observed in the other five groups of dataset GSE50856 (Fig. 2). In the SFM-smart data, the coefficient of variation (CV) of FCs increased from 0.18 to 0.33 as the quantity of the input RNA decreased from 1000 pg to 25 pg (Additional file 1: Figure 1Sa). Similar results were observed in the data for SFM-DP, SFM-CEL, AA100-Smart, AA100-DP, AA100-CEL, Raji and MCF-7 (Additional file 1: Figure 1Sa, b and c). Thus, a large amplification bias exists for the three amplification techniques even the amplification begins from 1000 pg mRNA input.

For the Raji group of dataset GSE17565, there were 12.02, 23.79, 41.73, 57.84% of genes with a FC value

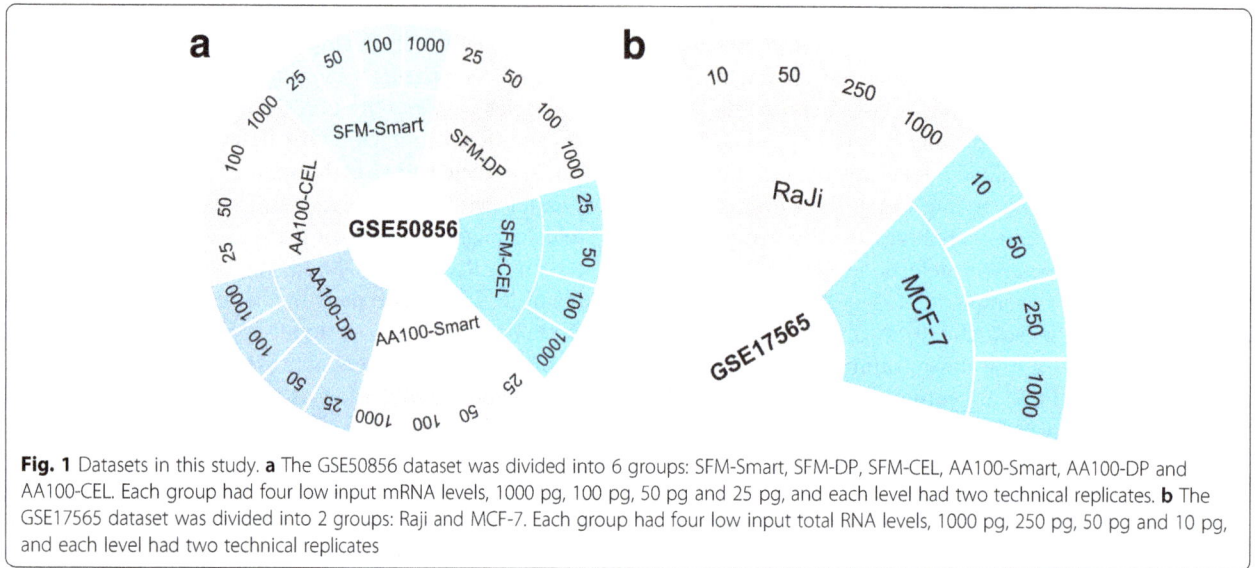

Fig. 1 Datasets in this study. **a** The GSE50856 dataset was divided into 6 groups: SFM-Smart, SFM-DP, SFM-CEL, AA100-Smart, AA100-DP and AA100-CEL. Each group had four low input mRNA levels, 1000 pg, 100 pg, 50 pg and 25 pg, and each level had two technical replicates. **b** The GSE17565 dataset was divided into 2 groups: Raji and MCF-7. Each group had four low input total RNA levels, 1000 pg, 250 pg, 50 pg and 10 pg, and each level had two technical replicates

larger than or equal to 2 in the expression values, respectively, in the 1000 pg, 250 pg, 50 pg, 10 pg samples compared with the paired high-input samples. Similar results were also observed for the MCF-7 group (Fig. 2). Obviously, the amplification procedure has a profound negative impact on the measurements of gene expression levels of the low-input samples.

Robustness of REOs against amplification bias

Using the same datasets, we evaluated the consistency scores between the low-input samples and the high-input samples, i.e. the proportions of the REOs of the gene pairs in the high-input RNA samples maintained in the low-input samples. All genes from the gene expression profiles were involved in the REO gene pairs.

In the SFM-Smart dataset GSE50856, 88.53 and 88.63% of the stable REOs in the high-input mRNA samples were respectively kept in the two 1000 pg input mRNA technical replicates. Obviously, the REOs of gene pairs with small rank differences (i.e., close expression levels) tend to be sensitive to random measurement variations [34]. After excluding 10% of pairs with the

smallest rank differences, the percentages increased to 91.36 and 91.46% in the two 1000 pg input mRNA technical replicates, respectively. The percentage of the stable REOs in the high-input samples that were kept in the low input technical replicates, termed the consistency scores for short, decreased gradually when the input mRNA decreased. The consistency scores for the two 100 pg input technical replicates decreased to 85.66 and 85.36%, respectively, and increased to 88.24 and 87.92% after excluding the bottom 10% of the gene pairs in the high-input mRNA samples. For the two 50 pg input technical replicates, the consistency scores were 83.84 and 83.29%, respectively, and increased to 86.27 and 85.67%, respectively, after excluding the bottom 10% of the stable gene pairs. For the two 25 pg input technical replicates, the consistency scores were 82.11 and 81.99%, respectively, and increased to 84.39 and 84.26% after excluding the bottom 10% of the gene pairs (Fig. 3a). Similar results were also found in the SFM-DP (Fig. 3b), SFM-CEL (Fig. 3c), AA100-Smart (Additional file 1: Figure S2a), AA100-DP (Additional file 1: Figure S2b) and AA100-CEL groups (Additional file 1: Figure S2c). As shown in

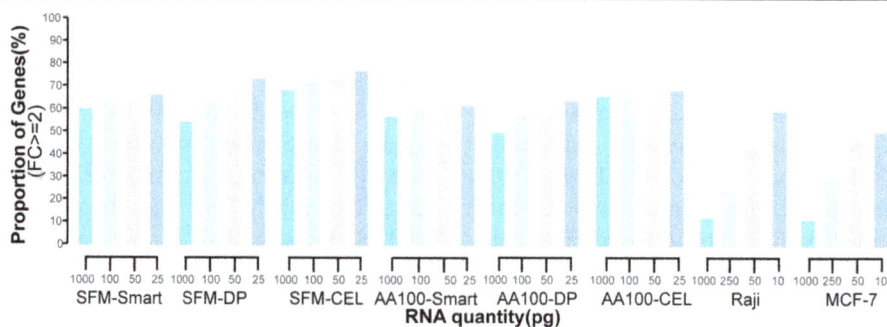

Fig. 2 Amplification bias. Proportion of genes with at least 2 folds-change of expression values

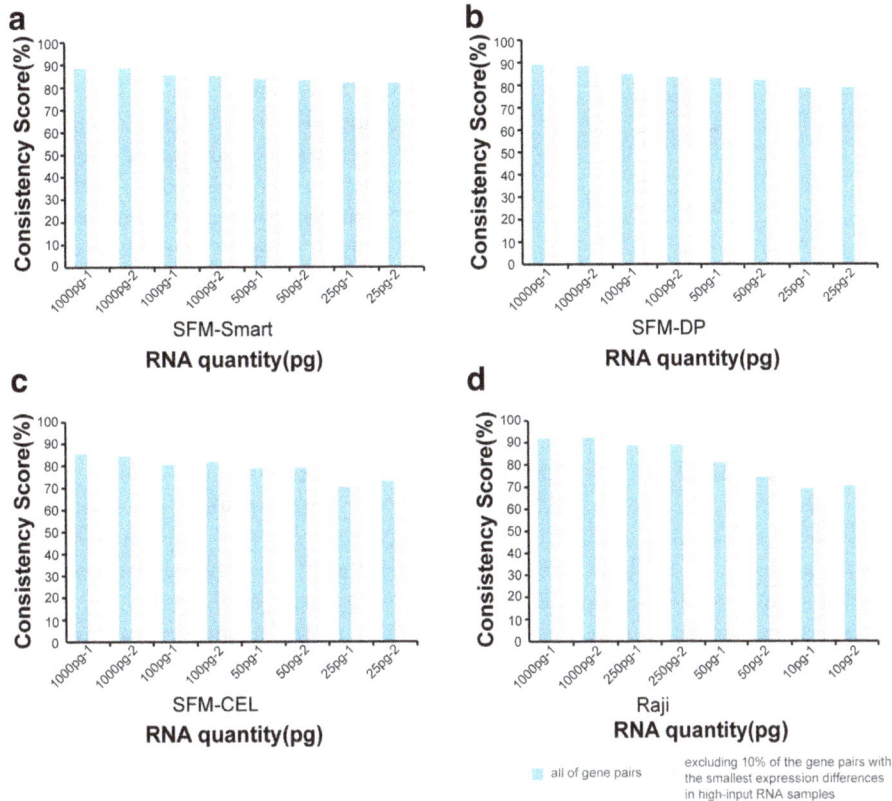

Fig. 3 Maintenance of REOs after excluding 0% to 10% gene pairs (**a**) The consistency scores between high-input RNA samples and low-input RNA samples of all gene pairs (blue) and after excluding 10% of the pairs with the smallest expression differences in the paired high-input RNA samples (pink) in the group of SFM-Smart (**b**) (**c**) (**d**) Similar as the Fig. a

the (Additional file 1: Figure S3, Figure S4), the percentage of the stable REOs in the high-input samples that were kept in each of the low input technical replicates increased when more gene pairs with small rank differences in the high-input samples were excluded.

For all the 164,238,402 gene pairs which had the same REOs among two technical replicates of the high-input samples in the Raji dataset GSE17565 measured by the Illumina HumanRef-8 v3.0 platform, 92.09 and 92.43% were respectively kept in the two technical replicates with 1000 pg input total RNA. The consistency scores increased to 94.83 and 95.08%, respectively, after excluding the bottom 10% gene pairs with the smallest rank differences. For the two 250 pg input technical replicates, the consistency scores for all the stable gene pairs were 88.87 and 89.20%, respectively, and increased to 91.49 and 91.76% after excluding the bottom 10% of the gene pairs. For the two 50 pg input technical replicates, the consistency scores were 81.06 and 74.52%, respectively, and increased to 83.39 and 76.42% after excluding the bottom 10% of the gene pairs. For the two 10 pg input technical replicates, the consistency scores were 69.32 and 70.58%, respectively, and increased to 70.87 and 72.23% after excluding the bottom 10% of the

gene pairs (Fig. 3d). Similar results were also found in the MCF-7 group (Additional file 1: Figure S4d).

Taken together, the above results showed that the REOs of gene pairs were robust against the amplification bias for the 1000 pg and 250 pg input samples but no longer stable in samples with less than 250 pg RNA input to a certain degree.

Performance of REOs-based signatures in low-input RNA samples

As a proof of principle that REOs-based signature identified from high-input RNA tissue samples are robust in low-input RNA samples, we collected 69 high-input RNA samples of lymphoma tissues from the GSE55267 dataset and 54 high-input RNA samples of breast cancer tissues from the GSE29431 dataset to search a REOs-based signature for discriminating the two types of tissues (Table 1). We obtained 106,213 highly stable gene pairs that have the same REOs in all lymphoma tissue samples and breast cancer tissue samples, respectively, but the REO patterns were reversal between the two tissue types. From these 106,213 gene pairs, we selected 3 gene pairs (Table 2) with the largest geometric mean of the average absolute rank difference in the lymphoma

Table 1 High-input RNA tissue samples used in this study

Tissue Sample Type	GEO ID	Sample Size
Lymphoma	GSE55267	69
Breast cancer	GSE29431	54
Lymphoma	GSE53820	81
Breast cancer	GSE10780	30
COAD	TCGA	41
Normal tissues paired with COAD	TCGA	41
Colon tumor tissues	GSE10950	25
Colon normal tissues	GSE10950	25
Colorectal tumors (CRC)	GSE81861	272
Normal mucosas paired with CRC	GSE81861	157

tissue samples and the average absolute rank difference in the breast cancer tissues samples (see Materials and Methods). The results showed that when $k = 3$ both sensitivity and specificity were 100%. Thus, these three gene pairs with the highest R_{ij} values, as described in Table 2, were selected as the classification signature. Using the 3 gene pairs as signature, we classified a given sample according to the majority vote rule. If 2 or 3 REOs of the 3 gene pairs in a sample were consistent with the REO patterns in the lymphoma tissue samples, the sample was identified as a lymphoma tissue sample; otherwise, the sample was identified as a breast cancer tissue sample. In the training datasets, obviously, all of the lymphoma tissues samples and the breast cancer tissue samples were correctly classified using the signature. In the independent validation dataset, consisting of 81 high-input RNA samples of lymphoma tissue from the GSE53820 dataset and 30 high-input RNA samples of breast cancer tissue from the GSE10780 dataset, all of the samples were correctly classified.

We further applied the REOs-based signature to distinguish Raji and MCF-7 cell lines profiled with high-input with 100 ng total RNA and low-input samples with as low as 50 pg total RNA from the GSE17565 dataset. All the 8 high-input Raji cell line samples, 8 high-input MCF-7 cell line samples, 12 low-input Raji cell line samples and 12 low-input MCF-7 cell line samples were correctly classified. This case study demonstrates that a REOs-based transcriptional signature identified from the high-input RNA tissue samples can be applied to classify low-input samples robustly.

Table 2 The 3 gene-pair signature

Gene pair No.	Gene A[a]	Gene B[a]
1	MMP3	RGS13
2	EPCAM	CD37
3	EPCAM	STAP1

[a]Gene A had a higher expression level than Gene B in breast cancer tissues and MCF-7 cell lines

As a second case study, we identified a REO-based signature from high-input RNA samples for discriminating primary colorectal tumors from normal colorectal tissues and showed that it can be robustly applied to low-input RNA samples summarized from single-cell RNA-seq data. Firstly, using the 41 colon adenocarcinoma samples and paired normal samples from TCGA, we identified two lists of gene pairs, each with identical REOs in all samples of the primary colorectal tumor tissue and the corresponding normal tissue, respectively. From the above two lists of gene pairs, 20,390 gene pairs were found to have reversal REOs between the tumor tissues and the normal tissues. Because there were an abundance of dropout events that led to zero expression values for approximately 90% of the genes measured in the single-cell data, it would be inappropriate to select only a few gene pairs as the diagnostic signature. Therefore, all the reversal gene pairs were directly used as the signature. In the training dataset, the 20,390 gene pairs correctly classified all the cancer and normal samples according to the majority voting rule. Then, we collected an independent dataset from GSE10950 with 25 high-input RNA samples of paired colon tumor tissues and colon normal tissues to validate this signature. Because only 18,227 gene pairs of the 20,390 gene pairs were measured in this dataset by the Illumina human Ref-8 v2.0 platform, these 18,227 gene pairs were used to classify the samples according to the majority voting rule and all the samples were correctly classified. However, with the same strategy, 272 colorectal tumor epithelial cells and 157 normal epithelial cells from the GSE81861 dataset could not be correctly classified. This result is not surprising since a cell contains only approximately 10 pg RNA and 90% of genes were measured with zero expression values. The REOs of gene pairs in such small input RNA samples would be unstable as demonstrated above. To address this issue, we constructed a pooled dataset from the single-cell RNA-seq results.

In the GSE81861 dataset, the 272 tumor epithelial cells and the 157 normal epithelial cells were extracted from 11 patients of primary colorectal tumors and paired normal tissues; however, there were no annotation on patients' information. We randomly assigned the 272 colorectal tumor epithelial cells into 11 samples with approximately equal number of cells: 10 samples each with 25 single cells and a sample with 22 single cells. Each simulated disease sample contains approximately 250 pg RNA. Similarly, the 157 normal epithelial cells were also randomly assigned into 10 samples each with 14 single cells and a sample with 17 single cells. Each sample approximately contains 140 pg RNA. In each sample, we calculated the sum of the measurement values for each gene to represent the expression levels of the genes [19]. Then, the REO signature with 20,390 gene pairs constructed from the high-input RNA samples was applied to classify the simulated low-input RNA samples from the single-cell data. Because only 18,308 gene pairs of the 20,390 gene pairs were measured in single cells by the Illumina HiSeq 2000 platform, we used the measured 18,308 gene pairs to classify the samples according to the majority voting rule. This random experiment was repeated for 100 times. The results showed that the average sensitivity and specificity were 100 and 73.55%, respectively. As demonstrated in the above Section, the REOs of gene pairs in the input samples with less than 250 pg RNA input tends to be less robust against the amplification bias. Therefore, for the 18,308 gene pairs, we respectively excluded 10 and 20% of pairs with the smallest average rank differences in either the normal samples or the disease samples, and used the remained gene pairs to classify the samples. For 100 random experiments, while the average sensitivity was kept at 100%, the average specificity increased to 91.82% (or 100%) when 10% (or 20%) of the gene pairs with the smallest average rank differences in either the normal samples or the disease samples were excluded.

Discussion

It is crucial to develop reliable analysis methods for the precise monitoring of global gene expression levels in limited clinical tissues in many research areas of biological and medical disciplines. For those methods based on quantitative gene expression values such as differential genes and risk score signatures, there exists large uncertainty for the low-input RNA samples due to inherent amplification bias and technical noise in the amplification procedures. However, the relative expression orderings of gene pairs are tolerant to these issues, which suggests us that we should take the advantage of the robustness of REOs to gain more reliable biological insight.

We compared serially diluted RNA samples to evaluate the impact of amplification techniques for low-input RNA samples on the gene expression profile measurements. As displayed in the study, thousands of genes had at least 2 folds-change of expression measurements in the low-input RNA samples compared with the paired high-input RNA samples due to the amplification procedure. Consequently, for the transcriptional signatures based on the quantitative expression levels, the risk threshold values determined from high-input RNA samples could not be applied to low-input RNA samples directly and vice versa. In contrast, we found that approximately 90% of REOs of gene pairs in high-input RNA samples were maintained in the diluted 1000 pg, low-input mRNA samples which were amplified and profiled by Smart-seq, DP-seq and CEL-seq techniques using the Illumina HiSeq 2000 platform. For the low-input total samples which were amplified and profiled by

the Whole-Genome DASL technique using the Illumina HumanRef-8 v3.0 platform, at least 90% of REOs of gene pairs in the high-input samples were maintained in the diluted 1000 pg and 250 pg input samples but unstable in the 50 pg and 10 pg input samples to a certain degree.

Our REO-based method facilitates gene expression profiling analysis in the context where the starting RNA material is extremely limited. A problem with the current study is that we cannot find appropriate data to verify the clinical value of the REOs-based signature. For the future study, it is worthwhile to further evaluate the method using clinically meaningful low-input RNA data such like tissues from minimally invasive tissue biopsy techniques and single-cell samples.

Conclusions

Thousands of genes have at least 2 folds-change of expression measurements in low-input mRNA and total RNA samples compared with the corresponding paired high-input samples. In contrast, most of the REOs of gene pairs in the high-input samples are maintained in the diluted low-input samples. Therefore, REOs-based disease signatures determined from high-input samples can be robustly applied to low-input samples.

Methods
Data sources and data preprocessing
All the gene expression data analyzed in this study were downloaded from the GEO database (http://www.ncbi.nlm.nih.gov/geo/), as described in details in Fig. 1 and Table 1. In Fig. 1, there are 2 datasets including mRNA sequencing data and whole genome gene expression data which were used to evaluate the amplification bias. In Table 1, there are, in total, 6 datasets of high-input RNA tissue samples, including 4 sets which were used to obtain the classification signature between breast cancer and lymphoma cancer and 2 datasets which were used to obtain the classification signature between colon tumor tissues and normal tissues.

The gene expression profiles of dataset GSE50856 were measured by the Illumina HiSeq 2000 platform for the low-input mRNA samples collected from day-4 embroid bodies of mouse embryonic stem cells (mESCs) differentiated in serum free media with and without Activin A treatment. The control samples were labeled with "SFM" and the Activin A-treated samples were labeled with "AA1000". The low-input samples were amplified and profiled by Smart-seq, DP-seq and CEL-seq techniques using the Illumina HiSeq 2000 platform, with those of the paired high-input (50 ng) mRNA samples. Based on the amplification methods and cell line treatment status, the dataset was divided into 6 groups: SFM-Smart, SFM-DP, SFM-CEL, AA100-Smart, AA100-DP and AA100-CEL. Each group had four input levels,

1000 pg, 100 pg, 50 pg and 25 pg, and each level had two technical replicates. The gene expression profile of dataset GSE17565 was measured by the Illumina HumanRef-8 v3.0 platform for two cell lines, Raji and MCF-7. The dataset was divided into 2 groups: Raji and MCF-7. The low-input total RNA samples were amplified and profiled by the Whole-Genome DASL technique using the Illumina HumanRef-8 v3.0 platform using the Illumina HumanRef-8 v3.0 platform, with those of the paired high-input (100 ng) total RNA samples. There were four input levels, 1000 pg, 100,250 pg, 50 pg and 25 10 pg as well for both cell lines, and every input level had two technical replicates (Fig. 1).

For the GSE50856 dataset, we downloaded the mappable reads that fell onto gene's exons. The experiments of standard RNA-seq, Smart-seq and DP-seq are single-end RNA-seq where every read corresponds to a single fragment. Thus, the RPKM (reads per kilobase of exon model per million mapped reads) and FPKM (fragments per kilobase of exon model per million mapped reads) metrics are conceptually analogous [35, 36], which could be used to quantify the gene expression level. The RPKM metric was estimated by the formula [37]: R = (10^9*C)/NL, where C is the number of mapped reads that fell onto the gene's exons, N is the total number of mapped reads in the experiment, and L is the sum of the exons in base pairs. On the other hand, The experiment of CEL-seq is paired-end sequencing where two reads correspond to a single fragment and only FPKM could be used to quantify the gene expression level. For the paired-end experiment, the FPKM value would be half of the RPKM value. This is not always true because in some cases only one of the two reads belonging to a fragment might be mapped. However, for most applications this simplification works [35]. The mouse mm9 genome was used for the genome annotation. By transforming the gene bank accession ID providing in the GSE50856 dataset into Entrez gene ID through the Source Batch Search database (http://source-search.princeton.edu/cgi-bin/source/sourceBatch-Search), 20,541 genes were analyzed in this dataset.

For dataset GSE17565 measured by Illumina HumanRef-8 v3.0 platform, dataset GSE10950 measured by Illumina humanRef-8 v2.0 platform and dataset GSE81861 measured by Illumina HiSeq 2000 platform, we directly downloaded the processed data. For 4 datasets of the expression profiles measured by Affymatrix microarrays, the raw expression data (.CEL files) were preprocessed using the Robust Multiarray Average algorithm [38]. For the data from TCGA, the level 3 RNA-seq datasets (RNAseqV2 RSEM) of mRNA were downloaded from the Broad Institute, Firehose (http://gdac.broadinstitute.org/runs/stddata__2016_01_28/).

Evaluation on amplification bias by fold change
For each of the measured genes, we calculated the average of its expression values in the technical replicates for

the low input and the paired high input RNA samples, respectively, and then calculated the fold changes (FCs) between the low input RNA samples and the paired high input RNA samples. We also calculated the FC between every paired low input RNA technical replicate and high input RNA technical replicates, and then calculated the coefficient of variation (CV) of the FCs.

Evaluation on REOs of gene pairs
Highly stable REOs of the gene pairs were obtained respectively from high-input RNA samples and low input RNA samples. We defined a REO as highly stable if the gene pair had identical REO direction in both technical replicates of one sample. The details are as following. The comparison of two genes in a gene pair (G_i, G_j) was viewed as an event with only two possible outcomes: the expression level of G_i was either higher or lower than that of G_j and the relative expression ordering was denoted as $G_i > G_j$ or $G_i < G_j$. If the REO of a pair was maintained in more than 99% of samples, the pair was called a highly stable gene pair. The REOs of two genes with small rank difference (i.e., close expression levels) tend to be unstable due to measurement variations.

To compare two lists of stable REOs, the consistency score, which was defined as k/n, was calculated, where n was the number of the gene pairs in the high-input RNA samples and k was the number of gene pairs with the consistent REOs in both the high-input RNA samples and low-input RNA samples.

REOs-based signature from high-input RNA samples for discriminating cancer tissues
First, we identified gene pairs each with identical REO in all samples of the two types of tissues, respectively, but with reversal REO patterns between the two types of samples. Then, we calculated the reversal degree for each gene pair as following equation,

$$\bar{R}_{ij} = \sqrt{\bar{R}_{ij(\text{lym})}\ \bar{R}_{ij(\text{bre})}}$$

where $\bar{R}_{ij(\text{lym})}$ and $\bar{R}_{ij(\text{bre})}$ are the arithmetic means of the absolute rank differences of the gene pair (i, j) in all samples of the two types of tissues, respectively.

Second, the gene pairs with reversal REOs were sorted in a descending order according to their reversal degrees. Obviously, the larger the R_{ij} value, the larger the reversal degree of the REO is between the two types of samples. Third, we selected the top k gene pairs, where k is an odd integer ranging from 1 to the total number of candidate gene pairs to classify the samples based on the majority vote rule. The value of k was chosen as the smallest number of gene pairs that reached the highest geometric mean of the sensitivity and specificity in the classification tests. Then, the selected gene-pair signature was tested in independent tissue samples measured with high-input RNA and in the cell line data measured with low-input RNA.

Performance evaluation

We called lymphoma tissue samples, colorectal tumor epithelial cells as positive samples, breast cancer tissue samples, Normal mucosa's epithelial cells paired with colorectal tumor as negative samples, and evaluated the performance of the classification signature using sensitivity and specificity which are calculated as follows:

$$Sensivity = \frac{TP}{TP + FN}$$

$$Specificity = \frac{TN}{TN + FP}$$

where TP, TN, FP and FN denote the number of true positives, true negatives, false positives and false negatives, respectively.

Statistical software for analysis

All statistical analyses were performed using the R 3.1.3 (http://www.r-project.org/). The main analyses codes are provided in the (Additional file 2).

Additional files

Additional file 1: Figure S1. The coefficient of variation (CV) of FCs. **(a)** The coefficient of variation (CV) of FCs in the three groups of SFM-DP, SFM-CEL and SFM-Smart respectively in the 25 pg, 50 pg, 100 pg and 1000 pg RNA quantity **(b) (c)** Similar as the Figure a. **Figure S2.** Maintenance of REOs after excluding 0 to 10% gene pairs. **(a)** The consistency scores between high-input RNA samples and low-input RNA samples of all gene pairs (blue) and after excluding 10% of the pairs with the smallest expression differences in the paired high-input RNA samples (pink) in the group of AA100-Smart **(b) (c) (d)** Similar as the Figure a. **Figure S3.** Maintenance of REOs after excluding 0 to 30% gene pairs. The consistency scores between high-input RNA samples and low-input RNA samples of all gene pairs, after excluding 0, 5, 10, 15, 20 and 30% of the pairs with the smallest expression differences in the paired high-input RNA samples in the group of AA100-Smart **(b) (c) (d)** Similar as the Figure a. (PDF 606 kb)

Additional file 2: The main analyses codes used in this research. (R 4 kb)

Abbreviations

AA100: Activin A treatment; AA100-CEL: Mouse embryonic stem cells differentiated in control serum free media (SFM) and the collected RNA amplified and profiled by CEL-seq using the Illumina HiSeq 2000 platform; AA100-DP: Mouse embryonic stem cells differentiated in control serum free media (SFM) and the collected RNA amplified and profiled by DP-seq using the Illumina HiSeq 2000 platform; AA100-Smart: Mouse embryonic stem cells subjected to Activin A treatment (AA100) and the collected RNA amplified and profiled by Smart-seq using the Illumina HiSeq 2000 platform; CEL-seq: Cell expression by linear amplification and sequencing; COAD: Colon adenocarcinoma; CRC: Colorectal tumors; DP-seq: Designed Primer-based RNA-sequencing strategy; FC: Fold change; GEO: Gene Expression Omnibus; REOs: Within-sample relative expression orderings; SFM: Serum free media; SFM-CEL: Mouse embryonic stem cells differentiated in control serum free media (SFM) and the collected RNA amplified and profiled by CEL-seq using the Illumina HiSeq 2000 platform; SFM-DP: Mouse embryonic stem cells differentiated in control serum free media (SFM) and the collected RNA amplified and profiled by DP-seq using the Illumina HiSeq 2000 platform; SFM-Smart: Mouse embryonic stem cells differentiated in control serum free media (SFM) and the collected RNA were amplified and profiled by Smart-seq using the Illumina HiSeq 2000 platform; SMART: Switching mechanism at 5'-end of the RNA transcript

Acknowledgements

Thank to all the individuals who participated in this study. We would also like to acknowledge the resources at GEO that facilitated this research.

Funding

This work was supported by National Natural Science Foundation of China (Grant numbers: 81,372,213, 81,572,935 and 21,534,008) and the Joint Technology Innovation Fund of Fujian Province (Grant number: 2016Y9044).

Authors' contributions

ZG, XLW and HPL conceived the project. HPL, YWL, JH and QZG performed computational experiments. HPL, HDY, RC and WCZ designed data analyses. HPL, KS, HC and YG interpreted data. HPL, XLW and ZG wrote the manuscript. All authors contributed to the preparation of the manuscript. All authors read and approved the final manuscript.

Competing interests

The authors declare that they have no competing interests.

Author details

[1]Department of Bioinformatics, Key Laboratory of Ministry of Education for Gastrointestinal Cancer, School of Basic Medical Sciences, Fujian Medical University, Fuzhou 350122, China. [2]Fujian Key Laboratory of Tumor Microbiology, Fujian Medical University, Fuzhou 350122, China. [3]Department of Systems Biology, College of Bioinformatics Science and Technology, Harbin Medical University, Harbin 150086, China. [4]Key Laboratory of Medical bioinformatics, Fujian Province, China.

References

1. De Rienzo A, Yeap BY, Cibas ES, Richards WG, Dong L, Gill RR, Sugarbaker DJ, Bueno R. Gene expression ratio test distinguishes normal lung from lung tumors in solid tissue and FNA biopsies. J Mol Diagn. 2014;16(2):267–72.
2. Libby DM, Smith JP, Altorki NK, Pasmantier MW, Yankelevitz D, Henschke CI. Managing the small pulmonary nodule discovered by CT. Chest. 2004; 125(4):1522–9.
3. Knudsen BS, Kim HL, Erho N, Shin H, Alshalalfa M, Lam LL, Tenggara I, Chadwich K, Van Der Kwast T, Fleshner N, et al. Application of a clinical whole-Transcriptome assay for staging and prognosis of prostate cancer diagnosed in needle Core biopsy specimens. J Mol Diagn. 2016;18(3):395–406.
4. Cabanski CR, Magrini V, Griffith M, Griffith OL, McGrath S, Zhang J, Walker J, Ly A, Demeter R, Fulton RS, et al. cDNA hybrid capture improves transcriptome analysis on low-input and archived samples. J Mol Diagn. 2014;16(4):440–51.
5. Soeda H, Sakudo F. NaCl and water responses across the frog tongue epithelium in vitro. Fukuoka Shika Daigaku Gakkai zasshi. 1990;17(3):251–9.
6. Wen L, Tang F. Single-cell sequencing in stem cell biology. Genome Biol. 2016;17:71.

7.   Liang J, Cai W, Sun Z. Single-cell sequencing technologies: current and future. J Genet Genomics. 2014;41(10):513–28.

8.   Datta S, Malhotra L, Dickerson R, Chaffee S, Sen CK, Roy S. Laser capture microdissection: big data from small samples. Histol Histopathol. 2015; 30(11):1255–69.

9.   Ramskold D, Luo S, Wang YC, Li R, Deng Q, Faridani OR, Daniels GA, Khrebtukova I, Loring JF, Laurent LC, et al. Full-length mRNA-Seq from single-cell levels of RNA and individual circulating tumor cells. Nat Biotechnol. 2012;30(8):777–82.

10. Bhargava V, Ko P, Willems E, Mercola M, Subramaniam S. Quantitative transcriptomics using designed primer-based amplification. Sci Rep. 2013;3:1740.

11. Hashimshony T, Wagner F, Sher N, Yanai I. CEL-Seq: single-cell RNA-Seq by multiplexed linear amplification. Cell Rep. 2012;2(3):666–73.

12. Bhargava V, Head SR, Ordoukhanian P, Mercola M, Subramaniam S. Technical variations in low-input RNA-seq methodologies. Sci Rep. 2014;4:3678.

13. Kolodziejczyk AA, Kim JK, Svensson V, Marioni JC, Teichmann SA. The technology and biology of single-cell RNA sequencing. Mol Cell. 2015;58(4): 610–20.

14. Boelens MC, te Meerman GJ, Gibcus JH, Blokzijl T, Boezen HM, Timens W, Postma DS, Groen HJ, van den Berg A. Microarray amplification bias: loss of 30% differentially expressed genes due to long probe - poly(a)-tail distances. BMC Genomics. 2007;8:277.

15. Spiess AN, Mueller N, Ivell R. Amplified RNA degradation in T7-amplification methods results in biased microarray hybridizations. BMC Genomics. 2003;4(1):44.

16. Oshlack A, Wakefield MJ. Transcript length bias in RNA-seq data confounds systems biology. Biol Direct. 2009;4:14.

17. van Haaften RI, Schroen B, Janssen BJ, van Erk A, Debets JJ, Smeets HJ, Smits JF, van den Wijngaard A, Pinto YM, Evelo CT. Biologically relevant effects of mRNA amplification on gene expression profiles. BMC Bioinformatics. 2006;7:200.

18. Degrelle SA, Hennequet-Antier C, Chiapello H, Piot-Kaminski K, Piumi F, Robin S, Renard JP, Hue I. Amplification biases: possible differences among deviating gene expressions. BMC Genomics. 2008;9:46.

19. Tariq MA, Kim HJ, Jejelowo O, Pourmand N. Whole-transcriptome RNAseq analysis from minute amount of total RNA. Nucleic Acids Res. 2011;39(18):e120.

20. Faherty SL, Campbell CR, Larsen PA, Yoder AD. Evaluating whole transcriptome amplification for gene profiling experiments using RNA-Seq. BMC Biotechnol. 2015;15:65.

21. Clement-Ziza M, Gentien D, Lyonnet S, Thiery JP, Besmond C, Decraene C. Evaluation of methods for amplification of picogram amounts of total RNA for whole genome expression profiling. BMC Genomics. 2009;10:246.

22. Qi L, Chen L, Li Y, Qin Y, Pan R, Zhao W, Gu Y, Wang H, Wang R, Chen X, et al. Critical limitations of prognostic signatures based on risk scores summarized from gene expression levels: a case study for resected stage I non-small-cell lung cancer. Brief Bioinform. 2016;17(2):233–42.

23. Xu H, Guo X, Sun Q, Zhang M, Qi L, Li Y, Chen L, Gu Y, Guo Z, Zhao W. The influence of cancer tissue sampling on the identification of cancer characteristics. Sci Rep. 2015;5:15474.

24. Cheng J, Guo Y, Gao Q, Li H, Yan H, Li M, Cai H, Zheng W, Li X, Jiang W, et al. Circumvent the uncertainty in the applications of transcriptional signatures to tumor tissues sampled from different tumor sites. Oncotarget. 2017;8(18):30265–75.

25. Freidin MB, Bhudia N, Lim E, Nicholson AG, Cookson WO, Moffatt MF. Impact of collection and storage of lung tumor tissue on whole genome expression profiling. J Mol Diagn. 2012;14(2):140–8.

26. Chen R, Guan Q, Cheng J, He J, Liu H, Cai H, Hong G, Zhang J, Li N, Ao L, et al. Robust transcriptional tumor signatures applicable to both formalin-fixed paraffin-embedded and fresh-frozen samples. Oncotarget. 2017;8(4): 6652–62.

27. Tan AC, Naiman DQ, Xu L, Winslow RL, Geman D. Simple decision rules for classifying human cancers from gene expression profiles. Bioinformatics. 2005;21(20):3896–904.

28. Geman D, d'Avignon C, Naiman DQ, Winslow RL. Classifying gene expression profiles from pairwise mRNA comparisons. Stat Appl Genet Mol Biol. 2004;3:Article19.

29. Zhao W, Chen B, Guo X, Wang R, Chang Z, Dong Y, Song K, Wang W, Qi L, Gu Y, et al. A rank-based transcriptional signature for predicting relapse risk of stage II colorectal cancer identified with proper data sources. Oncotarget. 2016;7(14):19060–71.

30. Qi L, Li Y, Qin Y, Shi G, Li T, Wang J, Chen L, Gu Y, Zhao W, Guo Z. An individualised signature for predicting response with concordant survival benefit for lung adenocarcinoma patients receiving platinum-based chemotherapy. Br J Cancer. 2016;115(12):1513–9.

31. Cai H, Li X, Li J, Ao L, Yan H, Tong M, Guan Q, Li M, Guo Z. Tamoxifen therapy benefit predictive signature coupled with prognostic signature of post-operative recurrent risk for early stage ER+ breast cancer. Oncotarget. 2015;6(42):44593–608.

32. Li X, Cai H, Zheng W, Tong M, Li H, Ao L, Li J, Hong G, Li M, Guan Q, et al. An individualized prognostic signature for gastric cancer patients treated with 5-fluorouracil-based chemotherapy and distinct multi-omics characteristics of prognostic groups. Oncotarget. 2016;7(8):8743–55.

33. Ao L, Song X, Li X, Tong M, Guo Y, Li J, Li H, Cai H, Li M, Guan Q, et al. An individualized prognostic signature and multiomics distinction for early stage hepatocellular carcinoma patients with surgical resection. Oncotarget. 2016;7(17):24097–110.

34. Guan Q, Chen R, Yan H, Cai H, Guo Y, Li M, Li X, Tong M, Ao L, Li H, et al. Differential expression analysis for individual cancer samples based on robust within-sample relative gene expression orderings across multiple profiling platforms. Oncotarget. 2016;7(42):68909–20.

35. Vikman P, Fadista J, Oskolkov N. RNA sequencing: current and prospective uses in metabolic research. J Mol Endocrinol. 2014;53(2):R93–101.

36. Trapnell C, Williams BA, Pertea G, Mortazavi A, Kwan G, van Baren MJ, Salzberg SL, Wold BJ, Pachter L. Transcript assembly and quantification by RNA-Seq reveals unannotated transcripts and isoform switching during cell differentiation. Nat Biotechnol. 2010;28(5):511–5.

37. Mortazavi A, Williams BA, McCue K, Schaeffer L, Wold B. Mapping and quantifying mammalian transcriptomes by RNA-Seq. Nat Methods. 2008;5(7):621–8.

38. Irizarry RA, Hobbs B, Collin F, Beazer-Barclay YD, Antonellis KJ, Scherf U, Speed TP. Exploration, normalization, and summaries of high density oligonucleotide array probe level data. Biostatistics. 2003;4(2):249–64.

4

Rapid evolutionary divergence of diploid and allotetraploid *Gossypium* mitochondrial genomes

Zhiwen Chen[1], Hushuai Nie[1], Yumei Wang[2], Haili Pei[1], Shuangshuang Li[1], Lida Zhang[3] and Jinping Hua[1*]

Abstract

Background: Cotton (*Gossypium* spp.) is commonly grouped into eight diploid genomic groups and an allotetraploid genomic group, AD. The mitochondrial genomes supply new information to understand both the evolution process and the mechanism of cytoplasmic male sterility. Based on previously released mitochondrial genomes of *G. hirsutum* (AD$_1$), *G. barbadense* (AD$_2$), *G. raimondii* (D$_5$) and *G. arboreum* (A$_2$), together with data of six other mitochondrial genomes, to elucidate the evolution and diversity of mitochondrial genomes within *Gossypium*.

Results: Six *Gossypium* mitochondrial genomes, including three diploid species from D and three allotetraploid species from AD genome groups (*G. thurberi* D$_1$, *G. davidsonii* D$_{3-d}$ and *G. trilobum* D$_8$; *G. tomentosum* AD$_3$, *G. mustelinum* AD$_4$ and *G. darwinii* AD$_5$), were assembled as the single circular molecules of lengths about 644 kb in diploid species and 677 kb in allotetraploid species, respectively. The genomic structures of mitochondrial in D group species were identical but differed from the mitogenome of *G. arboreum* (A$_2$), as well as from the mitogenomes of five species of the AD group. There mainly existed four or six large repeats in the mitogenomes of the A + AD or D group species, respectively. These variations in repeat sequences caused the major inversions and translocations within the mitochondrial genome. The mitochondrial genome complexity in *Gossypium* presented eight unique segments in D group species, three specific fragments in A + AD group species and a large segment (more than 11 kb) in diploid species. These insertions or deletions were most probably generated from crossovers between repetitive or homologous regions. Unlike the highly variable genome structure, evolutionary distance of mitochondrial genes was 1/6th the frequency of that in chloroplast genes of *Gossypium*. RNA editing events were conserved in cotton mitochondrial genes. We confirmed two near full length of the integration of the mitochondrial genome into chromosome 1 of *G. raimondii* and chromosome A03 of *G. hirsutum*, respectively, with insertion time less than 1.03 MYA.

Conclusion: Ten *Gossypium* mitochondrial sequences highlight the insights to the evolution of cotton mitogenomes.

Keywords: Mitochondrial genomes, Comparative genomics, Multiple DNA rearrangement, Unique segments, Repeat sequences, *Gossypium*

Background

Plant mitochondrial genomes (mtDNA) embrace notable characteristics, such as an extreme and highly diverse mitochondrial genome structure [1–4]. Plant mitochondrial genomes also possess highly branched and sigma-like structures [5–7] as well as multichromosomal genomes recently identified in three distantly-related angiosperm lineages [8–10]. The mitochondrial genome in plants is also noteworthy in that there is large variation in genome size (ranging from ~66 kb to 11.3 Mb) [8, 11] with highly variable intergenetic regions and a considerable proportion of repeated sequences [12], frequent rearrangements [13], massive gene loss [14], and frequent endogenous and foreign DNA transfer [15–17].

In terms of structure, angiosperm mitochondrial genomes are typically mapped as circular molecules with one or more larger (>1 kb) repetitive sequences, which promote active homologous inter- and intra-genomic recombination

* Correspondence: jinping_hua@cau.edu.cn
[1]Laboratory of Cotton Genetics, Genomics and Breeding /Key Laboratory of Crop Heterosis and Utilization of Ministry of Education/Beijing Key Laboratory of Crop Genetic Improvement, College of Agronomy and Biotechnology, China Agricultural University, Beijing 100193, China
Full list of author information is available at the end of the article

[4, 18, 19]. However, it is not clear how plant mitochondrial genomes rearrange so frequently or how the genome sizes can vary dramatically over relatively short evolutionary period. This dynamic organization of the angiosperm mitochondrial genome provides unique information as well as an appropriate model system for studying genome structure and evolution. More syntenic sequences will be helpful to interpret the evolutionary processes for diverse angiosperm mitochondrial structures.

Cotton (*Gossypium*) is the most important fiber crop plant in the world [20]. Four domesticated species remain as cultivated crops: the New World allopolyploid species *G. hirsutum* and *G. barbadense* (2n = 52), and the Old World diploid species *G. arboreum* and *G. herbaceum* (2n = 26) [20, 21]. The primary cultivated one is Upland cotton (*G. hirsutum* L.), accounting for more than 90% of global cotton fiber output. *Gossypium* includes 52 species: seven allotetraploid species and 45 diploids [21, 22]. The nascent allopolyploid species spread throughout the American tropics and subtropics, diverging into at least seven species, namely, *G. hirsutum* L. (AD$_1$), *G. barbadense* L. (AD$_2$), *G. tomentosum* Nuttalex Seemann (AD$_3$), *G. mustelinum* Miersex Watt (AD$_4$), *G. darwinii* Watt (AD$_5$), *G. ekmanianum* (AD$_6$), and *G. stephensii* (AD$_7$) [20–22]. The diploid *Gossypium* species comprise eight monophyletic genome groups, A, B, C, D, E, F, G and K group [20, 23]. With the rapid development of next-generation sequencing technologies [24, 25], cotton genomics research has rapidly progressed in recent years, such that nuclear genome sequences have now been published for model diploids (D$_5$-genome [26, 27], A$_2$-genome [28]), and for the allopolyploids (AD$_1$-*G. hirsutum* [29, 30], AD$_2$-*G. barbadense* [31, 32]). In addition, a large number of *Gossypium* organelle genome sequences have been released [33–39]. Compared to the highly conserved chloroplast genome structures [34–36], comparative analysis revealed rapid evolutionary divergence of *Gossypium* mitochondrial genomes [37–39], which proved that deep analyses of more mitochondrial genomes would provide new data to consider the evolutionary relationships and to explore the mechanism of cytoplasmic male sterility (CMS).

Cytoplasmic male sterility (CMS) is a maternally-conferred reproductive trait that relies on the expression of CMS-inducing mitochondrial sequences [40]. Many examples of CMS stem from the consequences of recombination [40–42]. Often, these chimeric CMS genes exhibit co-transcription with upstream or downstream functional genes, which typically affect the mitochondrial electron transfer chain pathways to fail to produce functional pollen [43]. Rearrangements in the mitochondrial DNA involving known mitochondrial genes as well as unknown sequences result in the creation of new chimeric open reading frames, which encode proteins containing transmembrane and lead to cytoplasmic male sterility by interacting with nuclear-encoded genes [43–45].

Here, six *Gossypium* mitochondrial genomes are reported, including three diploid species from D genome groups (*G. thurberi* D$_1$, *G. davidsonii* D$_{3-d}$ and *G. trilobum* D$_8$) and three allotetraploid species from AD genome groups (*G. tomentosum* AD$_3$, *G. mustelinum* AD$_4$ and *G. darwinii* AD$_5$). Comparative mitochondrial genome analysis then revealed rapid mitochondrial genome rearrangement and evolution between diploid and allotetraploid *Gossypium*. In addition, one of the most surprising outcomes of comparative analyses is how rapidly mitochondrial sequence segments altered within a single subspecies. Finally, the four mitogenomes of D group species provided the useful data resources for interpreting the CMS-related genes in *G. trilobum* D$_8$ cotton.

Methods

Plant materials and mitochondrial DNA extraction

Seeds of diploid and allotetraploid *Gossypium* species were acquired from the nursery on the China National Wild Cotton Plantation in Sanya, Hainan, China. Mitochondria were isolated from week-old etiolated seedlings, and the mitochondrial DNA samples were extracted from an organelle-enriched fraction isolated by differential and sucrose gradient centrifugation, essentially as described earlier [37–39, 46].

Mitochondrial genome sequencing and primary data processing

A total of ~5 million clean paired-end reads were sequenced from a ~500 bp library for each of three diploid species, respectively. We produced 300 bp read length with paired-end sequencing, using MiSeq sequencing method on Illumina platform at Beijing Biomarker Technologies Co, LTD. A total of ~11 million clean paired-end reads were sequenced from a ~500 bp library with paired-end, 300 bp read length, for each of three allotetraploid species, respectively, using the same method. Raw sequences were first evaluated by two quality control tools, Trimmomatic [47] and FilterReads module in Kmernator ([https://github.com/JGI-Bioinformatics/Kmernator]) to remove any potential undesirable artifacts in the data such as adapters or low quality or "N" bases and so on.

Genomes assembly and sequence verification

Six *Gossypium* draft mitogenomes were assembled de novo from the clean reads with velvet 1.2.10 [48] or combining FLASH [49] and Newbler (Version 2.53) methods, respectively. For the first assembly method, i.e., the 300-bp paired-end reads from six *Gossypium* species, we performed multiple velvet runs with different combinations of kmer values (for (kmer = 75; kmer <=209; kmer = kmer +2), (42 in total)). Three Kmer values (193, 195, 197), owning larger

N50 values, less contig number, were used to assemble the mitogenomes. For each velvet run, the minimum coverage parameter was set to 10× and scaffolding was turned off when the data sets contained paired-end reads. For each of assembly, mitochondrial contigs were identified by blastn [50] searches with known *Gossypium* mitochondrial genomes for scaffolding and gap filling [37–39]. The best draft assemblies for six *Gossypium* were chosen as the assembly that maximized total length of mitochondrial contigs after combining three Kmer values assembly. In another assembly method, we combined FLASH [49] and Newbler (Version 2.53) softwares together. First, FLASH provides the use of paired-end libraries with a fragment size (500 bp) shorter than twice the read length (300 bp) an opportunity to generate much longer reads (500 bp) by overlapping and merging read pairs [49]. The merging file was then assembled using Newbler (Version 2.53) software. Finally, the assembled mitochondrial scaffolds were aligned with known *Gossypium* mitochondrial genomes [37–39] for anchoring scaffold directions and gap filling. Thus, we combined two types of the assembly results to complete the six *Gossypium* mitogenomes. The final remaining gaps were filled by aligning individual pair-end sequence reads that overlapped the scaffolds or contig ends using Burrows-Wheeler Aligner (BWA 0.7.10-r789) software [51].

To evaluate six mitogenomes sequence assembly quality and accuracy, pair-end reads were mapped onto their respective consensus sequences with BWA 0.7.10-r789 [51]. The BWA mapping resulting SAM files were transformed into BAM files using samtools view program [52]. The BWA mapping results for these pair-end reads in BAM files were then used to calculate depth of sequencing coverage through samtools depth program [52]. For all six *Gossypium* species, the Illumina reads covered all parts of the genome consistently, with the average coverage ranging from 50× to 200 ×.

Genome annotations and sequence analyses

Gossypium mitochondrial genes from the six species were annotated using *G. hirsutum* and *G. barbadense* mitogenomes as references. Functional genes (other than tRNA genes) were identified by local blast searches against the database, whereas tRNA genes were predicted de novo using tRNA scan-SE [53]. Repeat-match program in MUMmer [54] was used to identify repeated sequences within six *Gossypium* mitogenomes. Their genome maps were generated using OGDRAW [55] and the repeat map was drawn by Circos [56].

Collinear blocks were generated among the ten mitochondrial genomes of *Gossypium* using the progressiveMauve program [57]. To determine the amount of *Gossypium* mitochondrial genome complexity shared between species, each pair of mitogenomes was aligned using blastn

[50] with an e-value cutoff of 1×10^{-5}. Using these parameters, the blastn searches should be able to detect homologous sequences as short as 30 bp. The unique segments in *Gossypium* mitogenomes identified in this study were summarized as follows: i) Paired-end reads were mapped onto their respective consensus sequences using the Burrows-Wheeler Aligner (BWA 0.7.10-r789) software [51]; ii) The BWA mapping resulting SAM files were transformed into BAM files using the samtools program [52] set to the default parameters; and iii) Structure variations (SVs) and InDels reported in this work were manually visualized using the Integrative Genomics Viewer (IGV) software [58].

RNA editing identification

RNA edit sites were computationally predicted using the batch version of the PREP-Mt. online server [59], with a cutoff value of 0.2.

Phylogenetic analyses and estimation of evolutionary divergence

For phylogenetic analyses, 36 protein-coding genes were extracted from 10 *Gossypium* species and two outgroups: *C. papaya* and *A. thaliana*. Sequence alignments for 36 concatenated genes, each chloroplast and mitochondrial coding exons were carried out by MAFFT [60]. Phylogenetic analyses were performed with the same methods to our previous studies [35, 36, 39]. P-distances for chloroplast and mitochondrial coding genes were calculated with MEGA5.05 [61].

Identifying nuclear mtDNAs in *Gossypium*, estimation of evolutionary divergence and divergence time between mitochondrial sequences and *numts*

Dot matrix comparisons were generated between the mitochondrial and nuclear chromosomes of four *Gossypium* species using the nucmer program of MUMmer with the parameters 100-bp minimal size for exact match and 500-bp minimal interval between every two matches [54]. The detailed comparison results were shown in Fig. 7: *G. raimondii* mitochondrial [39] and nuclear chromosomes [27] in Fig. 7a, *G. arboreum* mitochondrial [39] and nuclear chromosomes [28] in Fig. 7b, *G. hirsutum* mitochondrial [37] and nuclear chromosomes [30] in Fig. 7c, *G. barbadense* mitochondrial [38] and nuclear chromosomes [32] in Fig. 7d. Sequence alignments for each coding, intronic, and intergenic spacer regions were carried out by MAFFT [60] software. P-distances between mitochondrial sequences and numts were calculated with MEGA5.05 [61]. In order to estimate how old these insertions are, p-distance rates and some estimates of rate/million years were studied here. The divergence time of between mitochondrial native sequences and numts was calculated by the following Formula: $T = \text{p-distance}/(r_{nu} + r_{mt})$ [62]. Based on Gaut et al. (1996) and Muse et al. (2000), the r_{nu} and r_{mt} values were estimate as $r_{nu} = 6.5 \times 10^{-9}$ and $r_{mt} = 2 \times 10^{-10}$, respectively [63, 64]. It also has to be made clear that the underlying assumption is homogeneity in rate since their divergence from a common ancestor.

Results

Gossypium mitochondrial genomes from diploid and allotetraploid species

Six *Gossypium* mitochondrial genomes were obtained in present study, including three diploid D species and three allotetraploid AD species. Complete mitochondrial DNA sequences were deposited in the GenBank database respectively: *G. thurberi* D_1 (Accession No. KR736343), *G. davidsonii* D_{3-d} (Accession No. KR736344), *G. trilobum* D_8(Accession No. KR736346), *G. tomentosum* AD_3 (Accession No. KX388135), *G. mustelinum* AD_4 (Accession No. KX388136) and *G. darwinii* AD_5 (Accession No. KX388137) (Table 1). The six *Gossypium* mitogenomes were all assembled as single circular molecules of lengths about 644 kb in diploid (Additional file 1: Figure S1A) and 677 kb in allotetraploid (Additional file 1: Figure S1B), respectively. The genomic structures were identical within diploid group (Fig. 1) and allotetraploid group (Fig. 2), respectively, but differed between the two groups. The diploid group had six large repeats (>1 kb) whereas the allotetraploid group had four large repeats (Fig. 1; Fig. 2), which may be involved in the rearranged mitogenome organizations in diploid and allotetraploid *Gossypium*.

Comparison of mitochondrial gene content among *Gossypium* species reveals a conserved pattern of evolutionary stasis for diploid and allotetraploid species, respectively (Table 1). The *Gossypium* mitogenomes contain 36 protein coding genes with five genes (*rps1*, *rps2*, *rps11*, *rps13* and *rps19*) being lost during coevolution with

nucleus, compared to the common ancestor of seed plants [65, 66]. The repeat sequences confer some redundant gene copies (*nad4*, *nad9* and *mttB*) in three allotetraploid species with uncertain functions (Additional file 2: Table S1). These mitochondrial genomes (Table 1) show high identity in gene content but no similarity in genome organization (Fig. 1; Fig. 2) with each other or with previously published cotton mitochondrial genomes [37–39], with apparent major differences in genome organization and size.

Syntenic regions and rearrangement

After combining four other *Gossypium* mitochondrial genomes [37–39], totally, ten species of five allotetraploid and five diploid were used for analyses: one from A genome, four from D genome and five from AD genome groups, respectively (Table 1). Syntenic regions were identified between ten *Gossypium* mitochondrial genomes with eight large major syntenic blocks (Fig. 3). The genomic structures were totally identical in four species of D group, indicating that the mitochondrial genome structures may be highly conserved in D genome species. *G. trilobum* (D_8) contributed the cytoplasmic male sterility (CMS) cytoplasm in cotton [67–69], however, no genome rearrangement or large indel segments variations compared with mito-genomes of other D species, implying that the mitochondrial CMS-associated gene in cotton may function with different mechanism.

In addition, compared to D group species, the mitochondrial genomic structure in A group (A_2) was highly rearranged (Fig. 3). Interestingly, genome rearrangements also occurred among the five allotetraploid species, as was already reported in the *G. hirsutum* - *G. barbadense* comparison [38]. Despite the fact that three allotetraploid

Table 1 Main features of the ten assembled *Gossypium* mitogenomes

Genome Characteristics	G. thurberi	G. davidsonii	G. raimondii	G. trilobum	G. arboreum	G. hirsutum	G. barbadense	G. tomentosum	G. mustelinum	G. darwinii
Accession	KR736343	KR736344	KR736345	KR736346	KR736342	JX065074	KP898249	KX388135	KX388136	KX388137
Genome groups	D_1	D_{3-d}	D_5	D_8	A_2	AD_1	AD_2	AD_3	AD_4	AD_5
Genome size (bp)	644,395	644,311	643,914	644,460	687,482	621,884	677,434	677,295	677,306	677,286
Circular chromosomes	1	1	1	1	1	1	1	1	1	1
Percent G + C content	44.92	44.94	44.94	44.92	44.94	44.95	44.98	44.98	44.98	44.97
Protein genes	36	36	36	36	39	36	39	39	39	39
tRNA genes	27	27	27	27	30	29	30	30	30	30
Native	17	17	17	17	17	17	17	17	17	17
Plastid-derived	10	10	10	10	13	12	13	13	13	13
tRNAs with introns	3	3	3	3	4	3	4	4	4	4
rRNA genes	7*	7*	7*	7*	7*	4	6	6	6	6
Large repeats: >1 kb(number)	6	6	6	6	4	4	4	4	4	4

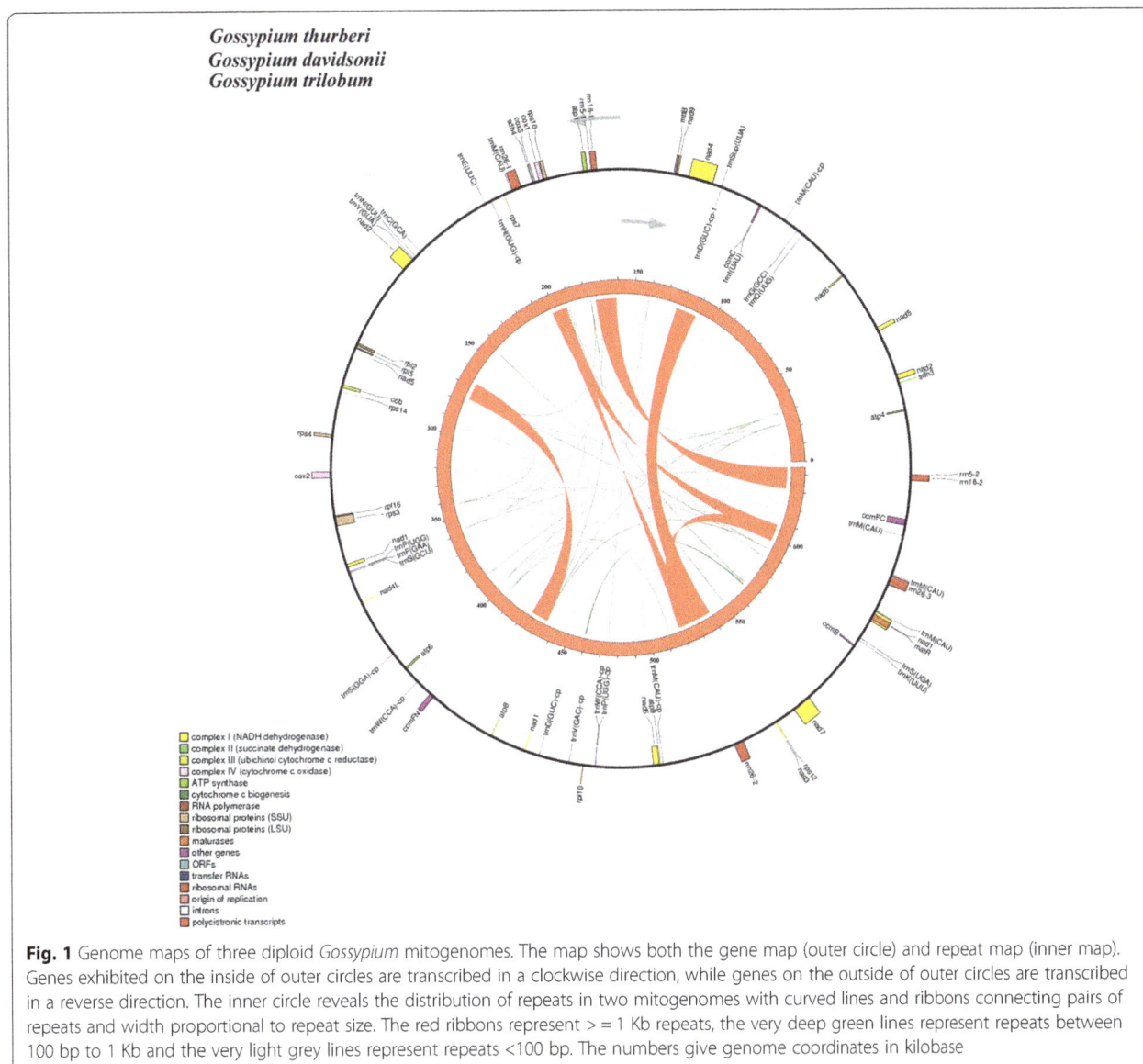

Fig. 1 Genome maps of three diploid *Gossypium* mitogenomes. The map shows both the gene map (outer circle) and repeat map (inner map). Genes exhibited on the inside of outer circles are transcribed in a clockwise direction, while genes on the outside of outer circles are transcribed in a reverse direction. The inner circle reveals the distribution of repeats in two mitogenomes with curved lines and ribbons connecting pairs of repeats and width proportional to repeat size. The red ribbons represent > = 1 Kb repeats, the very deep green lines represent repeats between 100 bp to 1 Kb and the very light grey lines represent repeats <100 bp. The numbers give genome coordinates in kilobase

species (*G. tomentosum* AD$_3$, *G. mustelinum* AD$_4$ and *G. darwinii* AD$_5$) exhibited the same genome organization, a disorder existed between the mitogenomes of *G. hirsutum* AD$_1$ and *G. barbadense* AD$_2$ (Fig. 3).

Gene order and repeat sequences

To uncover the formation mechanism of recombination generating multiple genomic arrangements in *Gossypium*, we presented the gene order with five major linear models and genes located in repeat regions shown in bold (Additional file 3: Figure S2). The gene orders in the D genome species are highly conserved but not identical to that in either *G. arboreum* (A$_2$) or the AD groups with six-seven gene clusters scattered. Though there exists a few changes in mitochondrial gene order within each of five models in the three *Gossypium* lineages as

shown by ten released mitogenomes, a minimum of two and three changes (inversions and translocations) need to be invoked to explain the differences of gene order among diploid and allotetraploid *Gossypium*, respectively (Fig. 3), and how these genomic rearrangements events happened are difficult to reconstruct. Repeat sequences have been suggested to serve as sites of homologous recombination, resulting in gene order changes in mitochondrial genomes [19].

The repeat sequences detected in the *Gossypium* mitogenomes in present and earlier studies [37–39] may be responsible for mitochondrial gene order changes between diploids and allotetraploids (Table 2). There mainly existed six or four large repeats in D group or (A + AD) groups, respectively. The repeat sizes were almost identical in D group but differed in (A + AD) groups. Despite a big

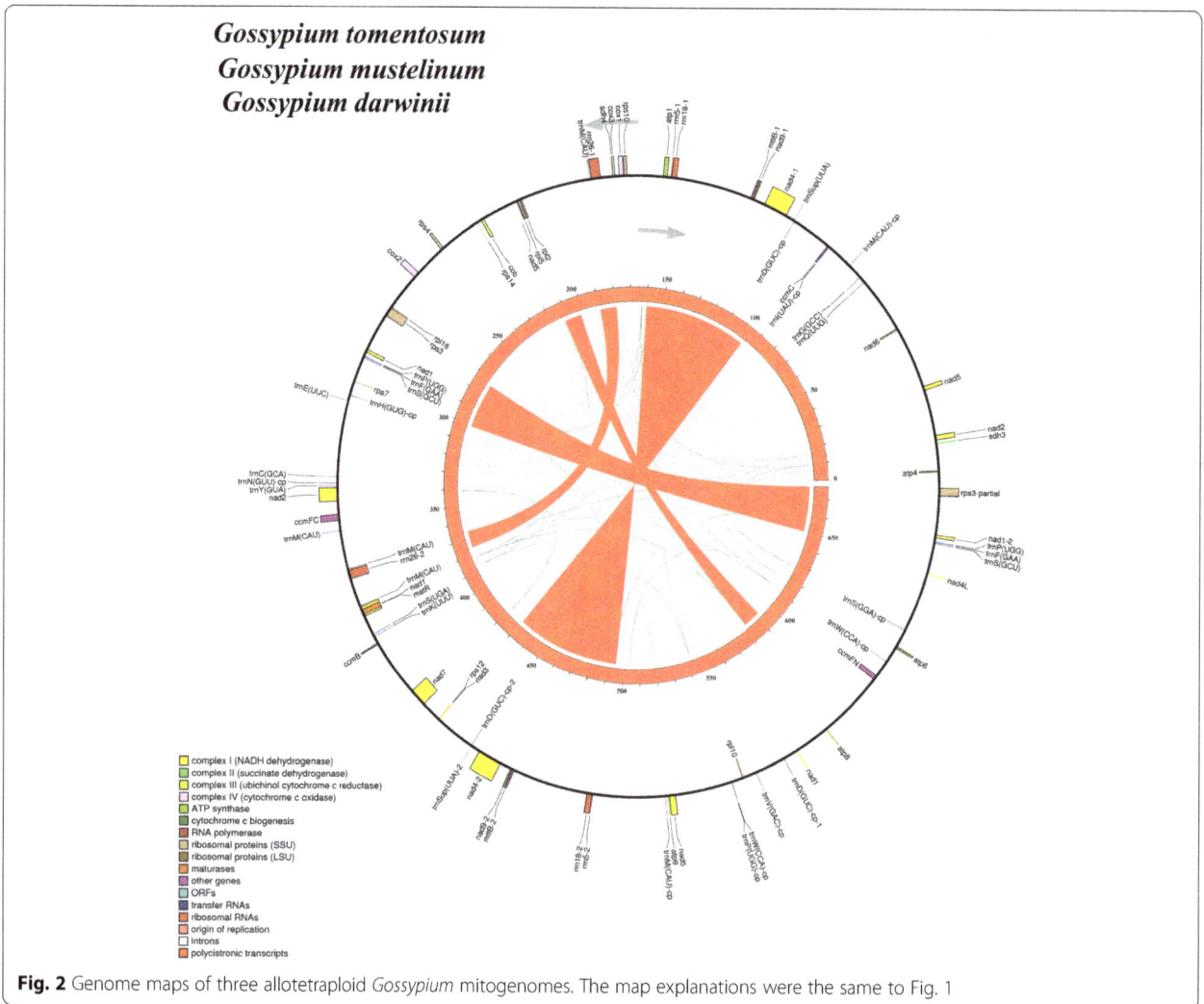

Fig. 2 Genome maps of three allotetraploid *Gossypium* mitogenomes. The map explanations were the same to Fig. 1

deletion, about 50 kb that occurred in R1 of *G. hirsutum* (AD₁), a 27 kb repeat was unique to the AD group. In addition, the repeat diverged considerably between the two diploid *Gossypium* groups (Table 2). In addition, genes in the border of the gene clusters in *Gossypium* were almost located in or close to the repeat sequences (Additional file 3: Figure S2). These variations in repeat sequences may perhaps cause the major inversions and translocations within the mitochondrial genome of the common ancestor shared by D-A and A-AD species after *Gossypium* had diverged. Evolution of gene order in diploid D group mitogenomes of *Gossypium* is overall quite conservative, but exists divergence between different diploid and allotetraploid lineages.

Conservation and variants in *Gossypium* mitochondrial genomes

Considering that all the *Gossypium* mitogenomes have similar genome complexity, comparative analysis were conducted to determine the proportion of the sequences

that each shared in common with the others (Table 3). One of the most surprising outcomes is how rapidly sequence segments were gained or lost. Genome specific fragments is not present in any two genomes of the D or AD groups, respectively (Table 3). While reciprocity is generally not seen in any other comparisons, even between the two diploid mitogenome groups: *G. arboreum* (A₂) lost 2.97% of the sequences present in D group, but D group lost 0.77% of the A species' sequences. *G. arboreum* (A₂) is attributed to be the putative maternal contributor to the progenitor of AD group [21, 34, 70], however, each of the five AD group genomes has lost substantial amounts of sequence that is present in the *G. arboreum* (A₂) genomes and vice versa (Table 3). The difference is more striking when comparing mitogenomes of D group and AD group: D group species lost only 0.64% of the AD group mitogenomes, but AD group species lost 4.41% of the D group mito-genomes. Reciprocal differences were more apparent in the comparisons between male-fertile and CMS (cytoplasmic male sterility) mitochondrial genomes [71–73].

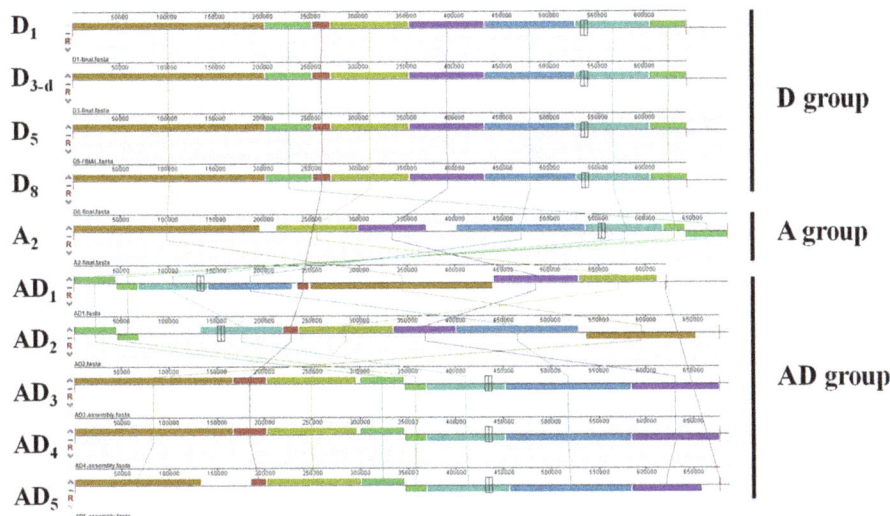

Fig. 3 Progressive Mauve show the genome size variation and the global rearrangement structure of 10 mitochondrial chromosomes among *Gossypium*. The mitogenome of *G. arboreum* (A_2) is the largest with a circular DNA molecule of 687,482 base pairs (bp) while the smallest mitogenome (from *G. hirsutum* AD_1) is only 621,884 bp. Each genome is laid out horizontally and homologous segments are shown as colored blocks connected across genomes. Blocks that are shifted downward in any genome represented segments that are inverted relative to the reference genome (*G. thurberi* D_1)

In fact, the genome complexity in *Gossypium* presented eight unique segments ranging from 108 bp to 7888 bp in length in D group mitochondrial genomes, comprising a total of 18,194 bp (Indels of <100 bp were not included) (Fig. 4a, showing one of the unique segments); while three specific fragments detected in (A + AD) group mitochondrial genomes with the largest size 3876 bp in length, 4315 bp in total (Fig. 4b). In addition, a large segment more than 11 kb in length that is present in the diploid mitogenome is not present in any of the other five allotetraploid mitogenomes (Fig. 4c). Despite the fact that the ancestor of A-genome group is the maternal source of extant allotetraploid species [20–23, 34], unique presence/absence variations existed as well (Fig. 4c; Additional file 4: Figure S3).

RNA editing in *Gossypium* mitochondrial genes

Post-transcriptional RNA-editing of mitochondrial genes is both ubiquitous and important for regulation [74]. Typically, RNA editing of mitochondrial transcripts in flowering plants occurs in coding regions of mitochondrial transcripts to convert specific cytosine residues to uracil (C → U) [75, 76]. For ten *Gossypium* species, we predicted sites of C-to-U editing using the PREP-Mt. online tool [59] with a cutoff of 0.2. The number of predicted C-to-U edits across the entire coding regions of their shared 36 protein genes is almost similar for ten *Gossypium* species (451), with one editing site lower in *G. hirsutum* (450 sites) caused in *nad3* gene (Table 4). The simplest interpretation of these results is that the

Table 2 Major repeat sizes (>1 kb) in *Gossypium* mitochondrial genomes

Gossypium species	Repeat 1 (kb)	Repeat 2 (kb)	Repeat 3 (kb)	Repeat 4 (kb)	Repeat 5 (kb)	Repeat 6 (kb)
G. thurberi	12,921	12,669	10,632	9121	8544	7317
G. davidsonii	12,937	12,669	10,624	9121	8544	7317
G. raimondii	12,741	12,670	10,624	9121	8544	7317
G. trilobum	12,921	12,669	10,650	9121	8544	7317
G. arboreum	63,789	32,975	15,029	10,246		
G. hirsutum	10,302	27,495	10,623	10,251		
G. barbadense	63,904	26,936	10,615	10,246		
G. tomentosum	63,888	27,425	10,614	10,246		
G. mustelinum	63,886	27,425	10,614	10,246		
G. darwinii	63,893	27,425	10,614	10,246		

Table 3 Percentage of *Gossypium* mitochondrial genome complexity that is absent in other genomes

	vs	D_1	D_{3-d}	D_5	D_8	A_2	AD_1	AD_2	AD_3	AD_4	AD_5
References genomes	D_1		0	0	0	0.77	0.64	0.64	0.64	0.64	0.64
	D_{3-d}	0		0	0	0.77	0.64	0.64	0.64	0.64	0.64
	D_5	0	0		0	0.77	0.64	0.64	0.64	0.64	0.64
	D_8	0	0	0		0.77	0.64	0.64	0.64	0.64	0.64
	A_2	2.97	2.97	2.97	2.97		0.12	0.12	0.12	0.12	0.12
	AD_1	4.41	4.41	4.41	4.41	1.63		0	0	0	0
	AD_2	4.41	4.41	4.41	4.41	1.63	0		0	0	0
	AD_3	4.41	4.41	4.41	4.41	1.63	0	0		0	0
	AD_4	4.41	4.41	4.41	4.41	1.63	0	0	0		0
	AD_5	4.41	4.41	4.41	4.41	1.63	0	0	0	0	

Note: Small deletions (< 100 bp) are not considered to be missing segments. D_1 = *G. thurberi*, D_{3-d} = *G. davidsonii*, D_5 = *G. raimondii*, D_8 = *G. trilobum*, A_2 = *G. arboreum*, AD_1 = *G. hirsutum*, AD_2 = *G. barbadense*, AD_3 = *G. tomentosum*, AD_4 = *G. mustelinum*, AD_5 = *G. darwinii*

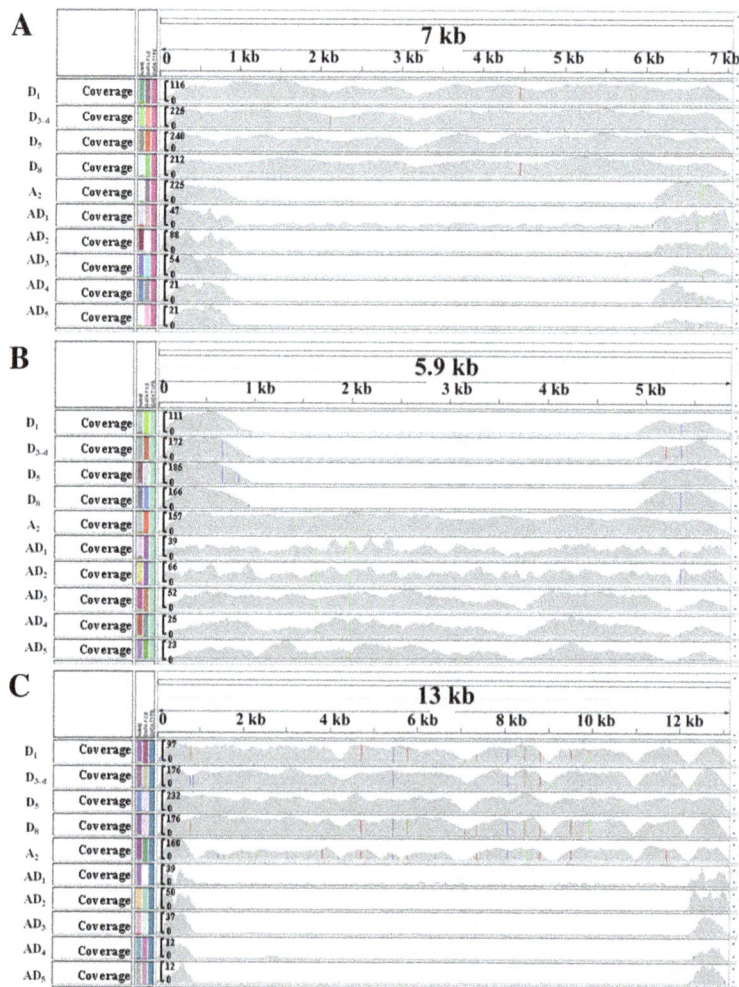

Fig. 4 Observed coverage of mapped paired-end reads supporting the existence of a large insertion or deletion in *Gossypium* species. IGV screenshot of the variability and coverage observed in ten *Gossypium* sequence samples. Upper panel represent the unique sequences coordinates. There are ten panels corresponding to the different *Gossypium* sequences. The track in each of these panels describes the density of read mapping or coverage depth. **a**: unique segment in D group. **b**: unique segment in A + AD group. **c**: unique segment in diploid groups

whole of edit sites in ten *Gossypium* species were present in their common ancestor, while the species-specific sites are less derived in cotton.

Mitochondrial genome evolution in *Gossypium*

Phylogenetic relationships among 10 *Gossypium* species with two outgroups, was generated using a concatenated analysis of 36 mitochondrial protein-coding genes (Fig. 5). The topology of the resulting tree supports *G. arboreum* as the maternal donor to polyploid cotton species, which further supports our former result [39]. We mapped these specific indels into the phylogenetic clades, as shown in Fig. 5, which implied an ongoing dynamic divergence process. First, eight mitogenome fragments (U1-U8) were involved in loss events after *G. arboreum* (A$_2$) diverged from a common ancestor shared with the D genome group. Subsequently, three genome fragments (U9-U11) were transferred from the nucleus to the mitochondrial genome in an A-genome ancestor or contributor/donator before the formation of the allopolyploidization event. Finally, a large genome fragment (U12), about 11 kb, was lost during the divergence process of allotetraploid ancestor species (Fig. 5). This 11 kb deletion (corresponding to U12 in diploid mitogenomes) was adjacent to the specific repeat sequence R2 in AD group, which might lead to the formation of R2 repeat sequences (unique to AD group species) during evolution. In addition, variations in repeat and Indels lengths also cause the great difference of *Gossypium* mitochondrial genome sizes (Fig. 3 and Fig. 5). MtDNA intergenic regions are known to possess more unique segments than genic regions, however, shorter repeats account for the relatively small size of the D-group mitochondrial genomes. Interestingly, a large deletion ~50 kb in length in R1 may lead to the small size of *G. hirsutum* (AD$_1$) [37], compared to the other four allotetraploid genomes. Comparative mitochondrial genome analysis revealed rapid mitochondrial genome rearrangement and evolution even within a single subspecies.

In addition, we calculated the p-distances representing evolutionary divergence from 78 chloroplast and 36 mitochondrial protein-coding exons among 10 *Gossypium* species, as shown in Fig. 6. Here, the average evolutionary divergence was 0.0031 in chloroplast genes but only 0.0005 in mitochondrial genes among 10 *Gossypium* species. The mitochondrial genes were highly conserved with low evolutionary divergence, however, their genome structures displayed the extremely rapid evolution of various changes, including repeat and large indels variations. Based on these results, the evolutionary distance of mitochondrial genomes are much lower than the chloroplast genomes in *Gossypium*, however, rapid varying mitogenome structures evolves much faster than the highly conserved chloroplast genomes [35–39].

MtDNAs insert into the nuclear chromosomes in *Gossypium*

In this study, four sets of mitochondrial and nuclear genomes of *Gossypium* species (two diploids and two allotetraploid) were analyzed. Numts in four *Gossypium* nuclear genomes were detected by whole-genome alignment. Dot matrix analysis of mitochondrial vs nuclear genomes in *G. raimondii* (D$_5$) show that there is a stretch of ~598 kb (92.91%) of sequence that is nearly identical to that of the *G. raimondii* mitochondrial genome (Fig. 7a) in chromosome 1. This insertion is at least 99.80% identical to the mitochondrial genome, suggesting that the transfer event was very recent. The organization of the assembled mitochondrial genome differs from that of the mitochondrial DNA in the nucleus with an internal deletion (Fig. 7a), which might occur during or after transferring and represent an alternate isoform of the *G. raimondii* mitochondrial genome. In addition, *G. hirsutum* also has a nearly complete NUMT on chromosome A03 (Fig. 7c), and small to median-large fragments of mitochondrial DNA have been identified in three *Gossypium* species nuclear genomes (Fig. 7b-d), showing apparently sporadic fragmentation compared to *G. raimondii*. So much noise in *Gossypium* nuclear chromosomes of (Fig. 7b-d) are just repetitive derived elements. These results may be caused by the insertion of retrotransposon elements into mitochondrial DNA insertions that may contribute significantly to their fragmentation process in the other three nuclear genomes.

In addition, most *numts* had >99% nucleotide identity to the homologous organelle sequences, so the lack of divergence in *G. raimondii* indicates that they must have been transferred to the nucleus recently. In order to estimate how old these insertions are, p-distance rates and some estimates of rate/million years between mitochondrial sequences and numts were studied here. We have dated 20 larger NUMTs in *G. raimondii* (Additional file 5: Table S2), 16 larger NUMTs in *G. arboreum* (Additional file 6: Table S3), 15 larger NUMTs in *G. hirsutum* (Additional file 7: Table S4) and 12 larger NUMTs in *G. barbadense* (Additional file 8: Table S5). These data showed that the insertion time of NUMTs was close among one chromosome, but with big divergence between different chromosomes. For example, the different insertion time for five larger NUMTs in chromosome A03 of *G. hirsutum* (ranging from 0.33–1.03 MYA), with other chromosomes (insertion time ranging from 0.91–11.43 MYA) (Additional file 7: Table S4).

The p-distance of larger NUMTs ranged from 0~0.0009 in chromosome 01 of *G. raimondii* (Fig. 8a; Additional file 5: Table S2) and 0.0022~0.0069 in chromosome A03 of *G. hirsutum* (Fig. 8a; Additional file 7: Table S4). We have dated these larger NUMTs in chromosome 01 of *G. raimondii* with insertion time ranging from 0~0.13 MYA

Table 4 The numbers of edit sites in *Gossypium* mitochondrial protein-coding genes

Genes	G. thurberi	G. davidsonii	G. raimondii	G. trilobum	G. arboreum	G. hirsutum	G. barbadense	G. tomentosum	G. mustelinum	G. darwinii
atp1	6	6	6	6	6	6	6	6	6	6
atp4	11	11	11	11	11	11	11	11	11	11
atp6	9	9	9	9	9	9	9	9	9	9
atp8	3	3	3	3	3	3	3	3	3	3
atp9	6	6	6	6	6	6	6	6	6	6
ccmB	29	29	29	29	29	29	29	29	29	29
ccmC	24	24	24	24	24	24	24	24	24	24
ccmFc	13	13	13	13	13	13	13	13	13	13
ccmFn	31	31	31	31	31	31	31	31	31	31
cob	9	9	9	9	9	9	9	9	9	9
cox1	16	16	16	16	16	16	16	16	16	16
cox2	12	12	12	12	12	12	12	12	12	12
cox3	5	5	5	5	5	5	5	5	5	5
matR	15	15	15	15	15	15	15	15	15	15
mttB	25	25	25	25	25	25	25	25	25	25
nad1	19	19	19	19	19	19	19	19	19	19
nad2	25	25	25	25	25	25	25	25	25	25
nad3	13	13	13	13	13	12	13	13	13	13
nad4	29	29	29	29	29	29	29	29	29	29
nad4L	11	11	11	11	11	11	11	11	11	11
nad5	24	24	24	24	24	24	24	24	24	24
nad6	12	12	12	12	12	12	12	12	12	12
nad7	23	23	23	23	23	23	23	23	23	23
nad9	8	8	8	8	8	8	8	8	8	8
rpl2	4	4	4	4	4	4	4	4	4	4
rpl5	10	10	10	10	10	10	10	10	10	10
rpl10	4	4	4	4	4	4	4	4	4	4
rpl16	6	6	6	6	6	6	6	6	6	6
rps3	10	10	10	10	10	10	10	10	10	10
rps4	17	17	17	17	17	17	17	17	17	17
rps7	2	2	2	2	2	2	2	2	2	2
rps10	5	5	5	5	5	5	5	5	5	5
rps12	4	4	4	4	4	4	4	4	4	4
rps14	3	3	3	3	3	3	3	3	3	3
sdh3	4	4	4	4	4	4	4	4	4	4
sdh4	4	4	4	4	4	4	4	4	4	4
Total	451	451	451	451	451	450	451	451	451	451
Edits/gene	12.5	12.5	12.5	12.5	12.5	12.5	12.5	12.5	12.5	12.5

(Fig. 8b; Additional file 5: Table S2), and from 0.33~1.03 MYA in chromosome A03 of *G. hirsutum* (Fig. 8b; Additional file 7: Table S4). These results revealed that two nearly full length insertion events in *G. raimondii* (Chr01, Fig. 7a) and *G. hirsutum* (chromosome A03, Fig. 7c) occurred recently.

Discussion

From the perspective of divergence, *Gossypium* originated from a common ancestor approximately ten million years ago and an allopolyploidization event occurred approximately 1.5 million years ago [35, 36]. Plant mitochondrial genomes have experienced myriad synteny-disrupting

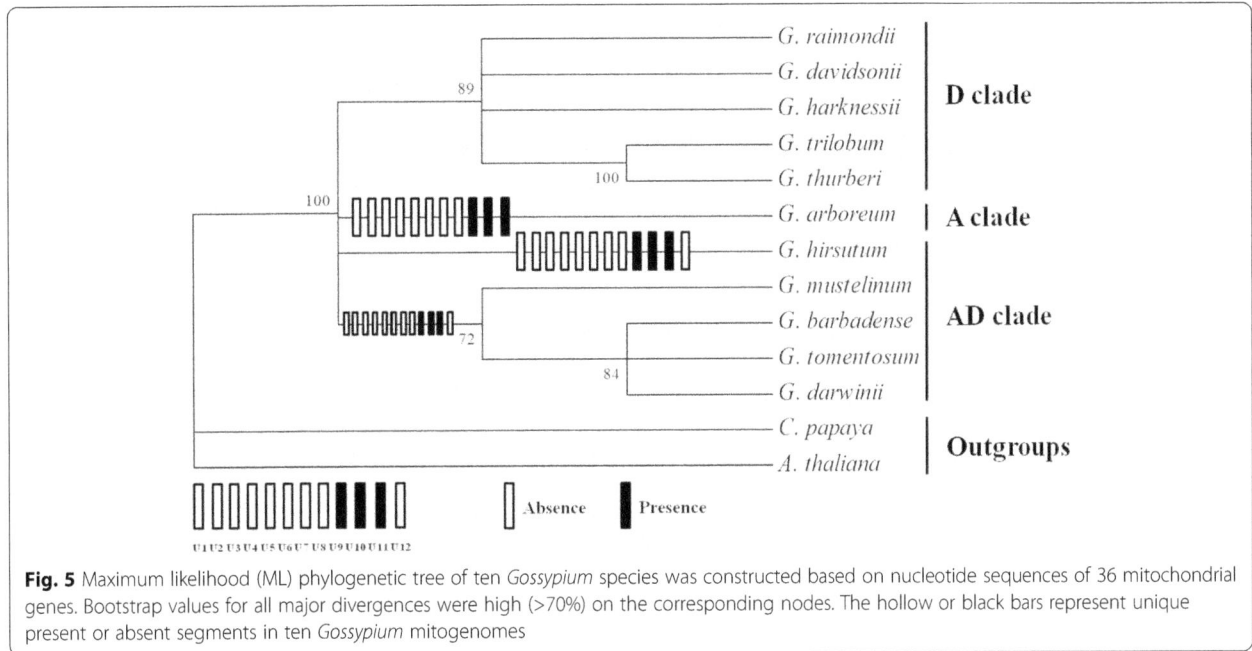

Fig. 5 Maximum likelihood (ML) phylogenetic tree of ten *Gossypium* species was constructed based on nucleotide sequences of 36 mitochondrial genes. Bootstrap values for all major divergences were high (>70%) on the corresponding nodes. The hollow or black bars represent unique present or absent segments in ten *Gossypium* mitogenomes

rearrangements even over a very short evolutionary timescale. Like most angiosperm mitogenomes abundant in repeat sequences with larger repeats mediating recombination at moderate to high frequency [19, 77], these recombination events generated multiple mito-genomic arrangements differed in *Gossypium* genome groups, which may be largely caused by both larger repeats and some key InDels or SVs during evolution, and quickly eroded synteny even among closely related plants [8, 72, 73, 78].

These cotton mitochondrial genomes diverged much, as indicated by the InDels events unique to A genome species, D and AD groups, respectively. All these structural variants (SVs) are located in intergenic regions of

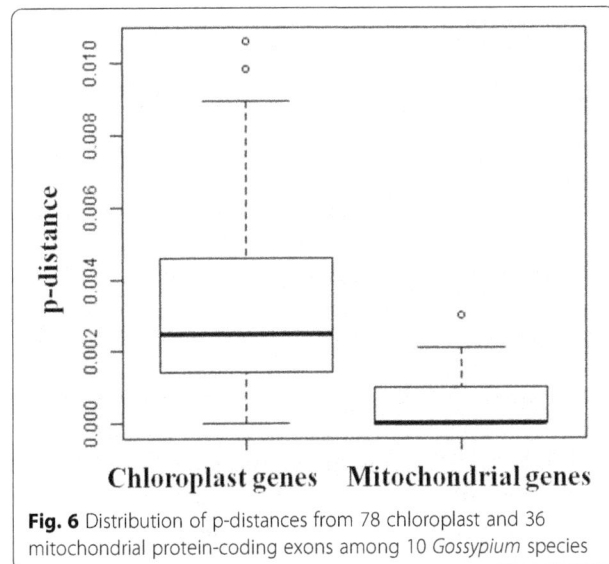

Fig. 6 Distribution of p-distances from 78 chloroplast and 36 mitochondrial protein-coding exons among 10 *Gossypium* species

mitogenomes. Some of them overlapped with their breakpoints and junctions occurring in repetitive and homologous genomic regions. And insertions or deletions were mostly generated from crossovers of repetitive regions or homologous regions [79].

There existed apparent inversions and translocations, which can offer clues to explain gene order differences of mitogenomes between different *Gossypium* groups. For examples, the mode of gene order changed by inversions and/or translocations was presented in early land plant mitochondrial genomes evolution of bryophytes [80] as well as rapidly rearranged mitochondrial genomes of vascular plants [71–73, 81, 82]. Apart from apparent rearranged mitogenome organizations in diploid and allotetraploid *Gossypium*, mitochondrial genome rearrangements have also been detected in diploid and allotetraploid species of *Brassica* [74, 83]. Generally, apparent variations in the mitogenome structures were always tested to be associated with cytoplasmic male sterility (CMS) and its maintainer lines [71–73], thus a new mitochondrial gene was produced by recombination and conferred CMS with its encoded protein interacted with the nuclear encoded mitochondrial protein to cause a detrimental interaction [43]. However, no genome rearrangement or large indel segments variations compared with mitogenomes of other D species, implying that the mitochondrial CMS-associated gene in cotton may function with different mechanism.

In addition, RNA-editing sites in *Gossypium* may not be in charge of cytoplasmic male sterility in D_8 cotton. RNA editing events have been compared in eight mitochondrial genes (*atp1, atp4, atp6, atp8, atp9,* and *cox1,*

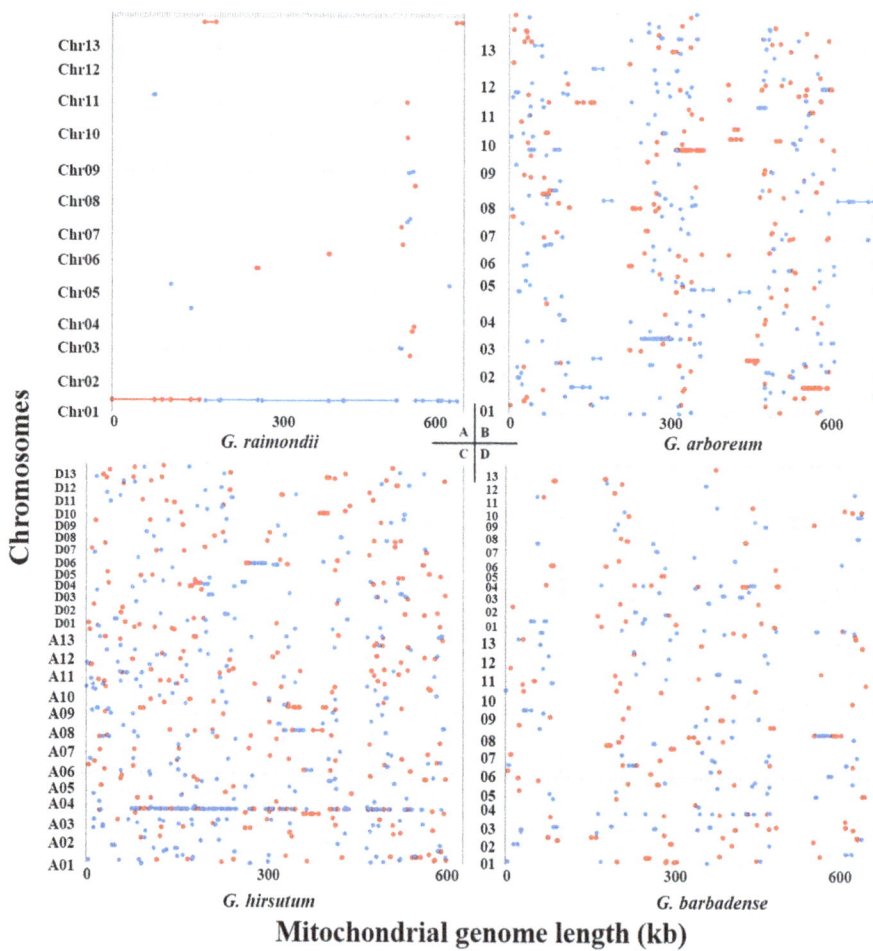

Fig. 7 Mitochondrial DNAs insertions into four *Gossypium* nuclear genomes detected by whole-genome alignment. The results were filtered to select only those alignments which comprise the one-to-one mapping between reference and query, and then display a dotplot of the selected alignments. The red and blue lines refer positive and reverse matches, respectively. **a**: Dot matrix analysis of *numts* in *Gossypium raimondii* (D₅) nuclear genome performed using MUMmer (Delcher et al., 2002). **b**: Dot matrix analysis of *numts* in *G. arboreum* (A₂) nuclear genome. **c**: Dot matrix analysis of *numts* in *G. hirsutum* (AD₁) nuclear genome. **d**: Dot matrix analysis of *numts* in *G. barbadense* (AD₂) nuclear genome

cox2, cox3) among CMS-D$_8$ three lines in cotton [75]. Although the frequencies of RNA editing events between mtDNA genes were different, no differences between cotton cytoplasms that could account for the CMS phenotype or restoration. In view of these results, the complete mitogenome sequences will provide the useful data resources for targeting the CMS-related genes in *G. trilobum* D$_8$ cotton in further studies.

As for MtDNAs insert into the nuclear chromosomes in *Gossypium*, Lin et al., (1999) and Stupar et al., (2001) also identified an intact mtDNA copy on chromosome 2 in the nucleus of Arabidopsis with more than 99% identity, which proved this type of mitochondrion-to-nucleus migration event [84, 85]. Second, these mitochondrion-to-nucleus migrations proved to be the independent events after the divergence of the *Gossypium* progenitors. These genome changes within the diploid and allotetraploid *Gossypium* species is worthy of more attention in future studies.

Conclusions

Plants mitochondrial genomes are evolutionarily intriguing because of the highly conserved genic content and slow rates of genic sequence evolution [18, 82]. These features contrasted sharply with the highly labile genomic structure, genome size, DNA repair mechanism and recombination induced by different types and origins of repeated sequences [82, 86–88]. Whole mitogenome sequences have been released in an ongoing process [9, 11, 38, 81, 89], which provide information for dissecting the evolutionary modifications in these genomes, such as gene loss [88], sequence acquisitions or loss [9], multiple sequence rearrangements [73] and dynamic structure evolution [38, 39]. Here, we presented six more cotton mitochondrial genomes, which showed apparently distinct divergence. Despite the short divergence time separating diploid and allotetraploid cotton species [35, 36], many of the hallmark features of mitochondrial genome evolution are evident,

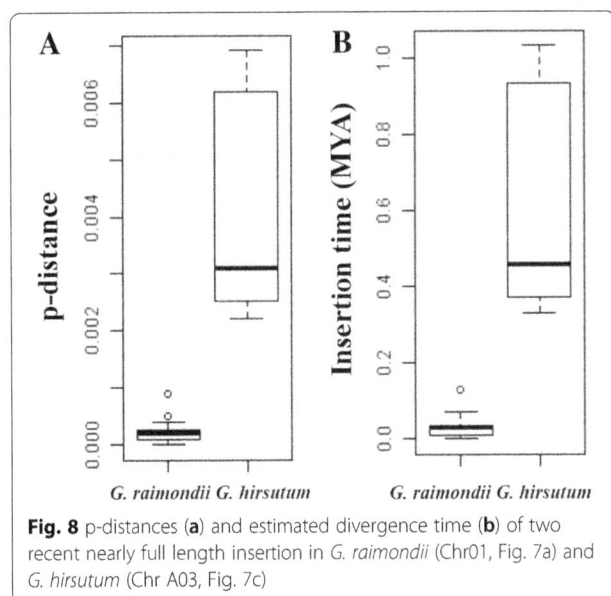

Fig. 8 p-distances (**a**) and estimated divergence time (**b**) of two recent nearly full length insertion in *G. raimondii* (Chr01, Fig. 7a) and *G. hirsutum* (Chr A03, Fig. 7c)

Additional file 4: Figure S3. Observed coverage of mapped paired-end reads supporting the existence of a small deletion (~500 bp) in *G. arboreum* compared to AD group species. IGV screenshot of the variability and coverage observed in ten samples of *Gossypium* sequence. Upper panel represent the unique sequences coordinates. There are five panels corresponding to the different *Gossypium* sequences. The track in each of these panels describes the density of read mapping or coverage depth. (JPEG 353 kb)

Additional file 5: Table S2. Nucleotide distances and divergence time (MYA) between mitochondrial sequences and corresponding *numts* in *G. raimondii*. Note: [a] twenty *numts* represent the largest mitochondrial fragments transferred into the nuclear chromosomes in *G. raimondii*. (DOCX 17 kb)

Additional file 6: Table S3. Nucleotide distances and divergence time (MYA) between mitochondrial sequences and corresponding *numts* in *G. arboreum*. (DOCX 17 kb)

Additional file 7: Table S4. Nucleotide distances and divergence time (MYA) between mitochondrial sequences and corresponding *numts* in *G. hirsutum*. Note: [a] fifteen *numts* represent the largest mitochondrial fragments transferred into the nuclear chromosomes in *G. hirsutum*. [b] represents five fragments from nearly full length mitochondrial fragments transferred into the nuclear A03 chromosomes in *G. hirsutum* (Fig. 7C). (DOCX 17 kb)

Additional file 8: Table S5. Nucleotide distances and divergence time (MYA) between mitochondrial sequences and corresponding *numts* in *G. barbadense*. Note: [a] twelve *numts* represent the largest mitochondrial fragments transferred into the nuclear chromosomes in *G. barbadense*. (DOCX 16 kb)

including differential genic content, genome rearrangements, inversion and translocation, gains/losses of multiple small and large repeats, presence/absence variations, and the mitogenome of *G. trilobum* D_8 cotton for targeting CMS-associated gene. Comparative analyses illustrated that four of the outcomes are quite surprising, including: 1) how rapidly mitochondrial genome rearrangements occur within a single subspecies (diverged ~ 10 mya), 2) how rapidly mitochondrial sequence segments are gained or lost, 3) RNA editing events were almost conserved in ten *Gossypium* mitogenomes, and 4) a previous unusual report of the integration of 93% of the mitochondrial genome of *G. raimondii* into chromosome 1 is confirmed with an estimation of insertion time 0.05 MYA. Increasing insight into the mechanisms and functional consequences of plant mitochondrial genome variation are expected to be helpful to elucidate the process of rapid evolutionary divergence mechanism between closely related mitochondrial genomes.

Additional files

Additional file 1: Figure S1. Genome maps of six diploid and allotetraploid *Gossypium* mitogenomes. Genes exhibited on the inside of outer circles are transcribed in a clockwise direction, while genes on the outside of outer circles are transcribed in a reverse direction. (JPEG 424 kb)

Additional file 2: Table S1. Gene contents of the six *Gossypium* mitogenomes. Note: Genes presented in multiple copies are denoted with a number (e.g., 2 or 3). (DOCX 18 kb)

Additional file 3: Figure S2. Gene order comparison among mitogenomes of *Gossypium*. Colored blocks represent regions of conserved gene clusters in the *Gossypium* genomes and genes in bold are located in the repeat regions. *Rpl2* and *atp8* are shown in red bold to indicate that they are just close to or partially overlapped with the repeat sequences. (JPEG 1006 kb)

Abbreviations
CMS: Cytoplasmic male sterility; G.: *Gossypium*; IGV: Integrative Genomics Viewer; Indels: Insertions and deletions; Mitogenome: Mitochondrial genome; mtDNA: Mitochondrial DNA; ORFs: Open reading frames; rRNAs: Ribosomal RNAs; tRNAs: Transfer RNAs

Acknowledgments
We are indebted to Dr. Shaoqing Li and late Prof. Yingguo Zhu (College of Life Sciences, Wuhan University, China) for helpful suggestions. We also thank Professor Jonathan F. Wendel (Iowa State University, Ames, USA), Dr. Corrinne E. Grover (Iowa State University, Ames, USA) and Professor Shu-Miaw Chaw (BRCAS, Taiwan, China) for helpful discussions. We are grateful to Dr. Kunbo Wang (Chinese Academy of Agricultural Sciences) for providing the seeds of wild *Gossypium* species used in present research.

Funding
This work was supported by grant from the National Natural Science Foundation of China (31671741) to J. Hua.

Authors' contributions
ZWC assembled the mitochondrial genome, annotated the mitochondrial genomes, performed the data analysis and prepared the manuscript. HSN, HLP, LDZ and SSL attended the data analyses and discussion. YMW maintained the experimental platform and participated in the bench work. JPH designed the experiments, provided research platform, guided the research and revised the manuscript. All authors approved the final manuscript.

Competing interests
The authors declare that they have no competing interests.

Author details
[1]Laboratory of Cotton Genetics, Genomics and Breeding /Key Laboratory of Crop Heterosis and Utilization of Ministry of Education/Beijing Key Laboratory of Crop Genetic Improvement, College of Agronomy and Biotechnology, China Agricultural University, Beijing 100193, China. [2]Institute of Cash Crops, Hubei Academy of Agricultural Sciences, Wuhan, Hubei 430064, China. [3]Department of Plant Science, School of Agriculture and Biology, Shanghai Jiao Tong University, Shanghai 200240, China.

References

1. Backert S, Nielsen BL, Borner T. The mystery of the rings: structure and replication of mitochondrial genomes from higher plants. Trends Plant Sci. 1997;2(12):477–83.
2. Chen Z, Zhao N, Li S, Grover CE, Nie H, Wendel JF, et al. Plant mitochondrial genome evolution and cytoplasmic male sterility. Crit Rev Plant Sci. 2017; 36(1):55–69.
3. Unseld M, Marienfeld JR, Brandt P, Brennicke A. The mitochondrial genome of *Arabidopsis thaliana* contains 57 genes in 366,924 nucleotides. Nat Genet. 1997;15(1):57–61.
4. Gualberto JM, Newton KJ. Plant mitochondrial genomes: dynamics and mechanisms of mutation. Annu Rev Plant Biol. 2017;68:225–52.
5. Oldenburg DJ, Bendich AJ. Size and structure of replicating mitochondrial DNA in cultured tobacco cells. Plant Cell. 1996;8(3):447–61.
6. Oldenburg DJ, Bendich AJ. Mitochondrial DNA from the liverwort *Marchantia polymorpha*: circularly permuted linear molecules, head-to-tail concatemers, and a 5′ protein. J Mol Biol. 2001;310(3):549–62.
7. Bendich AJ. The size and form of chromosomes are constant in the nucleus, but highly variable in bacteria, mitochondria and chloroplasts. BioEssays. 2007;29(5):474–83.
8. Sloan DB, Alverson AJ, Chuckalovcak JP, Wu M, McCauley DE, Palmer JD, et al. Rapid evolution of enormous, multichromosomal genomes in flowering plant mitochondria with exceptionally high mutation rates. PLoS Biol. 2012; 10(1):e1001241.
9. Rice DW, Alverson AJ, Richardson AO, Young GJ, Sanchez-Puerta MV, Munzinger J, et al. Horizontal transfer of entire genomes via mitochondrial fusion in the angiosperm *Amborella*. Science. 2013;342(6165):1468–73.
10. Alverson AJ, Rice DW, Dickinson S, Barry K, Palmer JD. Origins and recombination of the bacterial-sized multichromosomal mitochondrial genome of cucumber. Plant Cell. 2011;23(7):2499–513.
11. Skippington E, Barkman TJ, Rice DW, Palmer JD. Miniaturized mitogenome of the parasitic plant *Viscum scurruloideum* is extremely divergent and dynamic and has lost all *nad* genes. P Natl Acad Sci USA. 2015;112(27): E3515–24.
12. Kitazaki K, Kubo T. Cost of having the largest mitochondrial genome: evolutionary mechanism of plant mitochondrial genome. Journal of Botany. 2010;2010:1–12.
13. Galtier N. The intriguing evolutionary dynamics of plant mitochondrial DNA. BMC Biol. 2011;9:61.
14. Bonen L. Mitochondrial genes leave home. New Phytol. 2006;172(3):379–81.
15. Bergthorsson U, Adams KL, Thomason B, Palmer JD. Widespread horizontal transfer of mitochondrial genes in flowering plants. Nature. 2003;424(6945): 197–201.
16. Rodriguez-Moreno L, Gonzalez VM, Benjak A, Marti MC, Puigdomenech P, Aranda MA, et al. Determination of the melon chloroplast and mitochondrial genome sequences reveals that the largest reported mitochondrial genome in plants contains a significant amount of DNA having a nuclear origin. BMC Genomics. 2011;12:424.
17. Goremykin VV, Lockhart PJ, Viola R, Velasco R. The mitochondrial genome of *Malus domestica* and the import-driven hypothesis of mitochondrial genome expansion in seed plants. Plant J. 2012;71(4):615–26.
18. Davila JI, Arrieta-Montiel MP, Wamboldt Y, Cao J, Hagmann J, Shedge V, et al. Double-strand break repair processes drive evolution of the mitochondrial genome in Arabidopsis. BMC Biol. 2011;9:64.
19. Marechal A, Brisson N. Recombination and the maintenance of plant organelle genome stability. New Phytol. 2010;186(2):299–317.
20. Wendel JF, Cronn RC. Polyploidy and the evolutionary history of cotton. Adv Agron. 2003;78:139–86.

21. Wendel JF, Grover CE. Taxonomy and evolution of the cotton genus. In: Fang D, Percy R, editors. Cotton, Agronomy. Madison: Monograph 24, ASA-CSSA-SSSA; 2015.
22. Gallagher JP, Grover CE, Rex K, Moran M, Wendel JF. A New species of cotton from wake atoll, *Gossypium stephensii* (Malvaceae). Syst Bot 2017;42(1):115-123.
23. Wendel JF, Brubaker CL, Seelanan T. The origin and evolution of *Gossypium*. In: Stewart JM, Oosterhuis DM, Heitholt JJ, Mauney JR, editors. Physiology of cotton. Dordrecht: Springer Netherlands; 2010. p. 1–18.
24. Shendure J, Ji HL. Next-generation DNA sequencing. Nat Biotechnol. 2008; 26(10):1135–45.
25. Ansorge WJ. Next-generation DNA sequencing techniques. New Biotechnol. 2009;25(4):195–203.
26. Wang K, Wang Z, Li F, Ye W, Wang J, Song G, et al. The draft genome of a diploid cotton *Gossypium raimondii*. Nat Genet. 2012;44(10):1098–103.
27. Paterson AH, Wendel JF, Gundlach H, Guo H, Jenkins J, Jin D, et al. Repeated polyploidization of *Gossypium* genomes and the evolution of spinnable cotton fibres. Nature. 2012;492(7429):423–7.
28. Li F, Fan G, Wang K, Sun F, Yuan Y, Song G, et al. Genome sequence of the cultivated cotton *Gossypium arboreum*. Nat Genet. 2014;46(6):567–72.
29. Li F, Fan G, Lu C, Xiao G, Zou C, Kohel RJ, et al. Genome sequence of cultivated upland cotton (*Gossypium hirsutum* TM-1) provides insights into genome evolution. Nat Biotechnol. 2015;33(5):524–30.
30. Zhang T, Hu Y, Jiang W, Fang L, Guan X, Chen J, et al. Sequencing of allotetraploid cotton (*Gossypium hirsutum* L. acc. TM-1) provides a resource for fiber improvement. Nat Biotechnol. 2015;33(5):531–7.
31. Liu X, Zhao B, Zheng HJ, Hu Y, Lu G, Yang CQ, et al. *Gossypium barbadense* genome sequence provides insight into the evolution of extra-long staple fiber and specialized metabolites. Sci Rep-Uk. 2015;5:14139.
32. Yuan DJ, Tang ZH, Wang MJ, Gao WH, Tu LL, Jin X, et al. The genome sequence of Sea-Island cotton (*Gossypium barbadense*) provides insights into the allopolyploidization and development of superior spinnable fibres. Sci Rep-Uk. 2015;5:17662.
33. Lee SB, Kaittanis C, Jansen RK, Hostetler JB, Tallon LJ, Town CD, et al. The complete chloroplast genome sequence of *Gossypium hirsutum*: organization and phylogenetic relationships to other angiosperms. BMC Genomics. 2006;7:61.
34. Xu Q, Xiong GJ, Li PB, He F, Huang Y, Wang KB, et al. Analysis of complete nucleotide sequences of 12 *Gossypium* chloroplast genomes: origin and evolution of allotetraploids. PLoS One. 2012;7(8):e37128.
35. Chen Z, Feng K, Grover CE, Li P, Liu F, Wang Y, et al. Chloroplast DNA structural variation, phylogeny, and age of divergence among diploid cotton species. PLoS One. 2016;11(6):e0157183.
36. Chen Z, Grover CE, Li P, Wang Y, Nie H, Zhao Y, et al. Molecular evolution of the plastid genome during diversification of the cotton genus. Mol Phylogenet Evol. 2017;112:268–76.
37. Liu GZ, Cao DD, Li SS, AG S, Geng JN, Grover CE, et al. The complete mitochondrial genome of *Gossypium hirsutum* and evolutionary analysis of higher plant mitochondrial genomes. PLoS One. 2013;8(8):e69476.
38. Tang M, Chen Z, Grover CE, Wang Y, Li S, Liu G, et al. Rapid evolutionary divergence of *Gossypium barbadense* and *G. hirsutum* mitochondrial genomes. BMC Genomics. 2015;16:770.
39. Chen Z, Nie H, Grover CE, Wang Y, Li P, Wang M, et al. Entire nucleotide sequences of *Gossypium raimondii* and *G. arboreum* mitochondrial genomes revealed A-genome species as cytoplasmic donor of the allotetraploid species. Plant Biol. 2017;19(3):484–93.
40. Chen LT, Liu YG. Male sterility and fertility restoration in crops. Annu Rev Plant Biol. 2014;65:579–606.
41. Sloan DB, Muller K, McCauley DE, Taylor DR, Storchova H. Intraspecific variation in mitochondrial genome sequence, structure, and gene content in *Silene vulgaris*, an angiosperm with pervasive cytoplasmic male sterility. New Phytol. 2012;196(4):1228–39.
42. Tang H, Zheng X, Li C, Xie X, Chen Y, Chen L, et al. Multi-step formation, evolution, and functionalization of new cytoplasmic male sterility genes in the plant mitochondrial genomes. Cell Res. 2017;27(1):130–46.
43. Luo D, Xu H, Liu Z, Guo J, Li H, Chen L, et al. A detrimental mitochondrial-nuclear interaction causes cytoplasmic male sterility in rice. Nat Genet. 2013; 45(5):573–7.
44. Horn R, Gupta KJ, Colombo N. Mitochondrion role in molecular basis of cytoplasmic male sterility. Mitochondrion. 2014;19:198–205.
45. Hu J, Huang WC, Huang Q, Qin XJ, CC Y, Wang LL, et al. Mitochondria and cytoplasmic male sterility in plants. Mitochondrion. 2014;19:282–8.

46. Li SS, Liu GZ, Chen ZW, Wang YM, Li PB, Hua JP. Construction and initial analysis of five Fosmid libraries of mitochondrial genomes of cotton (*Gossypium*). Chinese Sci Bull 2013;58(36):4608-4615.

47. Bolger AM, Lohse M, Usadel B. Trimmomatic: a flexible trimmer for Illumina sequence data. Bioinformatics. 2014;30(15):2114–20.

48. Zerbino DR, Birney E. Velvet: algorithms for de novo short read assembly using de Bruijn graphs. Genome Res. 2008;18(5):821–9.

49. Magoc T, Salzberg SLFLASH. Fast length adjustment of short reads to improve genome assemblies. Bioinformatics. 2011;27(21):2957–63.

50. Altschul SF, Gish W, Miller W, Myers EW, Lipman DJ. Basic local alignment search tool. J Mol Biol. 1990;215(3):403–10.

51. Li H, Durbin R. Fast and accurate long-read alignment with burrows-wheeler transform. Bioinformatics. 2010;26(5):589–95.

52. Li H, Handsaker B, Wysoker A, Fennell T, Ruan J, Homer N, et al. The sequence alignment/map format and SAMtools. Bioinformatics. 2009;25(16):2078–9.

53. Lowe TM, Eddy SR. tRNAscan-SE: a program for improved detection of transfer RNA genes in genomic sequence. Nucleic Acids Res. 1997;25(5):955–64.

54. Delcher AL, Phillippy A, Carlton J, Salzberg SL. Fast algorithms for large-scale genome alignment and comparison. Nucleic Acids Res. 2002;30(11):2478–83.

55. Lohse M, Drechsel O, Bock R. OrganellarGenomeDRAW (OGDRAW): a tool for the easy generation of high-quality custom graphical maps of plastid and mitochondrial genomes. Curr Genet. 2007;52(5–6):267–74.

56. Krzywinski M, Schein J, Birol I, Connors J, Gascoyne R, Horsman D, et al. Circos: an information aesthetic for comparative genomics. Genome Res. 2009;19(9):1639–45.

57. Darling AE, Mau B, Perna NT. progressiveMauve: multiple genome alignment with gene gain, loss and rearrangement. PLoS One. 2010;5(6):e11147.

58. Robinson JT, Thorvaldsdottir H, Winckler W, Guttman M, Lander ES, Getz G, et al. Integrative genomics viewer. Nat Biotechnol. 2011;29(1):24–6.

59. Mower JP. The PREP suite: predictive RNA editors for plant mitochondrial genes, chloroplast genes and user-defined alignments. Nucleic Acids Res. 2009;37(Web Server issue):W253–9.

60. Katoh K, Standley DMMAFFT. Multiple sequence alignment software version 7: improvements in performance and usability. Mol Biol Evol. 2013;30(4):772–80.

61. Tamura K, Peterson D, Peterson N, Stecher G, Nei M, Kumar S. MEGA5: molecular evolutionary genetics analysis using maximum likelihood, evolutionary distance, and maximum parsimony methods. Mol Biol Evol. 2011;28(10):2731–9.

62. Chaw SM, Chang CC, Chen HL, Li WH. Dating the monocot-dicot divergence and the origin of core eudicots using whole chloroplast genomes. J Mol Evol. 2004;58(4):424–41.

63. Muse SV. Examining rates and patterns of nucleotide substitution in plants. Plant Mol Biol. 2000;42(1):25–43.

64. Gaut BS, Morton BR, McCaig BC, Clegg MT. Substitution rate comparisons between grasses and palms: synonymous rate differences at the nuclear gene *Adh* parallel rate differences at the plastid gene *rbcL*. Proc Natl Acad Sci U S A. 1996;93(19):10274–9.

65. Richardson AO, Rice DW, Young GJ, Alverson AJ, Palmer JD. The "fossilized" mitochondrial genome of *Liriodendron tulipifera*: ancestral gene content and order, ancestral editing sites, and extraordinarily low mutation rate. BMC Biol. 2013;11:29.

66. Guo W, Grewe F, Fan W, Young GJ, Knoop V, Palmer JD, et al. *Ginkgo* and *Welwitschia* mitogenomes reveal extreme contrasts in gymnosperm mitochondrial evolution. Mol Biol Evol. 2016;33(6):1448–60.

67. Zhang J, Turley RB, Stewart JM. Comparative analysis of gene expression between CMS-D$_8$ restored plants and normal non-restoring fertile plants in cotton by differential display. Plant Cell Rep. 2008;27(3):553–61.

68. Suzuki H, Rodriguez-Uribe L, Xu J, Zhang J. Transcriptome analysis of cytoplasmic male sterility and restoration in CMS-D$_8$ cotton. Plant Cell Rep. 2013;32(10):1531–42.

69. Suzuki H, Yu J, Wang F, Zhang J. Identification of mitochondrial DNA sequence variation and development of single nucleotide polymorphic markers for CMS-D$_8$ in cotton. Theor Appl Genet. 2013;126(6):1521–9.

70. Wendel JF. New world tetraploid cottons contain old world cytoplasm. Proc Natl Acad Sci U S A. 1989;86(11):4132–6.

71. Shearman JR, Sangsrakru D, Ruang-Areerate P, Sonthirod C, Uthaipaisanwong P, Yoocha T, et al. Assembly and analysis of a male sterile rubber tree mitochondrial genome reveals DNA rearrangement events and a novel transcript. BMC Plant Biol. 2014;14:45.

72. Tanaka Y, Tsuda M, Yasumoto K, Yamagishi H, Terachi T. A complete mitochondrial genome sequence of Ogura-type male-sterile cytoplasm and its comparative analysis with that of normal cytoplasm in radish (Raphanus sativus L.). BMC Genomics 2012;13:352.

73. Allen JO, Fauron CM, Minx P, Roark L, Oddiraju S, Lin GN, et al. Comparisons among two fertile and three male-sterile mitochondrial genomes of maize. Genetics. 2007;177(2):1173–92.

74. Grewe F, Edger PP, Keren I, Sultan L, Pires JC, Ostersetzer-Biran O, et al. Comparative analysis of 11 Brassicales mitochondrial genomes and the mitochondrial transcriptome of *Brassica oleracea*. Mitochondrion. 2014;19: 135–43.

75. Suzuki H, JW Y, Ness SA, O'Connell MA, Zhang JFRNA. Editing events in mitochondrial genes by ultra-deep sequencing methods: a comparison of cytoplasmic male sterile, fertile and restored genotypes in cotton. Mol Gen Genomics. 2013;288(9):445–57.

76. Takenaka M, Verbitskiy D, van der Merwe JA, Zehrmann A, Brennicke A. The process of RNA editing in plant mitochondria. Mitochondrion. 2008;8(1):35–46.

77. Woloszynska M. Heteroplasmy and stoichiometric complexity of plant mitochondrial genomes–though this be madness, yet there's method in't. J Exp Bot. 2010;61(3):657–71.

78. Palmer JD, Herbon LA. Plant mitochondrial DNA evolves rapidly in structure, but slowly in sequence. J Mol Evol. 1988;28(1–2):87–97.

79. Carbonell-Caballero J, Alonso R, Ibanez V, Terol J, Talon M, Dopazo JA. Phylogenetic analysis of 34 chloroplast genomes elucidates the relationships between wild and domestic species within the genus *Citrus*. Mol Biol Evol. 2015;32(8):2015–35.

80. Liu Y, Xue JY, Wang B, Li L, Qiu YL. The mitochondrial genomes of the early land plants *Treubia lacunosa* and *Anomodon rugelii*: dynamic and conservative evolution. PLoS One. 2011;6(10):e25836.

81. Wu ZQ, Cuthbert JM, Taylor DR, Sloan DB. The massive mitochondrial genome of the angiosperm *Silene noctiflora* is evolving by gain or loss of entire chromosomes. P Natl Acad Sci USA. 2015;112(33):10185–91.

82. Mower, J., D. Sloan, and A. Alverson, 2012 Plant mitochondrial genome diversity: the genomics revolution, pp. 123–144 in *Plant Genome Diversity Volume 1*, edited by J. F. Wendel, J. Greilhuber, J. Dolezel and I. J. Leitch. Springer Vienna.

83. Yang J, Liu G, Zhao N, Chen S, Liu D, Ma W, et al. Comparative mitochondrial genome analysis reveals the evolutionary rearrangement mechanism in *Brassica*. Plant Biol. 2016;18(3):527–36.

84. Lin XY, Kaul SS, Rounsley S, Shea TP, Benito MI, Town CD, et al. Sequence and analysis of chromosome 2 of the plant *Arabidopsis thaliana*. Nature. 1999;402(6763):761–8.

85. Stupar RM, Lilly JW, Town CD, Cheng Z, Kaul S, Buell CR, et al. Complex mtDNA constitutes an approximate 620-kb insertion on *Arabidopsis thaliana* chromosome 2: implication of potential sequencing errors caused by large-unit repeats. Proc Natl Acad Sci U S A. 2001;98(9):5099–103.

86. Christensen AC. Plant mitochondrial genome evolution can be explained by DNA repair mechanisms. Genome Biology and Evolution. 2013;5(6):1079–86.

87. Christensen AC. Genes and junk in plant mitochondria-repair mechanisms and selection. Genome Biology and Evolution. 2014;6(6):1448–53.

88. Adams KL, Qiu YL, Stoutemyer M, Palmer JD. Punctuated evolution of mitochondrial gene content: high and variable rates of mitochondrial gene loss and transfer to the nucleus during angiosperm evolution. P Natl Acad Sci USA. 2002;99(15):9905–12.

89. Eberhard JR, Wright TF. Rearrangement and evolution of mitochondrial genomes in parrots. Mol Phylogenet Evol. 2016;94:34–46.

5

Discovering the potential of *S. clavuligerus* for bioactive compound production: cross-talk between the chromosome and the pSCL4 megaplasmid

Rubén Álvarez-Álvarez[1][†], Yolanda Martínez-Burgo[1][†], Antonio Rodríguez-García[1,2] and Paloma Liras[1*]

Abstract

Background: *Streptomyces clavuligerus* ATCC 27064, the industrial producer of the β-lactamase inhibitor clavulanic acid, carries 49 putative secondary metabolite gene clusters. These secondary metabolite gene clusters are distributed between its linear chromosome and the 1.8 Mb-plasmid pSCL4, a rich reservoir of bioactive compound gene clusters.

Results: The transcriptome and metabolome of *S. clavuligerus* ATCC 27064 and the pSCL4⁻ derived strain, *S. clavuligerus* pSCL4⁻, were analysed. Construction of the *S. clavuligerus* pSCL4⁻ strain resulted in the excision of a 303 kb stretch in the right arm of the chromosome and its translocation to pSCL4, producing a 2.1 Mb plasmid named pSCL4* .
The absence of pSCL4* results in changes in the transcription level of genes encoding regulatory proteins or proteins with various functions. Lack of pSCL4* results in upregulation of three chromosomal gene clusters for secondary metabolites (SMC), SMC18, for holomycin and N-propionylholothin biosynthesis, SMC11b for tunicamycin biosynthesis (located between SMC10 and SMC11), and SMC5. The SMC10, SMC11 and SMC6 gene clusters were downregulated, resulting in lower production of clavulanic acid, cephamycin C and desferrioxamine E, respectively. Clusters SMC8, SMC12, SMC13 and SMC19 were also downregulated. Production levels of bioactive compounds, such as alkylresorcinol or thiol-derived compounds, were affected in the plasmid-less strain.

Conclusions: The excision and translocation to pSCL4 of 303 kb from the right arm of the chromosome confirms that the ends of the chromosome arms are regions of high instability and supports the hypothesis that pSCL4 might have been excised from *S. clavuligerus* chromosomal right arm end. Cysteine and methionine metabolism in *S. clavuligerus* lacking pSCL4* may differ from that of the wild type strain, given the absence of sulfur metabolism genes located either in pSCL4 or at the right end of the chromosome, which led to levels of dithiolopyrrolones (holomycin, N-propionylholothin) and acetylhomocysteine thiolactone (citiolone) higher than those of the wild type strain. *S. clavuligerus* pSCL4⁻ shows strong differences in its transcriptome and metabolome; however, the loss of 2.1 Mb DNA is dispensable in this strain.

Keywords: *Streptomyces clavuligerus*, pSCL4 megaplasmid, Plasmid evolution, Transcriptome analysis, Clavulanic acid, Secondary metabolites

* Correspondence: paloma.liras@unileon.es
[†]Equal contributors
[1]Microbiology Section, Faculty of Biological and Environmental Sciences, University of León, León, Spain
Full list of author information is available at the end of the article

Background

Streptomyces clavuligerus ATCC 27064 is the industrial producer of clavulanic acid, a β-lactamase inhibitor widely used in combination with the semisynthetic β-lactam amoxicillin to treat various bacterial infections. Forty-nine putative secondary metabolite gene clusters (SMC) have been identified in this species [1], one of the largest number of SMCs found in any bacterium. These SMCs are distributed between its lineal chromosome, with 24 SMCs, and a rich reservoir of secondary metabolic pathways, the 1.8 Mb-plasmid pSCL4, encoding 25 SMCs. *S. clavuligerus* ATCC 27064 has three additional plasmids: pSCL1, pSCL2, and pSCL3, of 11.7, 120, and 430 kb in length, respectively [2], but SMCs have not been identified within them so far.

Most of these gene clusters for bioactive compounds are silent and/or cryptic [1]. Charusanti et al. [3] exploited adaptative laboratory evolution by co-cultivation of *S. clavuligerus* with a methicillin-resistant *Staphylococcus aureus* strain; in this way they obtained fourteen *S. clavuligerus* isolates producing holomycin, a bioactive compound whose gene cluster is silent in the wild type strain in tested laboratory conditions. Two of these strains lack the whole pSCL4, and a third one lacks the leftmost stretch of pSCL4. Also, the holomycin producer strains *S. clavuligerus oppA2::aph* [4] and *S. clavuligerus claR::aph* [5] showed a lower copy number of pSCL4 than the wild type strain, suggesting that pSCL4 might be involved in the activation of some secondary metabolite gene clusters.

To obtain a *S. clavuligerus* strain devoid of pSCL4, the *parA-parB* genes for pSCL4 segregation were deleted in the wild type strain [6]. *S. clavuligerus* pSCL4⁻ grew slightly slower than the wild type strain, but was a viable strain, indicating that the pSCL4 plasmid is dispensable [6]; likewise, in accordance with previous studies [3, 6, 7], *S. clavuligerus* pSCL4⁻ produced a large amount of holomycin.

In addition to its large potential to produce a wide variety of bioactive compounds, the megaplasmid pSCL4 carries some regulatory genes [1]. Among them are the only gene for a butyrolactone receptor in *S. clavuligerus* genome, genes for serine/threonine kinases, genes for pairs or orphan two-component regulatory systems, and for other regulatory protein families. Therefore, given the in silico cross-talk prediction between chromosomal and pSCL4-encoded genes [1], to gain more in-depth knowledge of the potential of *S. clavuligerus* in bioactive compound production, the metabolome and transcriptome of *S. clavuligerus* ATCC 27064 and the derived *S. clavuligerus* pSCL4⁻ strain were analyzed.

The rapid evolution of the array of secondary metabolites in *Streptomyces* may be due to the instability of their chromosome ends that leads to amplifications and deletions in the genome and contributes to transmission of genetic information between *Streptomyces* plasmids and the chromosome. In this sense the double recombination of a smaller plasmid with the chromosome seems to be the most plausible scenario to explain the origin of pSCL4 [1].

Our studies confirm a megaplasmid-chromosome cross-regulation and support that pSCL4 may be originated by excision from the right arm of *S. clavuligerus* chromosome as previously suggested [1].

Results

Insights on the evolution of pSCL4

During the analysis of the transcriptome of both *S. clavuligerus* pSCL4⁻ and *S. clavuligerus* ATCC 27064, performed in the exponential and stationary growth phases, the genes along the chromosome in *S. clavuligerus* pSCL4⁻ were up- or downregulated in relation to the control strain; however, the coding DNA sequences (CDS) in the rightmost stretch of the right arm of the chromosome had systematically lower transcription levels in the three sampling times (Fig. 1a). These results might be due to a modulation of the transcription of these genes, located close to the right telomere, as a result of the lack of pSCL4. An alternative explanation is the absence of these genes in the pSCL4⁻ strain.

To determine the origin of this downregulation, the following genes were analyzed by qPCR in both strains: i) 10 genes located in the underexpressed region of the right end of the chromosome (between SCLAV_5482 and the right telomere), ii) genes located in the central region of the chromosome (SCLAV_5146, SCLAV_5308) or located in plasmid pSCL2 ($parA_{pSCL2}$), and iii) genes, such as $parB_{pSCL4}$, located in pSCL4, and therefore absent in *S. clavuligerus* pSCL4⁻ (Fig. 1b).

The qPCR analysis showed the same number of copies of SCLAV_5146, SCLAV_5308 and $parA_{pSCL2}$ in *S. clavuligerus* pSCL4⁻ as in the wild type strain. As expected, in *S. clavuligerus* pSCL4⁻ amplification of $parB_{pSCL4}$ was undetectable, and the same occurs to the 10 genes located in the right arm of the chromosome that were analyzed (Fig. 1b). In the DNA stretch showing lower transcriptional signal level (Fig. 1a), the relative amount of DNA ranged from 10^{-6} for SCLAV _5719 to 10^{-4} for SCLAV_5585 (Fig. 1b), suggesting that genes located downstream of SCLAV_5491, including the right telomere, are not present in *S. clavuligerus* pSCL4⁻. In order to determine when the translocation occurred, strains *S. clavuligerus* ATCC 27064, *S. clavuligerus parAB::aac*, derived from the wild type strain, and *S. clavuligerus* pSCL4⁻ constructed from *S. clavuligerus parAB::aac*, were analyzed by final time PCR to test every gene between SCLAV_5487 and SCLAV_5491 (Fig. 2). While all the genes were present in *S. clavuligerus* ATCC 27064 and *S. clavuligerus parAB::aac*, genes SCLAV_5489 to SCLAV_5491 were absent in *S. clavuligerus* pSCL4⁻. This suggests that the excision

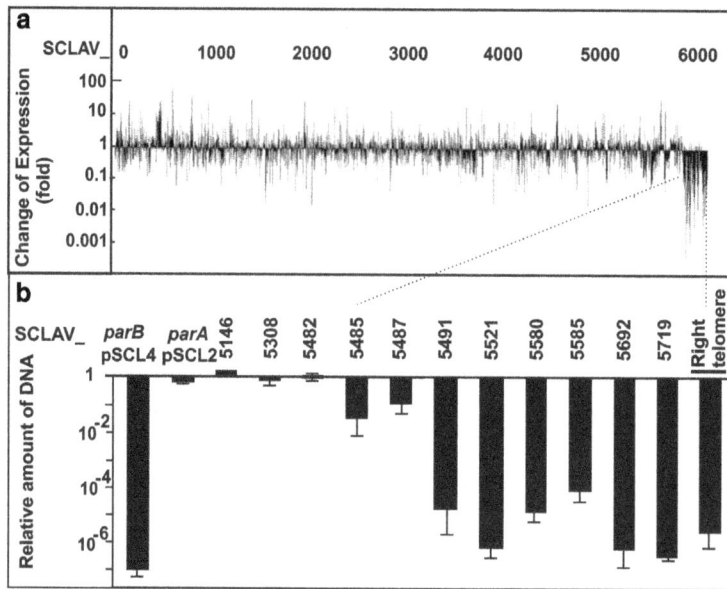

Fig. 1 Change of gene expression level and DNA quantification of genes in *S. clavuligerus* pSCL4- in relation to wild type strain. **a** Change of expression level of genes in *S. clavuligerus* pSCL4⁻ chromosome at 46.5 h in relation to the wild type strain. The pattern of the change of gene expression level was the same at 22.5 h and 60 h (not shown). **b** qPCR of: i) pSCL4 located genes (*parB*_pSCL4) deleted in *S. clavuligerus* pSCL4⁻, ii) genes located in plasmids different from pSCL4 (*parA*_pSCL2), iii) genes located in the central part of the chromosome (SCLAV_5146, SCLAV_5308), and iv) genes located in the 303 kb stretch of the right arm of the chromosome (SCLAV_5482, 5485, 5487, 5491, 5521, 5580, 5585, 5692, 5719 and the right telomere (nt 6,760,214 to 6,760,380)

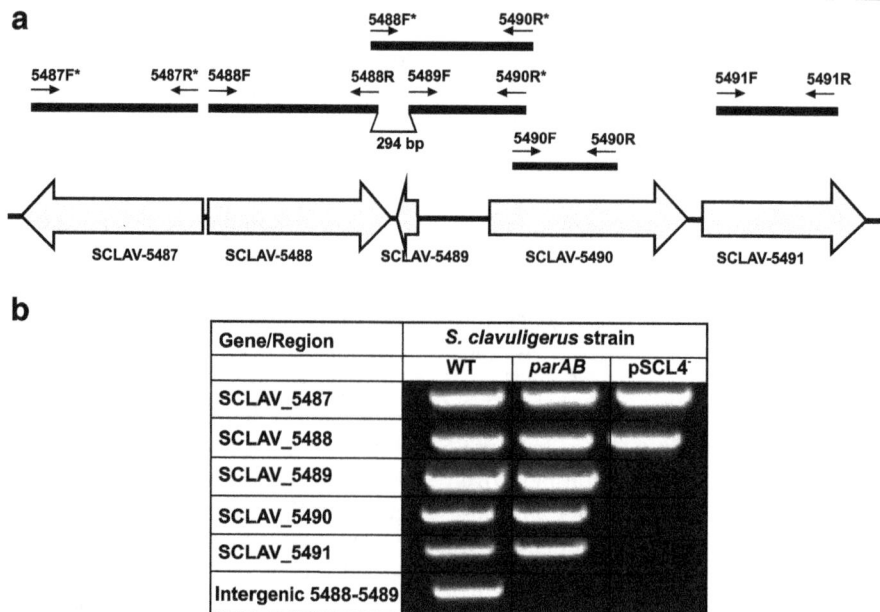

Fig. 2 PCR-based location of genes SCLAV_5487 to SCLAV_5491 in the *Streptomyces* studied strains. **a** The genes and regions indicated were analyzed in *S. clavuligerus* ATCC 27064, *S. clavuligerus parAB::aac* and *S. clavuligerus* pSCL4⁻ using the indicated oligonucleotides. **b** Regions amplified to determine the position in which the excision occurred in *S. clavuligerus parAB::aac*. The excision occurred at the 294 nt intergenic region shown in the upper part of the figure

occurred downstream of SCLAV_5488 (Fig. 2b); therefore, the intergenic SCLAV_5488 to SCLAV_5489 was analyzed; a qPCR amplification product was detected in the wild type strain, but not in *S. clavuligerus parAB::aac* or *S. clavuligerus* pSCL4⁻ (Fig. 2b).

The amplification of genes SCLAV_5487 to SCLAV_5491, and the lack of amplification of the SCLAV_5488 to SCLAV_5489 intergenic region in *S.clavuligerus parAB::aac* (Fig. 2b) suggests that those genes were already excised from the chromosome in this strain, and translocated to plasmid pSCL4 to form a 2.1 Mb megaplasmid named pSCL4*. The region in which the excision may take place is 294 nt in length (Fig. 2a) and may form a hairpin loop in all its length with a ΔG value of −85.98 kcal/mol (Additional file 1: Figure S1). This secondary structure might facilitate the excision of the 303 kb chromosomal DNA fragment, subsequently translocated to the megaplasmid and lost when pSCL4* is eliminated in *S. clavuligerus* pSCL4⁻. An alternative hypothesis is that the deletion of pSCL4 *parA-parB* segregation genes forces the integration of the megaplasmid in the right arm of the chromosome to avoid its loss, but due to the large size of this plasmid, the right arm of the chromosome might have split at one of the chromosomal hot spots during replication. Therefore, when we refer to the lack of pSCL4 in strain *S. clavuligerus* pSCL4⁻, the lack of the 303 kb chromosomal DNA fragment is always included.

Most of the genes in the translocated chromosomal DNA fragment encode hypothetical proteins, but others encode regulators, proteins involved in antibiotic resistance, genes with different functions and four gene clusters for secondary metabolism (SMC20 from SCLAV_5489, SMC21, SMC22 and SMC23).

Therefore, in the transcriptome studies presented below, we will consider only genes up to SCLAV_5488 since the lower expression of the final 231 CDS (SCLAV_5488 to SCLAV_5719) is due to their absence in the studied strain.

Transcriptome analysis

To analyze possible crosstalk regulation between genes located in *S. clavuligerus* chromosome and genes in the megaplasmid pSCL4, the transcriptome of *S. clavuligerus* pSCL4⁻ [6] was compared to that *S. clavuligerus* ATCC 27064, the wild type strain. Transcriptome analysis based on microarrays showed that, as expected, the 1570 CDS located in pSCL4 showed a negative M_g value (which is proportional to the abundance of the transcript for a particular gene), in the plasmid-less strain due to the absence of the transcripts for these genes. In total, 210 chromosomal genes were upregulated in the strain lacking pSCL4, 15 corresponding to regulatory genes and 34 included in secondary metabolites gene clusters. In addition, 335 chromosomal genes were downregulated,

including 30 genes encoding regulatory proteins and 25 genes located in secondary metabolites gene clusters. Furthermore,

by metabolomics analysis, differences in secondary metabolites produced between both strains were identified.

Production of antibiotics

The production of clavulanic acid, cephamycin C and holomycin was assessed in *S. clavuligerus* ATCC 27064 and *S. clavuligerus* pSCL4⁻ grown in SA media. The lack of pSCL4 resulted in lower cephamycin C production and a decrease in clavulanic acid production (about 50% along the fermentation) (Additional file 1: Figure S2). Holomycin, not detectable in the wild type strain, was produced by the pSCL4⁻ mutant at high levels (885 μg/ mg DNA at 70.5 h), as described previously [6].

Concomitantly, all the genes for clavulanic acid production (cluster SMC10) were downregulated (Fig. 3, upper panel). The effect was weaker for genes of the early steps (*ceaS2, bls2, pah2, cas2*) than for genes of the late steps of the pathway. This global downregulation of the gene cluster may be consequence of *claR* low expression level (2.9-fold decrease), as described by Martínez-Burgo et al. [5]. Lack of pSCL4 affected differently the genes in the clavams gene cluster (SMC9) (Fig. 3, central panel). Upregulation of *cvm7, cvm11, cvm12, cvmH* and *cvmP* was observed at all the sampling times but *cvm9, cvm1, cvm2* and *cvm13* were downregulated. The decrease in cephamycin C production observed in *S. clavuligerus* pSCL4, fits well with the lower expression of the genes for the biosynthetic early (*pcbAB, pcbC*) and late steps (*cefD, cefE, cefF, cmcH, cmcI, cmcJ*) of the cephamycin C pathway (cluster SMC11, Fig. 3 lower panel).

All genes from *hlmA* to *hlmI*, for holomycin biosynthesis (cluster SMC18), were strongly upregulated, although the effect of the lack of pSCL4 on *hlmK, hlmL* and *hlmM* was lower (not shown), as previously described [7].

Lack of pSCL4: effect on expression level of chromosome genes

Secondary metabolites gene clusters. Gene-clusters for secondary metabolites (SMC) in *S. clavuligerus* have been detected using the antiSMASH prediction method [1, 8, 9]. Transcription levels of genes for the secondary metabolites biosynthesis were heterogeneous. Chromosomal gene clusters SMC20 to SMC23 were not present in *S. clavuligerus* pSCL4⁻ and expression of clusters SMC14 to SMC17, for the formation of two NRPS, a type II PKS and a phytoene/squalene type of compound, respectively, were not affected by the lack of plasmid pSCL4.

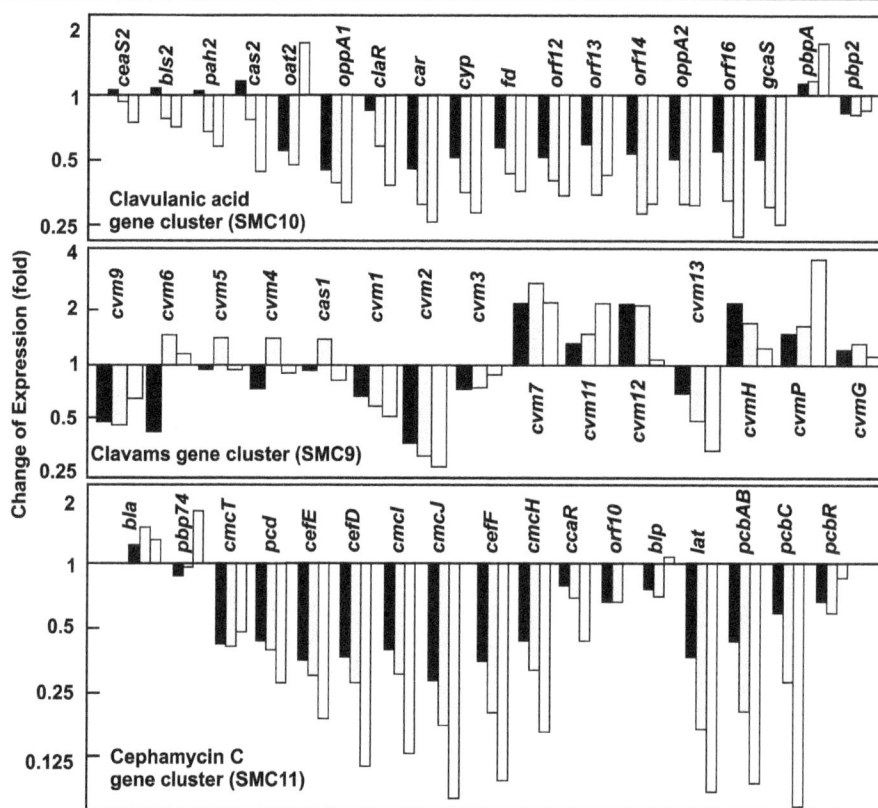

Fig. 3 Effect of the lack of pSCL4 on β-lactam biosynthesis gene clusters expression. Transcription level of genes of SMC10, SMC9 and SMC11 antibiotic gene clusters. Expression of the genes for the biosynthesis of clavulanic acid (SMC10), clavams 5S (SMC9), and cephamycin C (SMC11) in *S. clavuligerus* pSCL4⁻ is compared with those in *S. clavuligerus* ATCC 27064 (taken as 1.0). Bars represent the change of expression (fold) (base-2 logarithmic scale) at early exponential phase (black bars), exponential phase (grey bars) and stationary phase growth (white bars). The name of the genes is indicated over or below the bars

Tunicamycin and holomycin have been reported to be produced by *S. clavuligerus* [10]. Clusters SMC5, SMC11b and SMC18 for a type I PKS, for tunicamycin and for holomycin, respectively, showed all or most of their genes upregulated in the pSCL4-less strain (Table 1), as shown in detail for the tunicamycin and the SMC5 clusters (Fig. 4, lower left and upper panels).

Clusters SMC6, SMC8, SMC10, SMC11, SMC12, SMC13 and SMC19 encoding compounds with very different chemical structure (lantibiotics, β-lactams, PKS-NRPS, NRPS) or acting as siderophores [11], showed all or most of their genes downregulated in *S. clavuligerus* pSCL4⁻ (Table 1, shadowed in grey) as also shown with detail for the clavulanic acid or the cephamycin C gene clusters (Fig. 3, upper and lower panels) and for genes of the SMC12 cluster (Fig. 4, lower right panel). Other clusters SMC1, SMC2, SMC3, SMC4, SMC7, and SMC9 showed gene- and time-dependent up- or downregulation (Table 1, shadowed in dark grey).

The specific molecular structure of many of the compounds formed by the SMCs of *S. clavuligerus* is unknown and only the type of enzymes encoded by some genes allows to associate a type of chemical structure to these SMCs, but no correlation was detected between the transcription levels of the genes and the chemical structure of the compound produced by the clusters.

(ii) Regulators. Forty five chromosome regulatory genes were up- or downregulated in *S. clavuligerus* pSCL4⁻ at least 2-fold at all the sampling times (Table 2). SCLAV_2377, encoding the sigma factor SigE is one of the upregulated genes. Located in the operon *sigE-cseA-cseB-cseC,* this gene is part of a signal transduction pathway that allows *S. coelicolor* to sense and respond to changes in the integrity of its cell envelope [12]. Genes encoding the other three components of this pathway (CseA, a negative regulator; CseB, a response regulator; and CseC, a sensor histidine protein kinase) were also upregulated. A strong effect was also observed on the *ahpDC-oxyR* operon, which is activated by OxyR, as defense response to oxidative stress [13]. These genes were strongly upregulated at stationary growth phase (Table 2). SCLAV_4082 (*rpoE*), orthologous to *S. coelicolor* SigR [14], was upregulated at exponential growth phase. RpoE controls a regulon of 113 genes involved in oxidative

Table 1 Expression patterns in genes located in secondary metabolites gene clusters

Cluster	Genes	Product	Expression pattern
SMC5	SCLAV_0446-0497	Type I PKS	Most of the genes up-regulated at the two first sampling times
SMC11B	SCLAV_4276-4287	Tunicamycin	All the genes up-regulated at the three sampling times
SMC18	SCLAV_5267-5278	Holomycin	All the genes up-regulated at the three sampling times
SMC6	SCLAV_1942-1955	Siderophore	Most of the genes downregulated at the three sampling times
SMC8	SCLAV_2456-2469	Lantibiotic	All genes downregulated specially at the two first sampling times
SMC10	SCLAV_4178-4197	Clavulanic acid	All the genes downregulated at the three sampling times
SMC11	SCLAV_4198-4217	Cephamycin C	All the genes downregulated at the three sampling times
SMC12	SCLAV_4387-4392	Lantibiotic	All the genes downregulated at the three sampling times
SMC13	SCLAV_4460-4486	PKS-NRPS	Most of the genes downregulated at the three sampling times
SMC19	SCLAV_5325-5347	NRPS	Most of the genes downregulated, especially at the two final sampling times
SMC1	SCLAV_0001-0026	Macrolide type I PKS	Genes upregulated and genes downregulated at the three sampling times
SMC2	SCLAV_0082-0105	Unknown type	Genes upregulated and genes downregulated at the three sampling times
SMC3	SCLAV_0148-0152	Siderophore	Genes upregulated and genes downregulated at the three sampling times
SMC4	SCLAV_0153-0172	Terpene synthase	Genes upregulated and genes downregulated at the three sampling times
SMC7	SCLAV_2274-2302	NRPS	Genes upregulated and genes downregulated at the three sampling times
SMC9	SCLAV_2920-2935	Clavams	Genes upregulated and genes downregulated at the three sampling times
SMC20 to SMC23			These clusters are not present in *Streptomyces* pSCL4⁻

stress control. Sixty four genes of this regulon were slightly upregulated (equal to or higher than 1.5-fold in at least one of the three sampling times) as described for *S. clavuligerus* Δ*claR* [5]. The antisigma RsrA coding gene was also slightly upregulated at exponential growth phase.

The lack of pSCL4 resulted in thirty regulatory genes downregulated with a minimal 2-fold difference at all the sampling times (Table 2). A slight downregulation was observed on SCLAV_0691 and SCLAV_3814 genes, that encode *S. coelicolor* SigB and SigQ sigma-like factors, respectively, involved in morphological differentiation [15, 16] and on SCLAV_3146, encoding a transcriptional factor similar to *S. coelicolor* AtrA [17, 18]. Strongly downregulated was SCLAV_2573, for a transcriptional regulator similar to *S. coelicolor* WblA [19]. Two additional transcriptional regulators of unknown function shown in Table 2 were also downregulated.

(iii) Genes with diverse functions. In addition to the genes already mentioned, 446 genes encoding proteins

with different functions in the cell, were affected in the pSCL4⁻ strain, 161 upregulated and 285 downregulated (Fig. 5a). The most upregulated genes were SCLAV_0783 to SCLAV_0785, orthologous to SCO1557 to SCO1559, which encode a methionine transport system [20]. Expression of some of these genes will be discussed below in relation to the metabolites produced by *S. clavuligerus.*

(iv) Differentiation. S. clavuligerus pSCL4⁻ shows a very poor or null aerial mycelium or spores formation [6]. Concomitantly, several genes involved in differentiation were affected by the lack of pSCL4 (Additional file 1: Table S1). Seven genes involved in cell wall and membrane biogenesis were upregulated including SCLAV_5204, which encodes a *SsgA*-like protein (SALP) involved in peptidoglycan synthesis and the thickening of the spore cell-wall [21]. Also upregulated was SCLAV_1824, which encodes a small mechanosensitive channel involved in hypoosmotic stress protection [22] and, as previously mentioned,

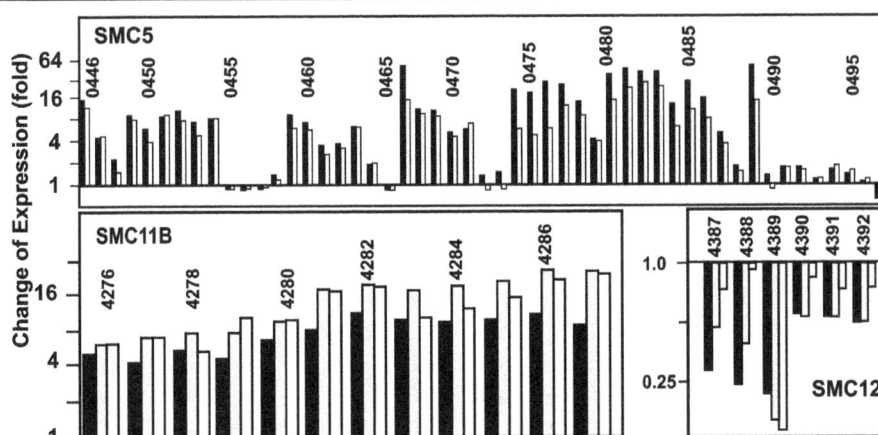

Fig. 4 Effects of the lack of pSCL4 on putative gene clusters for secondary metabolites biosynthesis. Change of expression (fold) of the upregulated SMC5 (upper panel) and SMC11b (lower left panel), clusters, and the downregulated SMC12 (lower right panel) gene clusters of *S. clavuligerus* pSCL4⁻ in comparison with those in *S. clavuligerus* ATCC 17064 (taken as 1.0). Only the exponential phase samples are shown for SCM5. Bars represent the change of expression (fold) (base-2 logarithmic scale) at early exponential phase (black bars), exponential phase (grey bars) and stationary phase growth (white bars). The gene SCLAV_0465 has no probe in the microarrays. The SCLAV_ number of the genes is indicated above the bars

Table 2 Genes encoding regulatory proteins[a] which are up- or down-regulated in S. clavuligerus pSCL4⁻

Gene	Product	M_c			BH-corrected P-value			Fold change		
		22.5 h	46.5 h	60 h	22.5 h	46.5 h	60 h	22.5 h	46.5 h	60 h
Up-regulated										
SCLAV_2377	ECF-subfamily RNA polymerase sigma factor	2.42	1.91	1.76	$1.80E^{-05}$	$1.43E^{-04}$	$2.61E^{-04}$	5.36	3.77	3.39
SCLAV_2378	Lipoprotein cseA	1.30	0.83	1.23	$1.55E^{-05}$	$1.08E^{-03}$	$8.98E^{-06}$	2.46	1.77	2.35
SCLAV_2379	Transcriptional regulatory protein cseB	1.34	1.08	1.71	$1.35E^{-03}$	$6.52E^{-03}$	$5.56E^{-05}$	2.54	2.11	3.26
SCLAV_2380	Sensor protein	1.60	1.22	2.19	$6.07E^{-04}$	$5.17E^{-03}$	$9.20E^{-06}$	3.04	2.33	4.56
SCLAV_3934	Alkyl hydroperoxide reductase ahpD	0.61	1.24	3.66	$3.34E^{-01}$	$2.15E^{-02}$	$2.27E^{-07}$	1.53	2.37	12.63
SCLAV_3935	Alkyl hydroperoxide reductase	0.57	1.27	3.90	$3.35E^{-01}$	$1.29E^{-02}$	$3.82E^{-08}$	1.49	2.41	14.89
SCLAV_3936	Putative LysR-family transcriptional regulator	1.11	1.48	2.06	$1.47E^{-02}$	$1.09E^{-03}$	$2.09E^{-05}$	2.16	2.79	4.17
SCLAV_4082	RNA polymerase sigma factor RpoE	1.23	1.64	0.66	$2.61E^{-03}$	$1.06E^{-04}$	$8.26E^{-02}$	2.35	3.11	1.58
SCLAV_4083	Putative anti-sigma factor	1.06	1.30	0.29	$4.63E^{-03}$	$5.29E^{-04}$	$4.56E^{-01}$	2.08	2.46	1.22
Down-regulated										
SCLAV_0691	RNA polymerase sigma factor	⁻1.35	⁻1.62	⁻1.21	$2.66E^{-03}$	$3.19E^{-04}$	$3.78E^{-03}$	2.55	3.08	2.32
SCLAV_2573	WhiB-family transcriptional regulator	⁻2.74	⁻3.47	⁻4.47	$6.05E^{-04}$	$3.12E^{-05}$	$9.92E^{-07}$	6.68	11.11	22.16
SCLAV_2833	Putative serine/threonine kinase anti-sigma factor	⁻1.16	⁻1.70	⁻2.54	$4.77E^{-02}$	$2.92E^{-03}$	$3.22E^{-05}$	2.24	3.24	5.81
SCLAV_3047	Predicted transcriptional regulator	⁻1.43	⁻2.74	⁻2.45	$2.53E^{-02}$	$5.41E^{-05}$	$1.40E^{-04}$	2.69	6.70	5.47
SCLAV_3146	TetR-family transcriptional regulator	⁻1.04	⁻1.20	⁻1.55	$5.19E^{-03}$	$1.16E^{-03}$	$5.07E^{-05}$	2.06	2.29	2.93
SCLAV_3422	Transcriptional regulator	⁻1.80	⁻2.65	⁻3.79	$3.07E^{-02}$	$1.37E^{-03}$	$2.01E^{-05}$	3.49	6.27	13.82
SCLAV_3814	ECF-subfamily RNA polymerase sigma factor	⁻1.27	⁻1.08	⁻1.71	$5.63E^{-03}$	$1.38E^{-02}$	$1.94E^{-04}$	2.42	2.11	3.27
SCLAV_4929	Regulatory protein	⁻1.71	⁻1.98	⁻3.90	$7.07E^{-03}$	$1.64E^{-03}$	$1.49E^{-06}$	3.28	3.94	14.92

[a]Only some of the most affected genes are shown

SCLAV_2573, which encodes a WhiB-like transcriptional regulator affecting morphological differentiation [19].

Twenty six genes related to environmental adaptation and differentiation, were downregulated in the pSCL4⁻ strain, including SCLAV_5177 and the block of genes from SCLAV_5181 to SCLAV_5192, downregulated in the three sampling times. These genes are orthologous to S. coelicolor mce operon, likely involved in survival in natural environment [23].

Aerial mycelium and sporulation of S. clavuligerus ΔclaR, which is defective in amfS expression [5], are extracellularly complemented by cross-feeding both by the wild type strain and by S. clavuligerus pSCL4⁻ (Fig. 5b). This indicates that S. clavuligerus pSCL4⁻ forms the SapB peptide encoded by amfS [24], as occurs with the wild type strain. The lack of extracellular complementation between the wild type strain and S. clavuligerus pSCL4⁻ (Fig. 5b) may be explained by the downregulation of atrA, wblA, sigB, sigQ and/or upregulation of sigE in the pSCL4-minus strain.

Validation of the transcriptome data
The transcriptome data previously showed was validated with 46.5 h RNA samples using RT-qPCR. Twelve genes were validated, including genes for the

biosynthesis of clavulanic acid (oppA2), cephamycin C (pcbC), holomycin (hlmA); several genes encoding regulatory proteins (SCLAV_3410, SCLAV_4464, SCLAV_4650, SCLAV_5308), as well as five genes for miscellaneous proteins (Additional file 1: Figure S3A). The RT-qPCR values consistently confirmed the transcriptome data obtained in the microarrays. A Pearson's correlation coefficient of 0.99 between the data from the two techniques was obtained (Additional file 1: Figure S3B).

Metabolomic analyses: Effect of the lack of pSCL4 on secondary metabolites production
The ability of S. clavuligerus pSCL4⁻ to produce secondary metabolites is lower than that of the wild type strain since 25 clusters for secondary metabolism are located in pSCL4 [1] and 3 clusters in the translocated 3'end region of the chromosome, all of which are deleted in S. clavuligerus pSCL4⁻. Indeed, concentrated broth extracts of S. clavuligerus pSCL4⁻ and S. clavuligerus ATCC 27064 analyzed by HPLC, showed notable differences in the 210 nm absorption pattern (Fig. 6a).

S. clavuligerus ATCC 27064 and S. clavuligerus pSCL4⁻ were grown in SA medium up to stationary phase as indicated above. For each strain, one 100 ml sample containing broth and mycelium was analyzed. Both strains produced compounds with structure of cyclic peptides, identical to those with siderophore activity (Fig. 6b). A

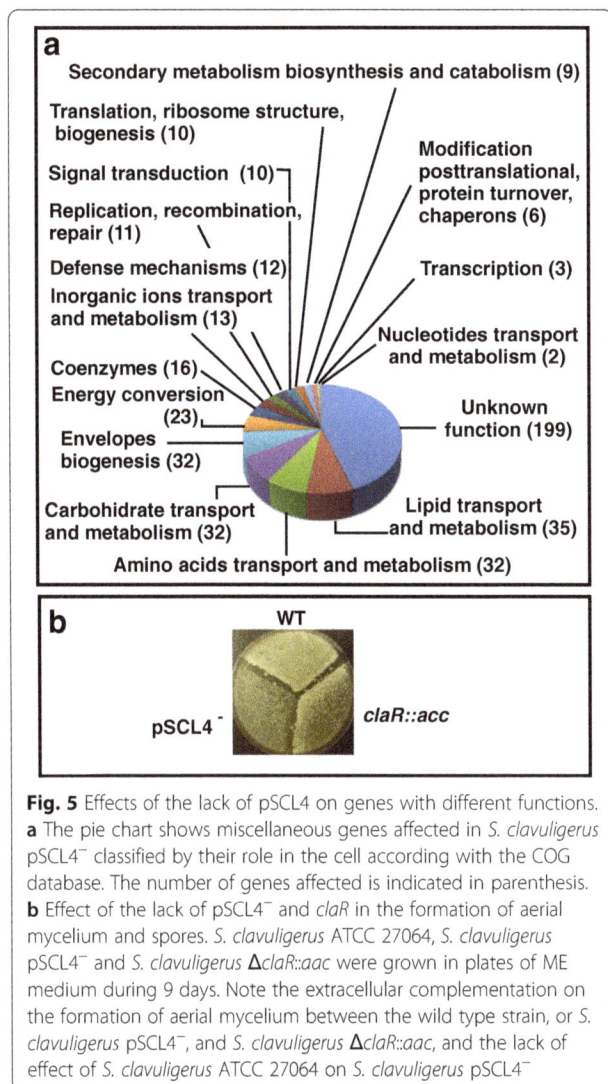

Fig. 5 Effects of the lack of pSCL4 on genes with different functions. **a** The pie chart shows miscellaneous genes affected in *S. clavuligerus* pSCL4⁻ classified by their role in the cell according with the COG database. The number of genes affected is indicated in parenthesis. **b** Effect of the lack of pSCL4⁻ and *claR* in the formation of aerial mycelium and spores. *S. clavuligerus* ATCC 27064, *S. clavuligerus* pSCL4⁻ and *S. clavuligerus* ΔclaR::aac were grown in plates of ME medium during 9 days. Note the extracellular complementation on the formation of aerial mycelium between the wild type strain, or *S. clavuligerus* pSCL4⁻, and *S. clavuligerus* ΔclaR::aac, and the lack of effect of *S. clavuligerus* ATCC 27064 on *S. clavuligerus* pSCL4⁻

peak of desferrioxamine E (nocardamine) was present in the broth extract of both strains (peaks 3 and 10, Fig. 6a) at different concentrations. In addition, the wild type strain showed small amounts of demethylenenocardamine and deoxynocardamine (peak 2, Fig. 6a) which were not detectable in the mutant. These desferrioxamine-related compounds have been previously described in a *Streptomyces* sp. isolated from a marine sponge [25].

Several organic acids present in extracts of the wild type strain, were not detectable in the *S. clavuligerus* pSCL4⁻ strain. These compounds derived from linear organic acids, i.e., 8-amino-2-methyl-7-oxononanoic acid (peak 4), or from resorcinol (1,3-benzenediol), i.e., 2-hexyl-5-methyl-resorcinol (with the same molecular mass as 2-butyl-5-propyl-resorcinol) that appears as mixtures in peaks 5 and 6 (Fig. 6b).

The most characteristic peaks found in *S. clavuligerus* pSCL4⁻ extracts were compounds containing sulfur, which are not detectable in the wild type strain. The

aromatic sulfur compounds, N-acetylhomocysteine thiolactone (citiolone) and the potentially toxic N-propionyl-3-aminodihydro-2(3H)-thiophenone, were identified in peaks 7 and 8, respectively (Fig. 6b). Both compounds have relatively similar structures and their biochemical origin is probably the lactone derived from homocysteine (2-amino-4-sulfanyl-butanoic acid) with N-acetyl or N-propionyl substitutions.

The other sulfur compounds found in *S. clavuligerus* pSCL4⁻ belong to the dithiolopyrrolone family (Fig. 6b), which corresponds well with the high production of holomycin (peak 9) and the high expression level of the holomycin biosynthesis genes in this strain (not shown). Surprisingly, the largest peak for a dithiolopyrrolone in our chromatogram corresponds to N-propionylholothin (peak 11), (with molecular weight identical to that of thiolutin). N-propionylholothin has been already described in an uncharacterized *Streptomyces* strain producing cephamycin C [26], supporting that peaks 11 corresponds to N-propionylholothin rather than to thiolutin, that has never been detected in *S. clavuligerus*. The MS and UV absorption spectra of all the compounds differentially produced by the wild type strain and the pSCL4⁻ mutant are shown in Additional file 1: Figure S4.

As far as we know, with the exception of the dithiolopyrrolones, all the other compounds found by HPLC-MS in this work, i.e., nocardamine and their derived analogs, homocysteine-derived lactones, linear organic acids or resorcinol-derived compounds, have never been previously described in *S. clavuligerus*.

Discussion

Medema et al. [1] proposed that pSCL4 might have evolved by recombination between a small plasmid and the *S. clavuligerus* chromosome. After excision, the megaplasmid would carry a large fragment of the chromosome. Excision of pSCL4 from the chromosome occurred mainly at a specific site to give a 1.8 Mb pSCL4 plasmid [1, 27], but different excision sites in the chromosome might exist. In the process of deleting the $parAB_{pSCL4}$ genes [6] we enriched and amplified a clone in which a chromosomal DNA fragment (303 kb, 231 CDS) was excised from the right arm of the chromosome and translocated to pSCL4 (1.8 Mb) resulting in the 2.1 Mb pSCL4* megaplasmid. The claim that all the genes in the chromosome are present in the megaplasmid-less *S. clavuligerus* strain obtained by adaptive evolution [3] suggest that these authors obtained a 1.8 Mb pSCL4⁻ in which no translocations ocurred.

The translocation of these chromosomal genes to the megaplasmid in *S. clavuligerus* parAB::aac is confirmed by their loss, as shown by PCR and by their low transcriptional signal level, in *S. clavuligerus* pSCL4⁻. Our finding confirms that the ends of the chromosome

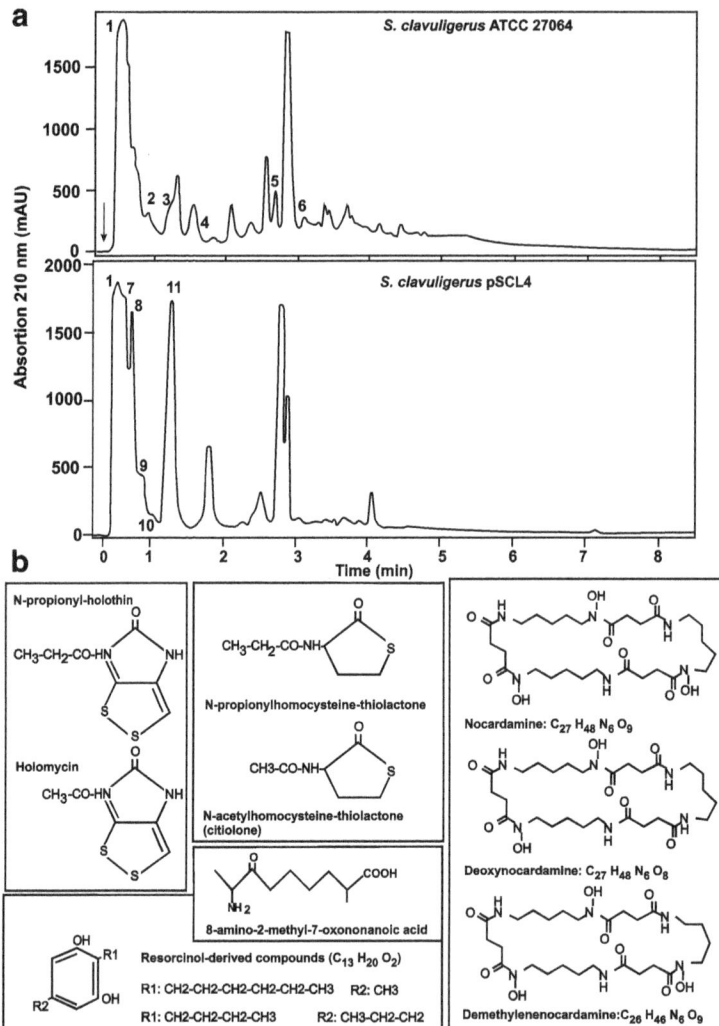

Fig. 6 Comparative HPLC analysis between *S. clavuligerus* ATCC 27064 and *S. clavuligerus* pSCL4⁻. **a** Components determined in *S. clavuligerus* ATCC 27064 extracts (upper pannel) and *S. clavuligerus* pSCL4⁻ extract (lower panel). 1:MOPS; 2: demethylenenocardamine and deoxynocardamine; 3: Nocardamine; 4: 8-amino-2-methyl-7-oxononanoic acid, 5: 2-hexyl-5-methyl-1,3-benzenediol (resorcinol) and 2-butyl-5-propyl-resorcinol, 6: 2-hexyl-5-methyl-1,3-benzenediol and 2-butyl-5-propyl-resorcinol, 1: MOPS, 7:citiolone, 8: N-propanoyl-3-aminodihydro-2(3H)-thiophenone (N-propionyl-homocysteine thiolactone); 9:holomycin, 10:nocardamine, 11: N-propionylholothin. **b** Chemical structure of the above indicated compounds

arms are regions of high instability [28] and supports the hypothesis that pSCL4 might have been excised from the chromosomal right arm end [1]. As a result of the lack of *S. clavuligerus parAB::aac* right telomere, this strain might have a circular chromosome, as described previously for other telomere-less linear *Streptomyces* genomes [29–31].

Deletion of only one regulator encoding gene, *claR*, located in the clavulanic acid cluster, produced multiple effects on the transcriptome of *S. clavuligerus* Δ*claR* [5] and affected several gene clusters for secondary metabolism in this strain. There are 132 regulatory genes located in pSCL4. It contains genes for two component systems (17), for regulators of the LysR-type (4), AraC-type (7), TetR-type (10), ArsR-type (7), MarR-type (6), SARP-type

(3), XRE-type (5) and genes for different types of additional regulators, including sigma factors (6), and the only butyrolactone-receptor protein (SCLAV_p0894) in *S. clavuligerus* [32, 33]. Lack of plasmid pSCL4 has, not surprisingly, a broad and large effect. These regulators might act directly, or in cascade, on multiple genes located in secondary metabolites clusters, or encoding miscellaneous genes or regulators located in the chromosome.

In agreement with the lower clavulanic acid production in *S. clavuligerus* pSCL4⁻, genes of the SMC10 cluster (for clavulanic acid formation) are slightly downregulated. Also downregulated are the clusters SMC6 (for desferrioxamine E), SMC11, SMC12 and SMC19.

In the upregulated gene clusters is especially important the upregulation of cluster SMC18, for holomycin biosynthesis [7], which correlates well with the production of this dithiolopyrrolone in *S. clavuligerus* pSCL4⁻ cultures. Also upregulated are SMC5, and SMC11b for tunicamycin biosynthesis. These results agree with the holomycin-tunicamycin overproduction described by Kenig and Reading [10] in the *S. clavuligerus* IT1 strain. However, tunicamycin was not detected in our samples.

A rich array of different compounds was found in the broths of the wild type and the pSCL4⁻ strain cultures. Desferrioxamine E was present in the supernatants of both strains, confirming that the siderophore biosynthesis cluster is located in the chromosome. Clusters SMC3 and SMC6 are candidates to encode nocardamine biosynthesis enzymes. Both clusters, but specially SMC6, are downregulated in the pSCL4-less strain, what may agree with the absence of desferrioxamine -related demethylenenocardamine and deoxynocardamine, and the lower amount of desferrioxamine, in this strain.

Alkylresorcinols are autoregulators in some Gram-positive bacteria reported to interact with DNA and modify its structure and viscosity [34]. To our knowledge this is the first time that they have been found in *Streptomyces*, and they appear to be encoded by genes located in pSCL4 since they were not detected in the *S. clavuligerus* pSCL4⁻ strain.

Cysteine and methionine metabolism in *S. clavuligerus* pSCL4⁻ differs from that of the wild type strain due to the absence of many sulfur metabolism genes that are located either in pSCL4 or at the right end of the chromosome (Additional file 1: Figure S5A, B). However, transcriptomic studies and the formation of metabolites by this strain confirms that enough flow exists to homocysteine and cysteine, the precursors of N-acylhomocysteine lactones and dithiopyrrolones, respectively. The lack of genes *metX* and *metY*, located in at the right end of the chromosome in *S. clavuligerus* pSCL4⁻, suggests that homocysteine formation from homoserine is very low or non-existing in this strain (Additional file 1: Figure S5A, B), but it can be still formed from methionine by the upregulated genes *metK* and *sahH* through S- adenosyl methionine and S-adenosylhomocysteine, respectively. *S. clavuligerus* pSCL4⁻ also lacks the pSCL4-located *metE* gene, to form methionine from homocysteine, but overexpresses the paralogous *metH* gene (Additional file 1: Figure S5A). Cysteine, is formed by the cysteine synthases encoded by SCLAV_4724 and SCLAV_2020, the last one being upregulated at stationary growth phase. Also, the cystathionine-γ-lyase (*cysA*) contributes to cysteine formation (Additional file 1: Figure S5B). However, the genes SCLAV_5668, encoding a cystathionine-γ-synthase, and SCLAV_p1477, encoding a cysteine synthase, do not exist in *S. clavuligerus* pSCL4-; in addition, a gene homologous to *metC*, for a cystathionine β-lyase [35] has not been detected in *S. clavuligerus* genome. Sulfur metabolism in *Streptomyces* is still poorly understood, and

more biochemical studies are required to know relationship between genes and enzymes and regulatory mechanisms in *Streptomyces* cysteine-methionine metabolism.

Dithiolopyrrolones production in *S. clavuligerus* pSCL4 fits well with the high expression of the holomycin cluster genes [7] of this strain. The metabolomic studies indicated that the largest peak in this group of compounds corresponds to N-propionylholothin. Holomycin was purified from an uncharacterized *S. clavuligerus* mutant by Kenig and Reading [10] and later by De la Fuente et al. [36] and Li and Walsh [37] from mutants lacking the *oppA2* gene; in both cases the purification was made from cultures grown in complex medium. The N-acetyltransferase encoded by *hlmA* has wide substrate specificity, an apparent Km of 15 μM for propionyl-CoA [37] and is able to form N-propionylholothin. The formation of detectable amounts of N-propionylholothin might respond to abundant propanoyl-CoA levels in the cells, in our culture conditions.

In summary, *S. clavuligerus* ATCC 27064 survival is not affected by the lack of 2.1 Mb of genetic information, although the amount and type of secondary metabolites produced and the expression of many genes located in the chromosome is altered. This study provides insight into the cross-talk between the chromosome and the pSCL4 megaplasmid.

Conclusions

The translocation of DNA fragments from the right arm of the chromosome to plasmids may be frequent in *S. clavuligerus*, as the deletion of the *parAB*$_{\text{pSCL4}}$ genes resulted in translocation of a 303 kb DNA fragment. The 1.8 Mb plasmid pSCL4 is dispensable and its lost originates the deletion of the translocated DNA chromosomal fragments, showing in a high plasticity of the *S. clavuligerus* genome. While *S. clavuligerus* strains cured of pSCL4 are viable, this genetic element carries information for secondary metabolites biosynthesis, and for regulatory elements that may modulate the expression of chromosomal genes. Therefore the number and production level of secondary metabolites in the cured strain differs from the wild type *S. clavuligerus*. Of special interest is the lower production of clavulanic acid by *S. clavuligerus* pSCL4⁻ and its high production of sulfur related metabolites, probably due to the alteration of the metabolic pathways leading to sulfur-containing amino acids in this strain.

Methods
Strains and culture conditions

S. clavuligerus ATCC 27064, as control strain, and the holomycin high producer *S. clavuligerus* pSCL4⁻ strain [6], lacking plasmid pSCL4, were used in transcriptome experiments. *S. clavuligerus* *parAB::aac*, the parental strain for *S. clavuligerus* pSCL4⁻, was used to locate the translocation of the right arm of the chromosome. The clavulanic acid non-producer, non-sporulating *S. clavuligerus* Δ*claR::aac* [5] was included in sporulation studies. The strains were pregrown in Trypticase Soy Broth (TSB) for 24 h at 28 °C and 220 rpm shaking to an optical density at 600 nm (OD$_{600}$) of 6.5. These seed cultures were used to inoculate (5%, *v/v*) duplicated 500-ml triple-baffled flasks containing 100 ml of defined SA medium [38], and cultures were grown for 72 h under the same conditions. ME medium [39] was used to analyze aerial mycelium formation and sporulation of the strains.

Antibiotic assays

Clavulanic acid and cephamycin C were quantified as indicated by Pérez-Redondo et al. [40] Holomycin was determined by bioassay against *Micrococcus luteus* ATCC 9341 as described by De la Fuente et al. [36].

Nucleic acid isolation and RT-qPCR

DNA was isolated as previously described [41]. Relative amount of DNA from the analyzed genes in the right arm of the chromosome, and in plasmids pSCL2 and pSCL4, was quantified by qPCR using 20 ng DNA as described by Lee et al. [42]. The genes analyzed were *parA*$_{pSCL2}$, *parB*$_{pSCL4}$, SCLAV_5146, 5308, 5482, 5485, 5487, 5491, 5521, 5580, 5585, 5692, 5719, and a region located in the right telomere including the right end of SCLAV_5719 (nucleotides 6,760,214 to 6,760,380), at 12 nt from the end of the chromosome. The chromosomal gene *hrdB* was used as control.

RNA isolation, purification and integrity analysis, and RT-qPCR were performed as indicated previously [5]. The oligonucleotides used in this work are shown in Additional file 1: Table S2.

Microarray design

The microarrays used in this work have been already described [5]. RNA was extracted from the culture samples at exponential (22.5 h, 46.5 h) and stationary (60 h) phase, and analysis were performed for two biological replicates for each condition (two strains and three growth times). Labeling of RNA preparations with Cy3-dCTP, labeling of genomic DNA as the reference sample with Cy5-dCTP (2.5 pmol/50 µl hybridization solution), and the purification procedures were carried out as described previously [43]. The hybridization conditions, washing, scanning with Agilent Scanner G2565BA, and the quantification of the images were accomplished as previously described [44].

Transcriptome analysis

Transcriptome analysis was performed as indicated by Martínez-Burgo et al. [5]. The M_g transcription values obtained are proportional to the abundance of the transcript for a particular gene [45] and correspond to the transcription values of the six experimental conditions, mutant versus wild type, corresponding to the three studied growth times. For each gene, M_c values and P values were calculated (three sets of values, one for each comparison). M_c values are the binary log of the differential transcription between the mutant and the wild-type strain. The Benjamini-Hochberg (BH) false-discovery rate correction was applied to the P values. For each comparison, a result was considered statistically significant if the BH-corrected P value was ≤0.05. A positive M_c value indicates upregulation, and a negative M_c value indicates downregulation. In this work, we study those genes with M_c value ≤ −1 in the three sampling times or M_c value ≥1 in the three sampling times.

Metabolomic analysis

S. clavuligerus ATCC 27064 and *S. clavuligerus* pSCL4$^-$ were grown in SA medium up to stationary phase as indicated above. For each strain, one 100 ml sample, containing broth and mycelium, was extracted with 1 volume of ethyl acetate with HCl 1%. Extracts from 200 ml of culture were concentrated, and dried samples were resuspended into 100 µL of methanol; 2 µL samples were analyzed in an HPLC Agilent 1200 Rapid Resolution connected to a mass spectrometer maXis from Bruker using a Zorbax SB-C8 (2.1 × 30 mm, 3.5 µm particle size) column. The mobile phase was composed by two solvents containing each ammonium formate 13 mM, and trifluoroacetic acid 0.01%: solvent A, water:acetonitrile 90:10; solvent B, water:acetonitrile 10:90. Elution was performed with a 0.3 ml min^{-1} flow, and the following gradient composition: 90:10 *v/v* 0 min; 0:100 v/v 6 min, 0:100 v/v 8 min, 90:10 v/v 8.1 min, 90:10 v/v 10 min.

Mass spectrometer was adjusted in positive mode ESI (Electrospray Ionization), using 4 kV in the capillary, a drying gas flow of 11 L min^{-1} at 200 °C and a nebulizer pressure of 2.8 bar. Equipment calibration before sample injection was performed using the ions cluster formed by the trifluoroacetic acid (TFA) in the presence of Na$^+$ ions. Before the chromatographic front was detectable, every injected sample was recalibrated using TFA-Na. Every chromatographic run was processed using the internal Bruker algorithm for the components extraction, and the more intense peaks, both by positive TIC and for 210 nm absorbance, were analyzed to interpret their exact mass and molecular formula. Both the retention time (RT) and exact mass were used as guideline to search in the High Performance Mass Spectrometry (HPMS) databases from MEDINA Foundation. When a match between the sample RT and mass, and the HPMS databases was found, a search in the Chapman & Hall Dictionary of Natural Products was performed to obtain the formula.

Bioinformatic analysis

The 294 nt intergenic region between SCLAV_5488 and SCLAV_5489, was analysed by the M. Zuker's DNA Fold Server (http://unafold.rna.albany.edu/).

Additional files

Additional file 1: Figure S1. Intergenic region from SCLAV_5488 to SCLAV_5489. Possible hairpin loop formed (ΔG = −85.98 kcal/mol) in the 294 nt excision region. The hairpin loop were predicted according to the M. Zuker's DNA Fold Server http://unafold. rna.albany.edu/). **Figure S2.** Antibiotics production. S. clavuligerus ATCC27064 (white circles) and S. clavuligerus pSCL4⁻ (black circles) were grown in SA medium, and production of clavulanic acid (left panel), cephamycin C (central panel) and holomycin (right panel) was quantified. **Figure S3.** RT-qPCR validation of the microarray data. (A) Genes tested: comparison of the data obtained for each gene analyzed in microarrays experiment (Mc values) and by RT-qPCR [$\log_2$ -2E($\Delta\Delta$Ct)]. (B) Representation of the correlation between the results showed in panel A. **Figure S4.** MS and UV absortion spectra of the compounds detected in S. clavuligerus and the pSCL4⁻ mutant. Only holomycin and N-propionyl holothin, detected in the mutant, have been described previously. **Figure S5.** Transcriptomic data of Methionine and Cysteine Metabolism Genes. A) Change of expression level of genes related to methionine or cysteine metabolism in S. clavuligerus pSCL4-. Genes not present in S. clavuligerus pSCL4⁻ since they are in the megaplasmid are indicated with double asterisk (**); those genes present among the 231 CDS absent at the right arm of the chromosome are indicated with single asterisk (*). B) Pathways of methionine and cysteine metabolism in Streptomyces. Open arrows indicated upregulated genes in the pSCL4- strain. Steps carried by enzymes encoded by genes not present are marked with a black circle. **Table S1.** Effect of lack of pSCL4⁻ in genes involved in cell envelope formation and morphological differentiation. **Table S2.** Oligonucleotides used in this work. (PDF 2056 kb)

Abbreviations
CDS: Coding sequences; SMC: Secondary metabolism gene cluster

Acknowledgements
The helpful scientific discussion with Prof. Juan F. Martín, the critical reading of the manuscript by Dr. Armando Á. Losada and the continous technical support of the company DISMED, are appreciated.

Funding
This work was supported by grant BIO2013–34723 from the Spanish Ministry of Economy and Competitiveness. Y. Martínez-Burgo and R. Álvarez-Álvarez received PFU fellowships from the Spanish Ministry of Education, Culture and Sport.

Authors' contributions
R.A-A obtained the S. clavuligerus pSCL4⁻ strain. Y.M-B and R.A-A performed the transcriptome studies and obtained the extracts for metabolomic studies. A.R-G analyzed the transcriptome data. PL directed the experiments, organized the data and write the manuscript. All authors read and approved the final manuscript.

Competing interests
The authors declare having not competing interest.

Author details
[1]Microbiology Section, Faculty of Biological and Environmental Sciences, University of León, León, Spain. [2]Institute of Biotechnology of León, INBIOTEC, León, Spain.

References
1. Medema MH, Trefzer A, Kovalchuk A, van den Berg M, Müller U, Heijne W, Wu L, Alam MT, Ronning CM, Nierman WC, Bovenberg RA, Breitling R, Takano E. The sequence of a 1.8-Mb bacterial linear plasmid reveals a rich evolutionary reservoir of secondary metabolic pathways. Genome Biol Evol. 2010;2:212–24.
2. Netolitzky DJ, Wu X, Jensen SE, Roy KL. Giant linear plasmids of beta-lactam antibiotic producing Streptomyces. FEMS Microbiol Lett. 1995;131:27–34.
3. Charusanti P, Fong NL, Nagarajan H, Pereira AR, Li HJ, Abate EA, Su Y, Gerwick WH, Palsson BO. Exploiting adaptive laboratory evolution of Streptomyces clavuligerus for antibiotic discovery and overproduction. PLoS One. 2012;7:e33727.
4. Lorenzana LM, Pérez-Redondo R, Santamarta I, Martín JF, Liras P. Two oligopeptide-permease-encoding genes in the clavulanic acid cluster of Streptomyces clavuligerus are essential for production of the β-lactamase inhibitor. J Bacteriol. 2004;186:3431–8.
5. Martínez-Burgo Y, Álvarez-Álvarez R, Rodríguez-García A, Liras P. The Pathway-Specific Regulator ClaR of Streptomyces clavuligerus Has a Global Effect on the Expression of Genes for Secondary Metabolism and Differentiation. Appl Environ Microbiol. 2015;81:6637–48.
6. Álvarez-Álvarez R, Rodríguez-García A, Martínez-Burgo Y, Robles-Reglero V, Santamarta I, Pérez-Redondo R, Martín JF, Liras P. A 1.8-Mb-reduced Streptomyces clavuligerus genome, relevance for secondary metabolism and differentiation. Appl Microbiol Biotechnol. 2014;98:2183–95.
7. Robles-Reglero V, Santamarta I, Álvarez-Álvarez R, Martín JF, Liras P. Transcriptional analysis and proteomics of the holomycin gene cluster in overproducer mutants of Streptomyces clavuligerus. J Biotechnol. 2013;163:69–76.
8. Blin K, Kazempour D, Wohlleben W, Weber T. Improved Lanthipeptide detection and prediction for antiSMASH. PLoS One. 2014;9:e103665.
9. Weber T, Blin K, Duddela S, Krug D, Kim HU, Bruccoleri R, Lee SY, Fischbach MA, Müller R, Wohlleben W, Breitling R, Takano E. Medema, MH. antiSMASH 3.0-a comprehensive resource for the genome mining of biosynthetic gene clusters. Nucleic Acids Res. 2015;43(W1):W237–43.
10. Kenig M, Reading C. Holomycin and an antibiotic (MM 19290) related to tunicamycin, metabolites of Streptomyces clavuligerus. J Antibiot. 1979;32:549–54.
11. Barona-Gómez F, Wong U, Giannakopulos AE, Derrick PJ, Challis GL. Identification of a cluster of genes that directs desferrioxamine biosynthesis in Streptomyces coelicolor M145. J Am Chem Soc. 2004;126:16282–3.
12. Paget MSB, Leibovitz E, Buttner MJ. A putative two-component signal transduction system regulates sigmaE, a sigma factor required for normal cell wall integrity in Streptomyces coelicolor A3(2). Mol Microbiol. 1999;33:97–107.
13. Hahn JS, Oh SY, Roe JH. Role of OxyR as a peroxide-sensing positive regulator in Streptomyces coelicolor A3(2). J Bacteriol. 2002;184:5214–22.
14. Kim MS, Dufour YS, Yoo JS, Cho YB, Park JH, Nam GB, Kim HM, Lee KL, Donohue TJ, Roe JH. Conservation of thiol-oxidative stress responses regulated by SigR orthologues in actinomycetes. Mol Microbiol. 2012;85:326–44.
15. Cho YH, Lee EJ, Ahn BE, Roe JH. SigB, an RNA polymerase sigma factor required for osmoprotection and proper differentiation of Streptomyces coelicolor. Mol Microbiol. 2001;42:205–14.
16. Shu D, Chen L, Wang W, Yu Z, Ren C, Zhang W, Yang S, Lu Y, Jiang W. AfsQ1-Q2-sigQ is a pleiotropic but conditionally required signal transduction system for both secondary metabolism and morphological development in Streptomyces coelicolor. Appl Microbiol Biotechnol. 2009;81:1149–60.
17. Uguru GC, Stephens KE, Stead JA, Towle JE, Baumberg S, McDowall KJ. Transcriptional activation of the pathway-specific regulator of the actinorhodin biosynthetic genes in Streptomyces coelicolor. Mol Microbiol. 2005;58:131–50.
18. Chen L, Lu Y, Chen J, Zhang W, Shu D, Qin Z, Yang S, Jiang W. Characterization of a negative regulator Avel for avermectin biosynthesis in Streptomyces avermitilis NRRL8165. Appl Microbiol Biotechnol. 2008;80:277–86.

19. Kang SH, Huang J, Lee HN, Hur YA, Cohen SN, Kim ES. Interspecies DNA microarray analysis identifies WblA as a pleiotropic down-regulator of antibiotic biosynthesis in *Streptomyces*. J Bacteriol. 2007;189:4315–9.

20. Gál J, Szvetnik A, Schnell R, Kálmán M. The *metD* D-methionine transporter locus of *Escherichia coli* is an ABC transporter gene cluster. J Bacteriol. 2002; 184:4930–2.

21. Noens EE, Mersinias V, Traag BA, Smith CP, Koerten HK, van Wezel GP. (2005). SsgA-like proteins determine the fate of peptidoglycan during sporulation of *Streptomyces coelicolor*. Mol Microbiol. 2005;58:929–44.

22. Pivetti CD, Yen MR, Miller S, Busch W, Tseng YH, Booth IR, Saier MHJr. Two families of mechanosensitive channel proteins. Microbiol Mol Biol Rev 2003; 67: 66-85.

23. Clark LC, Seipke RF, Prieto P, Willemse J, van Wezel GP, Hutchings MI, Hoskisson PA. Mammalian cell entry genes in *Streptomyces* may provide clues to the evolution of bacterial virulence. Sci Rep. 2013;3:1109.

24. Kodani S, Hudson ME, Durrant MC, Buttner MJ, Nodwell JR, Willey JM. The SapB morphogen is a lantibiotic-like peptide derived from the product of the developmental gene *ramS* in *Streptomyces coelicolor*. Proc Natl Acad Sci U S A. 2004;101:11448–53.

25. Lee HS, Shin HJ, Jang KH, Kim TS, Oh KB, Shin J. Cyclic peptides of the nocardamine class from a marine-derived bacterium of the genus *Streptomyces*. J Nat Prod. 2005;68:623–5.

26. Okamura K, Soga K, Shimauchi Y, Ishikura T, Lein J. Holomycin and N-propionylholothin, antibiotics produced by a cephamycin C producer. J Antibiot. 1977;30:334–6.

27. Song JY, Jeong H, Yu DS, Fischbach MA, Park HS, Kim JJ, Seo JS, Jensen SE, Oh TK, Lee KJ, Kim JF. *Draft genome sequence of Streptomyces clavuligerus NRRL 3585, a producer of diverse secondary metabolites*. J Bacteriol. 2010;192:6317–8.

28. Fischer G, Wenner T, Decaris B, Leblond P. Chromosomal arm replacement generates a high level of intraspecific polymorphism in the terminal inverted repeats of the linear chromosomal DNA of *Streptomyces ambofaciens*. Proc Natl Acad Sci U S A. 1998;95:14296–301.

29. Musialowski MS, Flett F, Scott GB, Hobbs G, Smith CP, Oliver SG. Functional evidence that the principal DNA replication origin of the *Streptomyces coelicolor* chromosome is close to the *dnaA-gyrB* region. J Bacteriol. 1994;176:5123–5.

30. Volff JN, Viell P, Altenbucher J. Artifitial circularization of the chromosome with concomitant deletion of its terminal inverted repeats enhances genetic instability and genome rearrangement in *Streptomyces lividans*. Mol Gen Genet. 1997;27:753–60.

31. Volff JN, Viell P, Altenbucher J. A new beginning with new ends, linearisation of circular chromosomes during bacterial evolution. FEMS Microbiol Lett. 2000;186:143–50.

32. Kim HS, Lee YJ, Lee CK, Choi SU, Yeo SH, Hwang YI, Yu TS, Kinoshita H, Nihira T. Cloning and characterization of a gene encoding the gamma-butyrolactone autoregulator receptor from *Streptomyces clavuligerus*. Arch Microbiol. 2004; 182:44–50.

33. Santamarta I, Pérez-Redondo R, Lorenzana LM, Martín JF, Liras P. Different proteins bind to the butyrolactone-receptor protein ARE sequence located upstream of the regulatory *ccaR* gene of *S. clavuligerus*. Mol Microbiol. 2005; 57:824–35.

34. Davidova OK, Deriabin DG, Nikiian AN, El'-Registan GI. Mechanisms of interaction between DNA and chemical analogues of microbial anabiosis autoinducers. Mikrobiologiia. 2005;74:616–25.

35. Kulkarni A, Zeng Y, Zhou W, VanLanen S, Zhang W, Chen S. A branch point of *Streptomyces* sulfur amino acid metabolism controls the production of albomycin. Appl Environ Microbiol. 2016;82:467–77.

36. De la Fuente A, Lorenzana LM, Martín JF, Liras P. Mutants of *Streptomyces clavuligerus* with disruptions in different genes for clavulanic acid biosynthesis produce large amounts of holomycin, possible crossregulation of two unrelated secondary metabolic pathways. J Bacteriol. 2002;184:6559–65.

37. Li B, Walsh CT. Identification of the gene cluster for the dithiolopyrrolone antibiotic holomycin in *Streptomyces clavuligerus*. Proc Natl Acad Sci U S A. 2010;107:19731–5.

38. Aidoo KA, Wong A, Alexander DC, Rittammer RA, Jensen SE. Cloning, sequencing and disruption of a gene from *Streptomyces clavuligerus* involved in clavulanic acid biosynthesis. Gene. 1994;147:41–6.

39. Sánchez L, Braña A. Cell density influences antibiotic biosynthesis in *Streptomyces clavuligerus*. Microbiology. 1996;142:1209–20.

40. Pérez-Redondo R, Rodríguez-García A, Martín JF, Liras P. The *claR* gene of *Streptomyces clavuligerus*, encoding a LysR-type regulatory protein controlling clavulanic acid biosynthesis, is linked to the clavulanate-9-aldehyde reductase (*car*) gene. Gene. 1998;211:311–21.

41. Pospiech A, Neumann B. A versatile quick-prep of genomic DNA from gram-positive bacteria. Trends Genet. 1995;11:217–8.

42. Lee C, Kim J, Shin SG, Hwang S. Absolute and relative qPCR quantification of plasmid copy number in *Escherichia coli*. J Biotechnol. 2006;123:273–80.

43. Álvarez-Álvarez R, Rodríguez-García A, Santamarta I, Pérez-Redondo R, Prieto-Domínguez A, Martínez-Burgo Y, Liras P. Transcriptomic analysis of *Streptomyces clavuligerus* Δ*ccaR, tsr*, effects of the cephamycin C-clavulanic acid cluster regulator CcaR on global regulation. Microb Biotechnol. 2014;7:221–31.

44. Yagüe P, Rodríguez-García A, López-García MT, Rioseras B, Martín JF, Sánchez J, Manteca A. Transcriptomic analysis of liquid nonsporulating *Streptomyces coelicolor* cultures demonstrates the existence of a complex differentiation comparable to that occurring in solid sporulating cultures. PLoS One. 2014;9:e86296.

45. Mehra S, Lian W, Jayapal KP, Charaniya SP, Sherman DH, Hu W-S. A framework to analyze multiple time series data, a case study with *Streptomyces coelicolor*. J Ind Microbiol Biotechnol. 2006;33:159–72.

6

Contrasting patterns of evolutionary constraint and novelty revealed by comparative sperm proteomic analysis in Lepidoptera

Emma Whittington[1], Desiree Forsythe[2], Kirill Borziak[1], Timothy L. Karr[3], James R. Walters[4] and Steve Dorus[1]*

Abstract

Background: Rapid evolution is a hallmark of reproductive genetic systems and arises through the combined processes of sequence divergence, gene gain and loss, and changes in gene and protein expression. While studies aiming to disentangle the molecular ramifications of these processes are progressing, we still know little about the genetic basis of evolutionary transitions in reproductive systems. Here we conduct the first comparative analysis of sperm proteomes in Lepidoptera, a group that exhibits dichotomous spermatogenesis, in which males produce a functional fertilization-competent sperm (eupyrene) and an incompetent sperm morph lacking nuclear DNA (apyrene). Through the integrated application of evolutionary proteomics and genomics, we characterize the genomic patterns potentially associated with the origination and evolution of this unique spermatogenic process and assess the importance of genetic novelty in Lepidopteran sperm biology.

Results: Comparison of the newly characterized Monarch butterfly (*Danaus plexippus*) sperm proteome to those of the Carolina sphinx moth (*Manduca sexta*) and the fruit fly (*Drosophila melanogaster*) demonstrated conservation at the level of protein abundance and post-translational modification within Lepidoptera. In contrast, comparative genomic analyses across insects reveals significant divergence at two levels that differentiate the genetic architecture of sperm in Lepidoptera from other insects. First, a significant reduction in orthology among Monarch sperm genes relative to the remainder of the genome in non-Lepidopteran insect species was observed. Second, a substantial number of sperm proteins were found to be specific to Lepidoptera, in that they lack detectable homology to the genomes of more distantly related insects. Lastly, the functional importance of Lepidoptera specific sperm proteins is broadly supported by their increased abundance relative to proteins conserved across insects.

Conclusions: Our results identify a burst of genetic novelty amongst sperm proteins that may be associated with the origin of heteromorphic spermatogenesis in ancestral Lepidoptera and/or the subsequent evolution of this system. This pattern of genomic diversification is distinct from the remainder of the genome and thus suggests that this transition has had a marked impact on lepidopteran genome evolution. The identification of abundant sperm proteins unique to Lepidoptera, including proteins distinct between specific lineages, will accelerate future functional studies aiming to understand the developmental origin of dichotomous spermatogenesis and the functional diversification of the fertilization incompetent apyrene sperm morph.

Keywords: Spermatogenesis, Lepidoptera, Fertility, Sexual selection, Testis, Mass spectrometry, Parasperm, Apyrene sperm, Positive selection, Genomic

* Correspondence: sdorus@syr.edu
[1]Center for Reproductive Evolution, Department of Biology, Syracuse University, Syracuse, NY, USA
Full list of author information is available at the end of the article

Background

Spermatozoa exhibit an exceptional amount of diversity at both the ultrastructure and molecular levels despite their central role in reproduction [1]. One of the least understood peculiarities in sperm variation is the production of heteromorphic sperm via dichotomous spermatogenesis, the developmental process where males produce multiple distinct sperm morphs that differ in their morphology, DNA content and/or other characteristics [2]. Remarkably, one sperm morph is usually fertilization incompetent and often produced in large numbers; such morphs are commonly called "parasperm", in contrast to fertilizing "eusperm" morphs. Despite the apparent inefficiencies of producing sperm morphs incapable of fertilization, dichotomous spermatogenesis has arisen independently across a broad range of taxa, including insects, brachiopod molluscs and fish. This paradoxical phenomenon, where an investment is made into gametes that will not pass on genetic material to the following generation, has garnered substantial interest, and a variety of hypotheses regarding parasperm function have been postulated [3]. In broad terms, these can be divided into three main functional themes: (**1**) facilitation, where parasperm aid the capacitation or motility of eusperm in the female reproductive tract, (**2**) provisioning, where parasperm provide nutrients or other necessary molecules to eusperm, the female or the zygote and (**3**) mediating postcopulatory sexual selection, where parasperm may serve eusperm either defensively or offensively by delaying female remating, influencing rival sperm, or biasing cryptic female choice. Despite experimental efforts in a number of taxa, a robust determination of parasperm function has yet to be attained.

Dichotomous spermatogenesis was first identified in Lepidoptera [4], the insect order containing butterflies and moths, over a century ago and is intriguing because the parasperm morph (termed apyrene sperm), is anucleate and therefore lacks nuclear DNA. Although it has been suggested that apyrene sperm are the result of a degenerative evolutionary process, several compelling observations suggest that dichotomous spermatogenesis is likely adaptive. First, it has been clearly demonstrated that both sperm morphs are required for successful fertilization in the silkworm moth (*Bombyx mori*) [5]. Second, phylogenetic relationships indicate ancestral origins of dichotomous spermatogenesis and continued maintenance during evolution. For example, dichotomous spermatogenesis is present throughout Lepidoptera, with the sole exception of two species within the most basal suborder of this group. Although multiple independent origins of sperm heteromorphism in Lepidoptera has yet to be formally ruled out, a single ancestral origin is by far the most parsimonious explanation [6].

Third, the ratio or eupyrene to apyrene varies substantively across Lepidoptera but is relatively constant within species, including several cases where apyrene comprise up to 99% of the sperm produced [7]. While variation in the relative production of each sperm morph is not in itself incompatible with stochastic processes, such as drift, it is nearly impossible to reconcile the disproportionate investment in apyrene without acknowledging that they contribute in some fundamental way to reproductive fitness. Although far from definitive, it has also been suggested that this marked variability across species is consistent with ongoing diversifying selection [6]. Arriving at an understanding of apyrene function may be further complicated by the possibility that parasperm are generally more likely to acquire lineage specific functionalities [8].

To better understand the molecular basis of dichotomous spermatogenesis, we recently conducted a proteomic and genomic characterization of sperm in *Manduca sexta* (hereafter *Manduca*) [9]. An important component of our analysis was to determine the taxonomic distribution of sperm proteins, which revealed an unexpectedly high number of proteins that possess little or no homology to proteins outside of Lepidoptera. This pattern is consistent with genetic novelty associated with dichotomous spermatogenesis in Lepidoptera, although we cannot formally rule out relaxation of purifying selection (on apyrene sperm proteins, for example) as an explanation for this marked divergence. Sperm proteins unique to Lepidoptera were also determined to be significantly more abundant than other sperm proteins. Given that apyrene spermatogenesis accounts for 95% of all sperm production in Manduca [7], these proteins are likely to be present and function in the more common apyrene sperm morph.

To provide a deeper understanding of the role of genetic novelty and genomic diversification in the evolution of dichotomous spermatogenesis, we have characterized the sperm proteome of the Monarch butterfly (*Danaus plexippus*; hereafter Monarch). In addition to its phylogenetic position and its continued development as a model butterfly species, we have pursued this species because of its distinct mating behavior. Unlike most other Lepidopteran species, male Monarch butterflies employ a strategy of coercive mating, as a consequence female Monarchs remate frequently [10]. In contrast, female remating is rare in *Manduca* and, as in many other Lepidoptera, females attract males via pheromonal calling behavior [11]. Interestingly, cessation of calling appears to be governed by molecular factors present in sperm or seminal fluid [12] and, as a consequence, non-virgin females rarely remate. Despite these behavioral differences, the proportion of eupyrene and apyrene produced is quite similar between these two species (~95–96%) [7, 13]. Thus, our focus on Monarch is motivated both by their disparate, polyandrous mating system and their utility as a representative butterfly species for comparative analyses with *Manduca*. Therefore, the overarching aims of this study were to (1) characterize the sperm proteome of the Monarch butterfly and compare it with the previously characterized sperm proteome of *Manduca*, (2) contrast patterns of orthology across diverse insect genomes between the sperm proteome and remainder of genes in the genome and (3) analyze genome-wide homology to assess the contribution of evolutionary genetic novelty to Lepidopteran sperm composition.

Methods

Butterfly rearing and sperm purification

Adult male Monarch butterflies, kindly provided by MonarchWatch (Lawrence, Kansas), were dissected between 5 and 10 days post eclosion. The sperm contents of seminal vesicles, including both apyrene and eupyrene sperm, were dissected via a small incision in the mid to distal region of the seminal vesicle. Samples were rinsed in phosphate buffer solution and pelleted via centrifugation (2 min at 15000 rpm) three times to produce a purified sperm sample. Sperm samples from 3 groups of 5 separate males were pooled to form three biological replicates [14].

Protein preparation and 1-dimensional SDS page

Samples were solubilized in 2X LDS sample buffer, as per manufacturers' instructions (Invitrogen, Inc) before quantification via the EZA Protein Quantitation Kit (Invitrogen, Inc). Protein fluorescence was measured using a Typhoon Trio + (Amersham Biosciences/GE Healthcare) with 488 nm excitation and a 610 nm bandpass filter. Fluorescence data was analyzed using the ImageQuant TL software. Three replicates of 25μg of protein were separated on a 1 mm 10% NuPAGE Novex Bis-Tris Mini Gel set up using the XCell SureLock Mini-Cell system (Invitrogen) as per manufacturer instructions for reduced samples. Following electrophoresis, the gel was stained using SimplyBlue SafeStain (Invitrogen, Inc) and destained as per manufacturer instructions. Each lane on the resulting gel (containing a sample from a single replicate) was sliced into four comparable slices, producing 12 gel fractions for independent tandem mass spectrometry analysis.

Tandem mass spectrometry (MS/MS)

Gel fractions were sliced into 1 mm^2 pieces for in-gel trypsin digestion. Gel fractions were reduced (DDT) and alkylated (iodoacetamide) before overnight incubation with trypsin at 37 °C. All LC-MS/MS experiments were performed using a Dionex Ultimate 3000 RSLC nanoUPLC

(Thermo Fisher Scientific Inc., Waltham, MA, USA) system and a QExactive Orbitrap mass spectrometer (Thermo Fisher Scientific Inc., Waltham, MA, USA). Separation of peptides was performed by reverse-phase chromatography at a flow rate of 300 nL/min and a Thermo Scientific reverse-phase nano Easy-spray column (Thermo Scientific PepMap C18, 2 μm particle size, 100A pore size, 75 mm i.d. × 50 cm length). Peptides were loaded onto a pre-column (Thermo Scientific PepMap 100 C18, 5 μm particle size, 100A pore size, 300 mm i.d. × 5 mm length) from the Ultimate 3000 autosampler with 0.1% formic acid for 3 min at a flow rate of 10 μL/min. After this period, the column valve was switched to allow elution of peptides from the pre-column onto the analytical column. Solvent A was water plus 0.1% formic acid and solvent B was 80% acetonitrile, 20% water plus 0.1% formic acid. The linear gradient employed was 2–40% B in 30 min. The LC eluant was sprayed into the mass spectrometer by means of an Easy-spray source (Thermo Fisher Scientific Inc.). All m/z values of eluting ions were measured in an Orbitrap mass analyzer, set at a resolution of 70,000. Data dependent scans (Top 20) were employed to automatically isolate and generate fragment ions by higher energy collisional dissociation (HCD) in the quadrupole mass analyzer and measurement of the resulting fragment ions was performed in the Orbitrap analyzer, set at a resolution of 17,500. Peptide ions with charge states of 2+ and above were selected for fragmentation. The mass spectrometry proteomics data have been deposited to the ProteomeXchange Consortium via the PRIDE partner repository with the dataset identifier PXD006454 [15].

MS/MS data analysis

MS/MS data was analyzed using X!Tandem and Comet algorithms within the Trans-Proteomic Pipeline (v 4.8.0) [16]. Spectra were matched against the *D. plexippus* official gene set 2 (OGS2) predicted protein set (downloaded from http://Monarchbase.umassmed.edu, last updated in 2012) with a fragment ion mass tolerance of 0.40 Da and a parent monoisotopic mass error of ±10 ppm. For both X!tandem and Comet, iodoacetamide derivative of cysteine was specified as a fixed modification, whereas oxidation of methionine was specified as a variable modification. Two missed cleavages were allowed and non-specific cleavages were excluded from the analysis. False Discovery Rates (FDRs) were estimated using a decoy database of randomized sequence for each protein in the annotated protein database. Peptide identifications were filtered using a greater than 95.0% probability based upon PeptideProphet [17] and the combined probability information from X!Tandem and Comet using Interprophet. Protein assignments were accepted if greater than 99.0%, as specified by the ProteinProphet [18] algorithms respectively. Proteins that contained identical peptides that could not be differentiated based on MS/MS analysis alone were grouped to satisfy the principles of parsimony. Protein inclusion in the proteome was based on the following stringent criteria: (**1**) identification in 2 or more biological replicates or (**2**) identification in a single replicate by 2 or more unique peptides. To identify post-translation modifications (PTMs) of proteins, X!Tandem and Comet were rerun allowing for variable phosphorylation of serine, threonine and tyrosine residues and acetylation of lysine residues. PTM locations were identified using PTMprophet in both the Monarch data presented here and a comparable dataset in *M. sexta* [19].

APEX protein quantitation and analysis
Relative compositional protein abundance was quantified using the APEX Quantitative Proteomics Tool [20]. The training dataset was constructed using fifty proteins with the highest number of uncorrected spectral counts (n_i), and identification probabilities. All 35 physicochemical properties available in the APEX tool were used to predict peptide detection/non-detection. Protein detection probabilities (O_i) were computed using proteins with identification probabilities over 99% and the Random Forest classifier algorithm. APEX protein abundances were calculated using a merged protXML file generated by the ProteinProphet algorithm and highly correlated (all pairwise p values $<9.3 \times 10^{-10}$). The correlation in APEX abundance estimates of orthologous proteins in Monarch and *Manduca* (abundance estimates from Whittington et al. [9]) were normalized, log transformed and assessed using linear regression. Differential protein abundance was analyzed using corrected spectral counts and the R (v 3.0.0) package EdgeR [21]. Results were corrected for multiple testing using the Benjamini-Hochberg method within EdgeR.

Lift-over between *D. plexippus* version 1 and 2 gene sets
Two versions of gene models and corresponding proteins are currently available for *D. plexippus*. Official gene set one (OGS1) was generated using the genome assembly as initially published [22], while the more recent official gene set 2 (OGS2) was generated along with an updated genome assembly [23]. While our proteomic analysis employs the more recent OGS2 gene models, at the time of our analysis only OGS1 gene models were included in publicly available databases for gene function and orthology (e.g. Uniprot and OrthoDB). In order to make use of these public resources, we assigned OGS2 gene models to corresponding OGS1 gene models by sequence alignment. Specifically, OGS2 coding sequences (CDS) were aligned to OGS1 CDS using BLAT [24], requiring 95% identity; the best aligning OGS1 gene model was

assigned as the match for the OGS2 query. In this way, we were able to link predictions of OGS1 gene function and orthology in public databases to OGS2 sequences in our analysis. Of the 584 OGS2 loci identified in the sperm proteome 18 could not be assigned to an OGS1 gene.

Functional annotation and enrichment analysis
Two approaches were employed for functionally annotating *D. plexippus* sperm protein sequences. First, we obtained functional annotations assigned by Uniprot to corresponding *D. plexippus* OGS1 protein sequences (Additional file 1) [25]. Additionally we used the Blast2GO software to assign descriptions of gene function and also gene ontology categories [26]. The entire set of predicted protein sequences from OGS2 were BLASTed against the GenBank non-redundant protein database with results filtered for $E < 10^{-5}$, and also queried against the InterPro functional prediction pipeline [27]. Functional enrichment of Gene Ontology (GO) terms present in the sperm proteome relative to the genomic background was performed using Blast2GO's implementation of a Fisher's exact test with a false discovery rate of 0.01%.

Orthology predictions and analysis
Two approaches were employed for establishing orthology among proteins from different species. First, we used the proteinortho pipeline [28] to assess 3-way orthology beween *D. plexippus* OGS2, *M. sexta* OGS1 [29], and *D. melanogaster* (flybase r6.12) gene sets. Proteinortho uses a reciprocal blast approach (>50% query coverage and >25% amino acid identity) to group genes with significant sequence similarity into clusters to identify orthologs and paralogs. For each species, genes with multiple protein isoforms were represented by the longest sequence in the proteinortho analysis. *D. melanogaster* and *M. sexta* ortholog predictions were then cross referenced to the published sperm of these two species [9, 30], allowing a three-way assessment of orthology in relation to presence in the sperm proteome. Using proteinortho allowed the direct analysis of the *D. plexippus* OGS2 sequences, which were not analyzed for homology in OrthoDB8 [31]. Potential annotation errors in the Monarch genome were investigated by identifying orthologs between Monarch and *Drosophila* which differed in length by at least 35%. These orthologs were manually curated using BLAST searches against available Lepidoptera and *Drosophila* genes to distinguish putative cases of misannotation from bona fide divergence in length.

A taxonomically broader set of insect ortholog relationships was obtained from OrthoDB8 and used to assess the proportion of orthologs among sperm proteins relative to the genomic background. A randomized sampling procedure was used to determine the null expectation for the proportion of orthologous proteins found between *D. plexippus* and the queried species. A set of 584 proteins, the number equal to detected *D. plexippus* sperm proteins, was randomly sampled 5000 times from the entire Monarch OGS2 gene set. For each sample, the proportion of genes with an ortholog reported in OrthoDB8 was calculated, yielding a null distribution for the proportion of orthologs expected between *D. plexippus* and the queried species. For each query species, the observed proportion of orthologs in the sperm proteome was compared to this null distribution to determine whether the sperm proteome had a different proportion of orthologs than expected and to assign significance. Comparisons were made to 12 other insect species, reflecting five insect orders: Lepidoptera (*Heliconius melpomene*, *M. sexta*, *Plutella xylostella*, *Bombyx mori*), Diptera (*Drosophila melanogaster*, *Anopheles gambiae*), Hymenoptera (*Apis mellifera*, *Nasonia vitripennis*), Coleoptera (*Tribolium castaneum*, *Dendroctonus ponderosae*), and Hemiptera (*Acyrthosiphon pisum*, *Cimex lectularius*).

Maximum likelihood phylogenetic analysis

The phylogenetic relationships (i.e. topology) among the 13 taxa considered here were taken from [32] (for Lepidoptera) and from [33] (among insect orders). Branch lengths for this topology were determined using maximum likelihood optimization with amino acid sequence data. Thirteen nuclear genes were selected from the set of 1-to-1 orthologous loci provided by the BUSCO Insecta listing from OrthoDB version 9 [34]. Genes were chosen for completeness among the focal species analyzed. The genes used in this analysis correspond to the following OrthoDB9 ortholog groups: EOG090W0153, EOG090W01JK, EOG090W059K, EOG090W05WH, EOG090W06ZM, EOG090W08E4, EOG090W08ZA, EOG090W09XZ, EOG090W0E59, EOG090W0EIQ, EOG090W0F8Q, EOG090W0JMT, EOG090W0JXV. Amino acid sequences were aligned using MUSCLE, with default parameters as implemented in the R package, "msa" [35]. Each alignment was then filtered with Gblocks to remove regions or poor alignment and low representation [36]. After filtering, the alignments yielded a total of 2618 amino acid positions for maximum likelihood analysis. Filtered alignments were concatenated and used as a single dataset for branch length estimation via the R package "phangorn" [37]. Model test comparisons for transition rate matrices were performed, with the optimal model (LG + gamma + invariant class) used for branch length optimization via the "pml.optim" function.

Phylogenetic distribution of sperm proteins

The taxonomic distribution of sperm proteins was determined by BLASTp analyses (statistical cut off of $e < 10^{-5}$ and query coverage of $\geq 50\%$) against the protein data sets of the following taxonomic groupings: butterflies (*Heliconius melpomene*, *Papilio xuthus*, *Lerema accius*), Lepidoptera (Butterflies with *M. sexta*, *Amyleios transitella*, and *Plutella xylostella*), Mecopterida (Lepidoptera with *D. melanogaster*), Mecopterida with *Tribolium casteneum*, and Insecta (all previous taxa as well as: *Apis mellifera*, *Pediculus humanus*, *Acyrthosiphon pisum*, and *Zootermopsis nevadensis*). Lepidopteran species were chosen to maximize species distribution across the full phylogenetic breadth of Lepidoptera, while also utilizing the most comprehensively annotated genomes based on published CEGMA scores (http://lepbase.org, [38]). Taxonomically restricted proteins were defined as those identified repeatedly across a given phylogenetic range but without homology in any outgroup species. Proteins exhibiting discontinuous phylogenetic patterns of conservation were considered unresolved.

Maximum likelihood analysis of molecular evolution

Orthology information for the four available *Papilionoidea* was obtained from OrthoDB v9 [39]. Coding sequences corresponding to protein entries for all orthology groups were obtained from Ensembl release 86 for *H. melpomene* and *M. cinxia*, and from lepbase v4 for *D. plexippus* and *P. glaucus*. Translated protein sequences were aligned using the linsi algorithm of MAFFT [40] and reverse translated in frame. Whole phylogeny estimates of dN and dS were obtained using the M1 model as implemented by the PAML software package [41]. Allowing for the absence of no more than one species, evolutionary analyses were conducted for a total of 10,258 orthology groups. Kolmogorov-Smirnov tests were used to compare the distribution of dN between groups of genes; dS was not utilized in these comparisons because synonymous sites were found to be saturated between all of the sequenced *Papilionoidea* genomes. Rapidly evolving sperm proteins were also identified as those in the top 5% of proteins based on dN after the removal of outliers exceeding twice the interquartile range genome-wide.

Results
Monarch sperm proteome

Characterization of the Monarch sperm proteome as part of this study, in conjunction with our previous analysis in *Manduca* [9], allowed us to conduct the first comparative analysis of sperm in Lepidoptera, and in insects more broadly, to begin to assess the origin and evolution of dichotomous spermatogenesis at the genomic level. Tandem mass spectrometry (MS/MS) analysis of Monarch sperm, purified in triplicate, identified 240 in all three replicates, 140 proteins in two replicates and 553 proteins identified by two or more unique peptides in at least a single replicate. Together this yielded a total of 584 high confidence protein identifications (Additional file 2). Of these, 41% were identified in all three biological replicates. Comparable with our previous analysis of *Manduca* sperm, proteins were identified by an average of 7.9 unique peptides and 21.1 peptide spectral matches. This new dataset thus provides the necessary foundation to refine our understanding of sperm composition at the molecular level in Lepidoptera. (Note: *Drosophila melanogaster* gene names will be used throughout the text where orthologous relationships exist with named genes; otherwise Monarch gene identification numbers will be used.)

Gene ontology analysis of molecular composition

Gene ontology (GO) analyses were first conducted to confirm the similarity in functional composition between the Monarch and other insect sperm proteomes. Analysis of Biological Process terms revealed a significant enrichment for several metabolic processes, including the tricarboxylic acid (TCA) cycle ($p = 2.22E-16$), electron transport chain ($p = 9.85E-18$), oxidation of organic compounds ($p = 1.33E-25$) and generation of precursor metabolites and energy ($p = 1.09E-30$) (Fig. 1a). GO categories related to the TCA cycle and electron transport have also been identified as enriched in the *Drosophila* and *Manduca* sperm proteomes [9]. Generation of precursor metabolites and energy, and oxidation of organic compounds are also the two most significant enriched GO terms in the *Drosophila* sperm proteome [30]. Thus, broad metabolic functional similarities exist between the well-characterized insect sperm proteomes.

An enrichment of proteins involved in microtubule-based processes was also observed, a finding that is also consistent with previously characterized insect sperm proteomes. Amongst the proteins identified are cut up (ctp), a dynein light chain required for spermatogenesis [42], actin 5 (Act5), which is involved in sperm individualization [43], and DPOGS212342, a member of the recently expanded X-linked *tektin* gene family in *Drosophila* sperm [44]. Although functional annotations are limited amongst the 10% most abundant proteins (see below), several contribute to energetic and metabolic pathways. For example, stress-sensitive B (sesB) and adenine nucleotide translocase 2 (Ant2) are gene duplicates that have been identified in the *Drosophila* sperm proteome and, in the case of Ant2, function specifically in mitochondria during spermatogenesis [45]. Also identified was Bellwether (blw), an ATP synthetase alpha chain which is required for spermatid development [46].

The widespread representation of proteins functioning in mitochondrial energetic pathways is consistent with the contribution of giant, fused mitochondria (i.e. nebenkern) in flagellum development and presence of mitochondrial derivatives in mature spermatozoa (Fig. 1a-b) [47]. In lepidopteran spermatogenesis, the nebenkern divides to form two derivatives, which flank the axoneme during elongation; ultrastructure and size of these derivatives varies greatly between species and between the two sperm morphs [7]. In *Drosophila*, the nebenkern acts as both an organizing center for microtubule polymerization and a source of ATP for axoneme elongation, however it is unclear to what extent these structures contribute to energy required for sperm motility. Of particular note is the identification of porin, a voltage-gated anion channel that localizes to the nebenkern and is critical for sperm mitochondrion organization and individualization [48]. Consistent with these patterns, Cellular Component analysis also revealed a significant enrichment of proteins in a broad set of mitochondrial structures and components, including the respiratory chain complex I ($p = 7.73E-09$), proton-transporting V-type ATPase complex ($p = 9.90E-08$) and the NADH dehydrogenase complex ($p = 7.73E-09$) (Fig. 1b). Aside from those categories relating to mitochondria, a significant enrichment was also observed amongst categories relating to flagellum structure, including microtubule ($p = 5.43E-18$) and cytoskeleton part ($p = 2.54E-12$). These GO categories included the two most abundant proteins in the proteome identified in both Monarch and *Manduca*, beta tubulin 60D (βTub60D) and alpha tubulin 84B (αTub84B). αTub84B is of particular interest as it performs microtubule functions in the post-mitotic spermatocyte, including the formation of the meiotic spindle and sperm tail elongation [49].

Molecular Function GO analysis revealed an enrichment of oxidoreductase proteins acting on NAD(P)H ($p = 7.06E-19$), as well as more moderate enrichments in several categories relating to peptidase activity or regulation of peptidase activity (data not shown). The broad representation of proteins involved in proteolytic activity is worthy of discussion, not solely because these classes of proteins are abundant in other sperm proteomes, but also because proteases are involved in the breakdown of the fibrous sheath surrounding Lepidoptera eupyrene sperm upon transfer to the female [7]. This process has been attributed to a specific ejaculatory duct trypsin-like arginine C-endopeptidase (initiatorin) in the silkworm (*B. mori*) [50] and a similar enzymatic reaction is needed for sperm activation in *Manduca* [51]. Blast2GO analyses identified three serine-type proteases in the top 5% of proteins based on abundance, including a chymotrypsin peptidase (DPOGS213461) and a trypsin

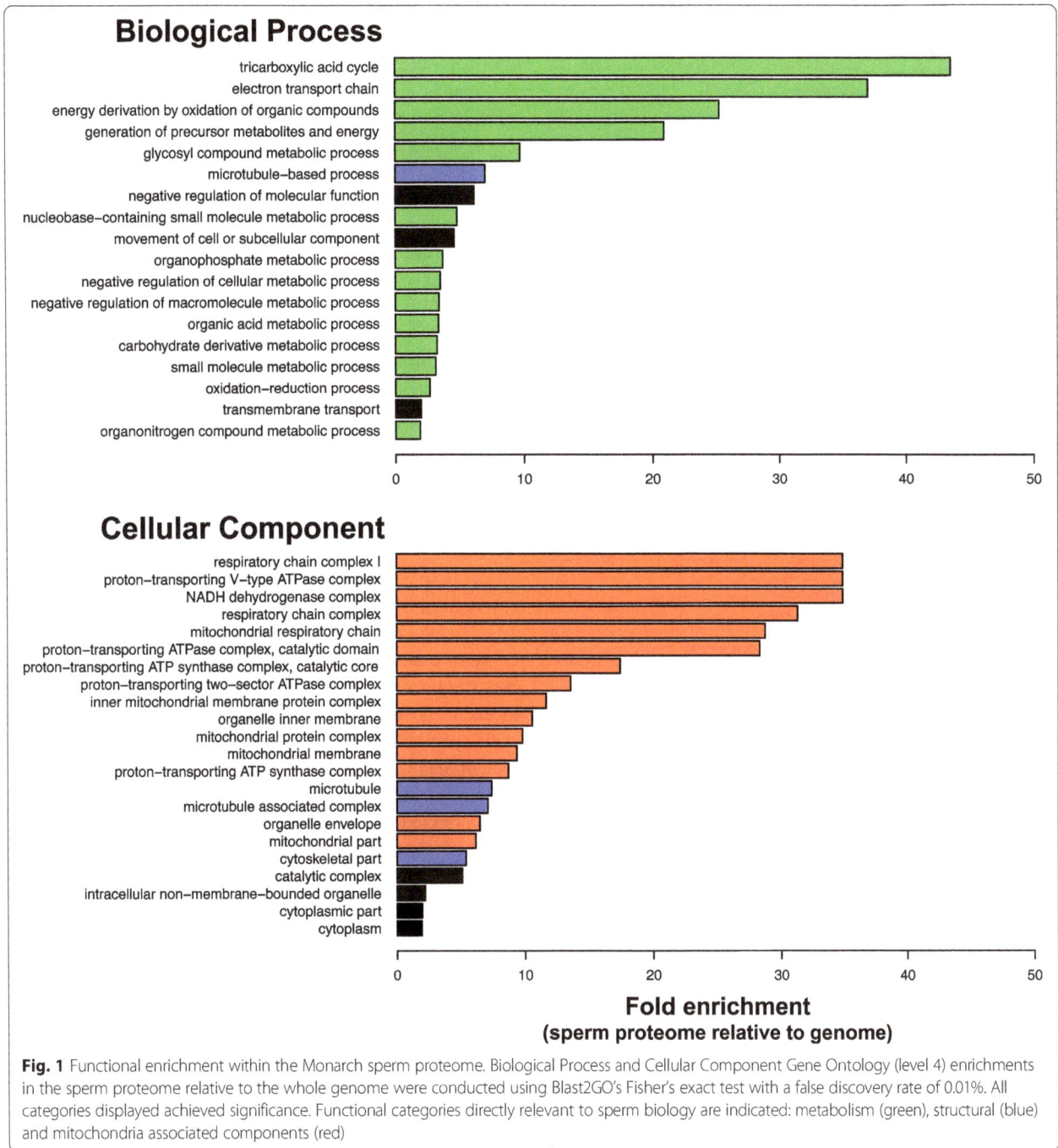

Fig. 1 Functional enrichment within the Monarch sperm proteome. Biological Process and Cellular Component Gene Ontology (level 4) enrichments in the sperm proteome relative to the whole genome were conducted using Blast2GO's Fisher's exact test with a false discovery rate of 0.01%. All categories displayed achieved significance. Functional categories directly relevant to sperm biology are indicated: metabolism (green), structural (blue) and mitochondria associated components (red)

precursor (DPOGS205340). These highly abundant proteases, particularly those that were also identified in *Manduca* (two of the most abundant proteases and 10 in total), are excellent candidates for a sperm activating factor(s) in Lepidoptera.

Conservation of Lepidoptera sperm proteomes
Our previous analysis of *Manduca* was the first foray into the molecular biology of Lepidopteran sperm and was motivated by our interest in the intriguing heteromorphic

sperm system that is found in nearly all species in this order [7]. Here we have aimed to delineate the common molecular components of lepidopteran sperm through comparative analyses. Orthology predictions between the two species identified relationships for 405 (69%) Monarch sperm proteins, of which 369 (91%) were within "one-to-one" orthology groups (Additional file 2). 298 of all orthologs (73.5%) were previously identified by MS/MS in the *Manduca* sperm proteome [9]. An identical analysis in *Drosophila* identified 203 (35%) Monarch proteins with orthology relationships, of which 166 (82%) were within "one-to-one" orthology groups (Additional file 2). 107 (52.7%) were previously characterized as components of the *Drosophila* sperm proteome [30, 52]. Thus there is a significantly greater overlap in sperm components between the two Lepidopteran species (two tailed Chi-square = 25.55, d.f. = 1, $p < 0.001$), as would be expected given the taxonomic relationship of these species. Additionally, gene duplication does not appear to be a widespread contributor to divergence relating to sperm form or function between Lepidoptera and *Drosophila*. It is also noteworthy that 27 orthologous proteins between Monarch and *Drosophila* were identified that differed substantially in length (>35%). Additional comparative analyses with gene models in other available Lepidoptera and *Drosophila* genomes indicated that 17 of these cases represent bona fide divergence in gene length, while the remainder are likely to represent gene model annotation errors in the Monarch genome. These issues were most commonly the result of inclusion/exclusion of individual exons with adjacent gene models and full gene model fusions (Additional file 2).

Recent comparative analyses of sperm composition across mammalian orders successfully identified a conserved "core" sperm proteome comprised of more slowly evolving proteins, including a variety of essential structural and metabolic components. To characterize the "core" proteome in insects, we conducted a GO analysis using *Drosophila* orthology, ontology and enrichment data to assess the molecular functionality of the 92 proteins identified in the proteome of all three insect species. This revealed a significant enrichment for proteins involved in cellular respiration ($p = 4.41e-21$), categories associated with energy metabolism, including ATP metabolic process ($p = 1.64e-15$), generation of precursor metabolites and energy ($p = 9.77e-21$), and multiple nucleoside and ribonucleoside metabolic processes. Analysis of cellular component GO terms revealed a significant enrichment for mitochondrion related proteins ($p = 3.72e-22$), respiratory chain complexes ($p = 8.25e-12$), dynein complexes ($p = 1.37e-5$), and axoneme ($p = 3.31e-6$). These GO category enrichments are consistent with a core set of metabolic, energetic, and structural proteins required for general sperm function. Similar sets of core sperm proteins have been identified in previous sperm proteome comparisons [9, 30, 52, 53]. Among this conserved set are several with established reproductive phenotypes in *Drosophila*. This includes proteins associated with sperm individualization, including cullin3 (Cul3) and SKP1-related A (SkpA), which acts in cullin-dependent E3 ubiquitin ligase complex required for caspase activity in sperm individualization [54], gudu, an

Armadillo repeat containing protein [55], and porin (mentioned previously) [48]. Two proteins involved in sperm motility were also identified: dynein axonemal heavy chain 3 (dnah3) [56] and an associated microtubule-binding protein growth arrest specific protein 8 (Gas8) [57].

Comparative analysis of protein abundance

Despite the more proximate link between proteome composition and molecular phenotypes, transcriptomic analyses far outnumber similar research using proteomic approaches. Nonetheless, recent work confirms the utility of comparative evolutionary proteomic studies in identifying both conserved [58] and diversifying proteomic characteristics [59]. We have previously demonstrated a significant correlation in protein abundance between *Manduca* and *Drosophila* sperm, although this analysis was limited by the extent of orthology between these taxa [9]. To further investigate the evolutionary conservation of protein abundance in sperm, a comparison of normalized abundance estimates between Monarch and *Manduca* revealed a significant correlation ($R^2 = 0.43$, p = <1 × 10^{-15}) (Fig. 2a). We note that this correlation is based on semi-quantitative estimates [20] and would most likely be stronger if more refined absolute quantitative data were available. Several proteins identified as highly abundant in both species are worthy of further mention. Two orthologs of *Sperm leucyl aminopeptidases* (S-LAPs) were identified. S-LAPs are members of a gene family first characterized in *Drosophila* that has recently undergone a dramatic expansion, is testis-specific in expression and encodes the most abundant proteins in the *D. melanogaster* sperm proteome [60]. As would be expected, several microtubule structural components were also amongst the most abundant proteins (top 20), including αTub84B and tubulin beta 4b chain-like protein, as well as succinate dehydrogenase subunits A and B (SdhA and SdhB), porin, and DPOGS202417, a trypsin precursor that undergoes conserved post translational modification (see below).

We next sought to identify proteins exhibiting differential abundance between the two species. As discussed earlier, Monarch and *Manduca* have distinct mating systems; female Monarch butterflies remate considerably more frequently than *Manduca* females, increasing the potential for sperm competition [10]. These differences may be reflected in molecular diversification in sperm composition between species. An analysis of differential protein abundance identified 45 proteins with significant differences after correction for multiple testing ($P < 0.05$; Fig. 2b), representing 7% of the proteins shared between species (Additional file 3). No directional bias was observed in the number of differentially abundant proteins (one-tail Binomial test; p

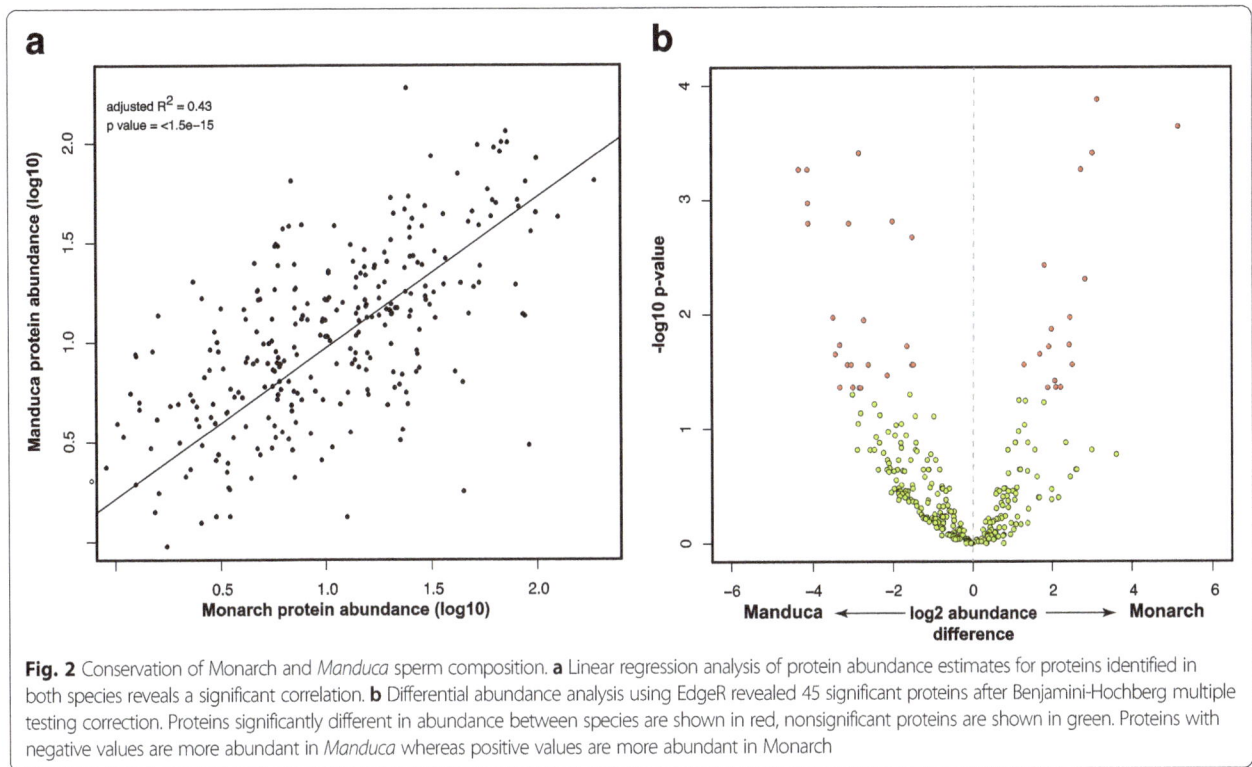

Fig. 2 Conservation of Monarch and *Manduca* sperm composition. **a** Linear regression analysis of protein abundance estimates for proteins identified in both species reveals a significant correlation. **b** Differential abundance analysis using EdgeR revealed 45 significant proteins after Benjamini-Hochberg multiple testing correction. Proteins significantly different in abundance between species are shown in red, nonsignificant proteins are shown in green. Proteins with negative values are more abundant in *Manduca* whereas positive values are more abundant in Monarch

value = 0.2757). Several of these proteins are worthy of further discussion given their role in sperm development, function or competitive ability. Proteins identified as more abundant in the Monarch sperm proteome were heavily dominated by mitochondrial NADH dehydrogenase subunits (subunits ND-23, ND-24, ND-39, and ND-51) and other mitochondria-related proteins, including ubiquinol-cytochrome c reductase core protein 2 (UQCR-C2), cytochrome C1 (Cyt-C1), and glutamate oxaloacetate transaminase 2 (Got2). Additionally, two proteins with established sperm phenotypes were identified as more abundant in *Manduca*. These included dynein light chain 90F (Dlc90F), which is required for proper nuclear localization and attachment during sperm differentiation [61], and cut up (ctp), a dynein complex subunit involved in nucleus elongation during spermiogenesis [42]. Serine protease immune response integrator (spirit) is also of interest considering the proposed role of endopeptidases in Lepidoptera sperm activation [50, 51]. Although it would be premature to draw any specific conclusions, some of these proteins play important mechanistic roles in sperm development and function and will be of interest for more targeted functional studies.

Post-translational modification of sperm proteins

During spermatogenesis, the genome is repackaged and condensed on protamines and the cellular machinery required for protein synthesis are expelled.

Consequently, mature sperm cells are considered primarily quiescent [62]. Nonetheless, sperm undergo dynamic molecular transformations after they leave the testis and during their passage through the male and female reproductive tract [63]. One mechanism by which these modifications occur is via post translational modification (PTM), which can play an integral part in the activation of sperm motility and fertilization capacity [64, 65]. Analysis of PTMs in Monarch identified 438 acetylated peptides within 133 proteins. Most notable among these are microtubule proteins, including alpha tubulin 84B (alphaTub84B), beta tubulin 60D (betaTub60D) and dyneins kl-3 and kl-5. Tubulin is a well-known substrate for acetylation, including the highly-conserved acetylation of N-terminus Lysine 40 of alphaTub84B. This modification is essential for normal sperm development, morphology and motility in mice [66]. A similar analysis in *Manduca* identified 111 acetylated peptides within 63 proteins. We found evidence for conserved PTMs within Lepidoptera in 19 proteins (36% of those identified in Monarch), including Lys40 of alphaTub84B.

In contrast to acetylation, only 75 Monarch sperm proteins showed evidence of phosphorylation, 53 of which were also modified in *Manduca* (71%). This included the ortholog of the Y-linked *Drosophila* gene WDY. Although a specific function for WDY in spermatogenesis has yet to be determined, WDY is expressed in a testis-specific manner and under positive selection in the *D. melanogaster* group [67]. The relative paucity of phosphorylation PTMs may reflect the fact that phosphorylation is one of the more difficult PTMs to identify with certainty via mass spectrometry based proteomics [68]. However, it is also noteworthy that sperm samples in this study were purified from the male seminal vesicle, and thus, before transfer to the female reproductive tract. Although far less is known about the existence of capacitation-like processes in insects, dynamic changes in the mammalian sperm phosphoproteome are associated with sperm capacitation and analogous biochemical alterations might occur within the female reproductive tract of insects [65]. We note that a similar extent of protein phosphorylation has been detected from *Drosophila* sperm samples purified in a similar manner (unpublished data; Whittington and Dorus). Lastly, identical acetylation and phosphorylation PTM patterns were identified for Monarch and *Manduca* HACP012 (DPOGS213379), a putative seminal fluid protein of unknown function previously identified in the Postman butterfly (*Heliconius melpomene*) [69, 70]. The identification of HACP012 in sperm, in the absence of other seminal fluid components, is unexpected but its identification was unambiguous as it was amongst the most abundant 10% of identified Monarch proteins. Seminal protein HACP020 (DPOGS203866), which exhibits signatures of recent adaptive evolution [70], was also identified as highly abundant (5th percentile overall); this suggests that some seminal fluid proteins may also be co-expressed in the testis and establish an association with sperm during spermatogenesis.

Rapid evolution of genetic architecture

Rapid gene evolution [71] and gene gain /loss [72], including de novo gene gain [73], are predominant processes that contribute to the diversification of male reproductive systems. Our previous study identified an enrichment in the number of Lepidoptera specific proteins (i.e. those without homology outside of Lepidoptera) in the sperm proteome relative to other reproductive proteins and non-reproductive tissues. We were unable, however, to determine from a single species whether novel genes contributed to sperm biology more broadly across all Lepidoptera. Here we employed two comparative genomic approaches to confirm and expand upon our original observation. First, we obtained whole-genome orthology relationships between Monarch and nine species, representing five insect orders, and compared the proportion of the sperm proteome with orthologs to the whole genome using a random subsampling approach. No significant differences were observed for three of the four Lepidoptera species analyzed and an excess of orthology amongst sperm proteins was identified in

the Postman butterfly ($p < 0.05$; Fig. 3). In contrast, we identified a significant deficit of sperm orthologs in all comparisons with non-Lepidopteran genomes (all $p < 0.01$). Orthology relationships in OrthoDB are established by a multi-step procedure involving reciprocal best match relationships between species and identity within species to account for gene duplication events since the last common ancestor. As such, the underrepresentation of orthology relationships is unlikely to be accounted for by lineage-specific gene duplication. Therefore, rapid evolution of sperm genes appears to be the most reasonable explanation for the breakdown of reciprocal relationships (see below). This conclusion is consistent with a diverse body of evidence that supports the influence of positive selection on male reproductive genes [71, 74], including those functioning in sperm [52, 75–78]. We note that we cannot rule out the influence of de novo gain but it is currently difficult to assess the contribution of this mechanism to the overall pattern.

The second analysis aimed to characterize the distribution of taxonomically restricted Monarch sperm proteins using BLAST searches across 12 insect species. Based on the analysis above, our a priori expectation was that a substantial number of proteins with identifiable homology amongst Lepidoptera would be absent from more divergent insect species. This analysis identified a total of 45 proteins unique to Monarch, 140 proteins (23.9% of the sperm proteome) with no detectable homology to proteins in non-Lepidopteran insect taxa and 173 proteins conserved across all species surveyed (Fig. 4a). Proteins with discontinuous taxonomic matches ($n = 171$) were considered "unresolved". Although the number of Monarch-specific proteins is considerably higher than the eight *Manduca*-specific proteins found in our previous study, the number of Lepidoptera specific is comparable to our previous estimate in *Manduca* ($n = 126$). These observations support the hypothesis that a substantial subset of lepidopteran sperm proteins are likely rapidly evolving and thus exhibit little detectable similarity. To pursue this possibility, we calculated nonsynonymous divergence (dN) for 10,212 genes across four species of butterfly and compared dN between Lepidoptera specific sperm proteins, sperm proteins with homology outside of Lepidoptera and the remainder of the genome (Fig. 4b). The average dN of Lepidoptera specific proteins was significantly higher than non-Lepidopteran specific proteins ($D = 0.34$, $p = 5.0 \times 10^{-9}$) and the remainder of the genome ($D = 0.28$, $p = 1.23 \times 10^{-7}$). Interestingly, sperm proteins with homology outside of Lepidoptera also evolve significantly slower than the genome as whole ($D = 0.30$. $p = 3.14 \times 10^{-6}$). Consistent with these trends, 17.7% of Lepidoptera specific sperm proteins where amongst the fastest evolving in the genome (top 5%), compared to only

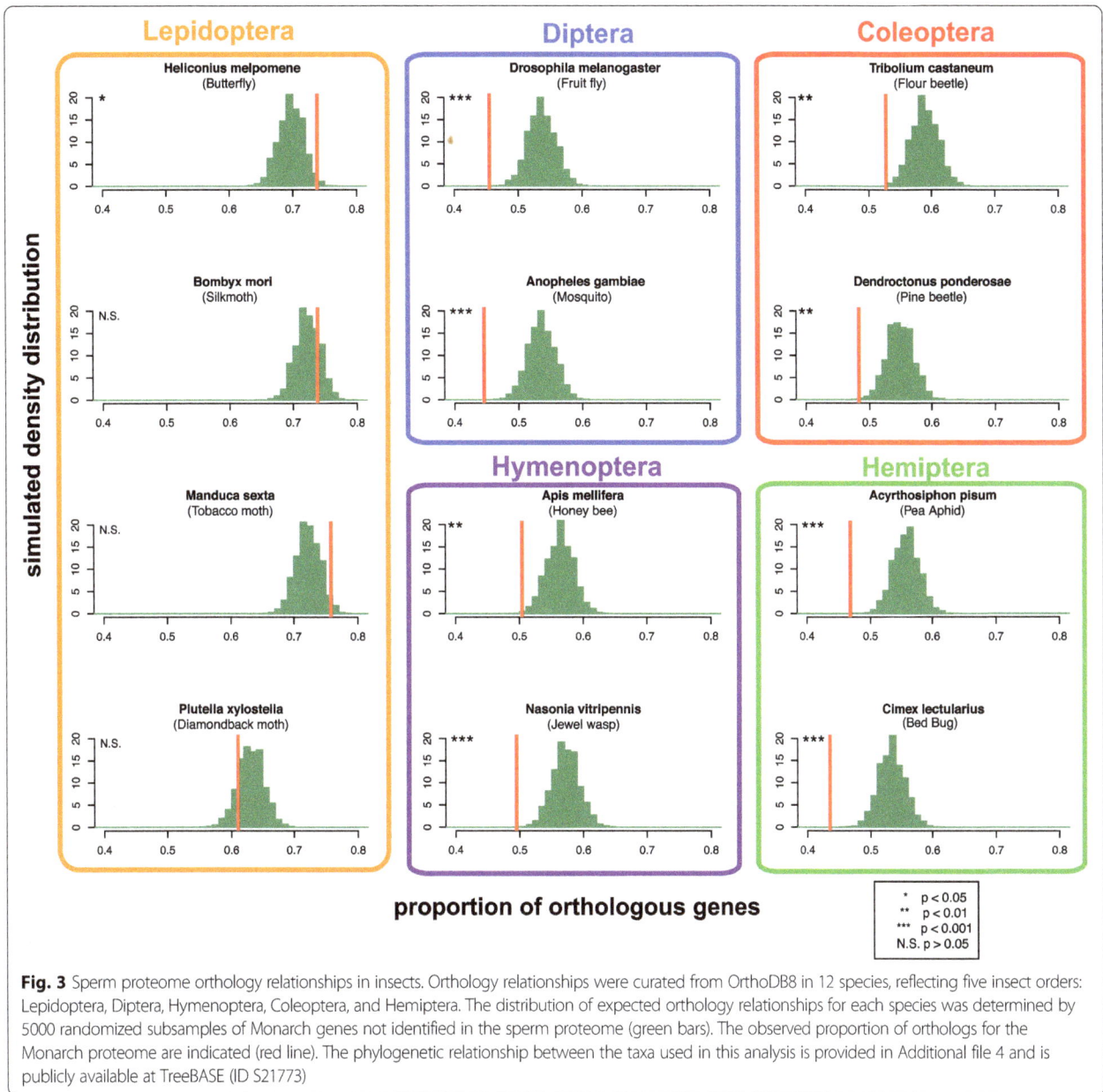

Fig. 3 Sperm proteome orthology relationships in insects. Orthology relationships were curated from OrthoDB8 in 12 species, reflecting five insect orders: Lepidoptera, Diptera, Hymenoptera, Coleoptera, and Hemiptera. The distribution of expected orthology relationships for each species was determined by 5000 randomized subsamples of Monarch genes not identified in the sperm proteome (green bars). The observed proportion of orthologs for the Monarch proteome are indicated (red line). The phylogenetic relationship between the taxa used in this analysis is provided in Additional file 4 and is publicly available at TreeBASE (ID S21773)

2.6% of sperm proteins with homology outside of Lepidoptera. In light of the rapid divergence of Lepidoptera specific proteins we next sought to assess their potential contribution to sperm function using protein abundance as a general proxy in the absence of functional annotation for nearly all of these proteins. As was observed in Whittington et al. [9], Lepidopteran specific proteins were found to be significantly more abundant than the remainder of the sperm proteome (D = 0.20, p = 0.0009, Fig. 4c).

Discussion

Dichotomous spermatogenesis in Lepidoptera, and in particular the production of sperm which do not fertilize oocytes, has intrigued biologists for over a century. Despite widespread interest, little is known about the functional roles fulfilled by apyrene sperm or why they have been retained in a nearly ubiquitous fashion during the evolution of Lepidoptera. Our comparative proteomic analysis of heteromorphic sperm, a first of its kind, provides important perspective and insights regarding the functional and evolutionary significance of this enigmatic reproductive phenotype. First, our analyses indicate that a substantial number of novel sperm genes are shared amongst Lepidoptera, thus distinguishing them from other insect species without dichotomous spermatogenesis, and suggest they are associated with heteromorphic spermatogenesis and the diversification

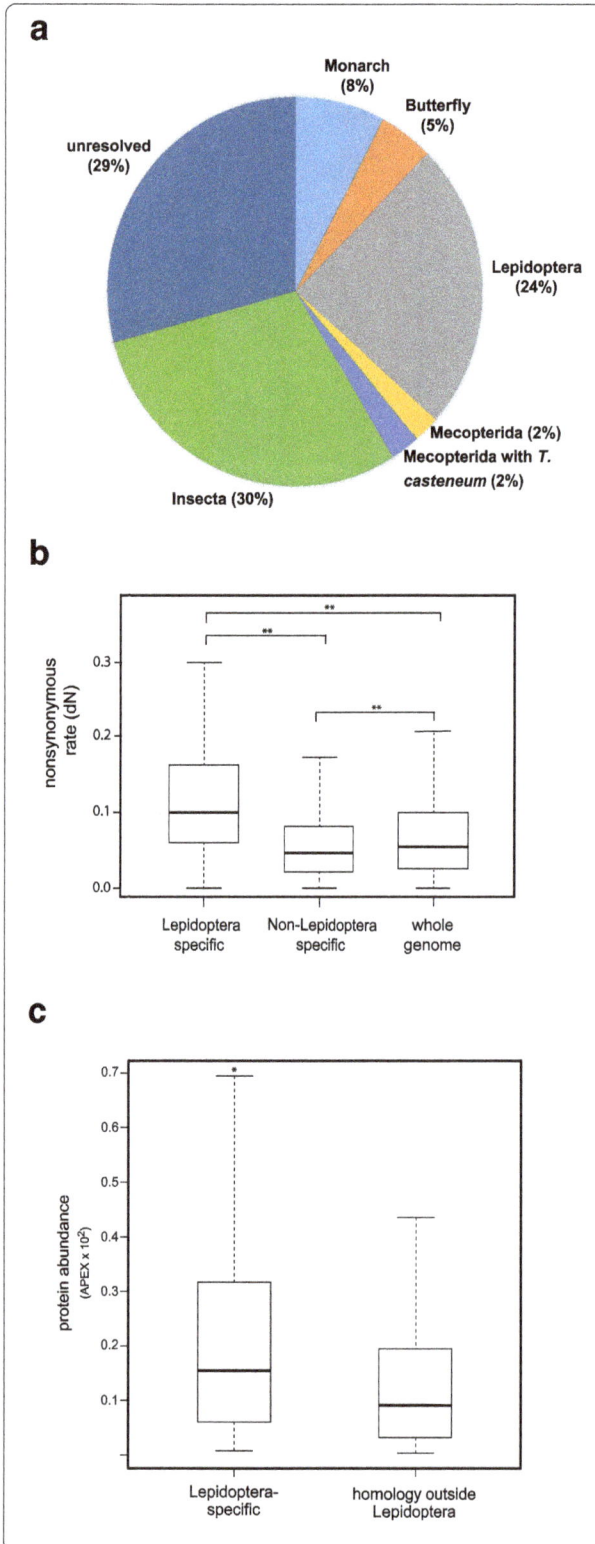

Fig. 4 Taxonomic distribution and evolution of Monarch sperm proteins. **a** Pie chart displaying the taxonomical distribution of proteins homologous to the Monarch sperm proteome and those unique to Monarch. BLAST searches were conducted beginning with closely related butterfly species and sequentially through more divergent species in Mecopterida, Mecopterida plus *Tribolium*, and Insecta. In order to be considered Lepidoptera specific, a protein was required to be present in at least at least one butterfly other than Monarch and at least one moth species. Proteins with discontinuous taxonomic patterns of homology are included in the category "unresolved". **b** Box plot showing nonsynonymous divergence (dN) of Monarch proteins across four species of butterfly ($n = 10,212$). Nonsynonymous divergence for sperm proteins identified as specific to Lepidoptera, sperm proteins with homology outside of Lepidoptera and the remainder of the genome are shown. Asterisks (**) indicate *p*-values less than 1.0×10^{-5}. **c** Box plot displaying the distribution of protein abundance estimates for proteins present only in Lepidoptera and those with homology in other insects. Asterisk (*) indicate p-values less than 0.001

of apyrene and eupyrene sperm. This observation can be attributed, at least in part, to the rapid evolution of Lepidoptera specific sperm genes. It is also possible that de novo gene gain may contribute to this observed genetic novelty, although it is not possible to assess this directly with the genomic and transcriptomic resources currently available in Lepidoptera. Our comparative and quantitative analyses, based on protein abundance measurements in both species, further suggests that some of these proteins contribute to apyrene sperm function and evolution. Given that apyrene sperm constitute the vast majority of cells in our co-mixed samples, it is reasonable to speculate that higher abundance proteins are either present in both sperm morphs or specific to apyrene cells. Confirmation of this will require targeted proteomic analysis of purified apyrene and eupyrene cell populations and will result in a refined set of candidates for further study in relation to apyrene sperm functionality. Ultimately, the comparative analysis of morph-specific sperm proteomes is critical to understanding the functional diversification of the fertilization incompetent apyrene sperm morph and the evolutionary maintenance of dichotomous spermatogenesis.

Conclusion

Our results indicate that the origin of heteromorphic spermatogenesis early in Lepidoptera evolution and/or the subsequent evolution of this system is associated with a burst of genetic novelty that is distinct from patterns of diversification across the remainder of the genome. The evolution of dichotomous spermatogenesis has therefore had a marked impact on Lepidoptera molecular evolution and suggests that focused studies of other reproductive transitions may inform our broader understanding of the evolution of reproductive genetic systems and their contribution to genomic novelty.

Additional files

Additional file 1: Functional Information- Predicted functions of Monarch proteins curated using Uniprot and Blast2GO. (XLSX 29 kb)

Additional file 2: Mass Spectrometry Data- Proteomic data including full Monarch sperm proteome, MS/MS results by replicate, PTM information, orthology relationships and rates of molecular evolution. (XLSX 100 kb)

Additional file 3: Protein Abundance and Quantitative Analyses- APEX protein abundance estimates and differential abundance analyses. (XLSX 54 kb)

Additional file 4: Phylogenetic Results- Phylogeny exhibiting the evolutionary relationship of the thirteen insect species utilized in this study. (PDF 129 kb)

Abbreviations

CDS: Coding Sequence; FDR: False Discovery Rate; GO: Gene Ontology; HCD: Higher energy Collisional Dissociation; LC: Liquid Chromatography; LC-MS/MS: Liquid Chromatography Tandem Mass Spectrometry; MS/MS: Tandem Mass Spectrometry; OGS1: Official Gene Set 1; OGS2: Official Gene Set 2; PTM: Post Translational Modification

Acknowledgements
We thank Monarch Watch and Channing Shives for support in rearing Monarch butterflies and Sheri Skerget for expert technical assistance. Computing for this project was performed on the Syracuse University Crush Virtual Research Cloud and the Community Cluster at the Center for Research Computing at the University of Kansas. We would also like to thank Eric Sedore and Larne Pekowsky of Information Technology Services at Syracuse University and Mike Deery, Renata Feret and Kathryn Lilley at the Cambridge Centre for Proteomics.

Funding
Funding for this study included Syracuse University support to SD, University of Kansas support to JW, and Syracuse University and Marilyn Kerr Fellowships to EW.

Authors' contributions
TLK, JW and SD designed the study; TLK and JW purified samples for MS analysis; EW, DF, KB, JW and SD analyzed the data; EW, TLK, JW and SD wrote the manuscript. All authors have read and approved the final version of this manuscript.

Competing interests
The authors declare that they have no competing interests.

Author details
[1]Center for Reproductive Evolution, Department of Biology, Syracuse University, Syracuse, NY, USA. [2]Science Education and Society, University of Rhode Island, Kingston, RI, USA. [3]Ecology and Evolutionary Biology, Kansas University, Lawrence, KS, USA. [4]Department of Genomics and Genetic Resources, Kyoto Institute of Technology. Saga Ippon-cho, Ukyo-ku, Kyoto, Japan.

References

1. Pitnick S, Birkhead TR, Hosken DJ. Sperm biology: an evolutionary perspective. 1st ed. Amsterdam: Academic Press/Elsevier; 2009.
2. Till-Bottraud I, Joly D, Lachaise D, Snook RR. Pollen and sperm heteromorphism: convergence across kingdoms? J Evol Biol. 2005;18:1–18.
3. Swallow JG, Wilkinson GS. The long and short of sperm polymorphisms in insects. Biol Rev Camb Philos Soc. 2002;77:153–82.
4. Meves F. Ueber oligopyrene und apyrene Spermien und über ihre Entstehung, nach Beobachtungen an Paludina und Pygaera. Arch Für Mikrosk Anat. 1902;61:1–84.
5. Sahara K, Kawamura N. Double copulation of a female with sterile diploid and polyploid males recovers fertility in *Bombyx mori*. Zygote Camb Engl. 2002;10:23–9.
6. Friedländer M. Control of the eupyrene–apyrene sperm dimorphism in Lepidoptera. J Insect Physiol. 1997;43:1085–92.
7. Friedländer M, Seth RK, Reynolds SE. Eupyrene and Apyrene sperm: dichotomous spermatogenesis in Lepidoptera. Adv Insect Physiol. 2005;32:206–308.
8. Snook RR, Hosken DJ, Karr TL. The biology and evolution of polyspermy: insights from cellular and functional studies of sperm and centrosomal behavior in the fertilized egg. Reproduction. 2011;142:779–92.
9. Whittington E, Zhao Q, Borziak K, Walters JR, Dorus S. Characterisation of the *Manduca sexta* sperm proteome: genetic novelty underlying sperm composition in Lepidoptera. Insect Biochem Mol Biol. 2015;62:183–93.
10. Oberhauser K, Frey D. Coercive mating by overwintering male monarch butterflies. In: Hoth J, Merino L, Oberhauser K, Pisanty I, Price S, Wilkinson T, editors. 1997 North American Conference on the Monarch Butterfly. Canada: Commission for Environmental Cooperation. 1997. pp. 67-78.
11. Sasaki M, Riddiford LM. Regulation of reproductive behaviour and egg maturation in the tobacco hawk moth, *Manduca sexta*. Physiol Entomol. 1984;9:315–27.
12. Stringer IAN, Giebultowicz JM, Riddiford LM. Role of the bursa copulatrix in egg maturation and reproductive behavior of the tobacco hawk moth, *Manduca sexta*. Int J Invertebr Reprod Dev. 1985;8:83–91.
13. Solensky MJ, Oberhauser KS. Male monarch butterflies, *Danaus plexippus*, adjust ejaculates in response to intensity of sperm competition. Anim Behav. 2009;77:465–72.
14. Karr TL, Walters JR. Panning for sperm gold: isolation and purification of apyrene and eupyrene sperm from lepidopterans. Insect Biochem Mol Biol. 2015;63:152–8.
15. Vizcaíno JA, Csordas A, del-Toro N, Dianes JA, Griss J, Lavidas I, et al. 2016 update of the PRIDE database and its related tools. Nucleic Acids Res. 2016;44:11033.
16. Deutsch EW, Mendoza L, Shteynberg D, Farrah T, Lam H, Tasman N, et al. A guided tour of the trans-proteomic pipeline. Proteomics. 2010;10:1150–9.
17. Keller A, Nesvizhskii AI, Kolker E, Aebersold R. Empirical statistical model to estimate the accuracy of peptide identifications made by MS/MS and database search. Anal Chem. 2002;74:5383–92.
18. Nesvizhskii AI, Keller A, Kolker E, Aebersold R. A statistical model for identifying proteins by tandem mass spectrometry. Anal Chem. 2003;75:4646–58.
19. Shteynberg DD., Mendoza L, Slagel J, Lam H, Nesvizhskii AI, Moritz R. PTMProphet: TPP software for validation of modified site locations on post-translationally modified peptides. 60th American Society for Mass Spectrometry (ASMS) Annual Conference, Vancouver, Canada, 2012.
20. Braisted JC, Kuntumalla S, Vogel C, Marcotte EM, Rodrigues AR, Wang R, et al. The APEX quantitative proteomics tool: generating protein quantitation estimates from LC-MS/MS proteomics results. BMC Bioinformatics. 2008;9: 529.
21. Robinson MD, McCarthy DJ, Smyth GK. edgeR: a bioconductor package for differential expression analysis of digital gene expression data. Bioinformatics. 2010;26:139–40.
22. Zhan S, Merlin C, Boore JL, Reppert SM. The monarch butterfly genome yields insights into long-distance migration. Cell. 2011;147:1171–85.
23. Zhan S, Reppert SM. MonarchBase: the monarch butterfly genome database. Nucleic Acids Res. 2013;41:D758–63.
24. Kent WJ. BLAT—the BLAST-like alignment tool. Genome Res. 2002;12:656–64.

25. The UniProt Consortium. UniProt: a hub for protein information. Nucleic Acids Res. 2015;43:D204–12.

26. Conesa A, Gotz S, Garcia-Gomez JM, Terol J, Talon M, Robles M. Blast2GO: a universal tool for annotation, visualization and analysis in functional genomics research. Bioinformatics. 2005;21:3674–6.

27. Zdobnov EM, Apweiler R. InterProScan–an integration platform for the signature-recognition methods in InterPro. Bioinforma Oxf Engl. 2001;17:847–8.

28. Lechner M, Findeiß S, Steiner L, Marz M, Stadler PF, Prohaska SJ. Proteinortho: detection of (co-)orthologs in large-scale analysis. BMC Bioinformatics. 2011;12:124.

29. Kanost MR, Arrese EL, Cao X, Chen Y-R, Chellapilla S, Goldsmith MR, et al. Multifaceted biological insights from a draft genome sequence of the tobacco hornworm moth, Manduca sexta. Insect Biochem Mol Biol. 2016;76:118–47.

30. Wasbrough ER, Dorus S, Hester S, Howard-Murkin J, Lilley K, Wilkin E, et al. The Drosophila melanogaster sperm proteome-II (DmSP-II). J Proteome. 2010;73:2171–85.

31. Waterhouse RM, Tegenfeldt F, Li J, Zdobnov EM, Kriventseva EV. OrthoDB: a hierarchical catalog of animal, fungal and bacterial orthologs. Nucleic Acids Res. 2013;41:D358–65.

32. Kawahara AY, Breinholt JW. Phylogenomics provides strong evidence for relationships of butterflies and moths. Proc R Soc B Biol Sci. 2014;281:20140970.

33. Misof B, Liu S, Meusemann K, Peters RS, Donath A, Mayer C, et al. Phylogenomics resolves the timing and pattern of insect evolution. Science. 2014;346:763–7.

34. Simão FA, Waterhouse RM, Ioannidis P, Kriventseva EV, Zdobnov EM. BUSCO: assessing genome assembly and annotation completeness with single-copy orthologs. Bioinformatics. 2015;31:3210–2.

35. Bodenhofer U, Bonatesta E, Horejš-Kainrath C, Hochreiter S. msa: an R package for multiple sequence alignment. Bioinformatics. 2015;31:3997–9.

36. Castresana J. Selection of conserved blocks from multiple alignments for their use in phylogenetic analysis. Mol Biol Evol. 2000;17:540–52.

37. Schliep KP. Phangorn: phylogenetic analysis in R. Bioinformatics. 2011;27:592–3.

38. Challis RJ, Kumar S, Dasmahapatra KKK, Jiggins CD, Blaxter M. Lepbase: the Lepidopteran genome database. bioRxiv 056994; doi: 10.1101/056994.

39. Zdobnov EM, Tegenfeldt F, Kuznetsov D, Waterhouse RM, Simão FA, Ioannidis P, et al. OrthoDB v9.1: cataloging evolutionary and functional annotations for animal, fungal, plant, archaeal, bacterial and viral orthologs. Nucleic Acids Res. 2017;45:D744–9.

40. Katoh K, Standley DM. MAFFT multiple sequence alignment software version 7: improvements in performance and usability. Mol Biol Evol. 2013;30:772–80.

41. Yang Z. PAML 4: Phylogenetic analysis by maximum likelihood. Mol Biol Evol. 2007;24:1586–91.

42. Joti P, Ghosh-Roy A, Ray K. Dynein light chain 1 functions in somatic cyst cells regulate spermatogonial divisions in Drosophila. Sci Rep. 2011;1:173.

43. Noguchi T. A role for actin dynamics in individualization during spermatogenesis in Drosophila Melanogaster. Development. 2003;130:1805–16.

44. Dorus S, Freeman ZN, Parker ER, Heath BD, Karr TL. Recent origins of sperm genes in Drosophila. Mol Biol Evol. 2008;25:2157–66.

45. Terhzaz S, Cabrero P, Chintapalli VR, Davies S-A, Dow JAT. Mislocalization of mitochondria and compromised renal function and oxidative stress resistance in Drosophila SesB mutants. Physiol Genomics. 2010;41:33–41.

46. Castrillon DH, Gönczy P, Alexander S, Rawson R, Eberhart CG, Viswanathan S, et al. Toward a molecular genetic analysis of spermatogenesis in Drosophila melanogaster: characterization of male-sterile mutants generated by single P element mutagenesis. Genetics. 1993;135:489–505.

47. Tokuyasu KT. Dynamics of spermiogenesis in Drosophila melanogaster. VI. Significance of "onion" nebenkern formation. J Ultrastruct Res. 1975;53:93–112.

48. Park J, Kim Y, Choi S, Koh H, Lee S-H, Kim J-M, et al. Drosophila Porin/VDAC affects mitochondrial morphology. PLoS One. 2010;5:e13151.

49. Hutchens JA, Hoyle HD, Turner FR, Raff EC. Structurally similar Drosophila Alpha-tubulins are functionally distinct in vivo. Mol Biol Cell. 1997;8:481–500.

50. Osanai M, Kasuga H, Aigaki T. Induction of motility of apyrene spermatozoa and dissociation of Eupyrene sperm bundles of the silkmoth, Bombyx mori, by initiatorin and trypsin. Invertebr Reprod Dev. 1989;15:97–103.

51. Friedländer M, Jeshtadi A, Reynolds SE. The structural mechanism of trypsin-induced intrinsic motility in Manduca sexta spermatozoa in vitro. J Insect Physiol. 2001;47:245–55.

52. Dorus S, Busby SA, Gerike U, Shabanowitz J, Hunt DF, Karr TL. Genomic and functional evolution of the Drosophila melanogaster sperm proteome. Nat Genet. 2006;38:1440–5.

53. Rettie EC, Dorus S. Drosophila sperm proteome evolution: insights from comparative genomic approaches. Spermatogenesis. 2012;2:213–23.

54. Arama E, Bader M, Rieckhof GE, Steller H. A ubiquitin ligase complex regulates caspase activation during sperm differentiation in Drosophila. PLoS Biol. 2007;5:e251. Bach E, editor

55. Cheng W, Ip YT, Xu Z. Gudu, an armadillo repeat-containing protein, is required for spermatogenesis in drosophila. Gene. 2013;531:294–300.

56. Karak S, Jacobs JS, Kittelmann M, Spalthoff C, Katana R, Sivan-Loukianova E, et al. Diverse roles of axonemal dyneins in Drosophila auditory neuron function and mechanical amplification in hearing. Sci Rep. 2015;5:17085.

57. Yeh S-D, Chen Y-J, Chang ACY, Ray R, She B-R, Lee W-S, et al. Isolation and properties of Gas8, a growth arrest-specific gene regulated during male gametogenesis to produce a protein associated with the sperm motility apparatus. J Biol Chem. 2002;277:6311–7.

58. Bayram HL, Claydon AJ, Brownridge PJ, Hurst JL, Mileham A, Stockley P, et al. Cross-species proteomics in analysis of mammalian sperm proteins. J Proteome. 2016;135:38–50.

59. Vicens A, Borziak K, Karr TL, Roldan ERS, Dorus S. Comparative sperm proteomics in mouse species with divergent mating systems. Mol Biol Evol. 2017;34:1403–16.

60. Dorus S, Wilkin EC, Karr TL. Expansion and functional diversification of a leucyl aminopeptidase family that encodes the major protein constituents of Drosophila sperm. BMC Genomics. 2011;12:177.

61. Li MG, Serr M, Newman EA, Hays TS. The Drosophila tctex-1 light chain is dispensable for essential cytoplasmic dynein functions but is required during spermatid differentiation. Mol Biol Cell. 2004;15:3005–14.

62. Hecht NB. Molecular mechanisms of male germ cell differentiation. BioEssays. 1998;20:555–61.

63. McDonough CE, Whittington E, Pitnick S, Dorus S. Proteomics of reproductive systems: towards a molecular understanding of postmating, prezygotic reproductive barriers. J Proteome. 2016;135:26–37.

64. Baker MA, Hetherington L, Weinberg A, Naumovski N, Velkov T, Pelzing M, et al. Analysis of phosphopeptide changes as spermatozoa acquire functional competence in the epididymis demonstrates changes in the post-translational modification of Izumo1. J Proteome Res. 2012;11:5252–64.

65. Platt MD, Salicioni AM, Hunt DF, Visconti PE. Use of differential isotopic labeling and mass spectrometry to analyze capacitation-associated changes in the phosphorylation status of mouse sperm proteins. J Proteome Res. 2009;8:1431–40.

66. Kalebic N, Sorrentino S, Perlas E, Bolasco G, Martinez C, Heppenstall PA. αTAT1 is the major α-tubulin acetyltransferase in mice. Nat Commun. 2013;4:1962.

67. Singh ND, Koerich LB, Carvalho AB, Clark AG. Positive and purifying selection on the Drosophila Y chromosome. Mol Biol Evol. 2014;31:2612–23.

68. Riley NM, Coon JJ. Phosphoproteomics in the age of rapid and deep proteome profiling. Anal Chem. 2016;88:74–94.

69. Walters JR, Harrison RG. Combined EST and proteomic analysis identifies rapidly evolving seminal fluid proteins in Heliconius butterflies. Mol Biol Evol. 2010;27:2000–13.

70. Walters JR, Harrison RG. Decoupling of rapid and adaptive evolution among seminal fluid proteins in Heliconius butterflies with divergent mating systems: seminal fluid proteins in Heliconius butterflies. Evolution. 2011;65:2855–71.

71. Swanson WJ, Vacquier VD. The rapid evolution of reproductive proteins. Nat Rev Genet. 2002;3:137–44.

72. Hahn MW, Han MV, Han S-G. Gene family evolution across 12 Drosophila genomes. PLoS Genet. 2007;3:e197.

73. Zhao L, Saelao P, Jones CD, Begun DJ. Origin and spread of de novo genes in Drosophila melanogaster populations. Science. 2014;343:769–72.

74. Haerty W, Jagadeeshan S, Kulathinal RJ, Wong A, Ravi Ram K, Sirot LK, et al. Evolution in the fast lane: rapidly evolving sex-related genes in Drosophila. Genetics. 2007;177:1321–35.

75. Vicens A, Lüke L, Roldan ERS. Proteins involved in motility and sperm-egg interaction evolve more rapidly in mouse spermatozoa. PLoS One. 2014;9:e91302.

76. Dorus S, Wasbrough ER, Busby J, Wilkin EC, Karr TL. Sperm proteomics reveals intensified selection on mouse sperm membrane and acrosome genes. Mol Biol Evol. 2010;27:1235–46.

77. Dean MD, Good JM, Nachman MW. Adaptive evolution of proteins secreted during sperm maturation: an analysis of the mouse epididymal transcriptome. Mol Biol Evol. 2008;25:383–92.

78. Vicens A, Gomez Montoto L, Couso-Ferrer F, Sutton KA, Roldan ERS. Sexual selection and the adaptive evolution of PKDREJ protein in primates and rodents. Mol Hum Reprod. 2015;21:146–56.

7

Identification of differentially expressed genes in flower, leaf and bulb scale of *Lilium* oriental hybrid 'Sorbonne' and putative control network for scent genes

Fang Du[1,2] [ID], Junmiao Fan[1], Ting Wang[1], Yun Wu[2], Donald Grierson[2,3], Zhongshan Gao[2*] and Yiping Xia[2*]

Abstract

Background: Lily is an economically important plant, with leaves and bulbs consisting of overlapping scales, large ornamental flowers and a very large genome. Although it is recognized that flowers and bulb scales are modified leaves, very little is known about the genetic control and biochemical differentiation underlying lily organogenesis and development. Here we examined the differentially expressed genes in flower, leaf and scale of lily, using RNA-sequencing, and identified organ-specific genes, including transcription factors, genes involved in photosynthesis in leaves, carbohydrate metabolism in bulb scales and scent and color production in flowers.

Results: Over 11Gb data were obtained and 2685, 2296, and 1709 differentially expressed genes were identified in the three organs, with 581, 662 and 977 unique DEGs in F-vs-S, L-vs-S and L-vs-F comparisons. By functional enrichment analysis, genes likely to be involved in biosynthetic pathways leading to floral scent production, such as *1-deoxy-D-xylulose-5-phosphate synthase* (*DXS*), *3-ketoacyl-CoA thiolase* (*KAT*), *hydroperoxide lyase* (*HPL*), *geranylgeranyl pyrophosphate* (*GGPP*) *4-hydroxy-3-methylbut-2-en-1-yl diphosphate* (*HDS*) and *terpene synthase* (*TPS*), and floral color genes, such as *dihydroflavonol 4-reductase* (*DFR*), *chalcone synthase* (*CHS*), *chalcone isomerase* (*CHI*), *flavonol synthase* (*FLS*) were identified. Distinct groups of genes that participate in starch and sucrose metabolism, such as *sucrose synthase* (*SS*), *invertase* (*INV*), *sucrose phosphate synthase* (*SPS*), *starch synthase* (*SSS*), *starch branching enzyme* (*SBE*), *ADP-glucose pyrophosphorylase* (*AGP*) and*β-amylase* (*BAM*) and photosynthesis genes (*Psa*, *Psb*, *Pet* and *ATP*) were also identified. The expression of six floral fragrance-related DGEs showed agreement between qRT-PCR results and RPKM values, confirming the value of the data obtained by RNA-seq. We obtained the open reading frame of the *terpene synthase* gene from *Lilium* 'Sorbonne', designated *LsTPS*, which had 99.55% homology to transcript CL4520. Contig5_All. In addition, 54, 48 and 50 differently expressed transcription factor were identified by pairwise comparisons between the three organs and a regulatory network for monoterpene biosynthesis was constructed.

Conclusions: Analysis of differentially expressed genes in flower, leaf and bulb scale of lily, using second generation sequencing technology, yielded detailed information on lily metabolic differentiation in three organs. Analysis of the expression of flower scent biosynthesis genes has provided a model for the regulation of the pathway and identified a candidate gene encoding an enzyme catalyzing the final step in scent production. These digital gene expression profiles provide a valuable and informative database for the further identification and analysis of structural genes and transcription factors in different lily organs and elucidation of their function.

Keywords: Gene expression, Organogenesis, *Lilium*, Transcriptome, RNA-seq

* Correspondence: gaozhongshan@zju.edu.cn; ypxia@zju.edu.cn
[2]Department of Horticulture, College of Agriculture & Biotechnology,
Zhejiang University, Hangzhou 310058, China
Full list of author information is available at the end of the article

Background

The processes whereby an organism develops its shape and the cells and organs take on specific metabolic roles and structural characteristics are essential aspect of development and differentiation. The analysis of mutants has been a very productive approach to understand these processes and identify important regulatory elements since the last Century and several key genes that control plant organogenesis have been identified from different crops using this approach. For example, the ABC floral organ identity genes [1], *LeHB-1*, that participates in regulating tomato floral organogenesis [2], *lic-1* contributing to plant architecture [3], and *ZmGSL* which plays a role in early lateral root development [4]. However, the analysis of mutants is time- and labor-consuming, and particularly difficult in plants with large genomes. With the development of biotechnology, the establishment of full genome-sequences has dramatically increased our knowledge of plant genetics and molecular biology [5]. Sequencing the expressed part of the genome is nowadays achievable, even for plants such as lily with large genomes (~36Gb) [6, 7], and this can reduce the complexity and provide useful information and tools for molecular analysis [7, 8], enabling the identification of important genes.

Lilies are highly prized monocotyledonous plants (family Liliaceae, genus *Lilium*) with prominent bulbs, linear or oval leaves with parallel veins, and ornamental flowers. Although the taxonomy of lilies is complex, the Illumina sequencing-based digital gene expression (DGE) profiling technology, also called RNA-seq technology, has recently been applied extensively in lily research. In the past three years a large volume of data about differentially expressed genes (DEGs) involved in vernalization [9], cold-stress response [10], carbohydrate metabolism [11], dwarfism [12], pollen germination [13], flower development [14], flower color biosynthesis [15, 16] have been generated and analyzed in lily. In 2011, a few genes for developmental traits of lily were located on the genetic map, for example for flower colour (*LFCc*), flower spots (*lfs*), stem colour (*LSC*), antherless phenotype (*lal*) and flower direction (*lfd*) [17].

Lily bulbs consist of imbricating scales, which are modified swollen leaf bases and the large flowers have six petaloid tepals, six stamens and a superior ovary. Flowers are characterized by showy color and fragrance which are both economically important and essential for attracting pollinators in the natural environment and bulb scales are rich in starch and are important storage organs in the dormant bulb. All organs of a lily originate from the basal plate of a bulb, which makes lily morphogenesis particularly interesting. However, there have been no reports of studies on gene expression during lily organogenesis. To develop SSR markers, we analyzed a hybrid assembly transcriptome database (L.-Unigene-All) from lily (accession number: SUB2623518), based on the Illumina HiSeq 2000 sequencing platform. Taking this as a reference sequence, we carried out a digital gene expression profiling of flowers, leaves and bulb scales of *Lilium* oriental 'Sorbonne', and identified DEGs, including transcription factors and structural genes, in the three organs for the first time, providing insights into the genes participated in the differentiation of these organs, focusing mainly on candidate genes putatively participating in flower color, scent production, leaf photosynthesis and bulb development. Quantitative real-time PCR (qRT-PCR) and gene cloning were carried out for selected candidate genes involved in scent production to verify the conclusions from RNAseq data and identify genes for future analysis.

Results

Summary of gene expression profiling

Three replicates for each organ were used for cDNA library construction. For one leaf sample 74.78% of the reads matched Tulip virus X (Additional file 1: Table S1) and the data for this sample were discarded. Overall, we obtained over 11 Gb data for lily flower, leaf and scale organs. From these reads, between 73.49 and 83.12% from each sample could be mapped to the reference transcriptome. There were approximately 6 Mb perfect reads for each organ, accounting for more than 67.05% of those that were mapped to the reference transcriptome (Table 1), excluding reads with more than 2 bp mismatches. Approximately 77.71% of these sequences matched to the reference transcriptome were unique matched reads, with some multi-position reads. These data, which lay a valuable cornerstone for future work, have been submitted to the NCBI SRA database (accession number: SRP084220).

Analysis of DEGs in each organ

Both RPKM value and false discovery rate were used for DEGs screening and original values of all transcripts are shown in Additional file 2: Table S2. Of the millions of gene sequences expressed in each biological sample, some transcripts were present in all three organs, with different expression levels but only a few thousand were significantly differentially expressed genes (DEGs) unique to a specific organ with undetectable expression in the other organs.

We obtained 2685 DEGs in the comparison of leaf-vs-flower (L-vs-F), 2296 in leaf-vs-scale (L-vs-S), and 1709 in flower-vs-scale (F-vs-S), with a false discovery rate value <0.001 [18]. From the F-vs-S comparison, there were more up-regulated genes (938) than down-regulated genes (771) in scales, taking flower as a reference organ, while with leaf as a reference organ,

Identification of differentially expressed genes in flower, leaf and bulb scale of Lilium oriental hybrid...

71

Table 1 Summary of gene expression profiling for three lily organs

Organ	Total reads	Total mapped Reads	Perf. match	≤2 mismatch	Unique match	Multiposition match
Flower	11,877,133	9,781,891 (82.36%)	6,639,595 (67.88%)	3,142,296 (32.12%)	8,163,994 (83.46%)	1,617,897 (16.54%)
	11,449,580	9,216,439 (80.50%)	6,350,723 (68.91%)	2,865,716 (31.09%)	7,136,152 (77.43%)	2,080,287 (22.57%)
	11,349,225	8,959,443 (78.94%)	6,040,856 (67.42%)	2,918,587 (32.58%)	7,444,467 (83.09%)	1,514,976 (16.91%)
Leaf	11,481,948	8,894,809 (77.47%)	5,680,563 (63.86%)	3,214,246 (36.14%)	6,766,819 (76.08%)	2,127,990 (23.92%)
	11,708,654	9,732,816 (83.12%)	6,914,703 (71.05%)	2,818,113 (28.95%)	6,977,857 (71.69%)	2,754,959 (28.31%)
Scales	11,682,968	8,902,142 (76.20%)	5,830,385 (65.49%)	3,071,757 (34.51%)	6,859,013 (77.05%)	2,043,129 (22.95%)
	11,101,394	8,158,589 (73.49%)	5,571,981 (68.30%)	2,586,608 (31.70%)	6,333,245 (77.63%)	1,825,344 (22.37%)
	10,552,892	8,566,038 (81.17%)	5,440,069 (63.51%)	3,125,969 (36.49%)	6,448,796 (75.28%)	2,117,242 (24.72%)

more genes were down-regulated in flower (1766) and scale (1534) (Fig. 1), indicating a greater number of transcripts of DEGs (65.77–66.81%) differentially expressed in leaf compared with flower and scale. There were 73 DEGs expressed in three comparisons (Additional file 3: Table S3). Thirty-six percent of these genes common had putative functions in photosynthesis and the others participated in basic processes, such as metabolism of carbohydrates, nitrogen and proteins. Two genes, CL1446.Contig13_All and Unigene6258_All, related to pathogene and defensin, were found in the three organs (Additional file 3: Table S3).

Taking uniquely expressed genes as those with an expression value of zero in two organs while having expression value more than zero in the third organ, 474, 825 and 404 unique genes were found for flower, leaf and scale, respectively. Of these, approximately 70% have been annotated (Additional file 4: Table S4, Additional file 5: Table S5 and Additional file 6: Table S6) and there are twice the number of unique genes in leaf compared to flower and scale (Table 2).

Gene ontology analysis of significantly DEGs

Approximately 17% of differentially expressed genes were unannotated in the database (296 in F-vs-S, 466 in L-vs-F and 393 in L-vs-S), suggesting that at least some might be new unknown transcripts, and further work will be necessary to investigate their roles. The majority (83%) of the DEGs had annotations indicating a likely

function, and the numbers and functions of genes in the three pairwise comparisons were similar (Fig. 2), with 24 functional groups classified under biology process, 9 under cellular components and 12 functional groups in molecular functions. In the biological process category, most genes were characterized as having functions in categories "cellular process" (GO: 0009987), "response to stimulus" (GO: 0050896) or "regulation of biological process" (GO: 0050789), and in the cellular component subgroup, most genes were classified as related to "cell" (GO: 0005623), "organelle" (GO: 0043226) and "membrane" (GO: 0016020). Most genes in the molecular function category encoded proteins with "catalytic activity" (GO: 0003824), "binding" (GO: 0005488) and "transporter activity" (GO: 0005215) (Fig. 2).

KEGG pathway analysis of significantly DEGs

KEGG (Kyoto Encyclopedia of Genes and Chromosomes) pathway enrichment analysis was used to identify DEGs significantly associated with specific metabolic pathways. A total of 1461 DEGs from the L-vs-F comparison were assigned to 117 KEGG pathways; for F-vs-S comparison, 885 DEGs had 109 pathway annotations; and in the L-vs-S comparison 1194 DEGs were related to 105 pathways. There were more DEGs with pathway annotations for L-vs-F than for the F-vs-S and L-vs-S comparisons. Genes for 95 pathways were expressed in all three organs and included housekeeping activities. More than 40% of the DEGs participated in

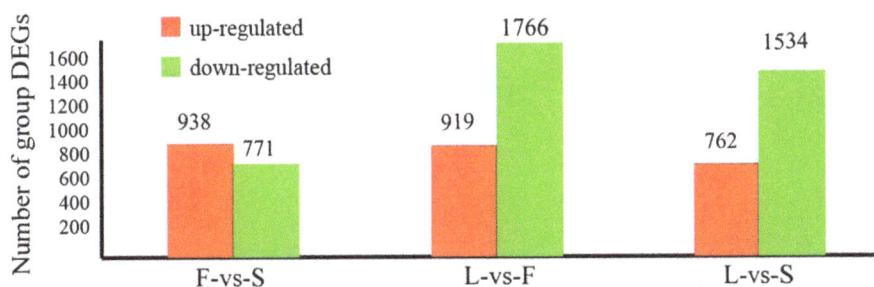

Fig. 1 Number of differentially expressed genes in leaf (L), flower (F) and bulb scale (S)

Table 2 Number and distribution of annotated unique organ genes

Organ	Unique gene	Annotated	Unannotated	Annotation percent (%)
Flower	474	324	150	68.4
Leaf	825	439	386	53.2
Scale	404	286	118	70.8

metabolic pathways, and around 20% in the three comparisons were related to biosynthesis of secondary metabolites. In addition, more than 10% of DEGs in three pairwise organ comparisons were involved in pathways of glycerophospholipid metabolism, endocytosis and ether lipid metabolism. Figure 3 lists 46 KEGG pathways where there were more than 10 DGEs between the three pairwise comparisons. There were no DEGs in 11 KEGG pathways in the F-vs-S comparison, and these were mainly pathways involved in photosynthesis and carbohydrate and protein metabolism. For L-vs-S comparison four KEGG pathways showed no DEGs, including nitrogen metabolism, alanine/glutamate metabolism, diterpenoid biosynthesis and galactose metabolism.

Mining of candidate genes related to flower color, fragrance, photosynthesis and bulb development

By GO and KEGG annotation, 20 up-regulated flower color genes were found with putative functions in flavonoid biosynthesis, including sequences homologous to *DFR*

(*dihydroflavonol 4-reductase*), *CHS* (*chalcone synthase*), *CHI* (*chalcone isomerase*), *FLS* (*Flavonol synthase*), *F3H* (*flavanone 3-hidroxylase*), *LAR* (*leucoanthocyanidin reductase*), *HCT* (*shikimate O-hydroxycinnamoyltransferase*), *LDOX* (*leucoanthocyanidin dioxygenase*), *TT7* (*flavonoid 3′-monooxygenase*) (Additional file 7: Table S7) and some of those genes, such as *DFR* and *CHS*, showed more than 10 fold higher expression level in flower compared to leaf and scale.

Monoterpenoids are the dominant classes of volatile compounds emitted from scented lilies [19–21] and genes encoding TPSs (terpene synthases) (Additional file 8: Table S8) involved in monoterpenoids biosynthesis were identified in both L-vs-F and F-vs-S comparisons. From comparisons of the predicted amino acid sequence with those of known genes from *Freesia hybrid* and *Litsea cubeba*, these were predicted to encode linalool and ocimene synthases, respectively. These *TPSs* were significantly up-regulated in flowers as were some other genes encoding enzymes participating in the 2-C-methyl-D-erythirtol 4-phosphate (MEP) pathway, which leads to the synthesis of monoterpenes, including *DXS* (*1-deoxy-D-xylulose-5-phosphate*), *GGPP* (*geranylgeranyl pyrophosphate*) and *HDS* (*4-hydroxy-3-methylbut-2-en-1-yl diphosphate*) (Additional file 5: Table S5). We also investigated *HPL* (*hydroperoxide lyase*) and *KAT* (*3-ketoacyl-CoA thiolase*) (Additional file 8: Table S8), which participate in the oxylipin metabolism and β-oxidation pathways, as these genes were highlighted

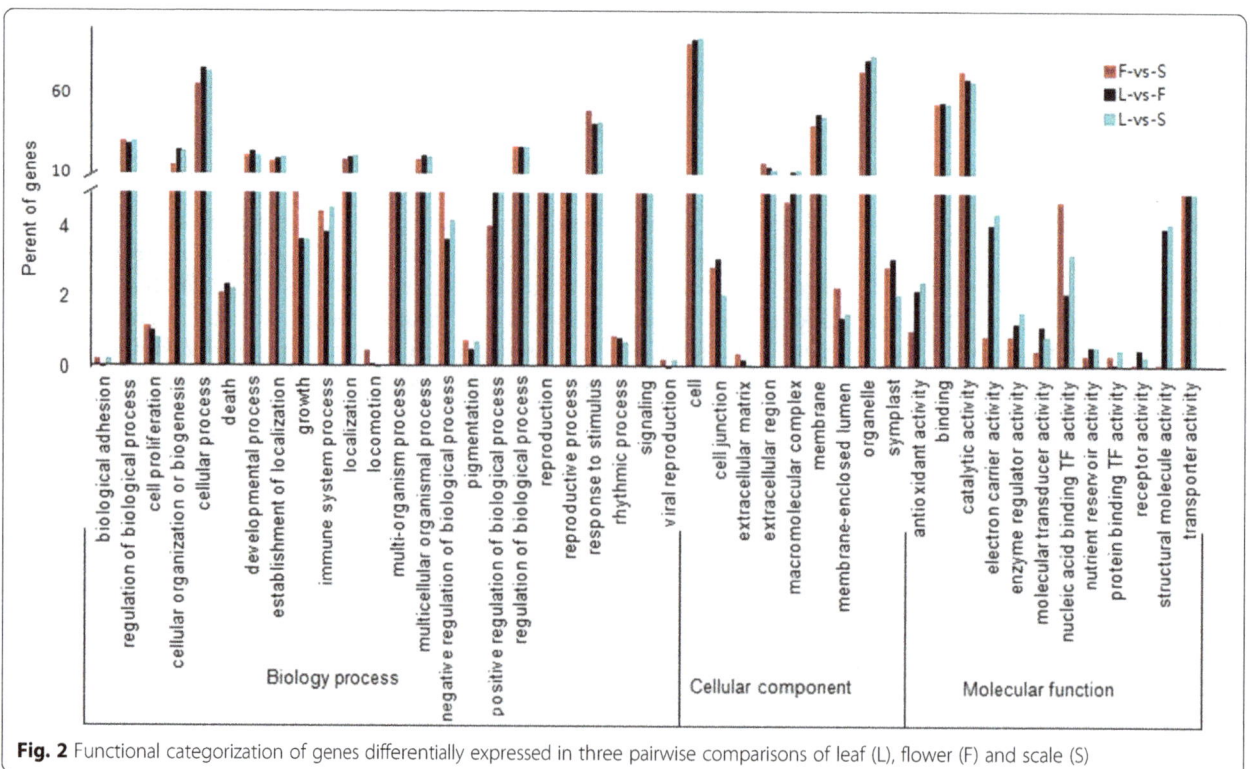

Fig. 2 Functional categorization of genes differentially expressed in three pairwise comparisons of leaf (L), flower (F) and scale (S)

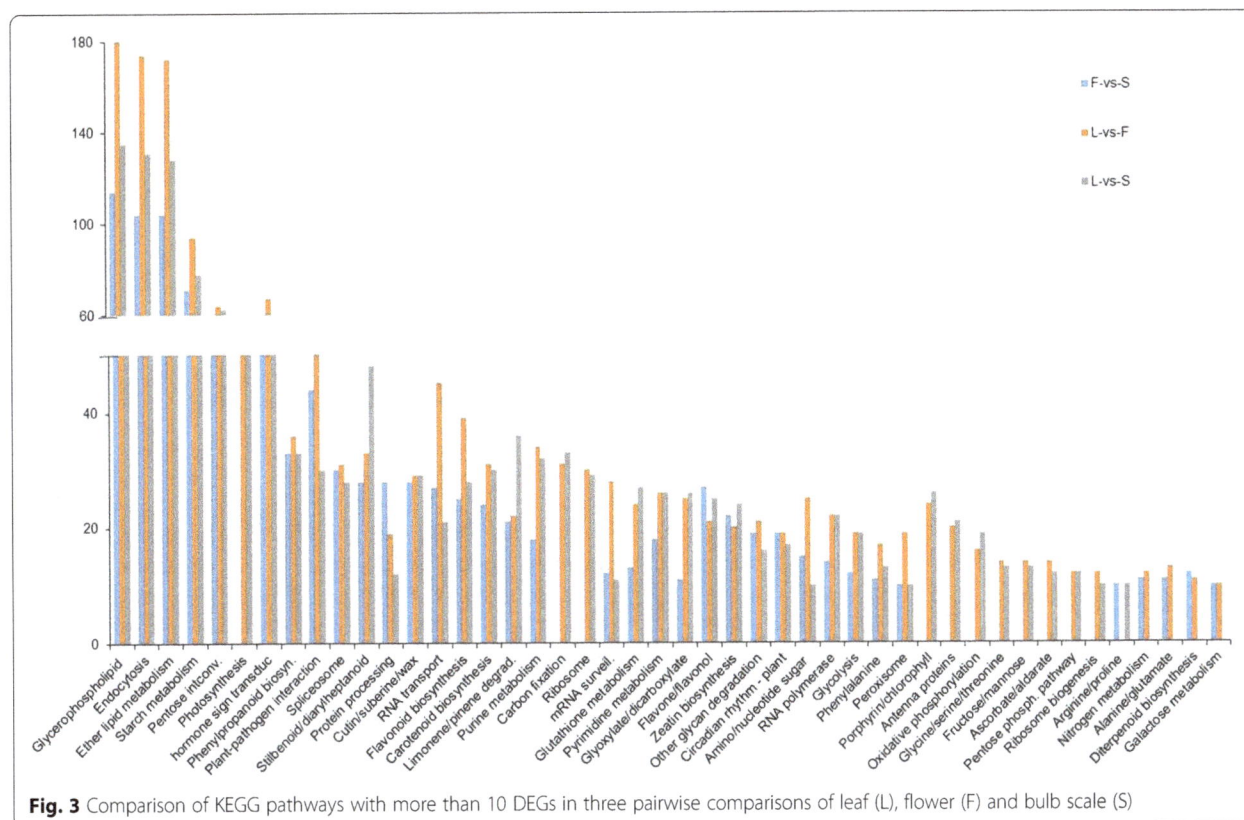

Fig. 3 Comparison of KEGG pathways with more than 10 DEGs in three pairwise comparisons of leaf (L), flower (F) and bulb scale (S)

in a recent lily floral fragrance study as being differentially expressed in lily cultivars [20]. Both *HPL* and *KAT* which were up-regulated in flower compared with scale and leaf.

Sixty-three DEGs involved in photosynthesis were identified as being up-regulated, in the transcriptome of mature leaves, related to photosystem, photosystem II, the cytochrome b6/f complex, photosystem electron transport and F-type ATPase (Additional file 9: Table S9).

Genes involved in carbohydrate metabolism, especially starch and sucrose, were of particular interest in this study because of their importance for bulb function. Twenty-one DEGs, 11 in L-vs-S and 11 in F-vs-S comparisons (with one in common for both L-vs-S and F-vs-S), were predicted to participate in metabolism of starch and sucrose (Additional file 10: Table S10). Key genes homologous to *SS (sucrose synthase), INV (invertase), SPS (Sucrose phosphate synthase), SSS (starch synthase), SBE (starch branching enzyme), AGP (ADP-glucose pyrophosphorylase)* and *BAM (amylase)* were identified. Most genes, putatively related to sucrose and starch hydrolysis, were down-regulated in scale relative to flower as reference. Different genes putatively related to sucrose or starch synthesis and hydrolysis were either up-regulated or down-regulated in scales with leaf as reference (Additional file 10: Table S10), presumably related to complicated metabolic differences in bulb at this developmental stage.

Expression patterns of genes involved in monoterpene biosynthesis

Flower scent production is of immense importance in plant biology and ecology. It plays a role in reproduction by promoting pollination through interactions with insects and other organisms, and it also contributes value to many commercially important horticultural flowers, such as *Lilium* cv. Sorbonne. Many scents are produce from the terpenoid pathway, as are many other biology important molecules. Therefore, we investigated genes encoding enzymes involved in monoterpene biosynthesis. In this study, five DEGs encoding four putative enzymes related to monoterpene biosynthesis were identified from the transcriptome (Additional file 8: Table S8). Genes *DXS* (Unigene8314_All), *HDS* (CL4079.Contig1_All), *GGPP* (CL1306.Contig1_All) and *TPS1* (Unigene1934_All) showed significantly higher expression in flower than in leaf and scale (Fig. 4), whereas *TPS2* showed extremely low expression in three organs.

Transcription factor analysis

Transcriptome unigenes of L.-Unigene-All were searched against the Plant Transcription Factor Database and 839 unigenes identified as transcription factors (TFs) sequences belonging to 53 putative transcription factor families, with the three largest groups being bHLH (75), MYB related (60) and C3H (55). These putative transcription factor

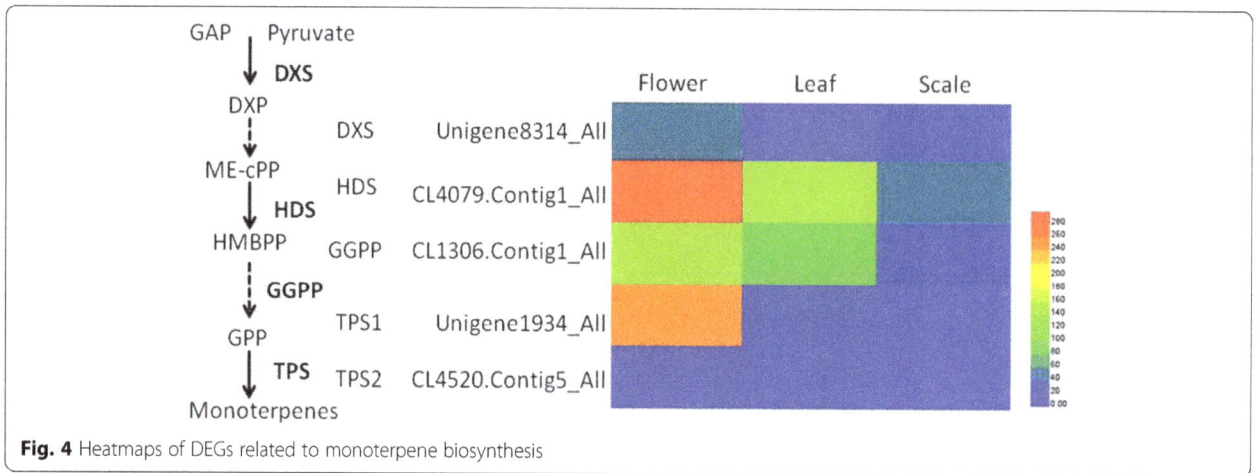

Fig. 4 Heatmaps of DEGs related to monoterpene biosynthesis

unigenes were subjected to three pairwise comparisons which identified 54, 48 and 50 TFs differentially expressed in the F-vs-S, L-vs-F and L-vs-S comparisons (Fig. 5). The bHLH transcription factors were the largest group for F-vs-S and L-vs-F comparison and ERFs were the largest group for the L-vs-S comparison (Fig. 5). These results provide a rich resource for future analysis of the role of transcription factors in lily organogenesis.

Construction of regulatory network of monoterpene biosynthesis

To provide a system view of the regulatory network responsible for controlling monoterpene analysis, networks were extracted based on correlation analysis between 839 putative TFs and seven putative DEGs related to flower fragrance. As a result, 31 putative TFs were identified as potentially involved in regulating the seven putative genes related to production of flower volatiles (Fig. 6a and Additional file 11: Table S11) ($P < 0.01$). These TFs were classified into 13 putative TF families, with the three largest

TF families being the bHLH (5 Unigenes), bZIP (5 unigenes) and MYB related (4 unigenes) families (Additional file 11: Table S11). The expression profiles of these TFs potentially related to flower volatile biosynthesis were hierarchically clustered and plotted in a heatmap (Fig. 6b). Three of these (Unigene15576_All, Unigene21249_All, Unigene1921_All) with extremely high expression levels in flowers (RPKM values of 1574.3, 1514.6 and 133.1, respectively) were excluded from the heatmap, but relevant data are shown in Additional file 11: Table S11. Most of those TFs were more highly expressed in flower than in leaf and scale. The expression patterns of all 31 TFs were positively correlated with those of flower scent genes (Additional file 11: Table S11).

qRT-PCR verification of DEGs identified by RNAseq

The expression profile of six flower fragrance-related DEGs from lily identified by the RNA-seq approach were tested, using quantitative RT-PCR, including key genes

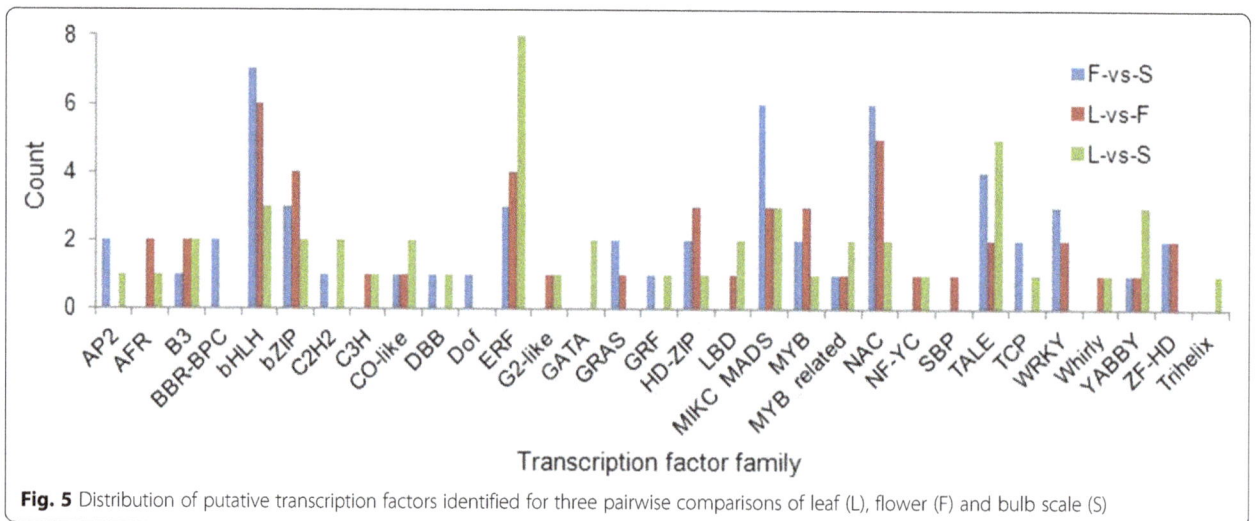

Fig. 5 Distribution of putative transcription factors identified for three pairwise comparisons of leaf (L), flower (F) and bulb scale (S)

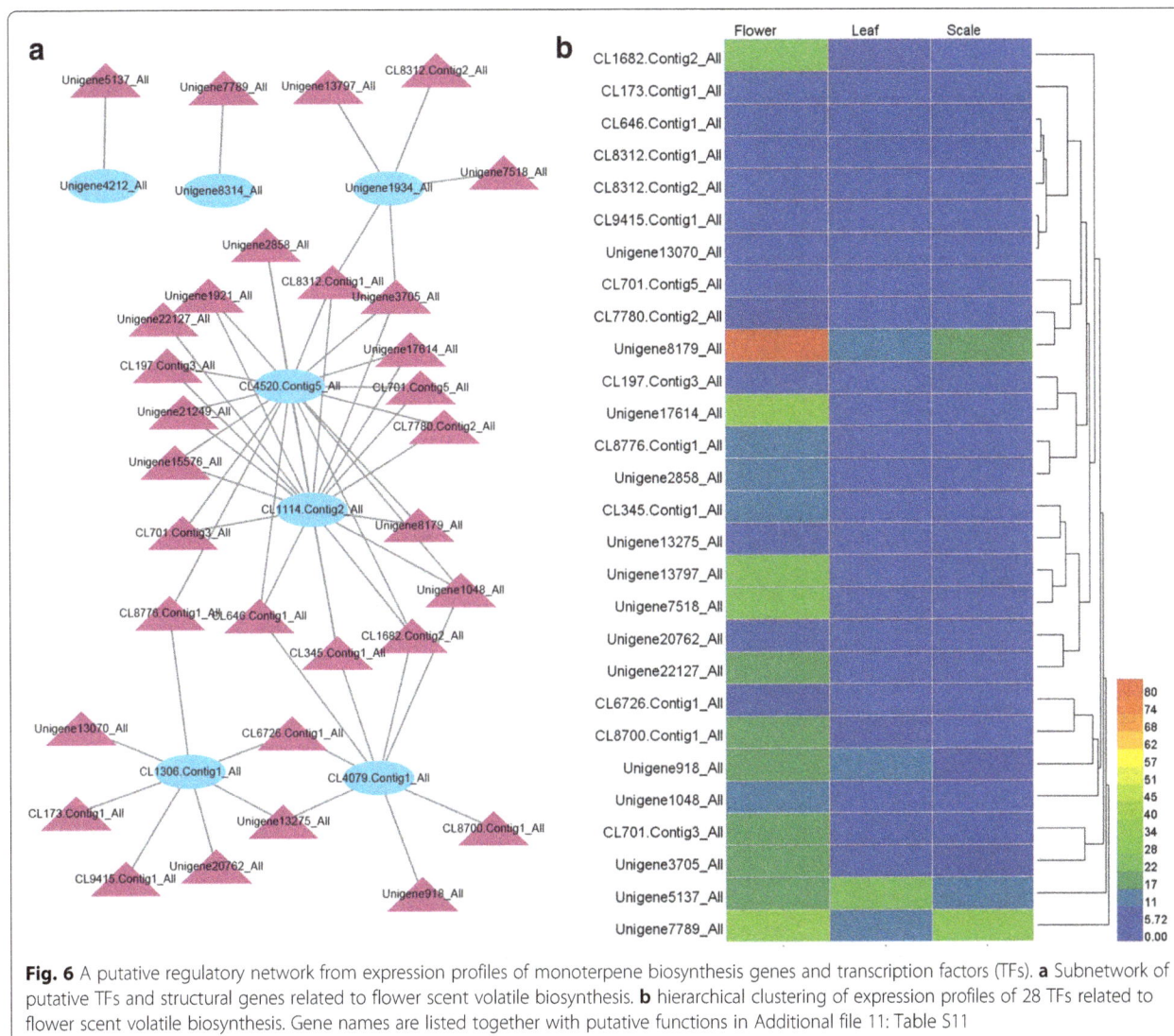

Fig. 6 A putative regulatory network from expression profiles of monoterpene biosynthesis genes and transcription factors (TFs). **a** Subnetwork of putative TFs and structural genes related to flower scent volatile biosynthesis. **b** hierarchical clustering of expression profiles of 28 TFs related to flower scent volatile biosynthesis. Gene names are listed together with putative functions in Additional file 11: Table S11

involved in synthesis of monoterpenoids biosynthesis: *TPS, DXS, GGPP, HDS*, and fatty acid derivatives and phenylpropanoid/benzenoid biosynthesis:*HPL* and *KAT*. The comparisons of expression measured by RNA-seq and qRT-PCR in the three organs were largely in agreement (Fig. 7). The expressions of these genes was highest in flowers, although some, such as *HPL* also had high expression in leaves.

Cloning of TPS genes from cDNA

Terpene synthases (TPS) are that key enzymes that catalyze the last step of the MEP pathway to produce terpenes. There are numerous terpenes and TPSs in plants and small changes in amino acid sequence generate the unique properties and diversity of these important compounds. Primers were designed based on the sequence of transcript CL4520.Contig5_All (accession number: SUB2623518), which had sequence homology with *TPS* genes, and attempts were made to clone the complete cDNA sequence from petal. However, these attempts failed with three different pairs of primers. When the sequence was checked against the NCBI database it was found to be a mixed clone, part of which was highly homologous (91.17% at the cDNA level) to the complete coding sequences of *LhTPS* from 'Siberia' (accession number: KF734591), together with a second unrelated open reading frame. Based on the *Lilium* oriental 'Sorbonne' sequence information, new primers were designed and a putative *TPS* sequence (NCBI accession: MF401556, designated *LsTPS*) was acquired with an open reading frame (ORF) of 1761 nucleotides, which has 99.55% sequence homology with the TPS coding sequence of transcript CL4520.Contig5_All. Phylogenetic analysis (Additional file 12: Figure S1) based on deduced amino acid sequences showed that *LsTPS* were highly homologous with CL4520.Contig5_All,

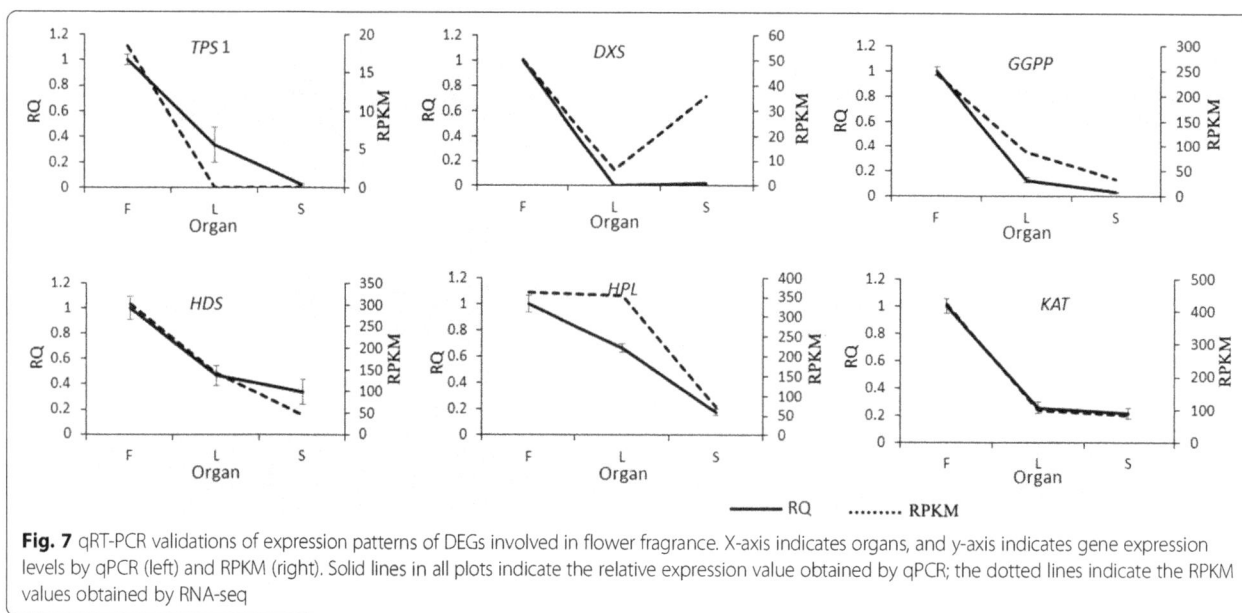

Fig. 7 qRT-PCR validations of expression patterns of DEGs involved in flower fragrance. X-axis indicates organs, and y-axis indicates gene expression levels by qPCR (left) and RPKM (right). Solid lines in all plots indicate the relative expression value obtained by qPCR; the dotted lines indicate the RPKM values obtained by RNA-seq

and with the other two amino acid sequences from *Lilium* 'Belladonna' and 'Siberia'. It was clustered with *LcTPS1*, which is a member of Tps-b group and was able to convert Geranyl diphosphate into *trans*-ocimene [22].

Discussion

Profiling of differentially expressed genes in lily organs

The aim of this study was to identify differences in the identities and expression patterns of specific genes and pathways operating in leaves, flowers and bulb scales. To our knowledge, this is the first report of the differentially expressed genes in these three major organs in lily. A large number of new genes and transcription factors involved into photosynthesis, bulb development, flower color and flower scents have been identified. As expected, RNA-seq generated a great deal of information, which will help identify targets to understand the factors controlling organ development and differentiation.

A greater number of DEGs were found in the L-vs-F and L-vs-S comparisons (2685, 2296) than in the F-vs-S group (1709), which indicated that compared to leaf, flower and scale had fewer differentially expressed genes (Fig. 1a). Each organ was also, as expected, characterized by expression of specific classes of genes. In leaves, for example these included genes involved in photosynthesis, carbon metabolism, nitrogen and protein metabolism. In contrast, 21 of the DEGs identified in scales were involved in starch and sucrose metabolism (Additional file 7: Table S7) and the importance of these pathways has previously been demonstrated at the transcriptional level in bulbous plants, such as *Lilium davidii* var. *unicolor* [11], *Gladiolus hybridus* [23] and *Tulipa gesneriana* [24].

Candidate genes related to flower color, fragrance, photosynthesis and bulb development

In this research, materials were harvested from plants at a stage when leaves were mature and flower color and scents were developing. We believe this is the best time to maximize the gene expression differences between the different organs and mine differentially expressed genes in organs of flower, leaf and scales. Flavonoids are accumulated in pink flowers of lily [25], and genes involved in the flavonoid biosynthesis pathway received special attention in this study with *Lilium* 'Sorbonne', which has pink flowers. A total of nine flower color-related genes, *DFR, CHS, CHI, FLS, F3H, LAR, HCT, LDOX, TT7*, were identified. Many of these belong to multi-gene families and several homologs were identified for each gene. Interestingly, a very recent de novo transcriptome sequencing of flower buds at different development phases of *Lilium* 'Sorbonne' also identified most of those genes [15], indicating the power of digital gene profiling in gene mining. All of the identified flower color genes have been cloned from different plants such as Arabidopsis and Petunia [26, 27], and an increasing number have been cloned from lily, such as *LhDFRs, LhCHS* [28, 29] and *LhF3S* [30]. *DFR* was identified from the pink petal of 'Sorbonne' in this study and *LhDFRs* have also been shown to be expressed in the Asiatic lily 'Montreux', which has pink flowers, but not in the Asiatic 'Connecticut King', which has yellow flowers [28].

Because oriental lily can emit complicated volatile compounds, it has been regarded as a potential model system for the elucidation of gene products responsible for the synthesis of volatiles [20]. The differential expression of monoterpene synthase genes has been suggested to be a key reason for differences in floral scent [31] and we identified a monoterpene synthase gene (*TPS*) predicted to code for enzymes synthesizing linalool and *trans*-ocimene. Three genes, *LcTPS1*, *LcTPS2* and *LcTPS3*, have been cloned from *Litsea cubeba* [22], encoding enzymes synthesizing *trans*-ocimene, α-thujene and (+)-sabinene respectively, and other TPS genes, encoding *myrcene synthase* (*LiMys*) and *linalool synthase* (*LiLis*), have also been cloned from *Lilium* 'Siberia', 'Novano' and *L. regale* [31–33]. Here, we cloned *LsTPS* from *Lilium* 'Sorbonne' and predicted the cloned sequence of *LsTPS* might encode enzymes synthesizing *trans*-ocimene by phylogenetic analysis (Additional file 8: Figure S1). Some other genes encoding enzymes involved in production of the terpenoid backbone in the MEP (2- C -methyl- D -erythritol 4-phosphate) pathway, *DXS, GGPP* and *HDS*, were also identified. DXS is a pivotal gene in the first step of MEP pathway and has been extensively studied in tomato, Arabidopsis and grape [34–36], and had been identified and cloned in lily by Johnson et al. [20]. However, functional analysis of these genes is still needed to confirm their role in lily flower scent production and a comparison of expression in scented lilies and those without fragrance [31] should also prove informative.

Unsurprisingly, DEGs participating in photosynthesis were identified as being up-regulated in leaves. Previously, homologous sequences of *psbA* and *atp* has been cloned in a study analyzing the complete plastid genomes of *L. nobilissimum* and *L. longiflorum* [37] and psbA-homologous proteins has been isolated from *L. superbum* [38], and this study has greatly increased the number of gene sequences identified in *Lilium*.

There have been several studies on the variation of carbohydrate compounds during bulblet development [39, 40]. Li et al. [11] revealed the variation in starch and sucrose content at different stages of bulblet initiation and enlargement in *Lilium davidii* var. *unicolor*. However, little is known about changes at the molecular level, especially between underground and above-ground stages. Kawagishi and Miura [41] divided the period of a lily development into four stages, and in this study we used the third stage, corresponding to flower bud expansion to flowering, when both above-ground and underground organs show a vigorous increase in dry weight, and bulbs can be both source and sink [42], underlying the complexities of metabolic activity in bulb scales. We found that putative sucrose- and starch-hydrolysis genes, such as *SS, INV* and *BAM* were down-regulated in scales compared to flowers as reference, and sucrose and starch synthesis genes, such as *SPS* and *SSS* were down-regulated also in scale with leaf as reference, which is consistent with the lack of rapid swelling and bulb growth for plants with 5.5 cm buds used in this study.

Identification of transcription factors and regulatory network

The largest families of transacting transcription factors modulate plant development [11, 43]. In this study, 839 unigenes were identified as TFs, and 31 of them were identified as potentially involved in regulating production of flower volatiles. Although terpenoid metabolism is very important in biochemical differentiation and tissue function in plants, not many TFs are known to be involved in the regulation of the pathway. In *Solanum lycopersicum*, bHLH and WRKY were identified activating terpene synthase promoters, with MYC synergistically transactivating the *SlTPS5* promoter [44]. and activated distinct terpeniod biosynthesis in *Catharanthus roseus* and *Medicago truncatula* [45]. Ethylene-related TFs have also been implicated in regulating TPSs, for example in *Artemisia annua*, AaERF1 and AaERF2 positively regulate artemisinin production [46] and in Newhall sweet orange CitAP2.10 activates the *CsTPS1* that produces valencene [47]. Based on this comparative information, bHLH, WRKY, ERF/AP2 family members would seem likely candidates for regulating lily flower TPS and studies on other pathways indicate that a complex network may be involved.

In this study, bHLH was the largest TF families in F-vs-S and L-vs-F comparison and in the putative flower volatile biosynthesis network (Fig. 5, Additional file 11: Table S11). CL4520.Contig5_All, a putative *TPS* gene (Additional file 12: Figure S1), were regulated by 17 TFs (Fig. 6a). Four of these were bHLH (Additional file 11: Table S11). However, flower volatile regulation is clearly complex (Fig. 6a), and many TFs, including bHLHs, probably regulate multiple structural genes synchronously during flower development, in concert with other TFs in the network. Candidate TFs have been identified in this study and their role can now be tested. Furthermore, the complete TF database identified from this study has additional potential for improving our understanding of the differential regulation of gene expression during development of leaves, bulbs and flowers.

Verification of the database

Although high-through sequencing technology has become a powerful tool to identify candidate genes and investigate gene expression patterns [48], further validation is needed, especially for non-model organisms without reference sequences, and for plants with huge genomes. Although in most cases, there was a good correlation between transcript abundance assayed by qRT-PCR and the transcription profile revealed by DGE profiling [11, 16, 49], there have also been reports of inconsistencies between the two methods [50, 51]. In this study, a general agreement was obtained by the two different methods. All verified genes except *HPL* had higher expression level in flower than in leaf and scale. A very recent study revealed *HPL* involvement in protecting the photosynthetic apparatus [52]. We deduce that this gene may play an important role in lily leaf and further functional studies are required to understand its role in lily leaf and flower. A putative error in the assembly of CL4520.Contig5_All was identified during subsequent attempts to clone the cDNA sequence although this did not effect the open reading frame of *LsTPS* and the complete sequence was subsequently obtained. To our knowledge, this is the first report of *TPS* gene from *Lilium* 'Sorbonne' and it is proposed that this is important for our understanding of scent production in this species. Overall, the extensive RNA-seq data provides a platform for candidate gene identification in lily organs and elucidation of their function.

Conclusions

In the present study, DEGs related to lily flower color, flower fragrance, photosynthesis and bulb development were identified by RNA-seq technology. Approximately 11 Mb data were generated for each lily organ, a few thousand DEGs were identified in the comparisons, and hundreds of genes specific for each of the three organs identified. By functional enrichment analysis, genes for floral scent (*TPS, DXS, KAT, HPL* et al.), floral color (*DFR, CHS, CHI* and *FLS* et al.), starch and sucrose metabolism (*SPS, SS, INV, SSS, SBE, AGP* and *BAM*) and photosynthesis (*Psa, Psb, Pet and ATP*) were identified. The expression of six floral fragrance-related DGEs showed a similar expression pattern between qRT-PCR results and RPKM values, confirming the value of the data obtained by RNAseq. *LsTPS* was cloned based on the sequence of transcript CL4520.Contig5_All with an ORF of 1761 nucleotides. The lily DGEs identified in this research include transcription factors, flower-specific genes with putative functions in flower color and scent biosynthesis, photosynthesis-related genes in leaf, and starch and sucrose metabolism-related genes in scales, and provide a valuable and informative database for understanding lily organogenesis, mining of important genes, and research into gene function. Based on these results a putative regulatory network for monoterpene biosynthesis is proposed.

Methods

Plant materials and RNA extraction

Lilium oriental hybrid 'Sorbonne' bulbs were planted in the field of the Institute of Landscape Architecture, Zhejiang University (ZJU), China in October, 2013 and they sprouted in March 30th, 2014. Three whole plants were harvested on a sunny day of June 13th at 9 am, 2014 (temperature 24 °C; 41,889.3 Lux) when leaves were mature and flower color and scent were developing

and brought to the lab immediately. 5.5 cm flower buds (named F), leaves from the centre of the stem (named L) and inner bulb scales (named S) were collected from each of the three plants. Three biological replicates were used for each organ. Samples were placed in 50 ml tubes and stored at –80 °C until use. The RNA isolation method was as described in Du et al. [53]. A total of nine RNA samples were isolated using a modified CTAB method [54] (three replicates for three organs) respectively. The quality and concentration of RNA were checked using an Angilent 2100. RNA with (RIN) ≥ 7, 28S:18S > 1, OD260/280 ≥ 2, and OD260/230 ≥ 2 were used for sequencing.

Construction of DGE database

Nine cDNA libraries were constructed respectively based on 9 samples from three replicates of three organs. Details for the construction was as described in Feng et al. [55]. RNA library processing and sequencing via Illumina HiSeqTM 2000 were carried out by staff of the Beijing Genome Institute (BGI) (Shenzhen, China). Clean reads were obtained by data filtering to remove reads with adaptor sequences, more than 10% unknown bases, and those with low quality bases above 50%. The clean reads were mapped to reference sequences of L.-Unigene-All (accession number: SUB2623518), a hybrid assembly comprehensive transcriptome acquired by our lab, using SOAP aligner/SOAP2 [56], and short sequences less than 200 were discarded. The parameters for SOAP aligner/SOAP2 were as follows: option = –m 0 -x 500 -s 40 -l 35 -v 5 -r 2. No more than two mismatches were allowed in the alignments. Once reads passed sequencing saturation and randomness assessment, a digital gene profiling database for each sample was set up.

Quantification of gene expression

The gene expression level was calculated by using RPKM (Reads per kb per million reads) which eliminates the influence of different sequence lengths and discrepancies due to sequencing depth. The differences in gene expression between samples can be compared directly by comparing RPKM values.

Screening of differentially expressed genes between two groups

The NOIseq method [57] was used to screen for differentially expressed genes between two groups. Two main steps were conducted: first, the noise distribution was calculated and then DEGs were divided into groups. The expression values for each gene in each group were used to calculate log2 (fold-change) M and the absolute difference value D of all pair conditions (gene expression value was scored as 0.001 for genes not expressed in a sample). For each gene, an average expression value across replicates was used to calculate M and D. Then, all these M/D values were pooled together to estimate the noise distribution. In the case where gene i is differentially expressed between two groups, Gi was set as one, otherwise as 0. This gave the P value, the probability of gene i being differentially expressed. Formulae for M, D, P calculation are as follows:

$$M^i = \log_2\left(x_1^i/x_2^i\right), D^i = \left|x_1^i - x_2^i\right|,$$

$$P\left(G^i = 1 | x_1^i, x_2^i\right) = P\left(G^i = 1 | M^i = m^i, D^i = d^i\right)$$
$$= P\left(|M^*| < |m^i|, D^* < d^i\right)$$

In this paper, genes with P greater than 0.8 and M greater than 2 were considered to be differentially expressed between groups.

Functional analysis of differentially expressed genes

All Unigene sequences were aligned to the protein databases NCBI non-redundant (NR), the Swiss-Prot protein database (Swiss-Prot, in UniProt), the Kyoto Encyclopedia of Genes and Genomes (KEGG, www.genome.jp/kegg/) and the Clusters of Orthologous Groups of proteins (COG, http://www.ncbi.nlm.nih.gov/COG/) (E value <10^{-5}) by BLASTx. To identify the main biological functions associated with the DEGs, all DEGs were mapped to GO terms in the database, the total numbers of genes for each term were calculated and a hypergeometric test were used. Taking corrected p-value ≤ 0.05 as a threshold, significantly enriched GO terms for DEGs were acquired. Then, a Blast2GO program [58] and WEGO software [59] were used to obtain GO functional classifications for DEGs following default parameters. Similarly, all DEGs were mapped to the KEGG database for KEGG pathway enrichment analysis of DEGs.

Validation DEGs by qRT-PCR

Real-time quantitative RT-PCR was used to validate the expression of a selected set of DEGs. Six primer pairs were designed with the Primer 3.0 (http://primer3.ut.ee/) program, to produce a 150 bp amplicon, based on lily transcriptome database L.-Unigene-All (Additional file 13: Table S12). After a preliminary experiment for evaluation of candidate reference genes, actin (forward primer sequence CACACTGGTGTCATGGTTGG; reverse primer sequence CACAATACC GTGCTCAATTGG), was used as an internal control. Real-time PCR reactions were performed using a 7500 Real Time PCR System (Thermo Fisher Scientific), in a total of 20 µl, with 1 µl cDNA, 0.8 µl forward primer (10 µM), 0.8 µl reverse primer (µM), 0.4 µl ROX, and 10 µl SYBR® Premix Ex TaqTM II(2×). The cycling conditions were as follows:

95 °C for 3 min, 40 cycles of 95 °C for 30 s, 55 °C for 30 s, and 72 °C for 1 min. Melting curves for each PCR product were analyzed at 95 °C for 15 s, cooling to 54 °C for 1 min, and then gradually heated at 0.1 °C/s to a final temperature of 95 °C. Relative quantitation (RQ) was calculated using the $2 - \Delta\Delta Ct$ method. Three RNA isolations and triplicate RT-PCR runs were implemented for each sample for biological and technical replication.

Cloning and nucleotide analysis of *TPS* gene

Gene cloning primers were designed taking sequences of transcript CL4520.Contig5_All from L.-Unigene-All as template. Forward primer (ATGGCAGCTATGAGCTGT) and reverse primer (TCATTCCAATGGGACATTATTG) were synthesized by Sangon Biotech. The PCR was performed with a Gene Amp kit (Transgen Biotech, China) according to the manufacturer's instructions, in a total of 50 µl, with 4 µl of petal cDNA, 1 µl forward primer (10 µM), 1 µl reverse primer (µM), 5 µl TransTaq-T buffer, 4 µl 2.5 mM dNTPs, and 1 µl TransTaq-T DNA Polymerase. The cycling conditions were as follows: 94 °C for 5 min, 35 cycles of 94 °C for 30 s, 55 °C for 30 s, and 72 °C for 1 min. Following detection and excision from the agarose, the fragment was subsequently cloned into pMD 19-T vector and transformed into Trans5α *E. coli* cell according to the manufacturer's protocol (Transgen Biotech, China). Three clones screened and vetted for amplicons of appropriate size were sequenced by Sangon Biotech. Multiple sequence alignment of plant *TPS* gene was performed using DNAMAN 8.0 (Lynnon Biosoft, USA) and a phylogenetic tree was constructed using MEGA5 with a bootstrap replication of 1000.

Identification of transcription factor

Transcriptome database L.-Unigene-All was used for transcription factor identification against the Plant Transcription Factor Database PlnTFDB (http://plntfdb.bio.uni-potsdam.de/v3.0/downloads.php) using BLASTX with an E-value cut-off of ≤ 10^{-5}.

Construction of gene expression profiles and gene co-expression network

HemI 1.0.3.7 [60] was used to construct a heatmap of DGEs related to monoterpene biosynthesis and expression profiles of transcription factors. Pearson correlation coefficient (PCC) were calculated between the two genes from two data sets, DEGs related to flower fragrance (Additional file 8: Table S8) and the expressed identified TFs. TFs with |PCC| ≥ 0.95 (p<0.01) were selected for co-expression network construction from the DEGs related to flower fragrance. The networks were visualized using Cytoscape _v.3.3.0 [61].

Additional files

Additional file 1: Table S1. Mapping reads alignment results for the discarded leaf sample. (XLSX 10 kb)

Additional file 2: Table S2. Original data for RPKM values and functional annotation of all transcripts. (XLS 33444 kb)

Additional file 3: Table S3. Review of common unigenes in flower, leaf and scale. (XLSX 13 kb)

Additional file 4: Table S4. Flower-specific genes. (XLSX 38 kb)

Additional file 5: Table S5. Leaf-specific genes. (XLSX 47 kb)

Additional file 6: Table S6. Scale-specific genes. (XLSX 36 kb)

Additional file 7: Table S7. Differentially expressed genes involved in flavonoid biosynthesis. (XLSX 17 kb)

Additional file 8: Table S8. Differentially expressed genes related to flower fragrance. (XLSX 16 kb)

Additional file 9: Table S9. Differentially expressed genes involved in photosynthesis. (XLSX 21 kb)

Additional file 10: Table S10. Differentially expressed genes involved in metabolism of starch and sucrose. (XLSX 18 kb)

Additional file 11: Table S11. Review of structural genes related to flower fragrance and their highly correlated transcript factors. (XLSX 23 kb)

Additional file 12: Figure S1. Alignment of deduced amino acid sequences of LsTPS with sequences from different fragrant plants. Constructed using MEGA5 with a bootstrap replications of 1000. (PDF 61 kb)

Additional file 13: Table S12. Primer sequences used for qRT-PCR validation of RNA-seq data. (PDF 135 kb)

Abbreviations

AGP: ADP-glucose pyrophosphorylase; BAM: β-amylase; CHI: Chalcone isomerase; CHS: Chalcone synthase; DEGs: Differentially expressed genes; DFR: Dihydroflavonol 4-reductase; DGE: Digital gene expression; DXS: 1-deoxy-D-xylulose-5-phosphate synthase; F3H: Flavanone 3-hdroxylase; FLS: Flavonol synthase; F-vs-S: Flowers-vs-scales; GGPP: Geranylgeranyl pyrophosphate; GO: Gene Ontology; HCT: Shikimate O-hydroxycinnamoyltransferase; HDS: 4-hydroxy-3-methylbut-2-en- 1-yl diphosphate; HPL: Hydroperoxide lyase; INV: Invertase; KAT: 3-ketoacyl-CoA thiolase; KEGG: Kyoto Encyclopedia of Genes and Genomes; LAR: Leucoanthocyanidin reductase; LDOX: Leucoanthocyanidin dioxygenase; L-vs-F: Leaf-vs-flower; L-vs-S: Leaves-vs-scales; qRT-PCR: quantitative real time PCR; RPKM: Reads per kb per million reads; SBE: starch branching enzyme; SPS: Sucrose phosphate synthase; SS: Sucrose synthase; SSS: Starch synthase; TPS: terpene synthase; TT7: Flavonoid 3′-monooxygenase

Acknowledgments

The authors thank Gene de novo Biotechnology (Guangzhou, China) for providing support on construction of regulatory networks.

Funding

This work was sponsored by the Natural Science Foundation of Shanxi, China (No. 201601D011077); Supporting Talent Research Project of Shanxi Agricultural University, China (No. 2014ZZ02); National Natural Science Foundation of China (No. 31772337). The funding agency had no role in the design of the study, collection, analysis, and interpretation of data, or writing the manuscript.

Availability of data and materials

The sequence datasets used as reference transcriptome, L.-Unigene-All, are available at the NCBI TSA database with accession number SUB2623518. The sequence datasets supporting the genes used in this article are available at the NCBI SRA database with accession number SRP084220. The sequence of *LsTPS* open reading frame cloned in this study are available at the NCBI Nucleotide database with accession number MF401556.

Author's contributions

ZSG and YPX conceived and designed the experiment. JMF and TW participated in RNA extraction, carried out the quantitative real-time PCR and gene clone. FD analyzed the sequence data and drafted the manuscript. YW participated in data analysis and manuscript drafting. DG, ZSG and YPX revised the manuscript. All authors read and approved the final manuscript.

Competing interests

The authors declare that they have no competing interests.

Author details

[1]College of Horticulture, Shanxi Agricultural University, Taigu 030801, China. [2]Department of Horticulture, College of Agriculture & Biotechnology, Zhejiang University, Hangzhou 310058, China. [3]Plant & Crop Sciences Division, School of Biosciences, University of Nottingham, Sutton Bonington Campus, Loughborough LE12 5RD, UK.

References

1. Bowman JL, Smyth DR, Meyerowitz EM. The ABC model of flower development: then and now. Development. 2012;139(22):4095–8.
2. Lin Z, Hong Y, Yin M, Li C, Zhang K, Grierson D. A tomato HD-zip homeobox protein, LeHB-1, plays an important role in floral organogenesis and ripening. Plant J. 2008;55(2):301–10.
3. Zhang C, Xu Y, Guo S, Zhu J, Huan Q, Liu H, Wang L, Luo G, Wang X, Chong K. Dynamics of brassinosteroid response modulated by negative regulator LIC in rice. PLoS Genet. 2012;8(4):e1002686.
4. Zimmermann R, Sakai H, Hochholdinger F. The Gibberellic acid stimulated-like gene family in maize and its role in lateral root development. Plant Physiol. 2010;152(1):356–65.
5. Leeggangers HA, Moreno-Pachon N, Gude H, Immink RG. Transfer of knowledge about flowering and vegetative propagation from model species to bulbous plants. Int J Dev Biol. 2013;57(6–8):611–20.
6. Leitch IJ, Beaulieu JM, Cheung K, Hanson L, Lysak MA, Fay MF. Punctuated genome size evolution in Liliaceae. J Evol Biol. 2007;20(6):2296–308.
7. Moreno-Pachon NM, Leeggangers HA, Nijveen H, Severing E, Hilhorst H, Immink RG. Elucidating and mining the *Tulipa* and *Lilium* transcriptomes. Plant Mol Biol. 2016;92(3):249–61.
8. Riesgo A, Andrade SCS, Sharma PP, Novo M, Pérez-Porro AR, Vahtera V, González VL, Kawauchi GY, Giribet G. Comparative description of ten transcriptomes of newly sequenced invertebrates and efficiency estimation of genomic sampling in non-model taxa. Front Zool. 2012;9:33.
9. Villacorta-Martin C, Haan JD, Huijben K, Passarinho P, Hamo LB, Zaccai M. Whole transcriptome profiling of the vernalization process in *Lilium longiflorum* (cultivar white heaven) bulbs. BMC Genomics. 2015;16(1):1–16.
10. Wang J, Wang Q, Yang Y, Liu X, Gu J, Li W, Ma S, Lu Y. *De novo* assembly and characterization of stress transcriptome and regulatory networks under temperature, salt and hormone stresses in *Lilium lancifolium*. Mol Biol Rep. 2014;41(12):8231–45.
11. Li X, Wang C, Cheng J, Zhang J, da Silva JA, Liu X, Duan X, Li T, Sun H. Transcriptome analysis of carbohydrate metabolism during bulblet formation and development in *Lilium davidii* Var. *unicolor*. BMC Plant Biol. 2014;14(1):358.
12. Zhu X, Chai M, Li Y, Sun M, Zhang J, Sun G, Jiang C, Shi L. Global transcriptome profiling analysis of inhibitory effects of paclobutrazol on leaf growth in lily (*Lilium Longiflorum*-Asiatic hybrid). Front Plant Sci. 2016;7

13. Lang V, Usadel B, Obermeyer G. *De novo* sequencing and analysis of the lily pollen transcriptome: an open access data source for an orphan plant species. Plant Mol Biol. 2015;87(1–2):69–80.

14. Liu X, Huang J, Wang J, Lu Y. RNA-Seq analysis reveals genetic bases of the flowering process in oriental hybrid lily cv. Sorbonne Russ J Plant Physl. 2014;61(6):880–92.

15. Zhang MF, Jiang LM, Zhang DM, Jia GX. *De novo* transcriptome characterization of *Lilium* 'Sorbonne' and key enzymes related to the flavonoid biosynthesis. Mol Genet Genomics. 2015;290(1):399–412.

16. Xu L, Yang P, Feng Y, Xu H, Cao Y, Tang Y, Yuan S, Liu X, Ming J. Spatiotemporal transcriptome analysis provides insights into bicolor tepal development in *Lilium* "tiny Padhye". Front Plant Sci. 2017;8:398.

17. Shahin A, Arens P, Van Heusden AW, Van Der Linden G, Van Kaauwen M, Khan N, Schouten HJ, Van De Weg WE, Visser RGF, Van Tuyl JM. Genetic mapping in *Lilium*: mapping of major genes and quantitative trait loci for several ornamental traits and disease resistances. Plant Breed. 2011;130(3):372–82.

18. Benjamini Y, Hochberg Y. Controlling the false discovery rate. A practical and powerful approach to multiple testing. J R Stat Soc. 1995;57(1):289–300.

19. Kong Y, Sun M, Pan HT, Zhang QX. Composition and emission rhythm of floral scent volatiles from eight lily cut flowers. J Amer Soc Hort Sci. 2012; 137(6):376–82.

20. Johnson TS, Schwieterman ML, Kim JY, Cho KH, Clark DG, Colquhoun TA. *Lilium* floral fragrance: a biochemical and genetic resource for aroma and flavor. Phytochemistry. 2016;122:103–12.

21. Zhang HX, Zeng-Hui HU, Leng PS, Wang WH, Fang XU, Zhao J. Qualitative and quantitative analysis of floral volatile components from different varieties of *Lilium* spp. Sci Agr Sinica. 2013;46(4):790–9. (in Chinese)

22. Chang YT, Chu FH. Molecular cloning and characterization of monoterpene synthases from *Litsea cubeba* (lour.) Persoon. Tree Genet Genom. 2011;7(4): 835–44.

23. He XL, Shi LW, Yuan ZH, Xu Z, Zhang ZQ, Yi MF. Effects of lipoxygenase on the corm formation and enlargement in *Gladiolus hybridus*. Sci Hortic. 2008; 118(1):60–9.

24. Yu Z, Chen LC, Suzuki H, Ariyada O, Erra-Balsells R, Nonami H, Hiraoka K. Direct profiling of phytochemicals in tulip tissues and in vivo monitoring of the change of carbohydrate content in tulip bulbs by probe electrospray ionization mass spectrometry. J Am Soc Mass Spectrom. 2009;20(12):2304–11.

25. Kong Y, Dou XY, Bao F, Lang LX, Bai JR. Advances in flower color mechanism of *Lilium*. Acta Hort Sinica. 2015;42(9):1747–59. (in Chinese)

26. Grotewold E. The genetics and biochemistry of floral pigments. Annu Rev Plant Biol Plant Biology. 2006;57(57):761–80.

27. Koes R, Verweij W, Quattrocchio F. Flavonoids: a colorful model for the regulation and evolution of biochemical pathways. Trends Plant Sci. 2005; 10(5):236–42.

28. Nakatsuka A, Izumi Y, Yamagishi M. Spatial and temporal expression of chalcone synthase and dihydroflavonol 4-reductase genes in the Asiatic hybrid lily. Plant Sci. 2003;165(4):759–67.

29. An L, Yang K, Zhang K, Zhao X, Wang W, Yang L, Wang J. Cloning and analysis of Chalcone synthase gene of lily. Acta Bot Boreal Occident Sin. 2011;31(3):0492–8. (in Chinese)

30. Yamagishi M. Oriental hybrid lily Sorbonne homologue of LhMYB12 regulates anthocyanin biosynthesis in flower tepals and tepal spots. Mol Breeding. 2010;28(3):381–9.

31. Tang B, Zenghui HU, Leng P, Yan J, Xiuyun WU. The expression of monoterpene synthase genes in *Lilium* with strong and light floral fragrance. J BJ Univ Agr. 2016;31(2):88–94. (in Chinese)

32. LI L, Wang H, Sun M, Zhang QX. Molecular cloning and expression analysis of a monoerpene synthase gene in *Lilium regale*. J FJ Agr For Univ. 2014; 43(4):397–401. (in Chinese)

33. Li TJ, Leng PS, Yang K, Zheng J, Hui HZ. Molecular cloning and characterization of monoterpene synthase gene in *Lilium* flowers. J BJ Univ Agr. 2014;29(3):6–10. (in Chinese)

34. Cordoba E, Salmi M, León P. Unravelling the regulatory mechanisms that modulate the MEP pathway in higher plants. J Exp Bot. 2009;60(10):2933–43.

35. Battilana J, Costantini L, Emanuelli F, Sevini F, Segala C, Moser S, Velasco R, Versini G, Grando MS. The 1-deoxy- d -xylulose 5-phosphate synthase gene co-localizes with a major QTL affecting monoterpene content in grapevine. Theor Appl Genet. 2009;118(4):653–69.

36. Lois LM, Rodríguezconcepción M, Gallego F, Campos N, Boronat A. Carotenoid biosynthesis during tomato fruit development: regulatory role of 1-deoxy-D-xylulose 5-phosphate synthase. Plant J Cell Molr Biol. 2000;22(6):503–13.

37. Kim JS, Kim JH. Comparative genome analysis and phylogenetic relationship of order Liliales insight from the complete plastid genome sequences of two lilies (*Lilium longiflorum* and *Alstroemeria aurea*). PLoS One. 2013;8(6): e68180.

38. Givnish TJ, Ames M, McNeal JR, McKain MR, Steele PR, dePamphilis CW, Graham SW, Pires JC, Stevenson DW, Zomlefer WB, et al. Assembling the tree of the monocotyledons: plastome sequence phylogeny and evolution of poales. Ann Mo Bot Gard. 2010;97(4):584–616.

39. Wang XN, Xiong L, Wu XW, Wang QG, Chen M, Bao LX. Relationship between starch saccharification and propagation of bulblets from scales in oriental hybrid lily (*Lilium* L.). SW China J Agr Sci. 2007;01: 115–9. (in Chinese)

40. Xia YP, Zheng HJ, Huang CH, Xu WW. Accumulation and distribution of ^{14}C-photosynthate during bulb developemnt of *Lilium* oriental hybrid. J Nucl Agril Sci. 2006;20(5):417–22. (in Chinese)

41. Kawagishi K, Miura T. Growth characteristics and effect of nitrogen and potassium topdressing on thickening growth of bulbs in spring-planted edible lily (*Lilium leichtlinii* Var. *maximowiczii* baker). JPN J Crop Sci. 1996; 65(1):51–7.

42. Wu SS, Wu JD, Jiao XH, Zhang QX, Lv YM. The dynamics of changes in starch and lipid droplets and sub-cellular localization of β-amylase during the growth of lily bulbs. J Integr Agri. 2012;11(4):585–92.

43. Hobert O. Gene regulation by transcription factors and microRNAs. Science. 2008;319(5871):1785–6.

44. Spyropoulou EA, Haring MA, Schuurink RC. RNA sequencing on *Solanum lycopersicum*, trichomes identifies transcription factors that activate terpene synthase promoters. BMC Genomics. 2014;15(1):402.

45. Mertens J, Moerkercke AV, Bossche RV, et al. Clade IVa basic helix–loop–helix transcription factors form part of a conserved jasmonate signaling circuit for the regulation of bioactive plant terpenoid biosynthesis. Plant Cell Physiol. 2016;57(12):2564–75.

46. Yu ZX, Li JX, Yang CQ, Hu WL, Wang LJ, Chen XY. The jasmonate-responsive AP2/ERF transcription factors AaERF1 and AaERF2 positively regulate artemisinin biosynthesis in *Artemisia annua* L. Mol Plant. 2012;5(2):353–65.

47. Shen SL, Yin XR, Zhang B, Xie XL, Jiang Q, Grierson D, et al. CitAP2.10 activation of the terpene synthase *CsTPS1* is associated with the synthesis of (+)-valencene in 'Newhall' orange. J Exp Bot. 2016;67(14): 4105–15.

48. Mutz KO, Heilkenbrinker A, Lonne M, Walter JG, Stahl F. Transcriptome analysis using next-generation sequencing. Curr opin biotech. 2013;24(1):22–30.

49. Yang F, Zhu G, Wang Z, Liu H, Xu Q, Huang D, Zhao C. Integrated mRNA and microRNA transcriptome variations in the multi-tepal mutant provide insights into the floral patterning of the orchid *Cymbidium goeringii*. BMC Genomics. 2017;18(1):367.

50. Li XW, Jiang J, Zhang LP, Yu Y, Ye ZW, Wang XM, Zhou JY, Chai ML, Zhang HQ, Arús P, et al. Identification of volatile and softening-related genes using digital gene expression profiles in melting peach. Tree Genet Genom. 2015;11(4):71.

51. Zu K, Li J, Dong S, Zhao Y, Xu S, Zhang Z, Zhao L. Morphogenesis and global analysis of transcriptional profiles of *Celastrus orbiculatus* aril: unravelling potential genes related to aril development. Genes Genom. 2017;39(6):623–35.

52. Savchenko T, Yanykin D, Khorobrykh A, Terentyev V, Klimov V, Dehesh K. The hydroperoxide lyase branch of the oxylipin pathway protects against photoinhibition of photosynthesis. Planta. 2017;245(6):1179–92.

53. Du F, Wu Y, Zhang L, Li XW, Zhao XY, Wang WH, Gao ZS, Xia YP. *De novo* assembled transcriptome analysis and SSR marker development of a mixture of six tissues from *Lilium* oriental hybrid 'Sorbonne'. Plant Mol Biol Rep. 2014;33(2):1–13.

54. Shan L, Li X, Wang P, Cai C, Zhang B, De Sun C, Zhang W, Xu C, Ferguson I, Chen K. Characterization of cDNAs associated with lignification and their expression profiles in loquat fruit with different lignin accumulation. Planta. 2008;227(6):1243.

55. Feng C, Chen M, Xu CJ, Bai L, Yin XR, Li X, Allan AC, Ferguson IB, Chen KS. Transcriptomic analysis of Chinese bayberry (*Myrica rubra*) fruit development and ripening using RNA-Seq. BMC Genomics. 2012;13:19.

56. Li R, Yu C, Li Y, Lam TW, Yiu SM, Kristiansen K, Wang J. SOAP2: an improved ultrafast tool for short read alignment. Bioinformatics. 2009; 25(15):1966–7.

57. Tarazona S, García-Alcalde F, Dopazo J, Ferrer A, Conesa A. Differential expression in RNA-seq: a matter of depth. Genome Res. 2011;21(12):2213–23.

58. Conesa A, Gotz S, Garcia-Gomez JM, Terol J, Talon M, Robles M. Blast2GO: a universal tool for annotation, visualization and analysis in functional genomics research. Bioinformatics. 2005;21(18):3674–6.
59. Ye J, Fang L, Zheng H, Zhang Y, Chen J, Zhang Z, Wang J, Li S, Li R, Bolund L, et al. WEGO: a web tool for plotting GO annotations. Nucleic Acids Res. 2006;34:293–7.
60. Deng WK, Wang YB, Liu ZX, Cheng H, Xue Y. Heml: a toolkit for illustrating heatmaps. PLoS One. 2014;9(11):e111988.
61. Shannon P, Markiel A, Ozier O, Baliga NS, Wang JT, Ramage D, et al. Cytoscape: a software environment for integrated models of biomolecular interaction networks. Genome Res. 2003;13(11):2498–504.

8

Chromosome level assembly and secondary metabolite potential of the parasitic fungus *Cordyceps militaris*

Glenna J. Kramer and Justin R. Nodwell*

Abstract

Background: *Cordyceps militaris* is an insect pathogenic fungus that is prized for its use in traditional medicine. This and other entomopathogenic fungi are understudied sources for the discovery of new bioactive molecules. In this study, PacBio SMRT long read sequencing technology was used to sequence the genome of *C. militaris* with a focus on the genetic potential for secondary metabolite production in the genome assembly of this fungus.

Results: This is first chromosome level assembly of a species in the *Cordyceps* genera. In this seven chromosome assembly of 33.6 Mba there were 9371 genes identified. *Cordyceps militaris* was determined to have the MAT 1-1-1 and MAT 1-1-2 mating type genes. Secondary metabolite analysis revealed the potential for at least 36 distinct metabolites from a variety of classes. Three of these gene clusters had homology with clusters producing desmethylbassianin, equisetin and emericellamide that had been studied in other fungi.

Conclusion: Our assembly and analysis has revealed that *C. militaris* has a wealth of gene clusters for secondary metabolite production distributed among seven chromosomes. The identification of these gene clusters will facilitate the future study and identification of the secondary metabolites produced by this entomopathogenic fungus.

Keywords: *Cordyceps militaris*, Entomopathogenic fungi, Genome, SMRT sequencing, Secondary metabolite

Background

Entomopathogenic fungi are a fascinating group of insect parasitic microbes, which include species from a variety of different fungal taxa including *Beauvaria*, *Hirsutella*, *Metarhizium Cordyceps* and *Ophiocordyceps* (Fig. 1). These entomopathogenic fungi typically have a lifecycle in which the host is infected and is killed in the process of fungal propagation. Two closely related genera of entomopathogemic fungi, *Cordyceps* and *Ophiocordyceps* (often times just referred to as cordyceps in common literature) are characterized by their unique lifecycle and a specific process by which they parasitize and reproduce using the insect host.

Fungal spores and hyphae are able to penetrate insect cuticle and then colonize and proliferate within their body cavity. As the insect's body is used as a nutrient reservoir for growth, the insect's behaviour is modified, eventually leading the host to die in an advantageous location for fungal spore dispersal. The fungus then emerges as a fruiting body from the corpse of the insect, which matures and disperses spores of the next generation. Spread globally, *Cordyceps* and *Ophiocordyceps* fungi have been described in climates across Asia, the Americas, Europe and Australia, with many of these species having not been characterized. Although these genera are believed to contain well over 400 species of fungi, there are a few standout examples which are revered for their medicinal potential or unusual host pathogenesis.

Ophiocordyceps sinensis, found in the mountains of Tibet, infects and kills ghost moth larvae to give the highly prized herbal remedy "dong chong xia cao," which is believed to treat a plethora of disorders [1]. This prized specimen is identified by the fruiting body growing from the ground, as the infected ghost moth larvae dies situated just below the surface of the soil with its head oriented upward, from which the fruiting body emerges. *Ophiocordyceps unilateralis*, also known as the zombie-ant fungus, is noted for its pathogenic process in ants, which is characterized by particular behaviour

* Correspondence: justin.nodwell@utoronto.ca
Department of Biochemistry, University of Toronto, MaRS Centre, West Tower, 661 University Avenue, Toronto, ON M5G 1M1, Canada

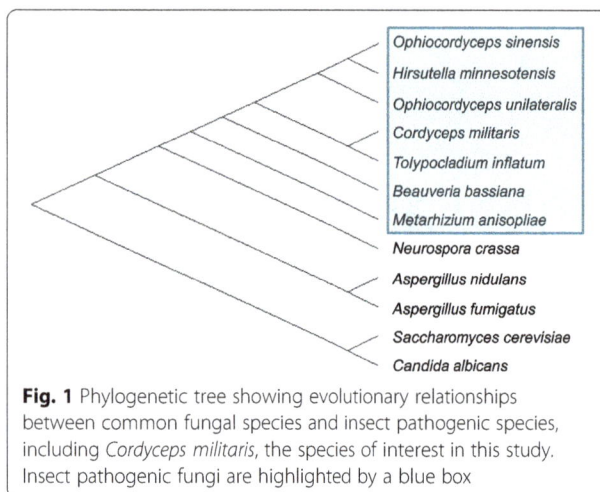

Fig. 1 Phylogenetic tree showing evolutionary relationships between common fungal species and insect pathogenic species, including *Cordyceps militaris*, the species of interest in this study. Insect pathogenic fungi are highlighted by a blue box

modifications in the host, that leaves the host ant perished with its jaw clamped to a leaf in prime location for spore dispersal [2]. *Cordyceps militaris*, which is a common component of supplements as it is also believed to have medicinal potential, is often used as a cheaper and more readily available version of *Ophiocordyceps sinensis* [3, 4].

The genome of a handful of these fungi have been sequenced, however, the available assemblies are often fragmented in over 500 contigs [5–8]. These assemblies do indicate that these fungi are capable to producing natural products, possibly over 30 distinct molecules per species. Only a few of these natural products from entomopathogenic fungi have been isolated and described, including the immunosuppressant, cyclosporine, from *Tolypocladium inflatum*, and fingolimoid, the immunomodulatory molecule derived from the *Isaria sinclarii* natural product myriosin, signifying that these fungi may be an underexplored source of novel molecules [9, 10].

Natural products have been established as a source of bioactive molecules, however, discovery has dwindled implying the need for new sources. Fungi have been shown to produce a wealth of diverse molecules [11] suggesting that *Cordyceps* and *Ophiocordyceps* could be an understudied and relevant avenue for the discovery of natural products. Furthermore, by studying the secondary metabolites in *Cordyceps and Ophiocordyceps* not only is there the potential for discovery of new bioactives, but for the identification and study of chemicals that have a role in the process of host pathogenesis, from behaviour modifying molecules to insecticides [12]. As genome sequencing becomes cheaper, faster, and more readily available, the possibility of taking a computational approach to genome mining for secondary metabolite discovery in fungi becomes a more realistic possibility, allowing for the study of secondary metabolites which may be cryptic under typical laboratory

conditions [13]. Indeed, laboratory culture of these organisms is a challenge due to their slow growth rates – a genome-based approach to their natural product genes is likely essential for this field to progress.

In this study, a new method of long read sequencing, Pacific Biosciences SMRT sequencing is applied to an exemplary sample from the *Cordyceps* genera, *Cordyceps militaris*, a strain isolated from butterfly pupa. The overarching goal is to provide a chromosome level genome assembly to serve as a model for the genera. Furthermore, as these fungi have the potential to produce many understudied natural products, this study is focused on the genetic potential for secondary metabolite expression in this organism.

Results
General genome features
Purified genomic DNA isolated from culture of *C. militaris* grown up from a single colony was sequenced using the Pacific Biosciences platform using a sheared large insert library [14]. Sequence data from 6 SMRT cells, providing approximately 180× coverage were assembled using two de novo assemblers, Celera with the PBcR protocol and the HGAP2 protocol from SMRT portal [15–17]. Both chosen assemblers were applied to self-correct the reads, a process in which the shorter PacBio reads were used to error correct the long PacBio reads. These corrected reads were then subsequently assembled into contigs. The PBcR-pipeline gave an assembly with 32 contigs, four of which had telomeric repeats (CCCTAA or TTAGGG)$_n$ on either the 5′ or 3′ end of the contig.

The second assembly protocol, the HGAP2 protocol from the SMRT portal software package, which also included an additional polishing step using the Quiver algorithm, yielded an even further improved assembly. This assembly, which contained 18 contigs, gave five assembled chromosomes, having telomeric repeats (CCCTAA or TTAGGG)$_n$ on both the 5′ and 3′ ends of the sequence and four having telomeric repeats on one of the 5′ or 3′ end of the sequence. After manually curating the assembly and submitting the curated assembly to the SMRT resequencing protocol, the resultant assembly contained 7 contigs, each with telomeres on both ends, indicative of 7 chromosomes. Coverage across these seven chromosomes, including regions where the assembly was manually curated is shown in Additional file 1: Figure S1. The coverage across the chromosomes is generally consistent in the assembly after manual curation and the SMRT resequencing protocol. The N_{50} (5.78kba) and N_{max} (8.29kba) remain unchanged when comparing the initial assembly and the curated and resequenced assembly. However, there is a spike in coverage in the contig corresponding to chromosome IV, possibly implying a collapsed repeat region within the assembly (Additional file 1: Figure S1D). When this area of the assembly is further inspected, it is noted that there are a large number of short low quality repeated reads, which align on top of >30× coverage of high quality reads that span this repetitive region. The repetitive sequence of this region and the overabundance of low quality reads in this region was noted prior to manual curation. The initial assembly consisted of 3 contigs containing these short repetitive reads, each about 30,000 bp in length, which overlapped in this region. With an overall coverage of approximately 150×, a genome size of 33.6 Mba, and a GC content of 50.9%. (Table 1), the BUSCO completeness of this assembly is 98.2% with only 0.4% missing [18]. Furthermore, the assembly is comparable to the previously sequenced *C. militaris* Cm01 (though that sequence is broken into a very large number of contigs) with a similar genome size of 32.2 Mba and a GC content of 51.4% [6].

The seven assembled *C. militaris* chromosomes range in size from 1.9 to 8.3 Mba. The sequenced haploid genomes of *Aspergillus niger* and *Neurospora crassa* contain eight and seven chromosomes, respectively [19, 20]. Furthermore, a karyotype analysis of *Tolypocladium inflatum* shows that this related species contains seven chromosomes ranging in size from 1.0 to 6.3 Mba [21], suggesting that the assembly with seven chromosomes is reasonable for *C. militaris*.

The MAKER genome annotation pipeline [22, 23] predicted 9907 genes for *C. militaris*. Passing the MAKER gene predictions through an additional evidence modeler using Funannotate gave a set of 9371 genes with a BUSCO analysis of the resulting gene set estimating a completeness of 93.7% with 3.9% of genes missing [18, 24]. Estimates of mean gene length, mean exon length, mean intron length and gene density (Table 2) are similar to those of *C. militaris* Cm01 and other filamentous ascomycete fungi (Table 3) [5–7, 12, 25, 26]. An Interpro analysis of the annotated genes using the Blast2GO suite was used to assign 8792 genes (93.8%) of genes InterPro IDs. A Gene Ontology (GO) annotation was assigned to 6453 of the genes (68.9%).

Mating type loci

The sequence of our isolate revealed the presence of only a MAT 1-2-1 mating type gene present on

Table 1 Main features of *C. militaris* genome assembly

Main Features of *C. militaris* Genome Assembly	
Genome size (Mba)	33.6
Number of chromosomes	7
Fold coverage	149.5×
GC content	50.9

Table 2 Features of *C. militaris* genome annotation

Main Features of *C. militaris* Genome Annotation	
Genome Size (Mba)	33.6
Number of Chromosomes	7
Number of genes	9371
Number of exons	26,128
Number of introns	16,759
Total gene length (Mba)	16.1
Mean gene length (bases)	1724
Gene density (genes per Mba)	278.9
Mean exon length (bases)	548
Mean intron length (bases)	111
Mean introns per gene (bases)	1.8
Genome coding (%)	48.0

chromosome VII, supporting the notion that this fungus is indeed heterothalic. Both MAT 1-1, MAT 1-2 and strains with hybrid mating loci have been reported [6]. The previously sequenced *C. militaris* Cm01 strain was determined to have both MAT 1-1-1 and MAT 1-1-2 mating type genes. The *C. militaris* Cm06 strain was reported to be a hybrid strain containing both the MAT 1-1 and MAT 1-2 mating types, with single spore isolates from this hybrid strain producing progeny with either the MAT 1-1 (93.3%) or MAT 1-2 (6.7%) loci [6]. Both the MAT 1-1 and MAT 1-2 containing isolates have been shown to fruit, but only the hybrid strain containing both the MAT 1-1 and MAT 1-2 loci was able to produce mature spores [6]. However, fruiting bodies were not observed under analogous conditions with the ATCC® 34164 strain. Comparison of the genomic regions containing the mating type genes in both our ATCC® 34164 strain and the Cm01 strain reveal that the genes in these regions are highly similar, with the exception of the MAT genes (Fig. 2).

Cordycepin

One hallmark molecule of interest in *C. militaris* is the nucleoside analogue cordycepin. Although this biosynthesis is unknown, it is proposed that this mechanism is dependent upon a reduction step, believed to be potentially catalyzed by a ribonucleoside diphosphate reductase (RNR) [6, 27, 28]. However, our sequenced *C. militaris* is similar to the sequenced Cm01 strain, in that it only seems to possess two type I RNRs (genes A9K55_000536 and A9K55_003140), both of which have homologues in non-cordycepin producing fungus, and have been identified in *C. militaris* Cm01, leaving the biosynthesis of cordycepin elusive [6].

Table 3 General features of ascomycetes related to *C. militaris*

	Cordyceps militaris ATCC® 34164	Cordyceps militaris Cm01	Tolypocladium inflatum	Fusarium graminearum	Ophiocordyceps sinensis	Ophiocordyceps unilateralis
Genome Size (MBa)	33.6	32.2	30.35	36.09	116.42	26.1
GC Content (%)	50.9	51.4	58.0	48.3	43.1	54.8
Predicted Genes	9371	9684	9998	13,321	7939	7831
Gene density (Genes/Mbp)	278.9	301	329	369	69	301
Mean gene length	1724	1742	1670	1583	1693	1420[a]
Mean exon length	548	507	570	508	NR	261[a]
Mean intron length	111	98	78	68	103	62[a]
Mean introns per gene	1.8	2.0	2.2	2.2	NR	3[a]

[a]Values presented as a median as opposed to a mean, NR is not reported in original publication

Secondary metabolite potential

Sequenced fungi from the *Cordyceps*, *Ophiocordypces*, and related genera have revealed the potential for production of over 30 diverse secondary metabolites per strain [7, 8, 26, 29, 30]. The limited number of prior systematic studies to identify bioactive secondary metabolites from *Cordyceps* and related fungi have shown that a number of novel molecules can be produced by these microbes [31–35]. However, these studies do not nearly capture the full secondary metabolite potential of these fungi, likely due to the fact that many of these metabolites may be cryptic and not expressed under the tested laboratory conditions. To determine whether this fungus could produce a wealth of secondary metabolites, the genetic potential for diverse metabolite production became the focus of

the study. Using two gene cluster predictors, Anti-SMASH and SMURF, all seven *C. militaris* chromosomes were profiled for the presence of genes that could be responsible for the biosynthesis of secondary metabolites [36–38]. The fungal version of Anti-SMASH, predicted 32 secondary metabolites and SMURF predicted 25. Taken together, the two algorithms predicted the presence of 36 unique gene clusters which could be responsible for secondary metabolite production in *C. militaris* (Table 4).

Distribution of these secondary metabolite producing genes were mapped on the 7 chromosomes (Fig. 3a). No gene clusters for secondary metabolites were noted in the presumably collapsed area of the genome shown in chromosome IV. The 36 metabolite producing gene clusters were from a variety of classes, including eight nonribosomal peptide synthetases (NRPS), seven type 1 polyketide synthases (T1PKS), six polyketide synthase-nonribosomal peptide synthetase (PKS-NRPS) hybrids, four terpenes, one indole and ten falling into other classes (Fig. 3b).

For comparison with the previously sequenced strain, the antiSMASH algorithm was also used to predict the presence of natural product producing

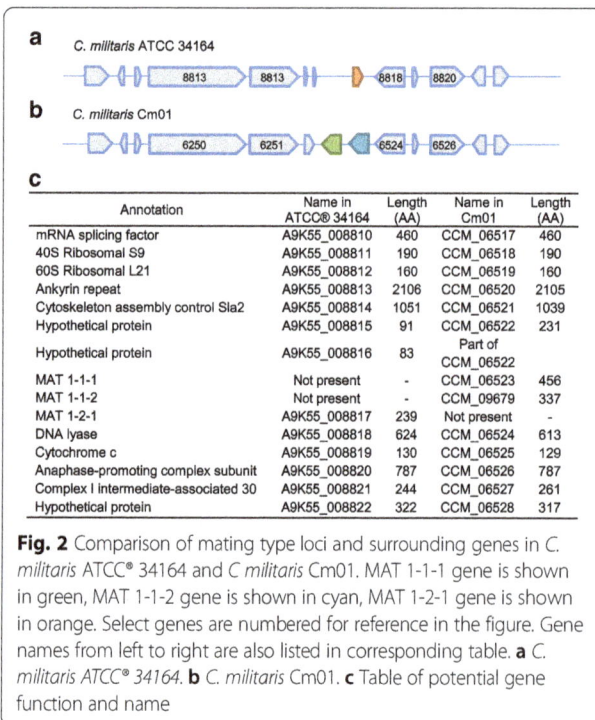

Fig. 2 Comparison of mating type loci and surrounding genes in *C. militaris* ATCC® 34164 and *C militaris* Cm01. MAT 1-1-1 gene is shown in green, MAT 1-1-2 gene is shown in cyan, MAT 1-2-1 gene is shown in orange. Select genes are numbered for reference in the figure. Gene names from left to right are also listed in corresponding table. **a** *C. militaris* ATCC® 34164. **b** *C. militaris* Cm01. **c** Table of potential gene function and name

Table 4 Number of natural product clusters predicted by the AntiSMASH and SMURF gene finding algorithms per chromosome and the number of unique natural products (NP) predicted in total by comparing the results of both algorithms

Chromosome	AntiSMASH	SMURF	Unique NP
I	1	1	1
II	3	1	4
III	4	4	5
IV	3	1	4
V	6	6	7
VI	8	6	8
VII	7	6	7
Total	32	25	36

a

b

	NRPS	T1PKS	NRPS-T1PKS	Indole	Other	Terpene	Total
I	-	1	-	-	-	-	**1**
II	-	2	-	-	1	1	**4**
III	2	-	1	-	2	-	**5**
IV	1	-	-	-	2	1	**4**
V	1	1	3	1	1	-	**7**
VI	2	3	-	-	2	1	**8**
VII	2	-	2	-	2	1	**7**
Total	**8**	**7**	**6**	**1**	**10**	**4**	**36**

Fig. 3 Putative natural products in *C. militaris*. The classes of natural products are denoted by the following colors NRPS (cyan), T1PKS (orange), T1PKS-NRPS (red), Indole (indigo), Other (green) Terpene (violet). **a** The distribution of natural product producing gene clusters on the chromosomes of *C. militaris*. **b** The number of natural product gene cluster on each chromosome, grouped by class

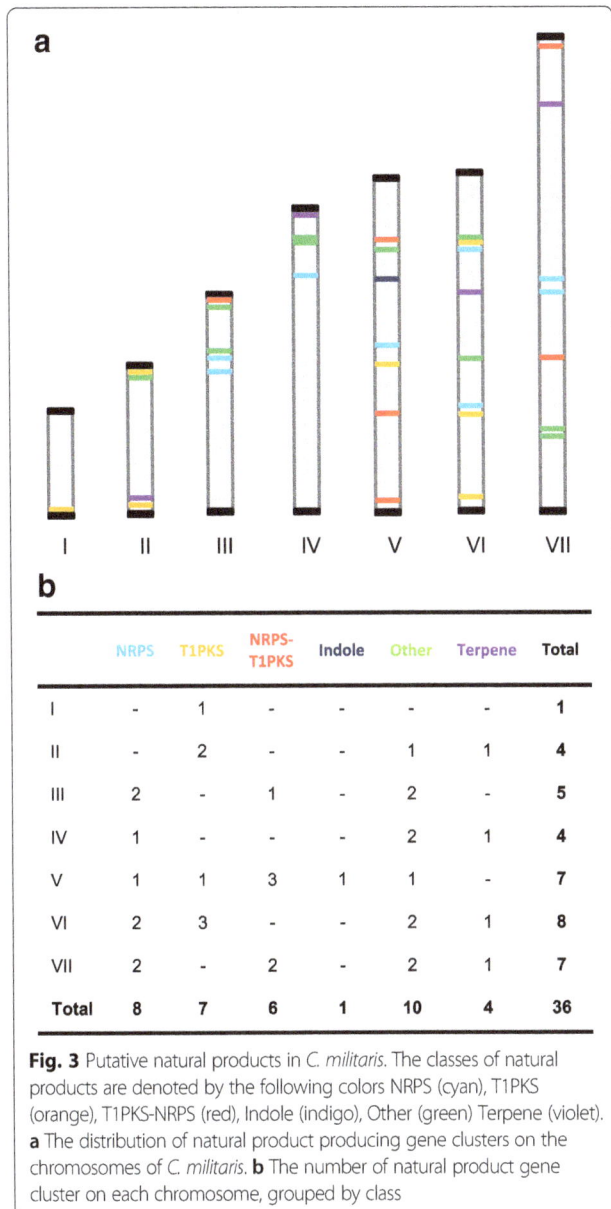

gene clusters in the in *C. militaris* Cm01. In Cm01 there were 28 natural product clusters identified by AntiSMASH, compared to the 32 in our strain of study. All of the 28 clusters were present in the ATCC® 34164 strain and the ATCC®34164 strain had four additional natural product clusters identified. These additional clusters are predicted to produce one indole (V-3), two T1PKS (VI-2, VI-8) and one T1PKS-NRPS (VII-5). Furthermore, using the ClusterFinder algorithm in the fungal version AntiSMASH, an additional 41 putative clusters were predicted, bringing the total to 73 predicted clusters from AntiSMASH with ClusterFinder, suggesting that the secondary metabolite potential of this organism is impressive.

Discussion

Herein is described the first chromosome level assembly of a *Cordyceps* genome. This seven chromosome assembly has revealed that this heterothallic strain, which contains 9371 genes, is capable of producing a wealth of secondary metabolites. Of the 36 gene clusters identified in ATCC®34164 by both the antiSMASH and SMURF algorithms (Additional file 2: Table S1), 3 clusters, III-1, V-6 and VII-5, are of particular interest as they have homology with gene clusters from other organisms that produce characterized natural products [39, 40].

It seems that a logical product of cluster III-1 could be a 2-pyridone alkaloid molecule. The hybrid NRPS-PKS central to this cluster (A9K55_001190) is similar to the NRPS-PKS responsible for the production of desmethyl-bassianin (70% identity) and tenellin (67% identity) from *Beauveria bassiana* and fumosorinone (66% identity) from *Isaria fumosoinone* [41–43]. Both the desmethyl-bassanin and tennellin gene clusters consist of 4 genes: the NRPS-PKS hybrid, an enoyl reductase and two cytochrome P450s (Fig. 4). In *C. militaris*, based on the sequence of gene A9K55_001191 it seems that the missing enoyl reductase may be fused with the cytochrome P450. Interestingly, a structurally related pigmented derivative, militarinone A, and the variants militarinone B–D, have been isolated from the possible *C. militaris* anamorph, *Paecilomyces militaris* [44, 45]. However, militarinone A–D was not identified by mass in extracts of the *C. militaris* strain of interest in this study.

On chromosome V, a cluster (V-6) with homology to the emercellamide producing cluster is present (Fig. 5). The emercellamide family molecules produced from the hybrid NRPS-PKS containing cluster have been described in the marine fungus *Emericella*, as well as the fungus *Aspergillus nidulans* [46, 47]. Other related molecules, scopularide A and W493-B have been isolated from *Scopulariopsis brevicaulis* and *Fusarium pseudogrami-nearum*, respectively [48–50]. The biosynthesis of emericellamide A in *Aspergillus* has been described and is shown to rely on four genes [47]. Comparing this emercellamide-like cluster in *C. militaris* to the gene cluster producing emercellamide in *Aspergillus* shows a conservation of 4 genes: an NRPS, a PKS, an acyl-transferase and a CoA ligase. The NRPS present in *C. militaris* (A9K55_005039) has 98% coverage and 43% identity with the NRPS in *Aspergillus nidulans*.

A cluster with similarities to the equisitin-producing cluster (VII-5) is also present in *C. militaris* on chromosome VII. This molecule, equisetin, was described as having structural similarities to the cholesterol lowering molecule lovastatin and was first isolated from *Fusarium equiseti* with a described bioactivity as a HIV-1 integrase inhibitor [51–53]. The biosynthesis, studied in *Fusarium*

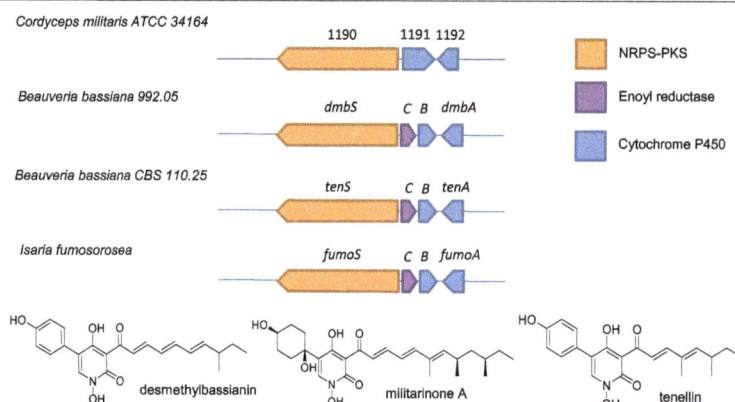

Fig. 4 Comparison of gene clusters producing a 2-pyridone alkaloid molecule to the genes A9K55_001190 to A9K55_001192 in cluster III-1 in *C. militaris* ATCC ® 34164. The desmethylbassianin cluster (*dmb*), tenellin (*ten*), and fumosorone (*fum*) clusters are shown. Genes are color coded by function. Structures of 2-pyridone alkaloids are displayed

heterosporum, reveals that the gene cluster consists of seven genes, two NRPS/PKS, two regulators, an oxidase, a methyltransferase and a transporter [53]. The comparable cluster in *C. militaris* consists of five genes, homologous to the genes present in the *F. heterosporum* minus one of the regulators and the oxidase (Fig. 6). The NRPS/PKS present in *C. militaris* (A9K55_008762) has 99% coverage and 50% identity with the NRPS/PKS in *Fusarium heterosporum* and seems to be well conserved among fungi in the *Aspergillus* and *Penicillum* genera.

This assembly of a genome from the *Cordyceps* genera, *Cordyceps militaris*, has shown the potential for production of an array of potentially novel natural products. This species is predicted to produce at least 36 secondary metabolites, three of which have significant similarity to characterized gene clusters. As fungal secondary metabolites can be cryptic under standard laboratory conditions, this assembled genome will allow for the application of genome mining techniques to guide the discovery and identification of new natural products. This can progress forward through a variety of techniques; one approach is to heterologously express gene clusters identified in the *C. militaris* genome. Alternatively, utilizing the genome to extrapolate potential natural products for expression can give important clues about the structure and favorable culture conditions of a secondary metabolite associated with a characterized

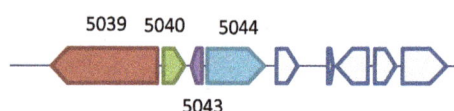

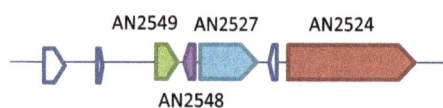

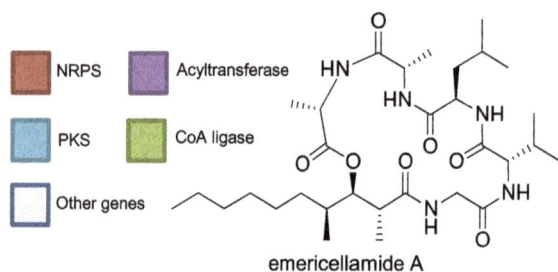

Fig. 5 Comparison of genes A9K55_005039, A9K55_005040, A9K55_005043 and A9K55_005044 in *C. militaris* ATCC® 34164 to an emericellamide producing gene cluster in *Aspergillus nidulans*. Genes are color coded by function. The structure of emericellamide is displayed

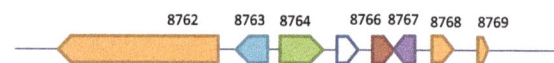

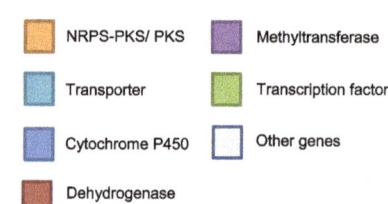

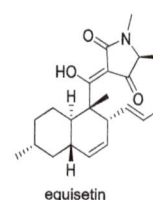

Fig. 6 Comparison of genes A9K55_008762 to A9K55_008769 in *C. militaris* ATCC® 34164 to an equisetin producing cluster in *Fusarium heterosporum*. Genes are color coded by function. The structure of equisetin is displayed

gene cluster. This knowledge can increase the likelihood of the production of the correlated molecule and simplify structural determination. Regardless, this assembly has shown that there is a great potential for the production of secondary metabolites in *C. militaris* and that this and other fungi from related *Cordyceps* and *Ophiocordyceps* genera could provide a wealth of molecular structural diversity.

Conclusions

Presented here is the first chromosome level assembly of a genome from the *Cordyceps* genera. This assembly and analysis has revealed that *C. militaris* has seven chromosomes containing a wealth of gene clusters for secondary metabolite production. Of the 36 gene clusters identified using the antiSMASH and SMURF algorithms, three clusters are found to have a high degree of similarity with clusters from other organisms that produce a known molecule. With this genome, further study and characterization of the secondary metabolites produced by *C. militaris* can be aided through genome based techniques including heterologous expression of gene clusters. As there is great potential for the production of secondary metabolites from *C. militaris*, this is one step towards discovering and characterizing the wealth of molecular structural diversity in this genera.

Methods
Phylogenetic tree construction
To compare the fungal species of interest, 18S rRNA sequences were obtained from the Silva database [54]. Sequence alignment was performed using ClustalW [55]. To construct the phylogenetic tree, The evolutionary history was inferred using the Neighbor-Joining method [56]. The optimal tree with the sum of branch length = 0.35499733 is shown. The evolutionary distances were computed using the Maximum Composite Likelihood method [57] and are in the units of the number of base substitutions per site. The analysis involved 12 nucleotide sequences. Codon positions included were 1st + 2nd + 3rd + Noncoding. All ambiguous positions were removed for each sequence pair. There were a total of 988 positions in the final dataset. Evolutionary analyses were conducted in MEGA7 [58].

Fungal strain and maintenance
Cordyceps militaris ATCC® 34164 was received from the American Type Culture Collection (ATCC). This strain, as described in the ATCC records was isolated from a butterfly pupa. Fungal cultures were maintained at 23.0 ° C on potato dextrose agar. The nrDNA of extracted genomic DNA was amplified using the ITS4 and ITS5 primer pairs, sequenced and compared against the BLAST database to determine sample validity [59–61].

Genomic DNA extraction and purification
Liquid cultures, containing 5 mL of seed media (10 g peptone, 40 g maltose, 10 g yeast extract, 1 g agar in 1 L DI water) in a culture tube were aseptically inoculated with a 3 mm square agar slab containing mycelial growth. The fungal mycelia were grown for 5 days at 23.0 °C. The mycelial mat was harvested, rinsed with sterile TE buffer and then frozen in liquid nitrogen. Frozen mycelia were macerated with a mortar and pestle with a spatula tip of aluminum oxide to aid in grinding the sample. The mycelial powder was transferred to a set of epitubes and 500 µL of CTAB DNA extraction buffer was added (100 mM Tris pH = 8.0, 10 mM EDTA, 2% CTAB, 2.8 M NaCl). The samples were incubated at room temperature for 5 min, then 2 µL of RNAse A (Thermo Scientific, 10 mg/mL) and 10 µL of Proteinase K were added (Invitrogen, 10 mg/mL) and inverted to mix. After centrifuging for 5 min, the pellet was ground in the epitube with a pellet pestle, then incubated for an additional 5 min before purification with phenol-chloroform. Each sample was washed twice with phenol-chloroform (50:50, phenol buffered with Tris pH 8.0, 600 µL) then twice with chloroform (600 µL). The resulting DNA containing aqueous portions were pooled and DNA was precipitated using cold ethanol (2.5× sample volume) and 3 M sodium acetate (0.1× sample volume). DNA was precipitated for at least 30 min by storing at –20 °C. The DNA precipitate was collected by centrifuging for 30 min, the pellet was washed with 70% ethanol and resuspended in TE buffer. The DNA was further purified with AMPure XP beads (Agencourt) by using an equal volume of beads to volume of DNA and eluting into TE. DNA was quantified using a PicoGreen assay (ThermoFisher) prior to sequencing.

Genome sequencing
The *Cordyceps militaris* DNA was sequenced using Pacific Biosciences RS II sequencing at the Genome Quebec Innovation Center (McGill University, Montreal, Canada). The sample was prepared using a large insert sheared DNA library and was sufficient for sequencing 6 SMRT cells.

Genome assembly
The sequencing reads were assembled using two different assemblers. The first assembler chosen was the PBcR pipeline from the Celera assembler (version 8.3rc2) using a genome size of 32 Mba [16]. The second assembler was SMRT portal (version 2.3.0) launched from an Amazon machine image. Assembly was performed on all 6 SMRT cells using the RS_HGAP Assembly.2 application with default settings and a genome size of 32 Mba [62]. The resulting assembly yielded 18 contigs, with five of these contigs containing characteristic telomeric

$(CCCTAA)_n$ or $(TTAGGG)_n$ repeats on both ends and four of these contigs containing telomeric repeats on one end. The 5 contigs with telomeres on both ends were taken to be fully sequenced chromosomes. The remaining 13 contigs were evaluated for overlapping regions that could possibly be used to join the contigs. Two of these 13 contigs were discarded due to low coverage (<30×), three of these remaining contigs were found to have overlapping regions that allowed them to be joined into 1 supercontig and the remaining 8 contigs were found to also contain overlapping regions that would allow them to be joined into a second supercontig. These overlapping regions were evaluated by subjecting the entire genome of 5 chromosomes initially assembled by SMRT portal, plus the two manually curated supercontig chromosomes, to the SMRT portal resequencing protocol as a reference genome, along with manually evaluating these overlapping regions by evaluating the reads that spanned the overlapping regions. The resulting assembly was in 7 contigs, with each end of the contig terminating in a telomeric repeat sequence.

Gene prediction, functional annotation and protein classification

Genome annotation was performed using the MAKER (version 2.31.8) pipeline using three *ab inito* gene prediction methods: Augustus trained for *Fusarium graminearum*, and GeneMark-ES and SNAP self-trained on the *C. militaris* genome [22, 23, 63–65]. Protein data from *Cordyceps brongniartii*, *C. militaris*, *Cordyceps confragosa*, *Ophiocordyceps sinensis*, *Ophiocordyceps unilateralis* and *Tolypocladium ophioglossoides* were used as protein evidence in MAKER. EST from *Cordyceps militaris* were downloaded from GeneBank and used as EST evidence in MAKER. Repeat elements were identified using Repeat Masker using the Repbase Library 20150807 [66]. A final set of consensus gene predictions was chosen using Exonerate [67]. The final gene set from MAKER was subjected to additional evidence modeling using Funannotate (0.7.0) [24]. The gene models were functionally annotated using the BLAST component of the Blast2GO software package and searching against the NCBI nr protein database (accessed July 2017) with the best hit being selected [68]. Gene families were established using the Interpro database using BlastProDOM, HMMPIR, HMMPfam, SuperFamily, SignalPHMM, HMMPanther [69]. The BLAST hits were mapped to the Gene Ontology database and KEGG analysis was carried out [70]. Secondary metabolite genes and gene clusters were predicted using both AntiSMASH, fungal version 4.0.0 and SMURF (accessed September 2016)[36, 38].

Attempt at fruiting body production

Using the protocol outlined in Zheng et al. fruiting body production was attempted with *C. militaris* ATCC® 34164 [6].

Attempt at militarinone production

Using the protocol in Schmidt et al. militiarone A–D production was attempted [44]. Specifically, precultures of *C. militaris* were used to inoculate 150 mL of medium (2% glucose, 2% neopeptone, 0.5% glycine, 0.2% K_2HPO_4, $MgPO_4$-$7H_2O$) in still 500 mL Erlenmeyer flasks at 25 °C. After 20 days the broth was removed and the mycelia were collected and freeze dried, then extracted with methanol. This methanol extract was treated with water (1.5 mL for 10 g of extract) and then partitioned in a 1:1:1:1 mixture of ethyl acetate/methanol/hexane/1% acetic acid. The lower phase was collected, concentrated, and then analyzed for militarinone production via UPLC-MS. No mass peaks corresponding to ionized militarinone A–D or sodium or acetate adducts of those natural products were apparent.

Coverage and identity

Coverage and identity of *C. militaris* genes compared to genes in known biosynthetic clusters was determined using BLAST [61].

Additional files

> **Additional file 1: Figure S1.** Coverage across chromosomes. Coverage across chromosomes from SMRT analysis resequencing protocol assembly. (DOCX 1935 kb)
>
> **Additional file 2: Table S1.** Predicted gene clusters in *C. militaris*. Predicted gene clusters are labeled, putative natural product class and the predicted length of each enzyme that is part of the putative cluster is given. (DOCX 106 kb)

Abbreviations
HGAP: Hierarchical genome assembly process; MAT: Mating type; PacBio: Pacific Biosciences; PBcR: PacBio corrected reads

Acknowledgements
We would like to acknowledge Purapharm Co. Ltd. Canada. for financial support for this project and the staff at the Génome Québec Innovation Centre for the guidance with experimental design for genome sequencing.

Funding
Funding was in part provided from CIHR grant MOP-57684.

Declarations
Not applicable.

Authors' contributions

GJK collected and analyzed data and wrote manuscript. JRN provided guidance with experimental design, data analysis and critical revision of the article. JRN and GJK authors have approved the final version to be published.

Competing interests

The authors declare that they have no competing interests.

References

1. Lo H-C, Hsieh C, Lin F-Y, Hsu T-H. A systematic review of the mysterious caterpillar fungus Ophiocordyceps sinensis in Dong-ChongXiaCao (Dōng Chóng Xià Cǎo) and related bioactive ingredients. J Tradit Complement Med. 2013;3:16–32.
2. Pontoppidan M-B, Himaman W, Hywel-Jones NL, Boomsma JJ, Hughes DP. Graveyards on the move: the spatio-temporal distribution of dead Ophiocordyceps-infected ants. PLoS One. 2009;4:e4835.
3. Paterson RRM. Cordyceps – a traditional Chinese medicine and another fungal therapeutic biofactory? Phytochemistry. 2008;69:1469–95.
4. Yue K, Ye M, Zhou Z, Sun W, Lin X. The genus Cordyceps: a chemical and pharmacological review. J Pharm Pharmacol. 2013;65:474–93.
5. Xia E-H, Yang D-R, Jiang J-J, Zhang Q-J, Liu Y, Liu Y-L, et al. The caterpillar fungus, Ophiocordyceps sinensis, genome provides insights into highland adaptation of fungal pathogenicity. Sci Rep. 2017;7:1806.
6. Zheng P, Xia Y, Xiao G, Xiong C, Hu X, Zhang S, et al. Genome sequence of the insect pathogenic fungus Cordyceps militaris, a valued traditional Chinese medicine. Genome Biol. 2011;12:R116.
7. Bushley KE, Raja R, Jaiswal P, Cumbie JS, Nonogaki M, Boyd AE, et al. The genome of tolypocladium inflatum: evolution, organization, and expression of the cyclosporin biosynthetic gene cluster. PLoS Genet. Public Library of Science. 2013;9:e1003496.
8. Agrawal Y, Khatri I, Subramanian S, Shenoy BD. Genome sequence, comparative analysis, and evolutionary insights into Chitinases of Entomopathogenic fungus Hirsutella thompsonii. Genome Biol Evol. 2015;7:916–30.
9. Adachi K, Chiba K. FTY720 story. Its discovery and the following accelerated development of sphingosine 1-phosphate receptor agonists as immunomodulators based on reverse pharmacology. Perspect Med Chem. 2007;1:11–23.
10. Traber R, Hofmann H, Kobel H. Cyclosporins. New analogues by precursor directed biosynthesis. J Antibiot (Tokyo). 1989;42:591–7.
11. Hoffmeister D, Keller NP. Natural products of filamentous fungi: enzymes, genes, and their regulation. Nat Prod Rep. 2007;24:393–416.
12. de Bekker C, Ohm RA, Loreto RG, Sebastian A, Albert I, Merrow M, et al. Gene expression during zombie ant biting behavior reflects the complexity underlying fungal parasitic behavioral manipulation. BMC Genomics. 2015; 16:620.
13. Keller NP, Turner G, Bennett JW. Fungal secondary metabolism – from biochemistry to genomics. Nat Rev Microbiol. 2005;3:937–47.
14. Rhoads A, Au KF. PacBio sequencing and its applications. Genomics Proteomics Bioinformatics. 2015;13:278–89.
15. Yandell M, Ence D. A beginner's guide to eukaryotic genome annotation. Nat Rev Genet. 2012;13:329–42.
16. Berlin K, Koren S, Chin C-S, Drake JP, Landolin JM, Phillippy AM. Assembling large genomes with single-molecule sequencing and locality-sensitive hashing. Nat Biotechnol. 2015;33:623–30.
17. De Novo Assembly – Pacific Biosciences [Internet]. Available from: http://www.pacb.com/products-and-services/analytical-software/smrt-analysis/analysis-applications/de-novo-assembly/. Cited 8 Jun 2017.
18. Simão FA, Waterhouse RM, Ioannidis P, Kriventseva EV, Zdobnov EM. BUSCO: assessing genome assembly and annotation completeness with single-copy orthologs. Bioinformatics. 2015;31:3210–2.
19. Baker SE. Aspergillus niger genomics: Past, present and into the future. Medical Mycology. 2006;44(Suppl 1):S17–S21.
20. Galagan JE, Calvo SE, Borkovich KA, Selker EU, Read ND, Jaffe D, et al. The genome sequence of the filamentous fungus Neurospora crassa. Nature. Nature Publishing Group. 2003;422:859–68.

21. Stimberg N, Walz M, Schörgendorfer K, Kiick U. Electrophoretic karyotyping from Tolypocladium inflatum and six related strains allows differentiation of morphologically similar species. Appl Microbiol Biotechnol. 1992;37:485–9.
22. Campbell MS, Holt C, Moore B, Yandell M. Genome annotation and curation using MAKER and MAKER-P. Curr Protoc Bioinformatics. 2014; 48:4.11.1–39.
23. Cantarel BL, Korf I, Robb SMC, Parra G, Ross E, Moore B, et al. MAKER: an easy-to-use annotation pipeline designed for emerging model organism genomes. Genome Res. 2008;18:188–96.
24. Palmer J. Funannotate: pipeline for genome annotation [Internet]. Available from: https://github.com/nextgenusfs/funannotate/ . Cited 1 May 2017.
25. Cuomo CA, Guldener U, Xu J-R, Trail F, Turgeon BG, Di Pietro A, et al. The Fusarium graminearum genome reveals a link between localized polymorphism and pathogen specialization. Science. 2007;317:1400–2.
26. Quandt CA, Bushley KE, Spatafora JW. The genome of the truffle-parasite Tolypocladium ophioglossoides and the evolution of antifungal peptaibiotics. BMC Genomics. 2015;16:553.
27. Lennon MB, Suhadolnik RJ. Biosynthesis of 3′-deoxyadenosine by Cordyceps militaris. Mechanism of reduction. Biochim Biophys Acta. 1976;425:532–6.
28. Reichard P. Ribonucleotide reductases: substrate specificity by allostery. Biochem Biophys Res Commun. 2010;396:19–23.
29. Hu X, Zhang YJ, Xiao GH, Zheng P, Xia YL, Zhang XY, et al. Genome survey uncovers the secrets of sex and lifestyle in caterpillar fungus. Chin Sci Bull. 2013;58:2846–54.
30. Larriba E, Jaime MDLA, Carbonell-Caballero J, Conesa A, Dopazo J, Nislow C, et al. Sequencing and functional analysis of the genome of a nematode egg-parasitic fungus, Pochonia chlamydosporia. Fungal Genet Biol. Elsevier Inc. 2014;65:69–80.
31. Asai T, Yamamoto T, Oshima Y. Aromatic Polyketide production in Cordyceps indigotica, an Entomopathogenic fungus, induced by exposure to a Histone Deacetylase inhibitor. Org Lett. 2012;14:2006–9.
32. Grudniewska A, Hayashi S, Shimizu M, Kato M, Suenaga M, Imagawa H, et al. Opaliferin, a new Polyketide from cultures of Entomopathogenic fungus Cordyceps sp. NBRC 106954. Org Lett. 2014;16:4695–7.
33. Khaokhajorn P, Samipak S, Nithithanasilp S, Tanticharoen M, Amnuaykanjanasin A. Production and secretion of naphthoquinones is mediated by the MFS transporter MFS1 in the entomopathogenic fungus Ophiocordyceps sp. BCC1869. World J Microbiol Biotechnol. 2015;31:1543–54.
34. Varughese T, Rios N, Higginbotham S, Elizabeth Arnold A, Coley PD, Kursar TA, et al. Antifungal depsidone metabolites from Cordyceps dipterigena, an endophytic fungus antagonistic to the phytopathogen Gibberella fujikuroi. Tetrahedron Lett. 2012;53:1624–6.
35. Krasnoff SB, Reátegui RF, Wagenaar MM, Gloer JB, Gibson DM, Cicadapeptins I. II: new Aib-containing peptides from the Entomopathogenic fungus Cordyceps h eteropoda. J Nat Prod. 2005;68:50–5.
36. Weber T, Blin K, Duddela S, Krug D, Kim HU, Bruccoleri R, et al. antiSMASH 3. 0–a comprehensive resource for the genome mining of biosynthetic gene clusters. Nucleic Acids Res. 2015;43:W237–43.
37. Fedorova ND, Moktali V, Medema MH. Bioinformatics approaches and software for detection of secondary metabolic gene clusters. Methods Mol Biol. 2012;944:23–45.
38. Khaldi N, Seifuddin FT, Turner G, Haft D, Nierman WC, Wolfe KH, et al. SMURF: genomic mapping of fungal secondary metabolite clusters. Fungal Genet Biol. 2010;47:736–41.
39. Fisch KM. Biosynthesis of natural products by microbial iterative hybrid PKS–NRPS. RSC Adv. 2013;3:18228.
40. Chooi Y-H, Tang Y. Navigating the fungal polyketide chemical space: from genes to molecules. J Org Chem. 2012;77:9933–53.
41. Liu L, Zhang J, Chen C, Teng J, Wang C, Luo D. Structure and biosynthesis of fumosorinone, a new protein tyrosine phosphatase 1B inhibitor firstly isolated from the entomogenous fungus Isaria fumosorosea. Fungal Genet Biol. 2015;81:191–200.
42. Eley KL, Halo LM, Song Z, Powles H, Cox RJ, Bailey AM, et al. Biosynthesis of the 2-Pyridone tenellin in the insect pathogenic fungus Beauveria bassiana. Chembiochem. WILEY-VCH Verlag. 2007;8:289–97.
43. Heneghan MN, Yakasai AA, Williams K, Kadir KA, Wasil Z, Bakeer W, et al. The programming role of trans-acting enoyl reductases during the biosynthesis of highly reduced fungal polyketides. Chem Sci The Royal Society of Chemistry. 2011;2:972.

44. Schmidt K, Gu W, Stoyanova S, Schubert B, Li Z, Hamburger M. Militarinone A, a Neurotrophic Pyridone alkaloid from *Paecilomyces militaris*. Org Lett. 2002;4(2):197–9.

45. Schmidt K, Riese U, Li Z, Hamburger M. Novel Tetramic acids and Pyridone alkaloids, Militarinones B, C, and D, from the insect pathogenic fungus *Paecilomyces militaris*. J Nat Prod. 2003;66:378–83.

46. D-C O, Kauffman CA, Jensen PR, Fenical W. Induced production of Emericellamides a and B from the marine-derived fungus *Emericella* sp. in competing co-culture. J Nat Prod. 2007;70:515–20.

47. Chiang Y-M, Szewczyk E, Nayak T, Davidson AD, Sanchez JF, Lo H-C, et al. Molecular genetic Mining of the Aspergillus Secondary Metabolome: discovery of the Emericellamide biosynthetic pathway. Chem Biol. 2008;15:527–32.

48. Lukassen M, Saei W, Sondergaard T, Tamminen A, Kumar A, Kempken F, et al. Identification of the Scopularide biosynthetic gene cluster in Scopulariopsis brevicaulis. Mar Drugs. 2015;13:4331–43.

49. Niheik K, Itoh H, Hashimoto K, Miyairi K, Okuno T. Antifungal Cyclodepsipeptides, W493 a and B, from *Fusarium* sp.: isolation and structural determination. Biosci Biotechnol Biochem. 1998;62:858–63.

50. Sørensen JL, Sondergaard TE, Covarelli L, Fuertes PR, Hansen FT, Frandsen RJN, et al. Identification of the biosynthetic gene clusters for the Lipopeptides Fusaristatin a and W493 B in *Fusarium graminearum* and *F. pseudograminearum*. J Nat Prod. 2014;77:2619–25.

51. Vesonder RF, Tjarks LW, Rohwedder WK, Burmeister HR, Laugal JA. Equisetin, an antibiotic from Fusarium equiseti NRRL 5537, identified as a derivative of N-methyl-2,4-pyrollidone. J Antibiot (Tokyo). 1979;32:759–61.

52. Campbell CD, Vederas JC. Biosynthesis of lovastatin and related metabolites formed by fungal iterative PKS enzymes. Biopolymers. 2010;93:755–63.

53. Sims JW, Fillmore JP, Warner DD, Schmidt EW. Equisetin biosynthesis in Fusarium heterosporum. Chem Commun (Camb). 2005:186–8.

54. Quast C, Pruesse E, Yilmaz P, Gerken J, Schweer T, Yarza P, et al. The SILVA ribosomal RNA gene database project: improved data processing and web-based tools. Nucleic Acids Res. 2013;41:D590–D6.

55. Thompson JD, Gibson TJ, Higgins DG. Multiple sequence alignment using ClustalW and ClustalX. Curr Protoc Bioinforma. 2002. doi: 10.1002/0471250953.bi0203s00.

56. Saitou N, Nei M. The neighbor-joining method: a new method for reconstructing phylogenetic trees. Mol Biol Evol. 1987;4:406–25.

57. Tamura K, Nei M, Kumar S. Prospects for inferring very large phylogenies by using the neighbor-joining method. Proc Natl Acad Sci U S A. 2004;101:11030–5.

58. Kumar S, Stecher G, Tamura K. MEGA7: molecular evolutionary genetics analysis version 7.0 for bigger datasets. Mol Biol Evol. 2016;33:1870–4.

59. Toju H, Tanabe AS, Yamamoto S, Sato H. High-coverage ITS primers for the DNA-based identification of ascomycetes and basidiomycetes in environmental samples. PLoS One. Public Library of Science. 2012;7:e40863.

60. Schoch CL, Seifert KA, Huhndorf S, Robert V, Spouge JL, Levesque CA, et al. Nuclear ribosomal internal transcribed spacer (ITS) region as a universal DNA barcode marker for fungi. Proc Natl Acad Sci. 2012;109:6241–6.

61. Altschul SF, Gish W, Miller W, Myers EW, Lipman DJ. Basic local alignment search tool. J Mol Biol. 1990;215:403–10.

62. Chin C-S, Alexander DH, Marks P, Klammer AA, Drake J, Heiner C, et al. Nonhybrid, finished microbial genome assemblies from long-read SMRT sequencing data. Nat Methods. 2013;10:563–9.

63. Stanke M, Keller O, Gunduz I, Hayes A, Waack S, Morgenstern B. AUGUSTUS: ab initio prediction of alternative transcripts. Nucleic Acids Res. 2006;34:W435–9.

64. Borodovsky M, Lomsadze A. Eukaryotic gene prediction using GeneMark. hmm-E and GeneMark-ES. Curr Protoc Bioinforma. 2011. doi: 10.1002/0471250953.bi0406s35.

65. Korf I. Gene finding in novel genomes. BMC Bioinformatics. BioMed Central. 2004;5:59.

66. Smit, A. F, Hubley, R., Green P. RepeatMasker Home Page [Internet]. Repeat Masker 3.0. Available from: http://www.repeatmasker.org/. Cited 30 Sep 2015.

67. Slater GSC, Birney E. Automated generation of heuristics for biological sequence comparison. BMC Bioinformatics. 2005;6:31.

68. O'Leary NA, Wright MW, Brister JR, Ciufo S, Haddad D, McVeigh R, et al. Reference sequence (RefSeq) database at NCBI: current status, taxonomic expansion, and functional annotation. Nucleic Acids Res. 2016;44:D733–45.

69. Hunter S, Apweiler R, Attwood TK, Bairoch A, Bateman A, Binns D, et al. InterPro: the integrative protein signature database. Nucleic Acids Res. Oxford University Press. 2009;37:D211–5.

70. Kanehisa M, Furumichi M, Tanabe M, Sato Y, Morishima K. KEGG: new perspectives on genomes, pathways, diseases and drugs. Nucleic Acids Res. Oxford University Press. 2017;45:D353–61.

9

Global gene expression reveals stress-responsive genes in *Aspergillus fumigatus* mycelia

Hiroki Takahashi[1,2]*, Yoko Kusuya[1], Daisuke Hagiwara[1], Azusa Takahashi-Nakaguchi[1], Kanae Sakai[1] and Tohru Gonoi[1]

Abstract

Background: *Aspergillus fumigatus* is a human fungal pathogen that causes aspergillosis in immunocompromised hosts. *A. fumigatus* is believed to be exposed to diverse environmental stresses in the host cells. The adaptation mechanisms are critical for infections in human bodies. Transcriptional networks in response to diverse environmental challenges remain to be elucidated. To gain insights into the adaptation to environmental stresses in *A. fumigatus* mycelia, we conducted time series transcriptome analyses.

Results: With the aid of RNA-seq, we explored the global gene expression profiles of mycelia in *A. fumigatus* upon exposure to diverse environmental changes, including heat, superoxide, and osmotic stresses. From the perspective of global transcriptomes, transient responses to superoxide and osmotic stresses were observed while responses to heat stresses were gradual. We identified the stress-responsive genes for particular stresses, and the 266 genes whose expression levels drastically fluctuated upon exposure to all tested stresses. Among these, the 77 environmental stress response genes are conserved in *S. cerevisiae*, suggesting that these genes might be more general prerequisites for adaptation to environmental stresses. Finally, we revealed the strong correlations among expression profiles of genes related to 'rRNA processing'.

Conclusions: The time series transcriptome analysis revealed the stress-responsive genes underlying the adaptation mechanisms in *A. fumigatus* mycelia. These results will shed light on the regulatory networks underpinning the adaptation of the filamentous fungi.

Keywords: Transcriptome, RNA-seq, *Aspergillus fumigatus*, Environmental stress, ESR

Background

Billions of people are infected with fungi every year in the world [1]. Filamentous fungus *Aspergillus fumigatus* is a ubiquitous fungus commonly found in soil, but is also reported as a major cause of invasive fungal aspergillosis infections in humans, especially in patients with compromised or suppressed host immunity [2–4].

Inhalation of airborne conidia is responsible for human infection [3, 4]. When *A. fumigatus* establishes an infection in the human lung, the mycelia must respond to highly variable conditions, which might impose stress on the fungal pathogen [4, 5]. Thus, the adaptation mechanism in mycelia plays an important role in terms of pathogenicity of *A. fumigatus*. Among environmental conditions, tolerance to higher temperatures is a critical trait for mammalian infections [6]. It is conceivable that oxidative stress needs to be overcome, as reactive oxygen species produced by host cells, such as neutrophils, damage the fungal pathogen [7–9]. Osmotic stress adaptation and sensitivity are also supposed to be important for conidial germination, growth and virulence [10–14]. It has been reported that MAP kinase SakA plays a pivotal role in responses and adaptation to osmotic stress [10].

* Correspondence: hiroki.takahashi@chiba-u.jp
[1]Medical Mycology Research Center, Chiba University, 1-8-1 Inohana, Chuo-ku, Chiba 260-8673, Japan
[2]Molecular Chirality Research Center, Chiba University, 1-33 Yayoi-cho, Inage-ku, Chiba 263-8522, Japan

Transcriptome analyses are particularly useful for obtaining a deeper understanding of gene regulation in organisms from bacteria to mammals. Microarray technology has been applied to A. fumigatus since whole genome sequencing was initiated in 2005 [6]. Nierman et al. (2005) have conducted time series microarray analyses upon exposure to heat shock of both 37 °C and 48 °C from 30 °C [6]. Do et al. (2009) have addressed the transcriptional networks by state space models using microarray data for more than 2000 genes studied by Nierman et al. (2005), and revealed the expression profiles of heat shock proteins, such as *hsp70* and *hsp30*, in response to heat stress [15]. Putative targets of *hsf1* were identified by integrative analysis of proteome and transcriptome [16].

In yeast, genome-wide gene expression analyses under various changes in the extracellular environment have been investigated by microarray [17–19]. Notably, Gasch et al. (2000) have proposed the environmental stress response (ESR) genes that responded to almost all the stressful conditions [18]. The ESR is a general adaptive response to suboptimal environments [18]. Indeed, Emri et al. (2015) identified 116 ESR genes in response to five different oxidative stress conditions in A. nidulans [20]. Some genes of A. fumigatus related to heat stress are orthologs of genes of S. cerevisiae [6], indicating that the stress response in A. fumigatus might be partially related to the response of the ESR genes.

It is noted that RNA-seq technology appears to be the most powerful tool for transcriptome analysis, and has great potential to investigate the transcriptome in A. fumigatus, e.g. developmental stages [21–27]. It is demonstrated that RNA-seq data were better correlated with proteome data than microarray data in A. fumigatus [24]. So far, no studies have investigated the integrative stress responses in A. fumigatus, such as seeking the ESR genes. In the present study, we explored the global transcriptome responses by RNA-seq in A. fumigatus mycelia upon exposure to heat, superoxide, and osmotic stresses.

Results

Time series transcriptome analyses in response to diverse environmental stresses

To determine the genes that respond to several stress conditions, we performed time series transcriptome analyses of mycelia of A. fumigatus Af293 using RNA-seq under four environmental conditions: heat stress from 30 °C to 37 °C (hereafter, HS1), heat stress from 30 °C to 48 °C (HS2), superoxide stress by adding menadione (SS), and osmotic stress by adding sorbitol (OS). Following each stress, RNA samples from the mycelia were

harvested at six time points: 0, 15, 30, 60, 120, and 180 min (Table 1). A total of 277,417,888 sequence reads were obtained by a MiSeq for HS1, and a HiSeq for HS2, SS, and OS (Additional file 1: Table S1, Additional file 2: Table S2).

FPKM values of 9840 genes were calculated as described in the Methods (Additional file 3: Table S3). To evaluate the expression levels in unshocked mycelia (0 min), we calculated Pearson correlation coefficients using 6932 genes that showed FPKM values higher than the median value in at least one condition (Additional file 4: Figure S1). There were high correlations between 0 min data of HS1, HS2, SS, and OS, e.g. $r = 0.81$ for HS1 and HS2, and $r = 0.98$ for SS and OS. Next, to elucidate transcriptome changes in the treated mycelia, the log2-transformed ratio values to the expression levels in unshocked mycelia were calculated, and used throughout this study. To illustrate global trends of transcriptome in response to HS1, HS2, SS, and OS, we performed principal component analysis (PCA) using 6932 genes (Additional file 5: Figure S2). The transcriptome of HS2 appeared to significantly differ from other three stresses. We observed similar behavior in the transcriptome of SS and OS. In addition, it appeared that the gene expression in response to SS and OS was transiently changed.

Furthermore, we conducted K-means clustering analysis (Fig. 1). The relatively low expressed genes were excluded, and the genes with FPKM values higher than the median value of at least one time point for each stress were used: 6092 genes in HS1, 6399 in HS2, 6001 in SS, and 5738 OS. The expression of 184 genes enriched in the 'oxidation-reduction process' was gradually up-regulated throughout HS1, and maximized at 180 min (Fig. 1, cluster 1). That of the 620 genes enriched in 'translation' was down-regulated (cluster 5). Meanwhile, the expression of the 413 (cluster 1) enriched in 'rRNA processing' gradually increased throughout HS2, and that of the 678 genes (cluster 3) enriched in 'transmembrane transport' gradually increased until 60 min upon exposure to HS2. The 735 genes (cluster 5) enriched in the 'oxidation-reduction process' decreased during exposure to HS2.

Table 1 Experimental conditions

Condition	Medium	Sampling point
Heat stress from 30 °C to 37 °C (HS1)	YG	0 (unshocked), 15, 30, 60, 120, and 180 min
Heat stress from 30 °C to 48 °C (HS2)		
Superoxide stress by menadione (SS)	AMM	
Osmotic stress by sorbitol (OS)		

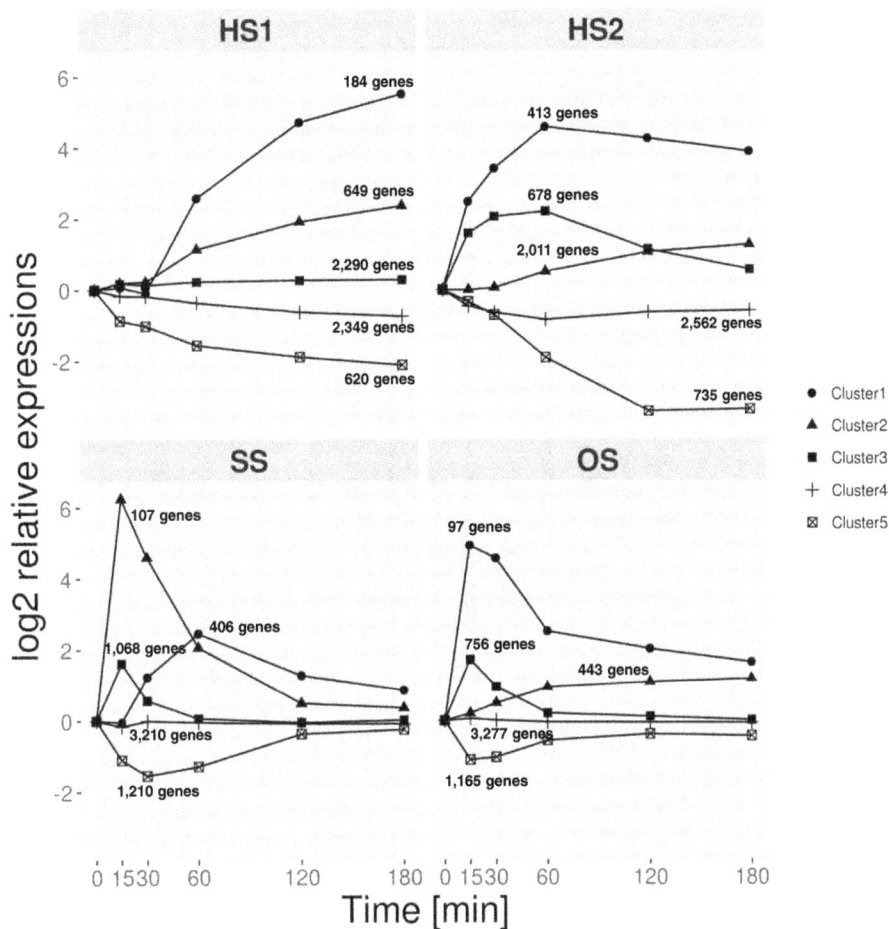

Fig. 1 K-means clustering of transcription profiles using five centroids. The average log2-transformed ratio values were plotted. Among 6092 genes in HS1, clusters 1, 2, 3, 4, and 5 consist of 184 (3.0%), 649 (10.7%), 2290 (37.6%), 2349 (38.6%), and 620 (10.2%) genes, respectively. Among 6399 genes in HS2, clusters 1, 2, 3, 4, and 5 consist of 413 (6.5%), 2011 (31.4%), 678 (10.6%), 2562 (40.0%), and 735 (11.5%) genes. Among 6001 genes in SS, clusters 1, 2, 3, 4, and 5 consist of 406 (6.8%), 107 (1.8%), 1068 (17.8%), 3210 (53.5%), and 1210 (20.2%) genes. Among 5738 genes in OS, clusters 1, 2, 3, 4, and 5 consist of 97 (1.7%), 443 (7.7%), 756 (13.2%), 3277 (57.1%), and 1165 (20.3%) genes

In contrast to heat stresses, the transient dynamics in response to SS and OS were observed. The expression of the 107 genes (cluster 2) enriched in the 'oxidation-reduction process' largely increased in SS at 15 min, and then reverted back to a similar level for the unshocked condition. The expression levels of catalase genes, cat1 (Afu3g02270) and cat2 (Afu8g01670), in cluster 2 were 30.1- and 241.8-fold up-regulated at 15 min. The 1210 genes (cluster 5) enriched in 'translation' transiently decreased at 30 min. Similarly, the 97 genes (cluster 1) transiently responded to OS. The genes related to the high-osmolarity glycerol response (HOG) pathway, e.g. dprA (Afu4g00860), dprB (Afu6g12180), dprC (Afu7g04520), ptcD (Afu5g13740), and atfD (Afu6g12150), were observed in cluster 1 [12, 28–30]. Consistent with previous reports, the expression levels of dprA and dprB were 99.4- and 71.3-fold up-regulated at 15 min [12]. In addition, we observed that the expression of MAPK sakA (Afu1g12940) in

cluster 2 as a major component of HOG pathway was 6.6-fold up-regulated. Taken together, two types of adaptation in response to suboptimal conditions were observed, namely gradual shift in HS1 and HS2, and transient shift in SS and OS.

Identification of the stress-responsive genes in response to environmental changes

We sought stress-responsive genes and identified 1598, 3383, 1735, and 1085 genes as the stress-responsive genes whose expression levels were differentially expressed in HS1, HS2, SS, and OS, respectively (Fig. 2). The thresholds used for judging whether genes responded to stress were >3-fold or <1/3-fold. In HS1, the maximum numbers of differentially expressed genes were measured at 180 min; that is, 730 up- and 553 down-regulated genes. Likewise, in HS2, 1152 and 1063 as up- and down-regulated genes, respectively, were

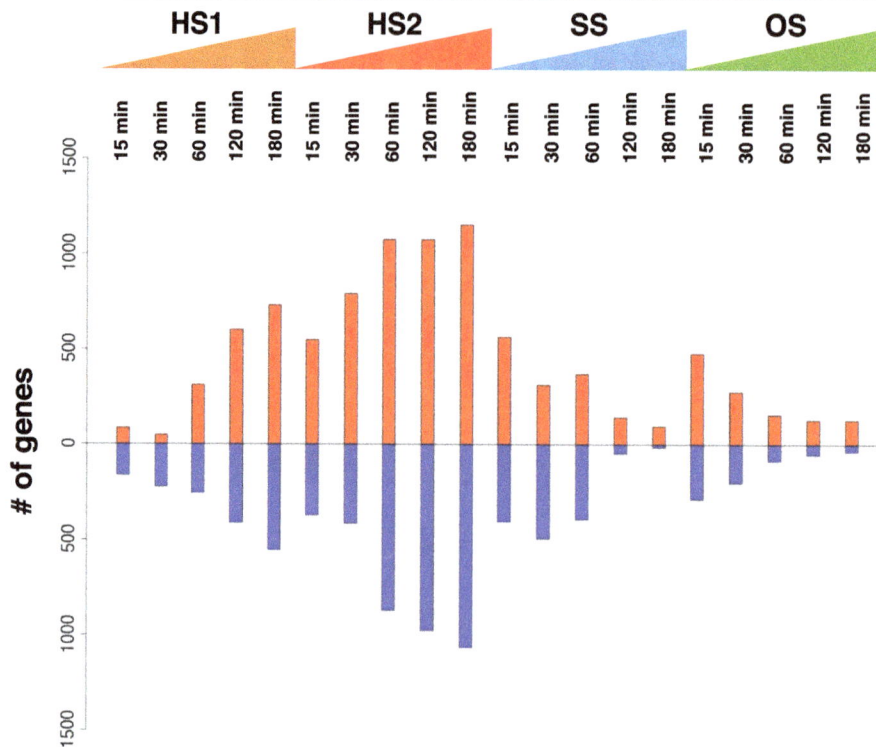

Fig. 2 Bar plots of the numbers of genes with altered expression at each time point. Red and blue indicate the numbers of up- and down-regulated genes, respectively

measured at 180 min. In SS, 564 up- and 491 down-regulated genes were detected at 15 and 30 min, respectively. Furthermore, in OS, 478 up- and 284 down-regulated genes were observed at 15 min. Consistent with the results by K-means analysis and PCA (Fig. 1, Additional file 5: Figure S2), the gene expression upon exposure to heat stresses was gradually changed, while transient changes were observed in SS and OS.

Comparison with microarray data of heat stress

We identified 1598 (HS1) and 3383 (HS2) genes as described above. Do et al. (2009) have reported 726 and 2200 genes for heat shock of 37 °C and 48 °C based on microarray data, respectively [6, 15]. Comparing identified genes, 268 and 1044 genes were observed in both experiments of 37 °C and 48 °C, respectively, and almost all the expression profiles of these genes appear to be consistent (Additional file 6: Figure S3). According to Do et al. (2009), six heat shock genes were identified as the hub nodes by network analysis, namely *hsp70* (Afu1g07440), *hsp78* (Afu1g11180), *hsp30* (Afu3g14540), *hsp90* (Afu5g04170), Afu6g06470, and *hscA* (Afu8g03930). In response to heat shock of 37 °C, the expression levels of *hsp78*, *hsp30*, *hsp90*, and Afu6g06470 were up-regulated at 15 min in our HS1 data, although those were up-regulated throughout the time series in Do et al. (2009) (Fig. 3a). The expression profile of *hscA*

consistent across both experiments. In response to heat shock of 48 °C, the expression profiles except for *hscA* were similar, while that of *hscA* exhibited the opposite trend (Fig. 3b).

Acute responses upon exposure to superoxide and osmotic stresses

Among the stress-responsive genes, 892 (51%) in SS and 728 (67%) in OS genes were up/down-regulated at one or both of 15 and 30 min, while 141 (9%) in HS1 and 332 (10%) in HS2 were observed. This indicated that the majority of the genes respond to SS and OS in a rapid and transient manner. As kinase and transcription factor (TF) play an important role in sensing superoxide and osmotic stresses [15, 31–33], we sought such regulator genes from the set of genes that responded to SS and OS. A total of 32 and 24 regulator genes, including kinase and transcription factor, were up-regulated at 15 min in SS and OS, respectively, while 14 and 20 genes were down-regulated at 15 min (Fig. 4a, b). Consistent with the observation that TF *yap1* (Afu6g09930) responded to H_2O_2 [9], we observed that the expression of *yap1* was 5.7-fold up-regulated at 15 min in response to SS. The expression of histidine kinase *phkB* (Afu3g12530) was detected as transiently responsive genes in OS. The expression of *sakA* was not changed in response to SS, although it was up-regulated in response

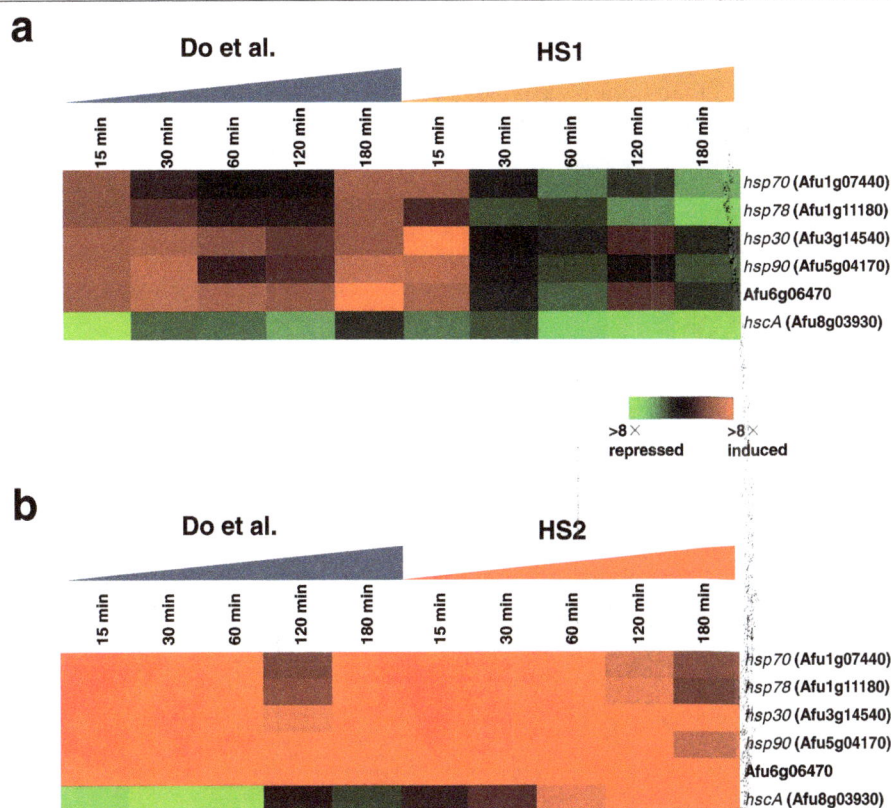

Fig. 3 Comparison of the gene expression of six heat shock proteins between microarray and RNA-seq data. **a** Expression profiles of six heat shock proteins in response to HS1. **b** Expression profiles of six heat shock proteins in response to HS2. Color scale is indicated

to OS, concordant with a previous study [34]. The expression of MAPKKK *sskB* (Afu1g10940) was 5.3-fold changed at 15 min. Interestingly, the expression of histidine kinase *tcsB* was transiently up-regulated at 15 min in response to HS1 and SS, while it was down-regulated at 15 min in response to OS. This is consistent with the report that *tcsB* is negatively regulated by osmotic stress in *A. nidulans* and *A. fumigatus* [31, 35]. The expression of *tcsB* was up-regulated at 15, 30, and 60 min in response to HS2. In addition, we observed the up-regulation of Atf family genes including *atfB* (Afu5g12960), *atfC* (Afu1g17360), and *atfD* under the OS condition, while the expression of *atfA* (Afu3g11330) was 2.6-fold changed at 15 min [28, 35]. Comparing the genes observed in SS and OS, three (Afu1g01560, Afu2g03490, and Afu5g08480) and seven (Afu1g05150, Afu1g10760, Afu2g01520, Afu2g10770, *azf1* (Afu6g05160), Afu6g09820, and Afu6g11110) uncharacterized genes were up- and down-regulated at 15 min under both SS and OS conditions, respectively.

Characterization of the stress-responsive genes

Next, to investigate which gene functions are associated with the individual stresses, Gene Ontology (GO) enrichment analysis was conducted. In total, 56, 67, and 89 GO terms for the categories Cellular Component,

Molecular Function, and Biological Process, respectively, were tested. Among 89 GO terms associated with Biological Process, 15 GO terms were overrepresented for up-regulated genes (Fig. 5a), while 24 GO terms were overrepresented for down-regulated genes (Fig. 5b). In HS2, GO terms related to ribosome, such as 'rRNA processing', 'endonucleolytic cleavage in ITS1 to separate SSU-rRNA from 5.8S rRNA and LSU-rRNA from tricistronic rRNA transcript (SSU-rRNA, 5.8S rRNA, LSU-rRNA)', 'maturation of SSU-rRNA from tricistronic rRNA transcript (SSU-rRNA, 5.8S rRNA, LSU-rRNA)', and 'ribosomal large subunit assembly', were up-regulated throughout the time series. The expression of genes annotated as 'protein folding' was up-regulated at 15 min in both HS1 and HS2. In SS, the expression of genes annotated as 'cell redox homeostasis' was up-regulated at 15 min. A total of 203 genes annotated as 'oxidation-reduction process' were up-regulated upon exposure to HS1, SS, and OS. Among them, 15 genes were up-regulated under all HS1, SS, and OS conditions, indicating that these are essential for oxidation reduction in *A. fumigatus*. Intriguingly, 'ergosterol biosynthetic process' was overrepresented in response to SS and OS (Figs. 5b, 6). The expression levels of 15 genes annotated as 'ergosterol biosynthetic process', including *erg6*, *erg2*, *erg3A*,

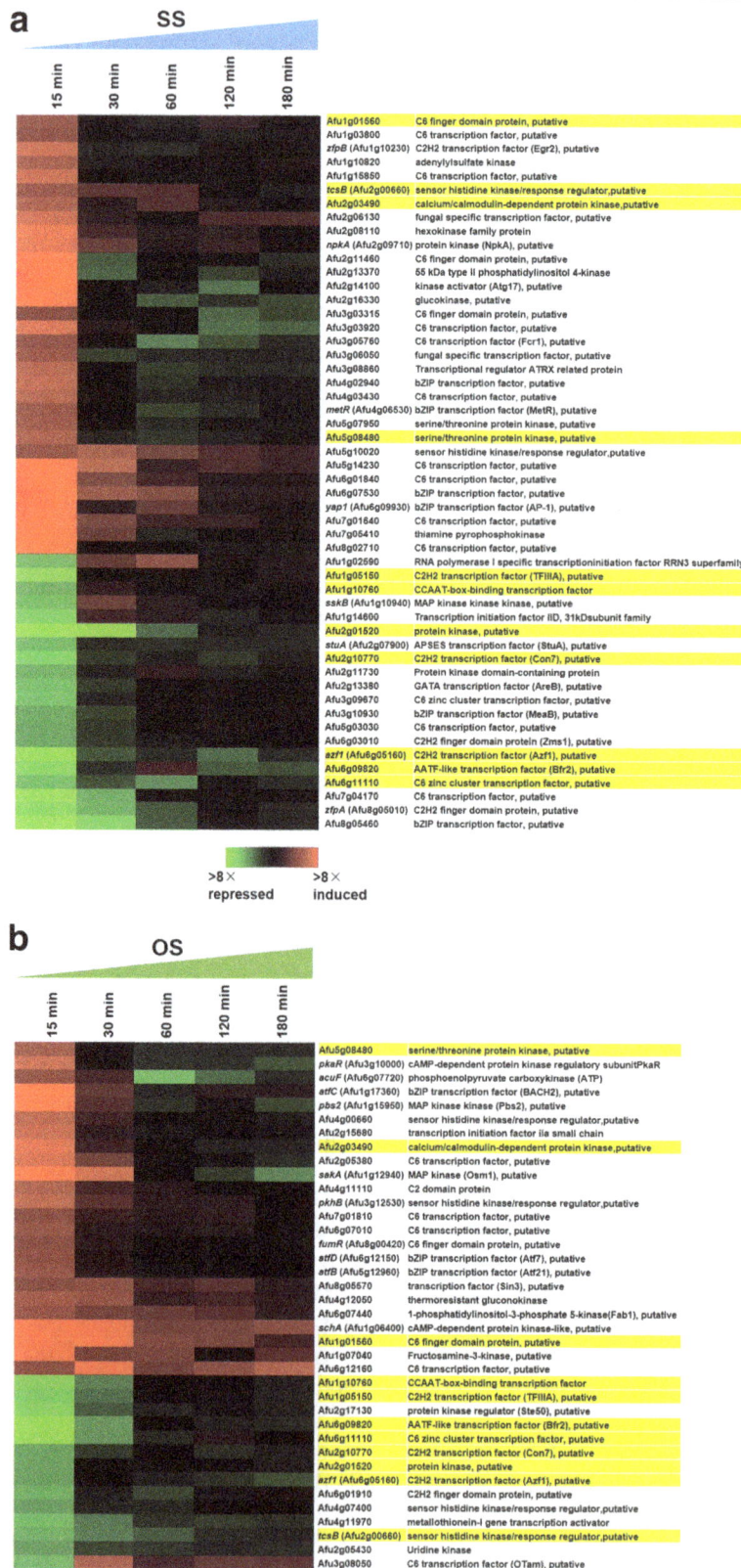

a SS

	15 min	30 min	60 min	120 min	180 min		

Afu1g01560 — C6 finger domain protein, putative
Afu1g03800 — C6 transcription factor, putative
zfpB (Afu1g10230) — C2H2 transcription factor (Egr2), putative
Afu1g10820 — adenylylsulfate kinase
Afu1g15850 — C6 transcription factor, putative
tcsB (Afu2g00660) — sensor histidine kinase/response regulator,putative
Afu2g03490 — calcium/calmodulin-dependent protein kinase,putative
Afu2g06130 — fungal specific transcription factor, putative
Afu2g08110 — hexokinase family protein
npkA (Afu2g09710) — protein kinase (NpkA), putative
Afu2g11460 — C6 finger domain protein, putative
Afu2g13370 — 55 kDa type II phosphatidylinositol 4-kinase
Afu2g14100 — kinase activator (Atg17), putative
Afu2g16330 — glucokinase, putative
Afu3g03315 — C6 finger domain protein, putative
Afu3g03920 — C6 transcription factor, putative
Afu3g05760 — C6 transcription factor (Fcr1), putative
Afu3g06050 — fungal specific transcription factor
Afu3g08860 — Transcriptional regulator ATRX related protein
Afu4g02940 — bZIP transcription factor, putative
Afu4g03430 — C6 transcription factor, putative
metR (Afu4g06530) — bZIP transcription factor (MetR), putative
Afu5g07950 — serine/threonine protein kinase, putative
Afu5g08480 — serine/threonine protein kinase, putative
Afu5g10020 — sensor histidine kinase/response regulator,putative
Afu5g14230 — C6 transcription factor, putative
Afu6g01840 — C6 transcription factor, putative
Afu6g07530 — bZIP transcription factor, putative
yap1 (Afu6g09930) — bZIP transcription factor (AP-1), putative
Afu7g01640 — C6 transcription factor, putative
Afu7g05410 — thiamine pyrophosphokinase
Afu8g02710 — C6 transcription factor, putative
Afu1g02690 — RNA polymerase I specific transcriptioninitiation factor RRN3 superfamily
Afu1g05150 — C2H2 transcription factor (TFIIIA), putative
Afu1g10760 — CCAAT-box-binding transcription factor
sskB (Afu1g10940) — MAP kinase kinase kinase, putative
Afu1g14600 — Transcription initiation factor IID, 31kDsubunit family
Afu2g01520 — protein kinase, putative
stuA (Afu2g07900) — APSES transcription factor (StuA), putative
Afu2g10770 — C2H2 transcription factor (Con7), putative
Afu2g11730 — Protein kinase domain-containing protein
Afu2g13380 — GATA transcription factor (AreB), putative
Afu3g09670 — C6 zinc cluster transcription factor, putative
Afu3g10930 — bZIP transcription factor (MeaB), putative
Afu5g03030 — C6 transcription factor, putative
Afu6g03010 — C2H2 finger domain protein (Zms1), putative
azf1 (Afu6g05160) — C2H2 transcription factor (Azf1), putative
Afu6g09820 — AATF-like transcription factor (Bfr2), putative
Afu6g11110 — C6 zinc cluster transcription factor, putative
Afu7g04170 — C6 transcription factor, putative
zfpA (Afu8g05010) — C2H2 finger domain protein, putative
Afu8g05460 — bZIP transcription factor, putative

>8× repressed — >8× induced

b OS

	15 min	30 min	60 min	120 min	180 min		

Afu5g08480 — serine/threonine protein kinase, putative
pkaR (Afu3g10000) — cAMP-dependent protein kinase regulatory subunitPkaR
acuF (Afu6g07720) — phosphoenolpyruvate carboxykinase (ATP)
atfC (Afu1g17360) — bZIP transcription factor (BACH2), putative
pbs2 (Afu1g15950) — MAP kinase kinase (Pbs2), putative
Afu4g00660 — sensor histidine kinase/response regulator,putative
Afu2g15680 — transcription initiation factor iia small chain
Afu2g03490 — calcium/calmodulin-dependent protein kinase,putative
Afu2g05380 — C6 transcription factor, putative
sakA (Afu1g12940) — MAP kinase (Osm1), putative
Afu4g11110 — C2 domain protein
pkhB (Afu3g12530) — sensor histidine kinase/response regulator,putative
Afu7g01810 — C6 transcription factor, putative
Afu6g07010 — C6 transcription factor, putative
fumR (Afu8g00420) — C6 finger domain protein, putative
atfD (Afu6g12150) — bZIP transcription factor (Atf7), putative
atfB (Afu5g12960) — bZIP transcription factor (Atf21), putative
Afu8g05570 — transcription factor (Sin3), putative
Afu4g12050 — thermoresistant gluconokinase
Afu6g07440 — 1-phosphatidylinositol-3-phosphate 5-kinase(Fab1), putative
schA (Afu1g06400) — cAMP-dependent protein kinase-like, putative
Afu1g01560 — C6 finger domain protein, putative
Afu1g07040 — Fructosamine-3-kinase, putative
Afu6g12160 — C6 transcription factor, putative
Afu1g10760 — CCAAT-box-binding transcription factor
Afu1g05150 — C2H2 transcription factor (TFIIIA), putative
Afu2g17130 — protein kinase regulator (Ste50), putative
Afu6g09820 — AATF-like transcription factor (Bfr2), putative
Afu6g11110 — C6 zinc cluster transcription factor, putative
Afu2g10770 — C2H2 transcription factor (Con7), putative
Afu2g01520 — protein kinase, putative
azf1 (Afu6g05160) — C2H2 transcription factor (Azf1), putative
Afu6g01910 — C2H2 finger domain protein, putative
Afu4g07400 — sensor histidine kinase/response regulator,putative
Afu4g11970 — metallothionein-I gene transcription activator
tcsB (Afu2g00660) — sensor histidine kinase/response regulator,putative
Afu2g05430 — Uridine kinase
Afu3g08050 — C6 transcription factor (OTam), putative

Fig. 4 (See legend on next page.)

(See figure on previous page.)
Fig. 4 Heat map of regulator genes upon exposure to SS and OS. **a** The expression profiles of 46 regulator genes. Thirty-two and 14 genes were up- and down-regulated at 15 min in SS, respectively. **b** The expression profiles of 44 regulator genes. Twenty-four and 20 genes were up- and down-regulated at 15 min in OS, respectively. The yellow boxes indicate 11 genes with altered expression at 15 min under both SS and OS conditions

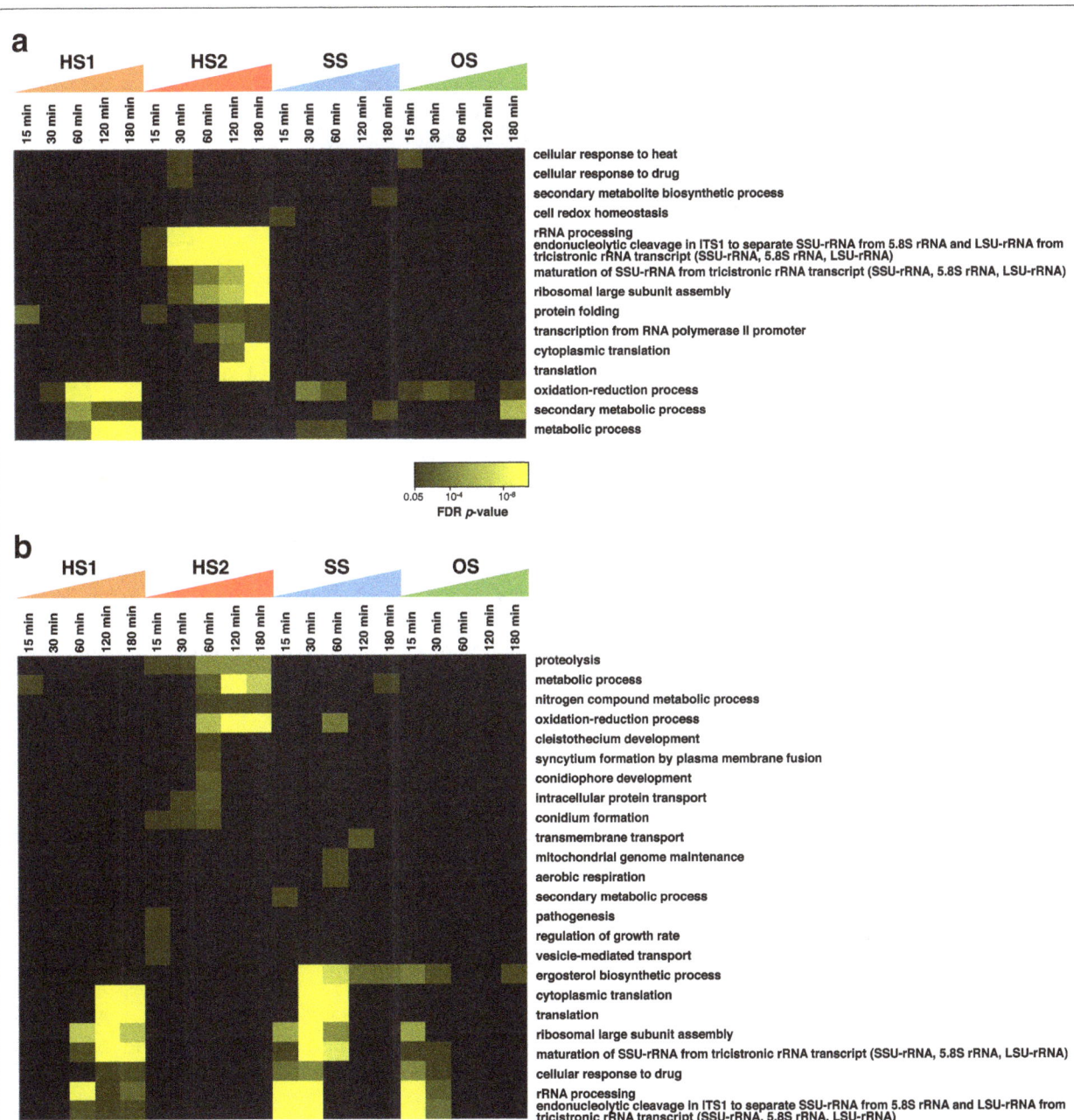

Fig. 5 Heat map of overrepresented GO terms based on FDR *p*-value. **a** 15 GO terms for up-regulated genes. **b** 24 GO terms for down-regulated genes. Color scale is indicated

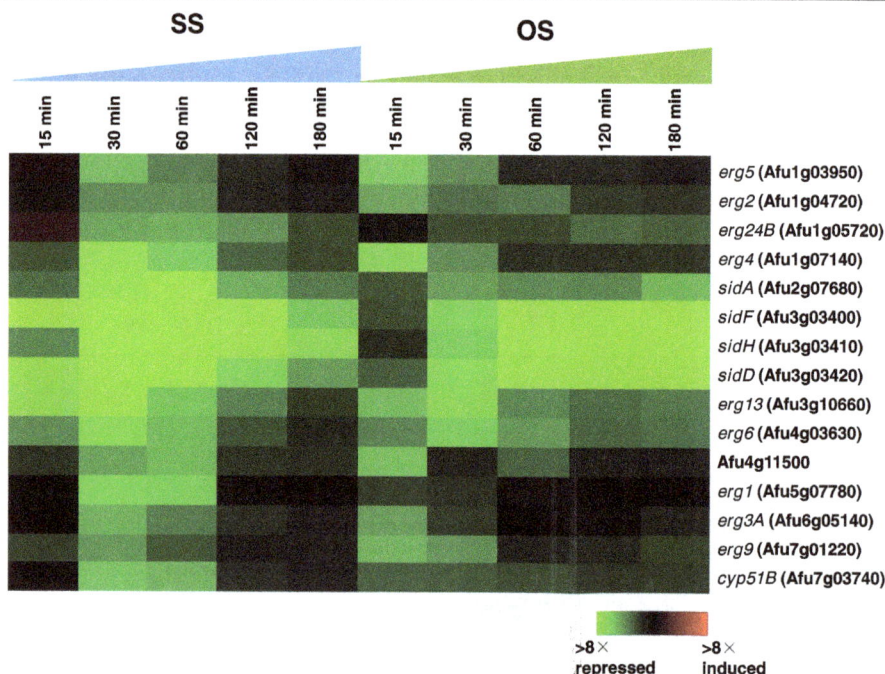

Fig. 6 Expression profiles of 15 genes annotated as 'ergosterol biosynthetic process' in response to SS and OS. Color scale is indicated

erg5, and *erg4*, were down-regulated under one or both of the SS and OS conditions (Fig. 6).

Identification of the stress-responsive genes upon exposure to all tested stresses

We sought the genes with altered expression under all four conditions (Fig. 7) [36]. Among 4647 genes, 266 were differentially expressed upon exposure to all four tested conditions. These genes were classified into eight clusters (Fig. 8, Additional file 7: Table S4).

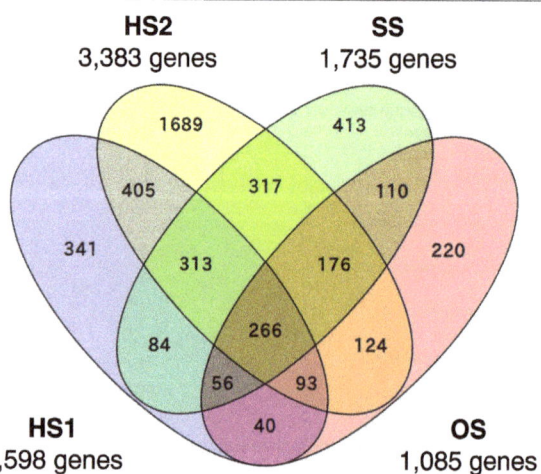

Fig. 7 Venn diagram of the stress-responsive genes in HS1 (1598 genes), HS2 (3383 genes), SS (1735 genes), and OS (1085 genes). A total of 266 genes responded to all four conditions. This figure was drawn using Venny 2.0 [36]

According to the functional annotations of genes in AspGD, 29 and 50 genes were annotated with 'conserved hypothetical protein' and 'hypothetical protein', respectively. Notably, the expression levels of 116 genes related to 'rRNA processing' were up-regulated in response to HS2, while they were transiently down-regulated in response to SS and OS (cluster 6). The expression levels of 40 genes related to 'metabolic process' were up-regulated under HS1, SS, and OS conditions, while they were down-regulated under the HS2 condition (cluster 1). Furthermore, we observed nine and 14 genes whose expression levels were up-regulated and down-regulated under all four conditions, respectively. The expression of Afu5g09910 encoding a putative *p*-nitroreductase-family protein was drastically activated upon exposure to all tested conditions. In particular, the expression was 709-fold changed at 15 min in SS. The expression of *msdS* (Afu1g14560) was transiently down-regulated upon exposure to HS1, SS, and OS. Especially, the expression was decreased to 1/22-fold at 120 min under the HS2 condition. *msdS* encodes a class I 1,2-α-mannosidase, and the cell wall integrity of the *msdS* mutant was slightly affected at a higher temperature [37]. In addition, five TFs, Afu1g04140 (cluster 1), Afu8g00420 (*fumR*) (cluster 1), Afu1g03800 (cluster 4), Afu3g11170 (cluster 4), and Afu6g12160 (cluster 7), were identified as stress-responsive genes.

Comparison with yeast ESR genes

In both *S. cerevisiae* and *Schizosaccharomyces pombe*, the ESR genes that respond to diverse types of environmental

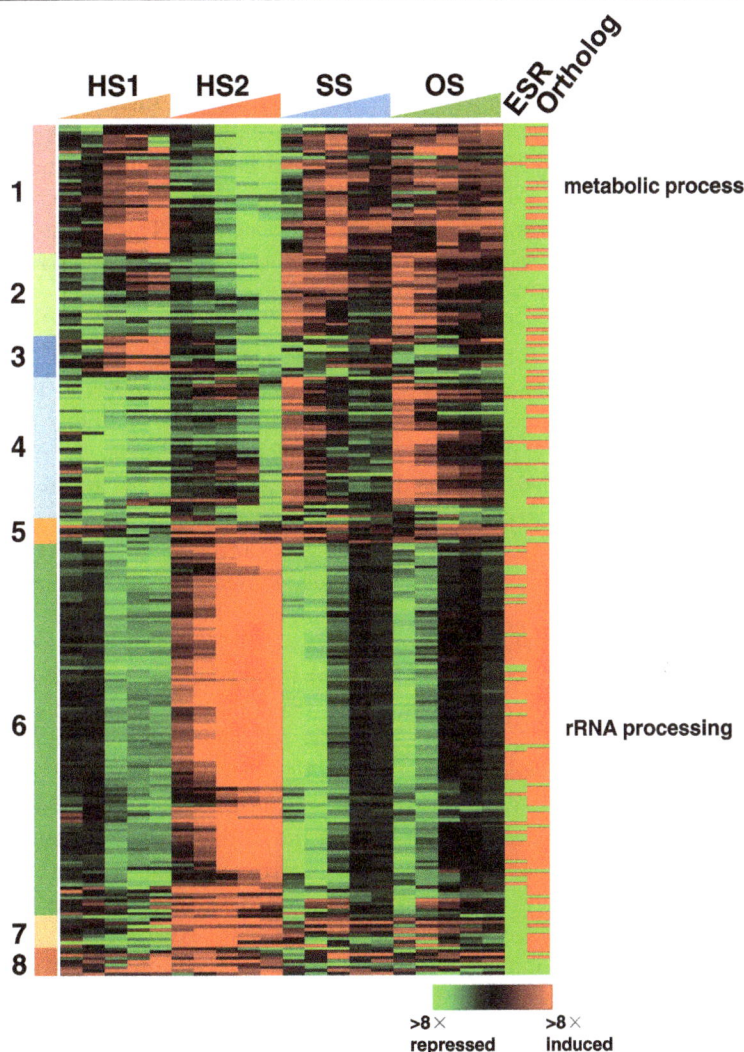

Fig. 8 Heat map of the 266 stress-responsive genes. Color scale is indicated. ESR and Ortholog indicate the existence of ESR and orthologous genes in *S. cerevisiae*, respectively. Red and green represent true and false, respectively, unlike the color scale of gene expression. The 'cutree' function in R was used to divide 266 genes into eight clusters. Clusters 1, 2, 3, 4, 5, 6, 7, and 8 consist of 40 (15.0%), 26 (9.8%), 13 (4.9%), 44 (16.5%), 8 (3.0%), 116 (43.6%), 10 (3.8%), and 9 (3.4%) genes, respectively

stresses have been identified [17, 18]. Particularly, in *S. cerevisiae*, Gasch et al. (2000) proposed 867 ESR genes including different functions, e.g. energy generation and storage, defense against reactive oxygen species, and DNA repair [18]. Here we attempted to identify *A. fumigatus* genes that are orthologous to the *S. cerevisiae* ESR (ScESR) genes. First, we conducted ortholog identification by reciprocal best-hit pairs against *S. cerevisiae* proteins, and consequently, 768 (768/867 ScESR genes, 88.6%) genes in *A. fumigatus* were identified as orthologs of ScESR genes, while among 9840 genes in *A. fumigatus*, 5399 (54.9%) genes corresponded to the genes of *S. cerevisiae*. Notably, the expression of 229 genes (29.8%) of the 768 *A. fumigatus* genes was not observed under any stress conditions tested. Among the 768 *A. fumigatus* genes, 77 (10.0%) were the

stress-responsive genes identified above (Additional file 7: Table S4). As shown in Fig. 8, 71 up-regulated genes in HS2 were orthologous to ScESR genes, such as chaperones and ribosomal processing protein (Fig. 8, cluster 6). The genes in *A. fumigatus* identified as the ScESR ortholog were significantly overrepresented in 266 stress-responsive genes (Fisher's exact test, *p*-value 2.35E − 25).

Correlation analysis reveals the clusters of co-expressed genes

Finally, we explored the co-expressed genes through four stress conditions. Co-expression relationships are useful for unravelling the transcriptional networks, such as TF-regulated genes [38–42]. By using normalized data, i.e. 4490 genes, correlation analyses were conducted. Gene-to-gene Pearson correlation coefficients were calculated, and the Gap statistic as described in the Methods was applied to estimate the cluster size [43]. We surveyed the hub gene, and observed Afu5g05710 annotated as pseudouridylate synthase family protein, whose expression correlated with 233 genes with $r > 0.95$. Seventy-three genes ($r > 0.98$) were estimated as co-expressed genes, 44 of which (cluster 6) were stress-responsive genes (Fig. 9a). These expression levels were up-regulated in HS2, while they were down-regulated in HS1, SS, and OS. The expression profiles of genes related to 'rRNA processing' were highly correlated to those of Afu5g05710 in response to all tested stresses. Furthermore, 18 genes were estimated as co-expressed genes with uncharacterized TF Afu3g11170 ($r > 0.87$) (Fig. 9b). These genes transiently responded to SS and OS at 15 min, while they were down-regulated upon exposure to HS1 and HS2.

Discussion

The aim of this study is to comprehensively elucidate the stress-responsive genes under diverse environmental stress conditions in pathogenic filamentous fungi *A. fumigatus*. We conducted time series transcriptome analyses by RNA-seq upon exposure to four different environmental stresses: heat stress from 30 °C to 37 °C (HS1) and from 30 °C to 48 °C (HS2), superoxide stress by adding menadione (SS), and osmotic stress by adding sorbitol (OS). The 0 min data were highly correlated (Additional file 4: Figure S1). We concluded that our RNA-seq data would be consistent in the initial growth stage, and the differences of growth time could have little impact. Furthermore, even though additional AMM was added in OS, the correlation between 180 min data of SS and OS was quite high ($r = 0.95$), and 180 min (T5) data of OS was positioned closely to that of SS (Additional file 5: Figure S2), indicating that the data of SS and OS could be compatible.

Comparison with microarray data

When compared with the previously published microarray data [15], we found that the expression profiles of heat shock genes were similar (Fig. 3). The profile of one of the heat shock genes, *hscA*, was different between two data sets, while the profiles of the other five genes were consistent. In addition, we comprehensively compared our HS1 and HS2 expression data with the microarray data (Additional file 6: Figure S3). The correlations between them for HS1 (37 °C) ($n = 676 \times 5$ time points) and HS2 (48 °C) were 0.30 (p-value, 1.22E − 70) and 0.39 (p-value, <1.00E − 70), respectively. It has been reported that the expression profile correlation between microarray and RNA-seq in *A. fumigatus* ranges from 0.14 to 0.41 [24]. Thus, we concluded that our RNA-seq data

are comparable with previous microarray data. We newly identified 1275 and 2254 genes that were up- or down-regulated in HS1 and HS2, respectively. It will be necessary to investigate the characterization of those genes in future. In our RNA-seq data, 36 and 94 genes that had been identified in microarray were missed because there were no mapping sequences in HS1 and HS2, respectively. This is most likely because these genes were misidentified by microarray based on the fluorescence intensity.

Stress-specific genes

We comprehensively revealed how the genes responded to specific stress conditions. A total of 39 GO terms were identified (Fig. 5a, b). In HS1, the expression of 117 genes annotated as 'oxidation-reduction process' was up-regulated, consistent with the observation that the genes involved in balancing the redox state were up-regulated upon heat stress treatment [16]. Although 'oxidation-reduction process' was not statistically overrepresented in HS2, the expression of 67 genes annotated as 'oxidation-reduction process' was up-regulated, indicating that the heat stress to some extent triggered the redox imbalance. Probably, enhancement of oxygen respiration by heat shock may cause the redox imbalance as previously reported in yeast [44]. It has been reported that the oxidative stress response leads to mitochondrial dysfunction in yeast [45] and *A. fumigatus* [9]. Consistent with these observations, the expression of five genes (Afu1g07450, Afu2g04270, Afu3g14490, Afu5g03640, and Afu5g07140) annotated as 'mitochondrial genome maintenance' was down-regulated at 60 min in SS (Fig. 5b). Interestingly, the expression of Afu3g14490, ketol-acid reductoisomerase, was differentially expressed upon exposure to airway epithelial cells in conidia [46], suggesting that Afu3g14490 may play important roles for environmental changes, e.g. oxidative damage, in both hyphae and conidia.

Upon heat stress, Hsf1 induces the expression of many genes, including chaperones as one of the key regulators in fungi including yeast and *A. fumigatus* [16]. In our transcriptome data, the expression of *hsf1* was largely induced in response to HS2 until 60 min, and slightly induced at 15 min in response to OS (2.5-fold). The expression levels of two genes, i.e. uncharacterized protein Afu4g10360 and MAP kinase kinase *pbs2* (Afu1g15950), were strongly correlated with those of *hsf1* ($r > 0.95$). The yeast ortholog of Afu4g10360, ETP1, is required for growth on ethanol and ethanol-induced transcriptional activation [47], suggesting that Afu4g10360 may sense environmental changes such as carbon source, heat stress, and osmotic stress, and relate to the transcriptional regulations. Although the growth defect of Δ*pbs2* strain was not observed at 48 °C [32], Δ*pbs2* strain was

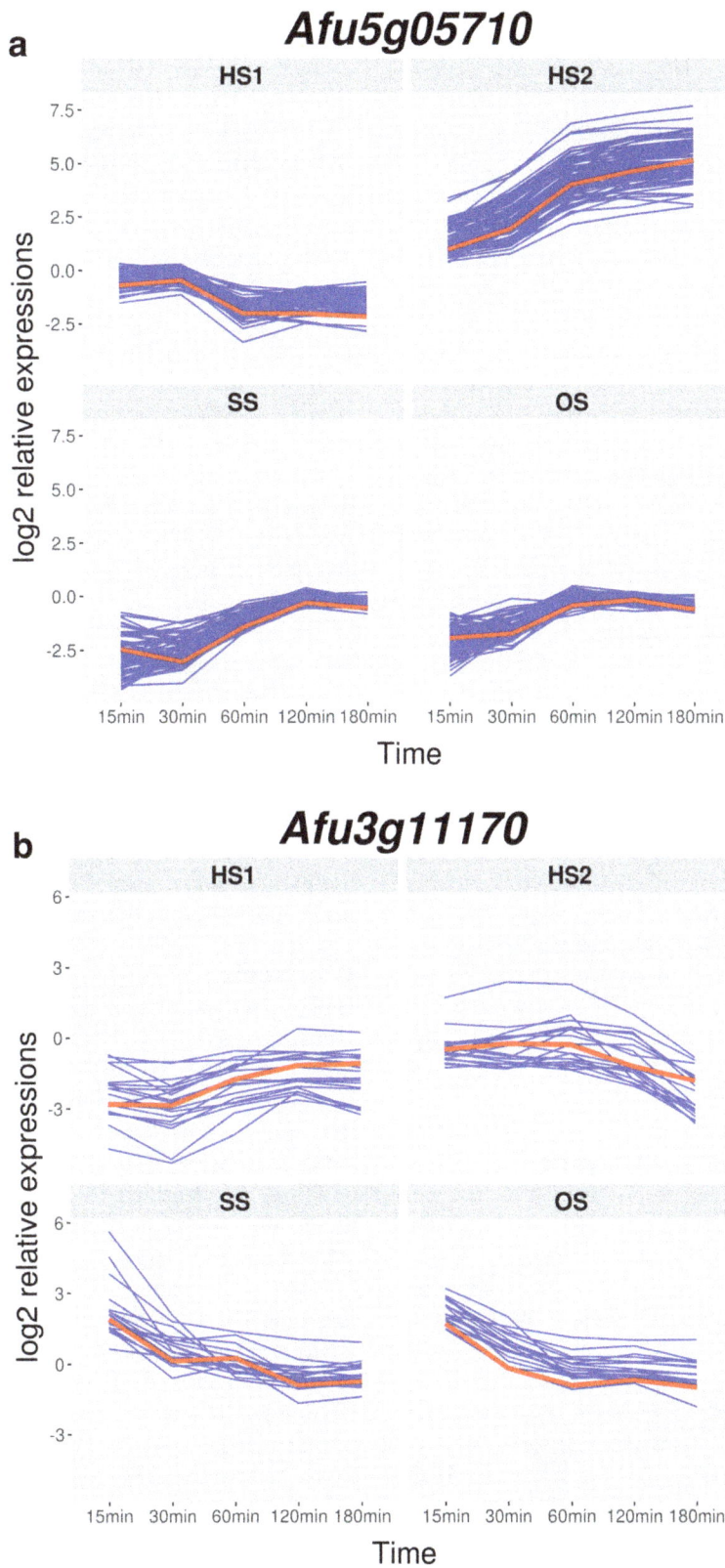

Fig. 9 Expression profiles of co-expressed genes. **a** Expression profiles of Afu5g05710 (red line) and 73 genes. **b** Expression profiles of Afu3g11170 (red line) and 18 genes. x-axis and y-axis correspond to the time after shocking the mycelia, 15, 30, 60, 120 and 180 min, and log2-transformed ratio values, respectively

hypersensitive to oxidative and osmotic stresses in *A. fumigatus* [13], indicating that the acute induction of *pbs2* may be a prerequisite for adaptation to oxidative and osmotic stresses.

The expression of the genes involved in 'ergosterol biosynthetic process' was down-regulated under SS and OS conditions (Figs. 5b, 6). This is consistent with the fact that repression of ergosterol biosynthesis is essential for stress resistance and important for salt stress in yeast [48]. Accordingly, the down-regulation of 'ergosterol biosynthetic process' might be prerequisite for adaptation to environmental challenges in *A. fumigatus* hyphae in terms of cell wall integrity. The expression of *srbB* (Afu4g03460) was 1/5-fold down-regulated at 30 min in SS. *srbB* is involved in the regulation of ergosterol biosynthesis upon hypoxia [49], suggesting that *srbB* would respond to not only hypoxia but also oxidative stress. Yap1 is known to function against H_2O_2 and menadione [9]. Consistent with the previous report, in our transcriptome analyses, *yap1* was identified as an SS-specific gene (Fig. 4a). The expression of *yap1* was up-regulated 5.7-fold at 15 min in response to SS, but did not change under other conditions (Additional file 8: Figure S4). Interestingly, the expression levels of some of the putative Yap1 target genes identified by Lessing et al. (2007) [9] were differentially expressed in not only SS but also HS2 and OS. For example, the expression of *cat2* responded to SS, HS2, and OS. The expression of *aspf3* (Afu6g02280) was induced in response to HS1, HS2, and SS, consistent with up-regulation upon exposure to heat stress [16]. This suggested that *cat2* and *aspf3* could be regulated not only by Yap1 but also other TFs. Among 28 putative Yap1 target genes, 14 genes, such as *aof2* (Afu4g13000), were not detected as SS-responsive genes.

The stress-responsive genes in *A. fumigatus*

We identified the 266 stress-responsive genes that were differentially expressed upon exposure to all four tested environmental changes, indicating that these genes are essential for mycelia of filamentous fungi to sense and respond to the environmental changes. We observed that most of these genes are involved in ribosome-related functions. The set of stress-responsive genes included 171 orthologs and 95 non-orthologs to *S. cerevisiae*. The proportion of *S. cerevisiae* ortholog genes was largely similar for the whole genome (54.9%) and the stress-responsive genes (64.3%). Among the 171 ortholog genes, 77 (45.0%) are orthologous to ScESR genes, which is relatively high compared with the genome-wide orthologs between *A. fumgiatus* and *S. cerevisiae* (768 orthologs to ScESR for 5399 ortholog genes, 14.2%). This suggested that the responses of ESR genes would be

more frequently conserved between *S. cerevisiae* and *A. fumigatus*. On the contrary, *A. fumigatus* stress-responsive genes included 94 genes (35.3%) that are orthologous to *S. cerevisiae* non-ESR genes. These genes might have evolved to respond to and sense the environmental changes in filamentous fungi, namely by transcriptional evolution. The remainder of the 95 genes (35.7%) have no ortholog in *S. cerevisiae*, suggesting that these genes have uniquely evolved in *A. fumigatus*, likely along with the spreading of its habitat.

Notably, the Afu5g09910 encoding a putative *p*-nitroreductase-family protein was drastically induced in response to all tested conditions. Its protein abundance is regulated by Yap1 against H_2O_2 [9]. In addition, it has been reported that the protein abundance and gene expression of Afu5g09910 are up-regulated upon exposure to human neutrophils and gliotoxin, respectively [50, 51]. This suggested that the expression of Afu5g09910 may be induced by diverse environmental challenges.

Correlation analysis

Finally, we conducted correlation analysis to estimate the co-expressed genes with a particular gene. The expression of Afu5g05710 was highly correlated with 73 genes enriched in 'rRNA processing'. Particularly, it was strongly induced under the HS2 condition, suggesting that these genes might be a prerequisite for protecting against sub-lethal exposures. The regulation of those genes, however, remains unknown.

Conclusion

Global gene expression analysis under diverse environmental conditions is important for understanding the pathogenicity of the opportunistic pathogen *A. fumigatus*. Here we performed time series transcriptome analysis under four different conditions, and identified stress-responsive genes for particular stresses. In addition, we identified the 266 genes whose expression levels were drastically changed upon exposure to heat, superoxide, and osmotic stresses in *A. fumigatus*. The 77 ESR genes common in filamentous fungi *A. fumigatus* and unicellular yeast *S. cerevisiae* responded to diverse environmental stresses, suggesting that these genes might be more general prerequisites for adaptation to environmental stresses. Further investigation of the molecular mechanisms will shed light on the adaptation in response to suboptimal conditions.

Methods

Fungal strain and growth condition

The strain used in this study was *A. fumigatus* Af293. Potato dextrose agar (PDA; Difco, Detroit, USA), YG [27], and AMM [52] were regularly used for culturing the strain. To collect fresh conidia, the stored conidia were incubated on a PDA plate at 30 °C, or an AMM at 37 °C for 1 week.

Environmental stress
Heat stress from 30 °C to 37 °C or 48 °C (HS1 or HS2)
Mycelia were grown in YG medium at 30 °C for 17 h before being transferred to a water bath of 37 °C or 48 °C. The samples in individual culture flasks were collected at 0, 15, 30, 60, 120, and 180 min after the transfer, and harvested mycelium was frozen in liquid nitrogen.

Superoxide stress (SS)
Mycelia were grown in AMM medium at 37 °C for 24 h. Following superoxide stress by adding menadione (Sigma-Aldrich, St. Louis, USA) for a final concentration of 10 μM menadione, the samples were collected at 0, 15, 30, 60, 120, and 180 min, and harvested mycelium was frozen in liquid nitrogen.

Osmotic stress (OS)
Mycelia were grown in AMM medium 37 °C for 24 h. Following the addition of AMM supplemented with 2 M sorbitol (Wako, Osaka, Japan) for a final concentration of 1 M sorbitol, the samples were collected at 0, 15, 30, 60, 120, and 180 min, and harvested mycelium was frozen in liquid nitrogen.

RNA extraction
After frozen mycelia were disrupted with zirconia beads using a multi-beads shocker (Yasui Kikai, Osaka, Japan), total RNAs were extracted using an RNeasy Mini Kit (Qiagen, Hilden, Germany) and contaminating genomic DNAs were removed with an RNase-Free DNase set (Qiagen, Hilden, Germany) according to the manufacturer's instructions.

RNA sequencing
TruSeq RNA Sample Prep Kit v2 (Illumina, San Diego, USA) and KAPA Stranded mRNA-Seq Kit (Kapa Biosystems, Wilmington, USA) were employed to prepare the libraries of mRNA samples for multiplexed sequencing according to the manufacturers' protocol (Additional file 1: Table S1). The qualities of all libraries were determined by an Agilent 2100 Bioanalyzer (Agilent Technologies, Santa Clara, USA). Paired-end 25 bp and single-end 60 bp were performed with the aid of a MiSeq (Illumina) and a HiSeq1500 (Illumina), respectively.

Expression analysis
Illumina data sets were trimmed using Trimmomatic (ver. 0.33), where sequencing adapters and sequences with low-quality scores were removed [53]. Cleaned reads were mapped to the genome sequence of *A. fumigatus* Af293 (29,420,142 bp; genome version: s03-m05-r04) from AspGD [54] using STAR (ver. 2.4.2a) with the default parameters other than '–alignIntronMax 1000' [55]. FPKMs were calculated using cuffdiff in Cufflinks (ver. 2.2.1) with default parameters [56]. Data analyses were conducted using the R programming language (https://www.r-project.org/), corrplot (ver. 0.84) [57], ggplot2 (ver. 1.0.1) [58] and cummerbund (ver. 2.18.0) software [59].

Identification of yeast orthologous genes
The amino acid sequences of 6713 proteins (downloaded on 13th January 2016) in *S. cerevisiae* were obtained from the Saccharomyces Genome Database (SGD, http://www.yeastgenome.org). A list of 867 ESR genes was obtained from Gasch et al. (2000) [18]. The reciprocal best-hit pairs between *A. fumigatus* and *S. cerevisiae* by BLASTP (ver. 2.2.28+) [60] analysis with '-evalue 1e-4' were used to identify the ortholog genes.

GO analysis
Genes were functionally categorized using their GO information [61] obtained from AspGD, and overrepresented GO terms were identified using Fisher's exact test. The one-tailed Fisher's exact p-value corresponding to the overrepresentation of GO categories with equal to or greater than 20 genes was calculated based on counts in 2×2 contingency tables. p-values were corrected by the false discovery rate method [62], and the threshold was set as 0.05.

Estimation of the cluster size using the gap statistic
Pearson correlation coefficient was applied to identify the co-expressed genes targeted by particular genes, e.g. TFs. To estimate cluster size, we adopted the Gap statistic proposed by Hastie et al. (2000) [43]. We assumed that gene expression levels in a cluster with k genes are highly coherent, and measured them by calculating $R^2(k)$.

$$R^2(k) = \frac{V_B}{V_T} \tag{1}$$

V_B and V_T are between variance and total variance, respectively. We obtained a randomized data matrix by permuting the elements within each gene expression, and calculated $R^2(k)^*$. We then defined the Gap statistic as the observed $R^2(k)$ minus the mean of $R^2(k)^*$.

$$Gap(k) = R^2(k) - mean(R^2(k)^*) \tag{2}$$

The estimated cluster size k is given by:

$$k = argmax_k \, Gap(k) \qquad (3)$$

In this study, we generated 1000 permutated matrixes and estimated the cluster size k.

Additional files

Additional file 1: Table S1. Information of RNA-seq. (XLSX 9 kb)

Additional file 2: Table S2. Mapping results. (XLSX 11 kb)

Additional file 3: Table S3. FPKM values of 9840 genes. (XLSX 2563 kb)

Additional file 4: Figure S1. Correlation matrix between 0 min data of HS1, HS2, SS, and OS. Size of the circles is proportional to the correlation coefficient [57]. (PDF 45 kb)

Additional file 5: Figure S2. Principal component analysis (PCA). Size of the circles indicates time after exposure to the stress, i.e. 15 (T1), 30 (T2), 60 (T3), 120 (T4), and 180 (T5) min. (PDF 91 kb)

Additional file 6: Figure S3. Overview of heat stress genes. (a) Comparison of 268 genes identified in Do et al. (2009) and this study for heat stress of 37 °C. (b) Comparison of 1044 genes for heat stress of 48 ° C. Color scale is indicated. (PDF 47 kb)

Additional file 7: Table S4. log2 transformed ratio values of the 266 genes. (XLSX 289 kb)

Additional file 8: Figure S4. Expression profiles of 27 putative Yap1 target genes reported by Lessing et al. (2007). yap1 was up-regulated only at 15 min in response to SS. (PDF 85 kb)

Acknowledgements
We would like to thank Atsushi Iwama, Motohiko Oshima, and Atsunori Saraya (Chiba University) for technical support with the Illumina HiSeq 1500, and Ryoko Mori for technical assistance. We thank the National BioResource Project - Pathogenic Microbes in Japan (http://www.nbrp.jp/).

Funding
This work has been partly supported by the Takeda Science Foundation, MEXT KAKENHI (16K18671 and 16H06279), Japanese Initiative for Progress of Research on Infectious Disease for Global Epidemic (J-PRIDE), and the Institute for Global Prominent Research, Chiba University to HT, and the Tenure Tracking System Program of MEXT to YK and HT.

Authors' contributions
HT initiated and supervised the project, carried out all data analyses, and wrote the manuscript. YK conceived the study, participated in designing and coordinating the study, carried out the RNA-seq, and wrote the manuscript. DH coordinated the study and wrote the manuscript. ATN performed the RNA-seq. SK and TG coordinated the study. All authors read and approved the final manuscript.

Competing interests
The authors declare that they have no competing interests.

References

1. Brown GD, Denning DW, Levitz SM. Tackling human fungal infections. Science. 2012;336:647.
2. Thom C, Raper KB. A manual of the Aspergilli. Baltimore: Williams & Wilkins; 1945.
3. Dagenais TR, Keller NP. Pathogenesis of Aspergillus fumigatus in invasive Aspergillosis. Clin Microbiol Rev. 2009;22:447–65.
4. Kousha M, Tadi R, Soubani AO. Pulmonary aspergillosis: a clinical review. Eur Respir Rev. 2011;20:156–74.
5. Brown AJ, Haynes K, Quinn J. Nitrosative and oxidative stress responses in fungal pathogenicity. Curr Opin Microbiol. 2009;12:384–91.
6. Nierman WC, Pain A, Anderson MJ, Wortman JR, Kim HS, Arroyo J, Berriman M, Abe K, Archer DB, Bermejo C, et al. Genomic sequence of the pathogenic and allergenic filamentous fungus Aspergillus fumigatus. Nature. 2005;438:1151–6.
7. Chauhan N, Latge JP, Calderone R. Signalling and oxidant adaptation in Candida albicans and Aspergillus fumigatus. Nat Rev Microbiol. 2006;4:435–44.
8. Philippe B, Ibrahim-Granet O, Prévost MC, Gougerot-Pocidalo MA, Sanchez Perez M, Van der Meeren A, Latgé JP. Killing of Aspergillus fumigatus by alveolar macrophages is mediated by reactive oxidant intermediates. Infect Immun. 2003;71:3034–42.
9. Lessing F, Kniemeyer O, Wozniok I, Loeffler J, Kurzai O, Haertl A, Brakhage AA. The Aspergillus fumigatus transcriptional regulator AfYap1 represents the major regulator for defense against reactive oxygen intermediates but is dispensable for pathogenicity in an intranasal mouse infection model. Eukaryot Cell. 2007;6:2290–302.
10. Xue T, Nguyen CK, Romans A, May GS. A mitogen-activated protein kinase that senses nitrogen regulates conidial germination and growth in Aspergillus fumigatus. Eukaryot Cell. 2004;3:557–60.
11. Brown NA, Goldman GH. The contribution of Aspergillus fumigatus stress responses to virulence and antifungal resistance. J Microbiol. 2016;54:243–53.
12. Winkelströter LK, Bom VL, de Castro PA, Ramalho LN, Goldman MH, Brown NA, Rajendran R, Ramage G, Bovier E, Dos Reis TF, et al. High osmolarity glycerol response PtcB phosphatase is important for Aspergillus fumigatus virulence. Mol Microbiol. 2015;96:42–54.
13. Ma D, Li R. Current understanding of HOG-MAPK pathway in Aspergillus fumigatus. Mycopathologia. 2013;175:13–23.
14. de Castro PA, Chiaratto J, Winkelströter LK, Bom VL, Ramalho LN, Goldman MH, Brown NA, Goldman GH. The involvement of the Mid1/Cch1/Yvc1 calcium channels in Aspergillus fumigatus virulence. PLoS One. 2014;9:e103957.
15. Do JH, Yamaguchi R, Miyano S. Exploring temporal transcription regulation structure of Aspergillus fumigatus in heat shock by state space model. BMC Genomics. 2009;10:306.
16. Albrecht D, Guthke R, Brakhage AA, Kniemeyer O. Integrative analysis of the heat shock response in Aspergillus fumigatus. BMC Genomics. 2010;11:32.
17. Causton HC, Ren B, Koh SS, Harbison CT, Kanin E, Jennings EG, Lee TI, True HL, Lander ES, Young RA. Remodeling of yeast genome expression in response to environmental changes. Mol Biol Cell. 2001;12:323–37.
18. Gasch AP, Spellman PT, Kao CM, Carmel-Harel O, Eisen MB, Storz G, Botstein D, Brown PO. Genomic expression programs in the response of yeast cells to environmental changes. Mol Biol Cell. 2000;11:4241–57.
19. Strassburg K, Walther D, Takahashi H, Kanaya S, Kopka J. Dynamic transcriptional and metabolic responses in yeast adapting to temperature stress. OMICS. 2010;14:249–59.
20. Emri T, Szarvas V, Orosz E, Antal K, Park H, Han KH, JH Y, Pócsi I. Core oxidative stress response in Aspergillus nidulans. BMC Genomics. 2015;16:478.
21. Gibbons JG, Beauvais A, Beau R, McGary KL, Latgé JP, Rokas A. Global transcriptome changes underlying colony growth in the opportunistic human pathogen Aspergillus fumigatus. Eukaryot Cell. 2012;11:68–78.
22. Hagiwara D, Takahashi H, Kusuya Y, Kawamoto S, Kamei K, Gonoi T. Comparative transcriptome analysis revealing dormant conidia and germination associated genes in Aspergillus species: an essential role for AtfA in conidial dormancy. BMC Genomics. 2016;17:358.
23. McDonagh A, Fedorova ND, Crabtree J, Yu Y, Kim S, Chen D, Loss O, Cairns T, Goldman G, Armstrong-James D, et al. Sub-telomere directed gene expression during initiation of invasive aspergillosis. PLoS Pathog. 2008;4: e1000154.

24. Müller S, Baldin C, Groth M, Guthke R, Kniemeyer O, Brakhage AA, Valiante V. Comparison of transcriptome technologies in the pathogenic fungus *Aspergillus fumigatus* reveals novel insights into the genome and MpkA dependent gene expression. BMC Genomics. 2012;13:519.

25. O'Keeffe G, Hammel S, Owens RA, Keane TM, Fitzpatrick DA, Jones GW, Doyle S. RNA-seq reveals the pan-transcriptomic impact of attenuating the gliotoxin self-protection mechanism in *Aspergillus fumigatus*. BMC Genomics. 2014;15:894.

26. Rokas A, Gibbons JG, Zhou X, Beauvais A, Latgé JP. The diverse applications of RNA-seq for functional genomic studies in *Aspergillus fumigatus*. Ann N Y Acad Sci. 2012;1273:25–34.

27. Takahashi-Nakaguchi A, Muraosa Y, Hagiwara D, Sakai K, Toyotome T, Watanabe A, Kawamoto S, Kamei K, Gonoi T, Takahashi H. Genome sequence comparison of *Aspergillus fumigatus* strains isolated from patients with pulmonary aspergilloma and chronic necrotizing pulmonary aspergillosis. Med Mycol. 2015;53:353–60.

28. Hagiwara D, Asano Y, Marui J, Yoshimi A, Mizuno T, Abe K. Transcriptional profiling for *Aspergillus nidulans* HogA MAPK signaling pathway in response to fludioxonil and osmotic stress. Fungal Genet Biol. 2009;46:868–78.

29. Hagiwara D, Suzuki S, Kamei K, Gonoi T, Kawamoto S. The role of AtfA and HOG MAPK pathway in stress tolerance in conidia of *Aspergillus fumigatus*. Fungal Genet Biol. 2014;73:138–49.

30. Hagiwara D, Sakamoto K, Abe K, Gomi K. Signaling pathways for stress responses and adaptation in *Aspergillus* species: stress biology in the post-genomic era. Biosci Biotechnol Biochem. 2016;80:1667–80.

31. Miskei M, Karányi Z, Pócsi I. Annotation of stress-response proteins in the aspergilli. Fungal Genet Biol. 2009;46:S105–20.

32. Ji Y, Yang F, Ma D, Zhang J, Wan Z, Liu W, Li R. HOG-MAPK signaling regulates the adaptive responses of *Aspergillus fumigatus* to thermal stress and other related stress. Mycopathologia. 2012;174:273–82.

33. Hagiwara D, Takahashi-Nakaguchi A, Toyotome T, Yoshimi A, Abe K, Kamei K, Gonoi T, Kawamoto S. NikA/TcsC histidine kinase is involved in conidiation, hyphal morphology, and responses to osmotic stress and antifungal chemicals in *Aspergillus fumigatus*. PLoS One. 2013;8:e80881.

34. Reyes G, Romans A, Nguyen CK, May GS. Novel mitogen-activated protein kinase MpkC of *Aspergillus fumigatus* is required for utilization of polyalcohol sugars. Eukaryot Cell. 2006;5:1934–40.

35. Pereira Silva L, Alves de Castro P, Dos Reis TF, Paziani MH, Von Zeska Kress MR, Riaño-Pachón DM, Hagiwara D, Ries LN, Brown NA, Goldman GH. Genome-wide transcriptome analysis of *Aspergillus fumigatus* exposed to osmotic stress reveals regulators of osmotic and cell wall stresses that are SakA(HOG1) and MpkC dependent. Cell Microbiol. 2017;19:e12681.

36. Oliveros JC. Venny. An interactive tool for comparing lists with Venn's diagrams. 2007–2015. http://bioinfogp.cnb.csic.es/tools/venny/index.html

37. Li Y, Zhang L, Wang D, Zhou H, Ouyang H, Ming J, Jin C. Deletion of the msdS/AfmsdC gene induces abnormal polarity and septation in *Aspergillus fumigatus*. Microbiology. 2008;154:1960–72.

38. Harbison CT, Gordon DB, Lee TI, Rinaldi NJ, Macisaac KD, Danford TW, Hannett NM, Tagne JB, Reynolds DB, Yoo J, et al. Transcriptional regulatory code of a eukaryotic genome. Nature. 2004;431:99–104.

39. Kobayashi H, Akitomi J, Fujii N, Kobayashi K, Altaf-Ul-Amin M, Kurokawa K, Ogasawara N, Kanaya S. The entire organization of transcription units on the *Bacillus subtilis* genome. BMC Genomics. 2007;8:197.

40. Redestig H, Weicht D, Selbig J, Hannah MA. Transcription factor target prediction using multiple short expression time series from *Arabidopsis thaliana*. BMC Bioinformatics. 2007;8:454.

41. Takahashi H, Morioka R, Ito R, Oshima T, Altaf-Ul-Amin M, Ogasawara N, Kanaya S. Dynamics of time-lagged gene-to-metabolite networks of *Escherichia coli* elucidated by integrative omics approach. OMICS. 2011; 15:15–23.

42. Wada M, Takahashi H, Altaf-Ul-Amin M, Nakamura K, Hirai MY, Ohta D, Kanaya S. Prediction of operon-like gene clusters in the *Arabidopsis thaliana* genome based on co-expression analysis of neighboring genes. Gene. 2012; 503:56–64.

43. Hastie T, Tibshirani R, Eisen MB, Alizadeh A, Levy R, Staudt L, Chan WC, Botstein D, Brown P. 'Gene shaving' as a method for identifying distinct sets of genes with similar expression patterns. Genome Biol. 2000;1: RESEARCH0003.

44. Sugiyama K, Izawa S, Inoue Y. The Yap1p-dependent induction of glutathione synthesis in heat shock response of *Saccharomyces cerevisiae*. J Biol Chem. 2000;275:15535–40.

45. Demasi AP, Pereira GA, Netto LE. Yeast oxidative stress response. Influences of cytosolic thioredoxin peroxidase I and of the mitochondrial functional state. FEBS J. 2006;273:805–16.

46. Oosthuizen JL, Gomez P, Ruan J, Hackett TL, Moore MM, Knight DA, Tebbutt SJ. Dual organism transcriptomics of airway epithelial cells interacting with conidia of *Aspergillus fumigatus*. PLoS One. 2011;6:e20527.

47. Snowdon C, Schierholtz R, Poliszczuk P, Hughes S, van der Merwe G. ETP1/YHL010c is a novel gene needed for the adaptation of *Saccharomyces cerevisiae* to ethanol. FEMS Yeast Res. 2009;9:372–80.

48. Montañés FM, Pascual-Ahuir A, Proft M. Repression of ergosterol biosynthesis is essential for stress resistance and is mediated by the Hog1 MAP kinase and the Mot3 and Rox1 transcription factors. Mol Microbiol. 2011;79:1008–23.

49. Chung D, Barker BM, Carey CC, Merriman B, Werner ER, Lechner BE, Dhingra S, Cheng C, Xu W, Blosser SJ, et al. ChIP-seq and in vivo transcriptome analyses of the *Aspergillus fumigatus* SREBP SrbA reveals a new regulator of the fungal hypoxia response and virulence. PLoS Pathog. 2014;10:e1004487.

50. Sugui JA, Kim HS, Zarember KA, Chang YC, Gallin JI, Nierman WC, Kwon-Chung KJ. Genes differentially expressed in conidia and hyphae of *Aspergillus fumigatus* upon exposure to human neutrophils. PLoS One. 2008;3:e2655.

51. Carberry S, Molloy E, Hammel S, O'Keeffe G, Jones GW, Kavanagh K, Doyle S. Gliotoxin effects on fungal growth: mechanisms and exploitation. Fungal Genet Biol. 2012;49:302–12.

52. Kusuya Y, Hagiwara D, Sakai K, Yaguchi T, Gonoi T, Takahashi H. Transcription factor Afmac1 controls copper import machinery in *Aspergillus fumigatus*. Curr Genet. 2017;63:777–89.

53. Bolger AM, Lohse M, Usadel B. Trimmomatic: a flexible trimmer for Illumina sequence data. Bioinformatics. 2014;30:2114–20.

54. Cerqueira GC, Arnaud MB, Inglis DO, Skrzypek MS, Binkley G, Simison M, Miyasato SR, Binkley J, Orvis J, Shah P, et al. The *Aspergillus* genome database: multispecies curation and incorporation of RNA-Seq data to improve structural gene annotations. Nucleic Acids Res. 2014;42:D705–10.

55. Dobin A, Davis CA, Schlesinger F, Drenkow J, Zaleski C, Jha S, Batut P, Chaisson M, Gingeras TR. STAR: ultrafast universal RNA-seq aligner. Bioinformatics. 2013;29:15–21.

56. Trapnell C, Williams BA, Pertea G, Mortazavi A, Kwan G, van Baren MJ, Salzberg SL, Wold BJ, Pachter L. Transcript assembly and quantification by RNA-Seq reveals unannotated transcripts and isoform switching during cell differentiation. Nat Biotechnol. 2010;28:511–5.

57. Wei T, Simko V. R package "corrplot": visualization of a correlation matrix (Version 0.84). 2017. https://github.com/taiyun/corrplot .

58. Wickham H. ggplot2: elegant graphics for data analysis. New York: Springer; 2009.

59. Goff L, Trapnell C, Kelley D. cummeRbund: analysis, exploration, manipulation, and visualization of cufflinks high-throughput sequencing data. R package version 2.18.0. 2013.

60. Altschul SF, Madden TL, Schäffer AA, Zhang J, Zhang Z, Miller W, Lipman DJ. Gapped BLAST and PSI-BLAST: a new generation of protein database search programs. Nucleic Acids Res. 1997;25:3389–402.

61. Ashburner M, Ball CA, Blake JA, Botstein D, Butler H, Cherry JM, Davis AP, Dolinski K, Dwight SS, Eppig JT, et al. Gene ontology: tool for the unification of biology. The gene ontology consortium. Nat Genet. 2000;25:25–9.

62. Benjamini Y, Hochberg Y. Controlling the false discovery rate: a practical and powerful approach to multiple testing. J R Statist Soc B. 1995;57:289–300.

10

Comparative analyses of plastid genomes from fourteen Cornales species: inferences for phylogenetic relationships and genome evolution

Chao-Nan Fu[1,2], Hong-Tao Li[3], Richard Milne[4], Ting Zhang[3], Peng-Fei Ma[3], Jing Yang[3], De-Zhu Li[1,2,3*] and Lian-Ming Gao[1*]

Abstract

Background: The Cornales is the basal lineage of the asterids, the largest angiosperm clade. Phylogenetic relationships within the order were previously not fully resolved. Fifteen plastid genomes representing 14 species, ten genera and seven families of Cornales were newly sequenced for comparative analyses of genome features, evolution, and phylogenomics based on different partitioning schemes and filtering strategies.

Results: All plastomes of the 14 Cornales species had the typical quadripartite structure with a genome size ranging from 156,567 bp to 158,715 bp, which included two inverted repeats (25,859–26,451 bp) separated by a large single-copy region (86,089–87,835 bp) and a small single-copy region (18,250–18,856 bp) region. These plastomes encoded the same set of 114 unique genes including 31 transfer RNA, 4 ribosomal RNA and 79 coding genes, with an identical gene order across all examined Cornales species. Two genes (*rpl22* and *ycf15*) contained premature stop codons in seven and five species respectively. The phylogenetic relationships among all sampled species were fully resolved with maximum support. Different filtering strategies (none, light and strict) of sequence alignment did not have an effect on these relationships. The topology recovered from coding and noncoding data sets was the same as for the whole plastome, regardless of filtering strategy. Moreover, mutational hotspots and highly informative regions were identified.

Conclusions: Phylogenetic relationships among families and intergeneric relationships within family of Cornales were well resolved. Different filtering strategies and partitioning schemes do not influence the relationships. Plastid genomes have great potential to resolve deep phylogenetic relationships of plants.

Keywords: Plastid genome, Phylogenomics, Cornales, Alignment, Partitioning schemes, Gene loss

Background

The Cornales is a relatively small but diverse group, representing the basal lineage of the largest angiosperm clade, the Asterids [1–4]. It comprises 42 genera and approximately 605 species in ten families, including two large families (Hydrangeaceae and Loasaceae) and eight small families. The latter contain few genera, mostly with isolated geographic ranges, i.e. Cornaceae (*Cornus*), Nyssaceae (*Camptotheca*, *Nyssa*), Curtisiaceae (*Curtisia*), Grubbiaceae (*Grubbia*), Hydrostachyaceae (*Hydrostachys*), Alangiaceae (*Alangium*), Davidiaceae (*Davidia*), and Mastixiaceae (*Diplopanax*, *Mastixia*) [3, 5–8]. Cornales contains many ecologically and economically important species, including ornamentals in Cornaceae, Davidiaceae and Hydrangeaceae; moreover Camptotheca (Nyssaceae), is the source of camptothecin. Species in the order possess different habits (evergreen, deciduous), diverse growth forms (e.g. trees, shrubs, lianas, rhizomatous and herbs) and occur in tropical, temperate and boreal ecosystems.

The circumscription and phylogenetic relationships of the order have been investigated by a number of

* Correspondence: dzl@mail.kib.ac.cn; gaolm@mail.kib.ac.cn
[1]Key Laboratory for Plant Diversity and Biogeography of East Asia, Kunming Institute of Botany, Chinese Academy of Sciences, Kunming 650201, China
Full list of author information is available at the end of the article

Comparative analyses of plastid genomes from fourteen Cornales species: inferences...

109

phylogenetic analyses, mostly based on plastid DNA, beginning from the early twentieth century such as Olmstead et al. [9] and Chase et al. [10]. Increasing the amount of molecular markers has progressively improved phylogenetic resolution and branch support in Cornales [3, 5, 7, 8]. For example, based on six cpDNA regions and broader taxon sampling, Xiang et al. [3] obtained well supported but not fully resolved intra-family relationships for some families (e.g. Hydrangeaceae, Cornaceae) in this order.

Integrating genomic data into plant phylogenetic investigations is developing rapidly due to the availability of new methods of sampling genomes (e.g. genome skimming, transcriptomes, hybrid capture) facilitated by next-generation sequencing (NGS) technologies [11–14]. Complete plastid genomes have rapidly accumulated in the NCBI databases over the last few years. However, phylogenomic studies remain to be conducted for Cornales, and only a few plastid genomes within this order have been released, sporadically, on NCBI databases. The plastome is usually uniparentally inherited in seed plants [15]. It can provide an abundance of variable sites across its entire length for phylogenetic analyses [16]. Thus, plastid genomes show the potential for resolving recalcitrant phylogenetic relationships, at both high taxonomic levels such as green plants [17–19], and low taxonomic levels [20–22]. The most widely used approach for plastome phylogenomics is to analyze the concatenated coding genes as a whole [14, 20, 23, 24], but the noncoding sequences are also useful for inferring phylogenies at lower taxonomic levels when the plastid genomes are conserved [25–27]. Because different regions of plastid genomes vary in their evolutionary rates, partitioning the genome by genes or regions might be preferable for phylogenomic analysis [20]. Moreover, fast-evolving sites of plastid genomes that cannot be aligned with confidence could possibly result in misleading phylogenetic inference, and therefore removing the most variable sites or problematic regions might improve accuracy in phylogenetic inference [28–30].

Plastid genomes of angiosperms generally contain 110 to 130 distinct genes, and range in size from 120 to 160 kb. They usually show a typical quadripartite circular structure of two copies of large inverted repeat (IR) separated by the small single-copy (SSC) and the large single-copy (LSC) regions [31]. Although the plastid genome is reported as highly conserved in most angiosperms [32], it is subject to structural alterations such as extension or contraction of the IR region [33], the presence of large inversions or deletions [34, 35], pseudogenization and gene loss [36, 37]. Besides their phylogenetic utility, whole plastid genomes could also be used to investigate other aspects of molecular biology such as genome evolution on the structural and molecular level, and to develop fast evolved molecular markers for investigations of phylogeny and phytogeography [17, 27, 38–41].

In the present study, a total of 15 complete plastid genomes of 14 species representing 10 genera and seven families of Cornales were obtained. The main objectives of this study were to 1) characterize and compare the structure and gene organizations of plastid genomes in Cornales; 2) explore the potential effects of different partitioning schemes and alignment strategies on phylogenetic inferences; and 3) assess the application of the complete plastid genome sequences in Cornales, and provide genetic resources for future research.

Methods

Taxon sampling

The circumscription of families of Cornales followed Xiang et al. [3], and taxonomy within families and genera followed Flora of China [42] or the Plant List (http://www.theplantlist.org/) (accessed 1st January, 2013). A total of 15 individuals representing 14 species of 10 genera from 7 families in Cornales mainly occurring in China were sampled. Samples of three families (Grubbiaceae, Hydrostachyaceae and Loasaceae) could not be obtained for this study. The sampled species hence represented four out of the five major lineages suggested by Xiang et al. [3]. Two individuals of *Cornus capitata* were sequenced here to investigate the intra-specific variability within plastid genome. As outgroups, the plastid genome of *Fouquieria diguetii* of Ericales was newly sequenced, and the plastomes of three species within Caryophyllales (*Basella alba*, *Talinella dauphinensis*, *Gisekia pharnaceoides*) were obtained from another parallel work (unpublished data). These two orders are phylogenetically closest to Cornales [1–4]. Fresh leaves were collected in the field or from botanic gardens with the permission of the land owners or the botanic gardens (Table 1) and transferred to the laboratory under cool conditions (~4 °C) for total genomic DNA extraction. Voucher specimens were collected for each species, and deposited at the Herbarium of Kunming Institute of Botany (KUN), Chinese Academy of Sciences or the herbarium of the Royal Botanic Garden Edinburgh (E). Detailed information of the pecies sampled in this study is provided in Table 1.

DNA sequencing and genome assembly

Total genomic DNA was isolated from about 100 mg fresh leaf material with a modified CTAB method [43] in which 4% CTAB was used instead of 2% CTAB and with approximately 0.1% DL-dithiothreitol (DTT) added. Subsequently, plastid DNA was selectively amplified through long-range PCR using nine or fifteen primer pairs [44, 45]. All PCR products were pooled and diluted

Table 1 Taxa sampled in this study

Taxa	Family	Order	Locality	Voucher	Voucher specimen	GenBank accession number
Nyssa wenshanensis	Nyssaceae	Conales	China, Yunnan, Kunming Botanical Garden	Cai J. & Zhang T.	14CS9047	MG524995
Nyssa sinensis	Nyssaceae	Conales	China, Yunnan, Wenshan	Liu C., et al.	14CS8436	MG525000
Camptotheca acuminata	Nyssaceae	Conales	China, Yunnan, Yuxi	Cai J., et al.	13CS7273	MG525005
Davidia involucrata	Davidiaceae	Conales	China, Yunnan, Kunming Botanical Garden	Cai J. & Zhang T.	14CS9049	MG525002
Mastixia caudatilimba	Mastixiaceae	Conales	China, Yunnan, Xishuangbannan,	Guo Y.J., et al.	14CS9459	MG525001
Diplopanax stachyanthus	Mastixiaceae	Conales	China, Yunnan, Wenshan,	Zhang T. & Liu C.	14CS8795	MG524991
Hydrangea heteromalla	Hydrangeaceae	Conales	China, Yunnan, Kunming	Guo Y.J. & Liu C.	10CS1923	MG524994
Hydrangea aspera	Hydrangeaceae	Conales	China, Yunnan, Wenshan	Liu C., et al.	14CS8432	MG524992
Deutzia crassifolia	Hydrangeaceae	Conales	China, Yunnan, Chuxiong	Guo Y.J., et al.	14CS8216	MG524993
Alangium alpinum	Alangiaceae	Conales	China, Yunnan, Kunming Botanical Garden	Yang J.D.	14CS9086	MG525003
Alangium chinense	Alangiaceae	Conales	China, Yunnan, Wenshan	Cai J., et al.	14CS9130	MG524996
Cornus capitate #1	Cornaceae	Conales	China, Yunnan, Kunming	Ya J.D., et al.	14CS9213	MG524990
Cornus capitate #2	Cornaceae	Conales	China, Yunnan, Kunming Botanical Garden	Liu C. & Ya J.D.	14CS8464	MG524998
Cornus controversa	Cornaceae	Conales	China, Yunnan, Kunming Botanical Garden	Liu C. & Ya J.D.	14CS8466	MG525004
Curtisia dentata	Curtisiaceae	Conales	UK, Royal Botanic Garden Edinburgh	Möller M.	RBGE 19240177	MG524999
Fouquieria diguetii	Fouquieriaceae	Ericales	UK, Royal Botanic Garden Edinburgh	Möller M.	RBGE 19800074	MG524997
Basella alba	Basellaceae	Caryophyllales	China, Yunnan, Kunming	Yang J.D. et al.	14CS9526	Unpublished
Talinella dauphinensis	Talinaceae	Caryophyllales	UK, Royal Botanic Garden, Kew	Yi T.S.	Yi14363	Unpublished
Gisekia pharnaceoides	Gisekiaceae	Caryophyllales	China, Hainan, Lingshui,	Zhang T., et al.	14CS8741	Unpublished

to 0.2 ng/μL for library preparation. A short-insertion (500 bp) sequencing library was prepared following the Nextera XT Sample Preparation procedure (Illumina). The paired-end reads of 250 bp or 300 bp were generated using Illumina Miseq at the Laboratory of Molecular Biology of Germplasm Bank of Wild Species, Kunming Institute of Botany, Chinese Academy of Sciences. Four species could not be amplified through long range PCR: *Camptotheca acuminata*, *Cornus controversa*, *Mastixia caudatilimba* and *Hydrangea heteromalla*. These were sequenced instead from total DNA using Illumina Hiseq4000 after short-insert (500 bp) libraries constructed following the manufacturer's protocol (Illumina HiSeq 4000) and 143 bp paired-end reads for *Hydrangea heteromalla* and 90 bp paired-end reads for the other three species, which were generated at BGI Shenzhen, China.

The raw sequence reads were assembled using following steps. First, all reads were de novo assembled into contigs with CLC Genomics Workbench 8.0.2 (CLC Bio) under a word size of 60 bp, minimum contig length of 500 bp and map reads back to contigs with default settings. Second, a closely related genome of *Camellia sinensis* (NC_020019.1) was used as a reference, and contigs of each individual sample were aligned to it using local BLAST, from which the contigs of plastid genome can be selected. For *Cornus controversa* and *Hydrangea heteromalla*, this process produced two and

three long plastid contigs respectively, which were easily assembled into a complete genome by overlaps using Geneious v 8.1 [46]. Among the remaining 14 samples, parts of the genome were covered only by short contigs, which were hard to assemble directly. These were analyzed using the two successfully assembled species as reference sequences, and then manually concatenated by their overlaps in Geneious v 8.1.

Verification of the assembly was performed in three ways: 1) by mapping the reads to the assembled plastid genome sequences, 2) by comparing the 14 manually assembled genomes with two easily assembled ones, and 3) by obtaining the four boundary regions using newly designed primers under Sanger sequencing, which were showed in Additional file 1: Table S1.

Genome annotation and comparison

The complete genome sequences were annotated using the online program DOGMA [47] to predict protein-coding genes, transfer RNA (tRNA) genes, and ribosome RNA (rRNA) genes. Start and stop codons of protein-coding genes were determined using plastid/bacterial genetic codes, with the most closely matching reference genome as a guide. Graphical maps with annotation of genomes were drawn using OrganellarGenomeDRAW-tool (OGDraw) [48].

The 15 whole plastid genomes were aligned with Mauve v 2.3.1 [49] plugin in Geneious v 8.1, including only one copy of the IR, assuming collinear genomes for the full alignment. To compare the overall similarities among different plastid genomes, pairwise alignments of the 15 genomes of Cornales were performed in the mVISTA program [50], under LAGAN mode using the annotations of *Cornus controversa* as reference. Plastomes of Cornales were also aligned using MAFFT [51] and manually edited in Geneious v 8.1. To observe the plastid genome divergence and determine parsimony informative sites, sliding window analysis was conducted after alignment. In order to identify some mutational hotspots, the proportion of mutational events was calculated following a modified version of the formula used by Gielly and Taberlet [52]: the proportion of mutation events = $[(NS + ID)/L] * 100\%$, where NS is the number of nucleotide substitutions, ID is the number of indels and L is the aligned sequence length of each region. Hotspots were here defined as those regions with a value >20%. The step size was set to 200 bp, with a 600 bp window length as described by Xu et al. [27].

To test whether the abnormal gene of *rpl22* is disabled or not, the ratio of nonsynonymous and synonymous (ω, d_N/d_S) of *rpl22* for different branches was calculated in PAML v4.7 [53] using the codeml module.

Alignment and subdivision of plastid genomes

The whole plastid genomes of the 15 individuals of Cornales and the four outgroup species were aligned using the program MAFFT v 7.22 with default settings. Three primary data sets were generated for phylogenetic inference. The first data set comprised coding regions, i.e. exons of protein-coding genes, tRNAs and rRNAs; the second comprised all noncoding regions, i.e. intergenic regions and introns; the third comprised the entire plastid genome. Each gene and intergenic or intron was realigned using MAFFT v 7.22 with G-INS-i algorithm plugin in Geneious v 8.1. One of the IR regions was removed for all data sets to reduce overrepresentation of duplicated sequences.

Some regions in the whole plastome data set are highly variable and poorly aligned. So, in order to assess the effect of alignment quality on phylogeny, we compared the results from three different analysis strategies. First, the unfiltered alignment included all sequence positions of the plastomes in the alignment. Second, the lightly filtered alignment was created using the program Gblocks [54] to remove those regions that were identified as highly variable or ambiguously aligned, using the program's default parameters; only positions where 50% or more of the sequences had a gap were retained. Third, the strictly filtered alignment was generated using the same approach as the lightly filtered alignment, but excluding all those positions that had at least one gap.

Phylogenetic analyses

For the unfiltered, lightly filtered and strictly filtered alignments of coding, noncoding and complete plastome data sets, jModeltest v2.1.6 [55] was used, as implemented on the Cyberinfrastructure for Phylogenetic Research (CIPRES) cluster (http://www.phylo.org/), to estimate the optimal model of molecular evolution with the Akaike Information Criterion (AIC). Maximum likelihood (ML) analyses were conducted using RAxML v8.1.11 [56] as implemented on the CIPRES cluster. These RAxML searches relied on the general time reversible model of nucleotide substitution, with the gamma model of rate heterogeneity (GTR + G) as suggested (see RAxML manual). The ML trees were inferred using the rapid bootstrap with 1000 replicates, and the best-scoring ML tree was sought. Bayesian inference (BI) analyses were conducted with MrBayes v3.2.3 [57] as implemented on the CIPRES cluster with the models estimated for the different data sets (Additional file 1: Table S2). Two runs were conducted in parallel with four Markov chains (one cold and three heated), with each running for 2,000,000 generations from a random tree and sampled every 100 generations. The convergence was checked using the average standard deviation of split frequencies (ASDFs) (<0.01). The first 25% of the trees were discarded as burn-in, and the remaining trees were used to construct majority-rule consensus trees.

To investigate the issues of data partitioning for the plastid phylogenomic analysis, an algorithmic method for estimating an optimal partitioning scheme was conducted for the complete unfiltered data set. It was partitioned into the maximum possible number of data blocks based on genomic composition. We divided the whole plastid genome into 174 subsets: each gene, intergenic region or intron was regarded as a distinct subset, while subsets of less than 200 bp, or regions that only contained invariable nucleotide sites, were combined into large data subsets according to their function (see details in Additional file 1: Table S3). Subsequently, the program PartitionFinder v1.1.1 [58] was used to identify the best partitioning schemes of these 174 subsets according to the Bayesian information criterion (BIC) using a heuristic search (search = rcluster).

For partitioned ML phylogenetic analysis, a partitioned model was used to specify the regions of alignment, for which an individual model of nucleotide substitution was estimated. Individual per-partition branch lengths were estimated using RAxML v8.1.11 software. For partitioned BI phylogeny estimation, each partition was given its own optimal model (GTR + G or GTR + G + I) (Additional file 1: Table S3). All parameters were set to be unlinked across partitions except those for branch lengths and topology; branch length rate multipliers were unlinked in MrBayes v3.2.3.

Results
Characteristics of the plastid genomes

Fifteen complete plastid genomes of Cornales, plus *Fouquieria diguetii* of Ericales were newly generated in this study; these genome sequences have been submitted to GenBank (Table 1). The mean coverage depth of these plastomes ranged from 383× (*Alangium chinense*) to 2757× (*Camptotheca acuminata*). Henceforth, all text describing plastid genomes refers only to Cornales unless stated otherwise. The size of the 15 Cornales plastid genomes ranged from 156,567 bp in *Nyssa sinensis* to 158,715 bp in *Diplopanax stachyanthus*, and both individuals of *Cornus capitata* examined had the same plastid genome size (157,200 bp) (Table 2). All of the 16 sequenced plastid genomes displayed a typical quadripartite structure (Fig. 1), comprising a pair of IRs (25,859–26,451 bp) separated by the LSC (86,089–87,835 bp) and the SSC (18,250–18,856 bp) regions (Table 2). The LSC regions exhibited the greatest standard deviation in sequence length (s.d. = 586 bp), followed by SSC regions (s.d. = 188 bp) and the IR regions (s.d. = 147 bp). The full genomes encoded 114 unique genes, which included 31 tRNA genes, four rRNA genes and 79 protein-coding genes with the same gene order. There were 16 genes duplicated in the IR regions, resulting in a total of 130 genes (Additional file 1: Table S4). Seventeen of those genes contained one intron, and two genes (*ycf3* and *clpP*) contained two

introns. The length and GC content of coding, noncoding and complete plastid genome data sets are shown in Table 2. Noncoding regions (s.d. = 789 bp) showed more variation in sequence length than coding regions (s.d. = 115 bp). Among Cornales species, the percentage of the coding regions varied from 57.2% to 58.2%. The overall GC content is similar across individuals in coding and noncoding regions of Cornales, but a little higher than that of *Fouquieria diguetii* of Ericales (Table 2).

Boundaries between the IR and SSC/LSC regions were verified by Sanger sequencing; the results were identical with the NGS sequencing. Variation in the positions of the boundaries between IR and SSC/LSC are usually considered to be the primary mechanism causing length variation among the plastid genomes of higher plants (Kim and Kim, [59]), but only slight variation was detected within Cornales (Fig. 2). The IRa/LSC junction was located within the *rps19* gene in all but two species (*Hydrangea davidii* and *Deutzia crassifolia*), resulting in the presence of a part of the *rps19* gene in the IRb. In *Hydrangea davidii* and *Deutzia crassifolia*, the junction was located in the *rps19-rpl2* spacer. The IRb/SSC boundary positions in all species were located in the *ycf1* gene, with part of this gene duplicated from 972 to 1246 bp. The *ndhF* gene in seven species was completely located in the SSC region, whereas in the others it extended fractionally into the IRa region (Fig. 2).

Table 2 The plastid genome features of the sequenced species

Taxon	Full		LSC length (bp)	SSC length (bp)	IR length (bp)	Gene Number	Protein-coding	RNAs	Coding region		Noncoding region		Mean Coverage
	Length (bp)	GC (%)							Length (bp)	GC (%)	Length (bp)	GC	
Nyssa wenshanensis	156,598	37.9	86,109	18,261	26,114	114	79	35	91,073	40.3	65,525	34.6	974
Nyssa sinensis	156,567	37.9	86,089	18,250	26,114	114	79	35	91,073	40.3	65,494	34.6	802
Camptotheca acuminata	157,811	37.8	87,333	18,760	25,859	114	79	35	91,078	40.3	66,772	34.4	2757
Davidia involucrata	158,131	37.8	87,335	18,856	25,970	114	79	35	91,037	40.3	67,094	34.4	1026
Mastixia caudatilimba	158,221	37.8	87,418	18,797	26,003	114	79	35	90,962	40.3	67,259	34.4	1889
Diplopanax stachyanthus	158,715	37.8	87,679	18,632	26,202	114	79	35	90,944	40.2	67,771	34.6	1758
Hydrangea heteromalla	157,889	37.8	86,907	18,738	26,122	114	79	35	91,138	40.1	66,751	34.7	937
Hydrangea aspera	157,637	37.8	86,815	18,646	26,088	114	79	35	91,189	40.2	66,448	34.5	766
Deutzia crassifolia	157,035	37.6	86,583	18,714	25,869	114	79	35	91,099	40.1	65,936	34.1	549
Alangium alpinum	156,673	37.7	86,181	18,592	25,950	114	79	35	90,842	40.2	65,831	34.2	2121
Alangium chinense	156,684	37.7	86,185	18,603	25,948	114	79	35	90,824	40.2	65,860	34.2	383
Cornus capitate #1	157,200	38.2	86,564	18,412	26,112	114	79	35	90,928	40.5	66,272	35.0	1068
Cornus capitate #2	157,200	38.2	86,564	18,412	26,112	114	79	35	90,928	40.5	66,272	35.0	2523
Cornus controversa	158,668	37.8	87,835	18,705	26,064	114	79	35	90,823	40.4	67,845	34.3	573
Curtisia dentata	158,548	37.7	87,158	18,490	26,450	114	79	35	91,018	40.2	67,530	34.3	538
Fouquieria diguetii	157,895	37.3	87,321	18,482	26,046	114	79	35	91,244	39.9	66,651	33.7	1195

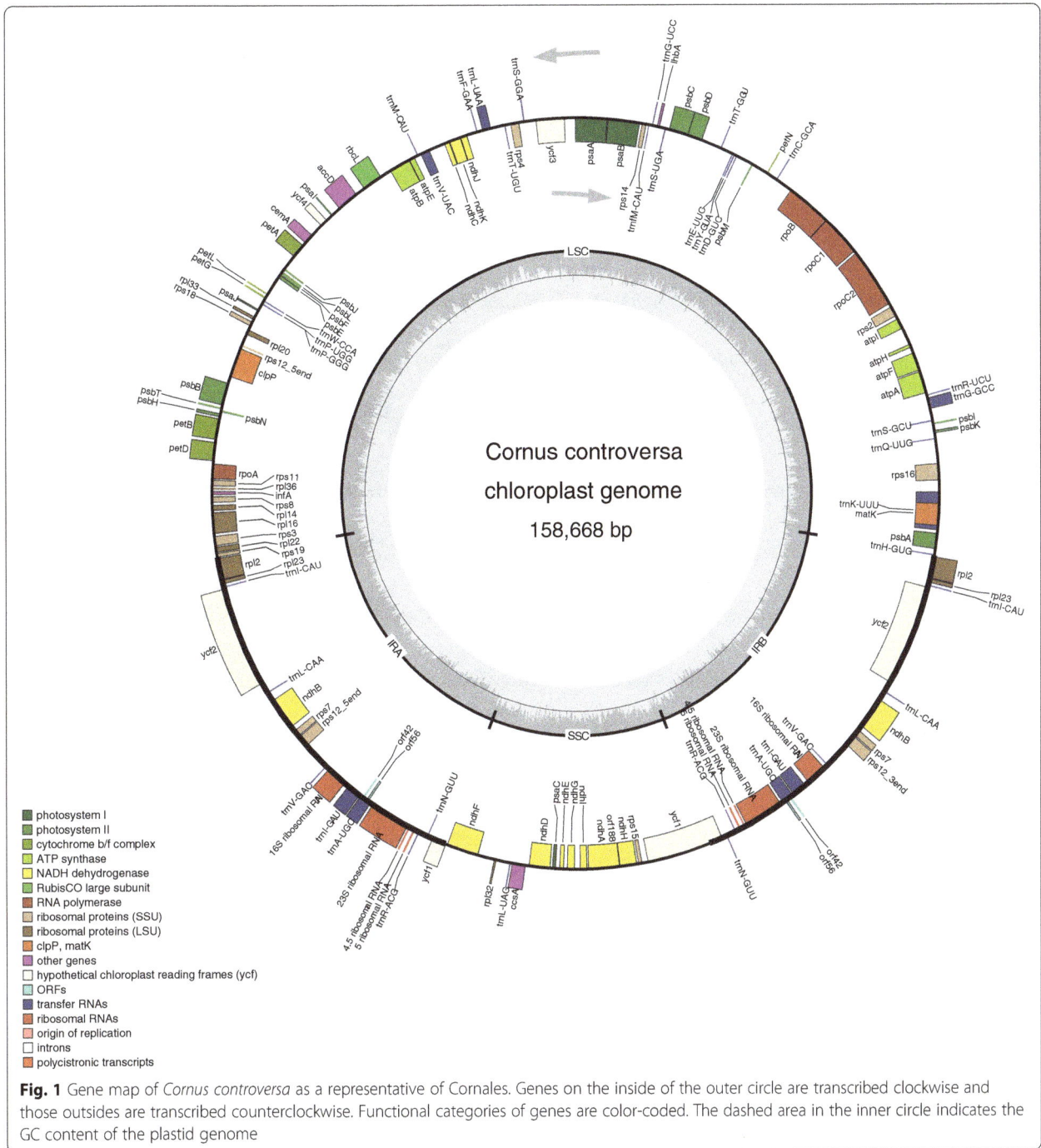

Fig. 1 Gene map of *Cornus controversa* as a representative of Cornales. Genes on the inside of the outer circle are transcribed clockwise and those outsides are transcribed counterclockwise. Functional categories of genes are color-coded. The dashed area in the inner circle indicates the GC content of the plastid genome

Genome sequence divergence among Cornales

The plastid genomes within Cornales showed high sequence similarities with identities of only a few regions below 90% (Additional file 2: Fig. S1), suggesting a high conservatism of plastid genomes within Cornales. The IR regions and coding regions were more conserved than the single-copy regions and noncoding regions (Additional file 2: Fig. S1).

Slide window analysis also showed much higher proportions of both mutation events and parsimony-informative sites in single-copy regions than in the IR region. From this, nine relatively highly variable regions (mutational hotspots) were identified from the plastid genomes, which might be undergoing more rapid nucleotide substitution. These comprised 2 gene regions and 7 intergenic regions: *matK, ndhF, trnK-rps16, rpoB-*

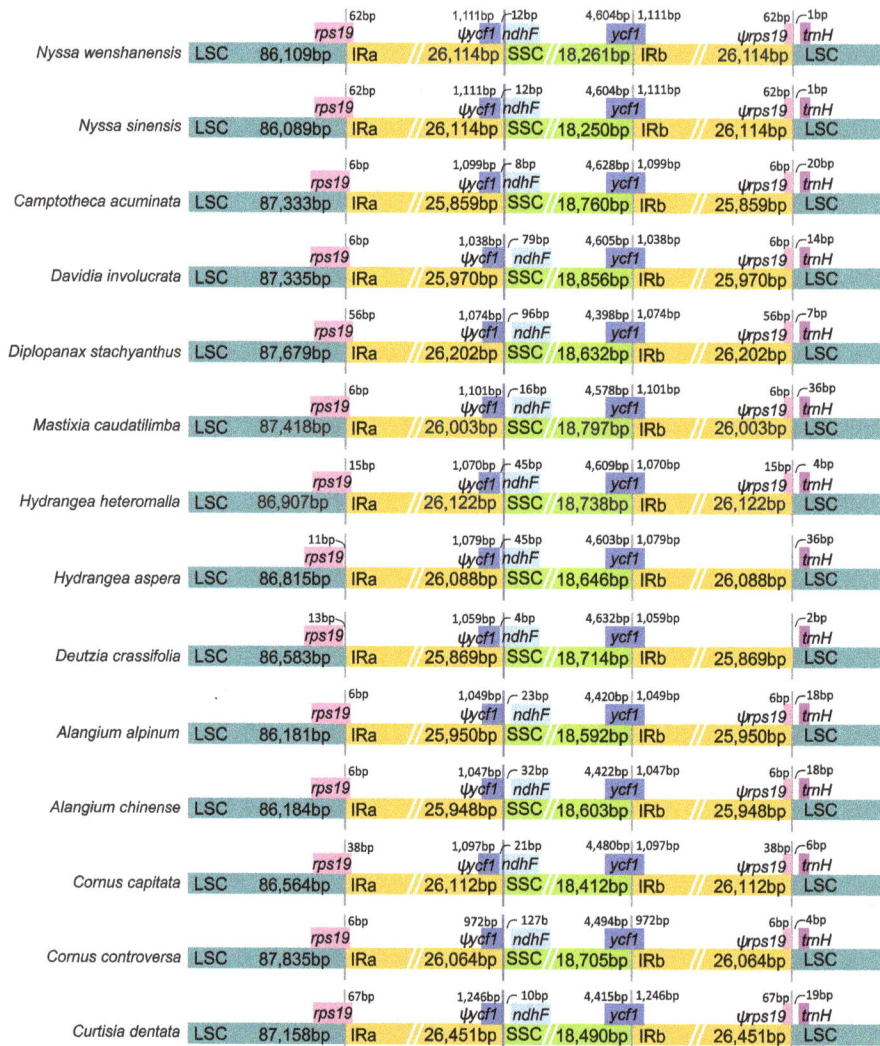

Fig. 2 Sliding window analysis of the whole chloroplast genomes of Cornales taxa

trnC, trnT-trnE, petA-psbJ, psbE-petL, rpl32-trnL, and *rps15-ycf1* regions (Fig. 3). These regions are potential molecular markers for application in phylogeny and phytogeography investigations.

Internal stop codons and putative loss of gene function

The genes *rpl22* and *ycf15* were interrupted by internal stop codons in seven and five Cornales species respectively. Both of them were further verified by Sanger sequencing using newly designed primers (Additional file 1: Table S1); the results were identical to the NGS-based plastid genome sequences.

For all species from Cornaceae, Alangiaceae and Curtisiaceae, a frameshift mutation generated premature termination codons within *rpl22*. Furthermore, *Mastixia caudatilimba* had one base change from G to A within *rpl22*, resulting in an internal termination codon (TGG to TGA) (Fig. 4a). Furthermore, this gene had a 19-bp and 5-

bp insertion in *Cornus capitata* and *C. controversa* respectively, plus a 1 bp deletion in both *Alangium* species, and a 1 bp insertion in *Curtisa dentata*, all occurring upstream of the internal stop codon (Fig. 4a). *rpl22* was found to be truncated in some species, with considerable length variation (384 bp to 474 bp). Despite this, the gene still exhibited nearly 80% nucleotide identities between species, with no big difference between those species with and those without internal stop codons. Furthermore, the ratio of nonsynonymous and synonymous (ω, d_N/d_S) of *rpl22* for different branches showed similar values in both the Cornaceae-Alangiaceae-Curtisiaceae clade ($\omega = 0.34569$) and Mastixiaceae clade ($\omega = 0.35594$), and no significant difference with background ($\omega = 0.36549$, $P > 0.33$) was found. This indicated that those genes containing stop codons have not accumulated mutations at an increased rate, and hence may not have lost their functions.

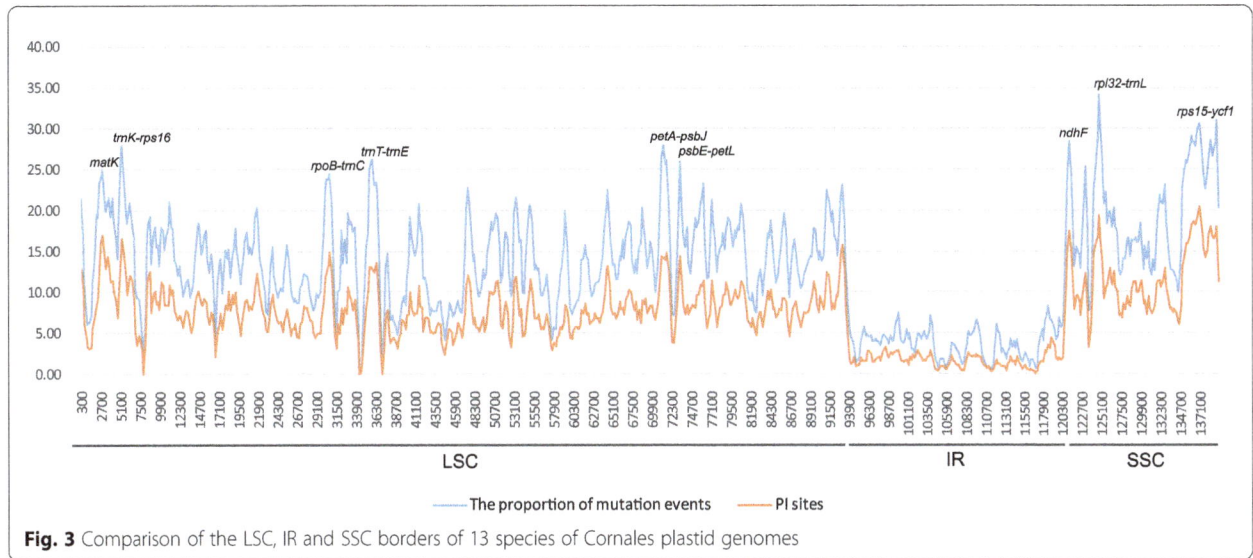

Fig. 3 Comparison of the LSC, IR and SSC borders of 13 species of Cornales plastid genomes

The gene *ycf15* varied from 102 bp to 249 bp among the 15 sequenced individuals of Cornales. For two species of *Alangium*, *ycf15* contained a large deletion (84 bp) at the 5′ end and a 10-bp deletion near the 3′ end, potentially causing a loss of function. Additional 4-bp and 5-bp deletions within *ycf15* led to internal stop codons in *Davidia involucrata* and *Hydrangea aspera*, respectively. Furthermore, in *Deutzia crassifolia*, a single substitution (G to A) within *ycf15* likewise resulted in an internal stop codon (TAG). In the remaining nine Cornales species, *ycf15* did not contain stop codons, and there was no evidence of loss of function (Fig. 4b). Because of these parallel function losses, *ycf15* was not annotated in this study.

Phylogenetic analyses

The unfiltered whole plastid genome data set, with one copy of the IR region excluded, was 148,838 bp in length. Variable and parsimony informative sites of this data set were 25.5% and 15.5%, respectively. The noncoding regions were more variable than the coding regions (33.9% vs 17.5% variable sites and 20.3% vs 11.1% parsimony informative sites) (Table 3). Compared to the unfiltered alignment, a total of 22,247 sites (14.9%) in the lightly filtered data set and a total of 42,513 sites (28.6%) in the strictly filtered alignment were removed. The unfiltered and lightly filtered data sets showed similar percentages of variable and parsimony informative sites, irrespective of calculation for the different regions or the complete genome. However, the strictly filtered alignment exhibited a somewhat decreased percentage of variable and parsimony informative sites in the all data sets (Table 3).

Using both ML and BI methods without data partitioning, the phylogenetic inference of Cornales from the whole unfiltered data set provided complete resolution of relationships among all species sampled, with maximum support (100%/1.0) for all nodes (Fig. 5). Nyssaceae was monophyletic and sister in turn to Davidiaceae, then a monophyletic Mastixiaceae, then a monophyletic Hydrangeaceae. Cornales comprised this clade plus another, in which Cornaceae and Alangiaceae (both monophyletic) were together sister to Curtisiaceae (Fig. 5). The phylogenetic topology of Cornales based on unfiltered coding and noncoding regions were consistent with that from the complete plastome data set. Only the sister relationship of Curtisiaceae and Cornaceae-Alangiaceae received support values below 99% or 1.0 from unfiltered data sets, with 81%/1.0 support from coding regions (Table 3, Fig. 5, Additional file 2: Fig. S2).

Likewise, using lightly filtered and strictly filtered data sets, both the topology and support values were almost identical (Additional file 2: Fig. S3, S4). However, when only coding data sets are used, the bootstrap support value for the sister relationship of Curtisiaceae and Cornaceae-Alangiaceae drops to 78% with strictly filtered alignment (Additional file 2: Fig. S4A).

When partitioning was applied using the program PartitionFinder, the whole unfiltered data set was divided into 13 partitions (Table S3). Topology and support values obtained from this analysis were consistent with unpartitioned analysis, except for a decrease in BS support value from 100% to 95% for the sister relationship of Curtisiaceae to Cornaceae-Alangiaceae in ML analysis (Additional file 2: Fig. S2C).

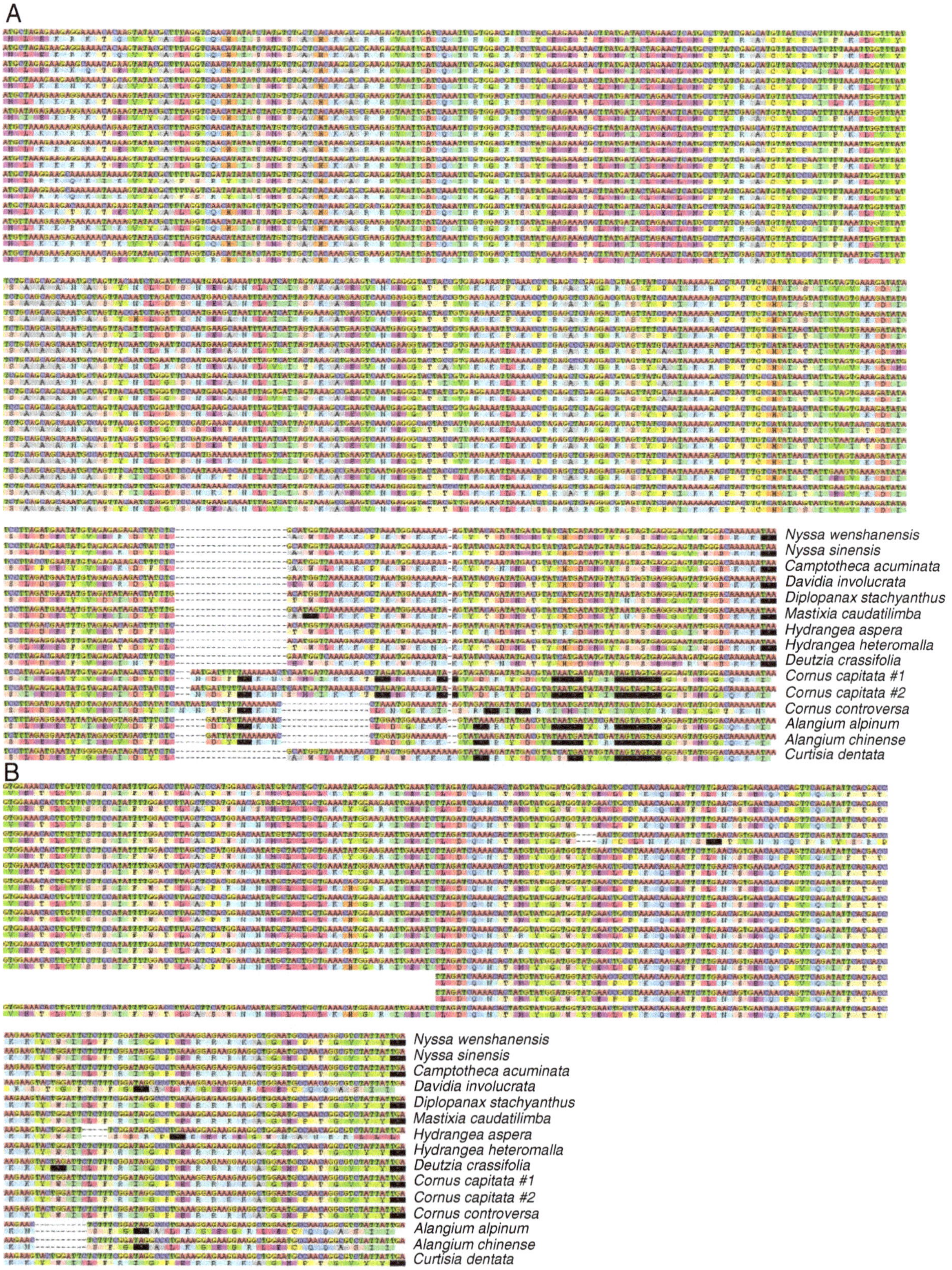

Fig. 4 Alignment of two abnormal genes among Cornales and outgroups. **a** *rpl22* gene; (**b**) *ycf15* gene

Table 3 Sequence alignment information and support values for key nodes under different alignment strategies

Data set	Blocks	Number of sites	Variable sites	Parsimony informative sites	Support value (LB/PP)	
					(Cornaceae-Alangiaceae)-Curtisiaceae	(Mastixiaceae-Davidiaceae-Nyssaceae) - Hydrangeaceae
Coding	Unfiltered	75,334	13,147(17.5%)	8336(11.1%)	81/1.0	99/1.0
	Light filtered	74,352	13,040(17.5%)	8305(11.2%)	82/1.0	100/1.0
	Strict filtered	72,369	12,267(17.0%)	7809(10.8%)	78/1.0	100/1.0
Noncoding	Unfiltered	72,056	24,406(33.9%)	14,643(20.3%)	99/1.0	100/1.0
	Light filtered	51,343	18,974(37.0%)	12,232(23.8%)	99/1.0	100/1.0
	Strict filtered	32,852	10,167(30.9%)	6545(19.9%)	100/1.0	100/1.0
Complete	Unfiltered	148,838	37,928(25.5%)	23,136(15.5%)	100/1.0	100/1.0
	Light filtered	126,591	32,266(25.5%)	20,678(16.3%)	99/1.0	100/1.0
	Strict filtered	106,325	22,745(21.4%)	14,551(13.7%)	100/1.0	100/1.0

Discussion

Structure of plastome and comparative analyses

In the present study, the complete plastid genomes of 14 species of Cornales were obtained for the first time. They showed the typical quadripartite structure of most angiosperms, including a pair of IR regions, separated by an LSC and an SSC region. The Cornales plastid genome was highly conserved in structure compared to most angiosperms [32], with all sampled species encoding the same set of 114 unique genes in same gene order (Table 2). The GC content was around the average for plant plastomes (GC = 37%) [60], but was slightly higher than

that of the outgroup taxa. The length variation of the Cornales plastid genomes observed here was low (156–159 kpb), and differences were mainly due to variation in noncoding regions (65–68 kbp). Length variation of plastid genomes was previously shown to result from expansion and contraction of the inverted repeat regions [61]. Here, we also found that the IR/SSC boundary located differently among the 14 species, but the location of boundary and length of IR regions only showed moderate variation (Fig. 2). Furthermore, there was no obvious phylogenetic implication of extension/ contraction of IRs among the Cornales plastomes.

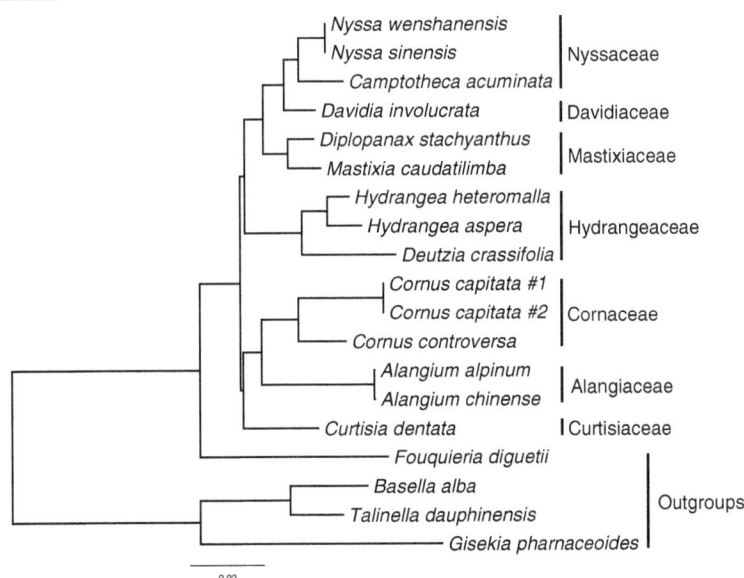

Fig. 5 Phylogenetic relationships of Cornales based on unfiltered whole plastid genome alignment. Nodes without values represent maximal support in both ML and BI methods

Premature stop codons in two genes, but no apparent loss of function in *rpl22*

The gene content is highly conserved among plastid genomes of land plants, although gene loss has been reported in several angiosperm lineages [17]. Two genes, *rpl22* and *ycf15*, contained premature termination codons in several species in the present study. *rpl22* showed premature termination codons in *Alangium*, *Cornus*, *Curtisia*, and *Mastixia*, making it about 20% shorter in these species compared to others. This gene appears to be absent from plastids in some taxa such as legumes [62] and was reported to have been transferred to the nucleus in both *Pisum* and Fagaceae [63, 64]. In Cornales, *rpl22* in those species with internal stop codons have not undergone a detectable increase in mutation rate compared to those without them, whereas such an increase would be expected if the gene was disabled in the former but functional in the latter. Moreover, the former group still contain nearly 80% of the normal gene sequence. This implies that *rpl22* in the plastid either functions as a gene in all examined species, or in none of them. If the former, truncation does not remove its function. If the latter, then it might be a pseudogene in all Cornales, as it is in *Citrus sinensis* [65]; if so it might have a non-coding function in the plastid. Possibly, a functional copy of *rpl22* might exist in the nucleus, as in *Pisum* and Fagaceae [63, 64], removing any selective disadvantage to loss of function in the chloroplast. Therefore, Cornales might be in the early stages of a process of losing *rpl22* from the plastid. Hence more data is needed, regarding function of *rpl22* in the chloroplast and whether a full copy exists in the nucleus.

The nucleotide sequence of *ycf15* has been shown to vary among angiosperm plastid genomes, with conserved motifs at 5′ to 3′ ends in some taxa (like tobacco) and an intervening region of about 250 bp in some other taxa (like *Eucalyptus globulus*) that renders it as a pseudogene [66]. A comparative study of *ycf15* transcripts in taxa with or without the insertion suggested that this gene may not be a protein-coding gene even when it is intact [67]. Although transcripts of *ycf15* were detected in some taxa like *Camellia*, it may have been removed from the pre-mRNA after transcription in order to activate the function of other genes, thus *ycf15* is possibly an intergenic sequence without function [68]. The non-coding *ycf15* hypothesis to some extent is supported by data from Cornales, within which four independent mutation events within *ycf15* either inserted stop codons (*Davidia*, *Hydrangea* and *Deutzia*) or deleted parts of the gene (*Alangium*). The evolutionary patterns of *ycf15* showed that they evolved in a discontinuous fashion across angiosperms [68] [69]. It shows an intact and conserved structure in nine Cornales

species, but cannot be translated normally in species of *Alangium*, *Davidia* and *Deutzia*. These three genera belong to distinct clades, implying separate and independent alterations in each case; hence *ycf15* might not provide phylogenetic implication (Fig. 5).

Influence of data set subdivision, alignment filtering, and data partitioning on phylogeny

In addition to the complete plastid genome, two data subsets were generated, one comprising all coding genes, and the other only noncoding regions. We conducted three filtering strategies (none, light and strict) on each of these three data sets. Phylogenetic inference from BI and ML analyses based on all data sets provided the same topology. All data sets supported the sister relationship between Curtisiaceae and Cornaceae-Alangiaceae, support for this clade from the coding region and its filtered data subsets was relatively low, e.g. 81%/1.0 for unfiltered (Additional file 2: Fig. S2A). Although the noncoding regions are usually excluded for phylogenomic analyses at high taxonomic levels [14, 24], the phylogenetic resolution within Cornales obtained from noncoding regions in all three strategies was high (Additional file 2: Fig. S2, S3, S4; Table 3). This might because plastid genomes within the order have a conserved collinear structure, and the noncoding regions can provide more phylogenetic signals. The treatment of problematic or ambiguous regions in alignments can affect the final phylogenetic relationships, and for alignments that are long enough, removal of problematic regions leads to better phylogenetic resolution [69, 70]. Conversely, in this study, alignment filtering has no influence in any of the coding, noncoding and complete alignment data sets (Table 3), which may also be due to the conservation of plastid genomes within Cornales.

When a genome-scale approach is adopted in phylogenetic analyses, partitioning is one of the most popular methods used to model the heterogeneity of molecular evolution among regions in an alignment for phylogenetic inference [71]. In the present study, however, data partitioning by PartitionFinder had no effect on the topology of the resulting phylogenetic trees compared to unpartitioned plastid genome data set. The phylogenetic relationships of Cornales were robustly resolved based on both partitioned and unpartitioned datasets. It was indicated that the longer the data set was, the less likely that the results will be affected by partitioning scheme [71]. This is perhaps because the whole plastid genome contains sufficient amount of phylogenetic signals (while noise is randomly dispersed) and may converge on the correct phylogenetic tree, irrespective of partitioning. It was also observed, in a previous phylogenetic study of Cornales with six plastid fragments, that the partitioned data sets presented the same topology as the unpartitioned ones with only some differences among the branch support values [3]. Irrespective of different regions or data subsets, partitioned or unpartitioned data sets used, our results suggested that the plastid genome as a whole contains sufficient phylogenetic signals in different regions within Cornales to fully resolve the phylogenetic relationships. The conservatism in genome structure and gene content along with abundance of phylogenetic signal of Cornales plastid genomes, both coding and noncoding regions, make it a valuable phylogenetic tool, at and below family level.

Phylogenetic implication among Cornales with plastid genome

Preliminary phylogenetic frameworks for Cornales have previously been provided based on a few molecular markers, but relationships among families of the order tended to be poorly resolved [5, 7, 72]. A later study [3], based on six plastid loci (*rbcL*, *matK*, *ndhF*, *atpB*, *trnL-F* and *trnH-K*) and a broader taxon sampling, recovered five major clades: Cornaceae-Alangiaceae, Curtisiaceae-Grubbiaceae, Mastixiaceae-Nyssaceae-Davidiaceae, Hydrostachyaceae, and Hydrangeaceae-Loasaceae; relationships between these clades were well supported. However, cpDNA-based relationships were contradicted by 26S rDNA data in the study, and relationships within some families (e.g. Hydrangeaceae, Cornaceae) were not fully recovered [3]. Therefore, further work is needed on relationships within these families. Our own work recovered identical relationships with even higher support, demonstrating that greater genome coverage can compensate for reduced taxon sampling, at least in some cases. The sequence and structure of the whole plastid genome has been recognized for its great potential to resolve relationships for phylogenetically recalcitrant plant groups [14, 21, 40, 73, 74]. Given that resources will seldom permit full genome sequencing across large numbers of taxa, the best strategy for wide taxonomic sampling is to identify marker regions that contain a high proportion of phylogenetically useful information. To this end, our study identified two genes (*matK*, *ndhF*) and seven regions *trnK-rps16*, *rpoB-trnC*, *trnT-trnE*, *petA-psbJ*, *psbE-petL*, *rpl32-trnL*, and *rps15-ycf1* that are mutational hotspots, and are hence recommended as phylogenetic markers within Cornales, and perhaps beyond it.

Conclusions

Phylogenomic data have rapidly accumulated and been broadly used for resolving phylogenetic relationships in the last few years. In the present study, fifteen full plastid genomes of 14 Cornales species were sequenced to investigate the phylogenetic relationships and plastome evolution of Cornales. Comparative analysis of the plastid genomes revealed that plastomes of the order have a conserved collinear structure with identical gene content and order. Two genes (*rpl22* and *ycf15*) contained premature stop codons in seven and five species respectively. Plastid genomes showed strong potential for resolving phylogenetic relationships within Cornales, both for the interfamily and intrafamily relationships, with very strong support. Different partitioning schemes and filtering strategies (none, light and strict) of sequence data sets have no effect on phylogenetic relationships. The topology recovered from coding and noncoding data sets was likewise identical to that for the whole plastome. However, the coding data set provided lower support values than the latter two data sets. Mutational hotspots and highly informative regions of Cornales were identified. All data presented here are fundamental to phylogenomic analyses of Cornales, and will be a useful genomic resource for future studies of evolutionary biology.

Additional files

> **Additional file 1: Table S1.** The primers newly designed in this study for four junctions and two genes (*rpl22*, *ycf15*). **Table S2.** Molecular models selected for all the data sets of the three alignment strategies. **Table S3.** Model selected for each data partition identified by software PartitionFinder for unfiltered complete plastid genomes. **Table S4.** Gene category and gene contained in plastid genomes of Cornales. (DOC 147 kb)
>
> **Additional file 2: Fig. S1.** A percent identity plot showing the overall sequence similarity of the fourteen Cornales plastid genomes. **Fig. S2.** Phylogenetic relationships of Cornales based on three different data sets with light filtered alignment. **Fig. S3.** Phylogenetic relationships of Cornales based on three different data sets with strict filtered alignment. (PDF 1121 kb)

Abbreviations

AIC: Akaike information criterion; BI: Bayesian inference; BIC: Bayesian information criterion; GC: Guaninecytosine; IR: Inverted repeat region; LB: Likelihood bootstrap; LSC: Large single copy region; ML: Maximum likelihood; PP: Posterior probability; SSC: Small single region

Acknowledgements

We are grateful to Jun-Bo Yang, Xiang-Qin Yu, Jian-Jun Jin, Gang Yao and Zhi-Rong Zhang for their help with data analysis and laboratory work. We also thank Dr. Michael Möller from Royal Botanic Garden Edinburgh for the constructive comments and English editing for the early version of the MS. We thank Jie Cai, Ting-Shuang Yi, Cheng Liu, Yong-Jie Guo and Ji-Dong Ya from the CAS Kunming Institute of Botany, and the Royal Botanic Garden Edinburgh and Royal Botanic Gardens, Kew for providing samples.

Funding

This study was supported by the National Key Basic Research Program of China (2014CB954100), the Large-scale Scientific Facilities of the Chinese Academy of Sciences (Grant No: 2017-LSFGBOWS-01), Major program of CAS Kunming Institute of Botany (2014KIB02), and the Applied Fundamental Research Foundation of Yunnan Province (2014GA003).

Authors' contributions

LG and DL designed the research, TZ, CF, LG and DL collected samples, CF, JY and HL collected the data, CF and PM analyzed the data, CF, LG, PM, RM and DL wrote and revised the manuscript. CF performed the study, participated in the data analysis and wrote the manuscript. HL participated in the data analysis. RM revised the manuscript. TZ participated in samples collection. PM participated in the data analysis and manuscript revising. JY participated in the DNA sequencing. DL designed the research and revised the manuscript. LG designed the research, collected study materials and revised the manuscript. All authors read and approved the final manuscript.

Competing interests

The authors declare that they have no competing interests.

Author details

[1]Key Laboratory for Plant Diversity and Biogeography of East Asia, Kunming Institute of Botany, Chinese Academy of Sciences, Kunming 650201, China. [2]University of Chinese Academy of Sciences, Beijing 100049, China. [3]Germplasm Bank of Wild Species in Southwest China, Kunming Institute of Botany, Chinese Academy of Sciences, Kunming 650201, China. [4]Institute of Molecular Plant Sciences, University of Edinburgh, King's Buildings, Edinburgh, Scotland EH9 3JH, UK.

References

1. Soltis DE, Soltis PS, Chase MW, Mort ME, Albach DC, Zanis M, et al. Angiosperm phylogeny inferred from 18S rDNA, *rbcL*, and *atpB* sequences. Bot J Linn Soc. 2000;133:381–461.
2. The Angiosperm Phylogeny Group. An update of the angiosperm phylogeny group classification for the orders and families of flowering plants: APG IV. Bot J Linn Soc. 2016;181(1):1–20.
3. Xiang QY, Thomas DT, Xiang QP. Resolving and dating the phylogeny of Cornales–effects of taxon sampling, data partitions, and fossil calibrations. Mol Phylogenet Evol. 2011;59(1):123–38.
4. Zeng L, Zhang N, Zhang Q, Endress PK, Huang J, Ma H. Resolution of deep eudicot phylogeny and their temporal diversification using nuclear genes from transcriptomic and genomic datasets. New Phytol. 2017;214(3):1338–54.
5. Fan CZ, Xiang QY. Phylogenetic analyses of Cornales based on 26S rRNA and combined 26S rDNA-*matK-rbcL* sequence data. Am J Bot. 2003;90(9):1357–72.
6. Xiang QY. Cornales (dogwood). In: eLS. Chichester: John Wiley & Sons, Ltd; 2005. http://www.els.net
7. Xiang QY, Moody ML, Soltis DE, Fan CZ, Soltis PS. Relationships within Cornales and circumscription of Cornaceae-*matK* and *rbcL* sequence data and effects of outgroups and long branches. Mol Phylogen Evol. 2002;24(1):35–57.
8. Xiang QY, Manchester SR, Thomas DT, Zhang W, Fan CZ. Phylogeny, biogeography, and molecular dating of cornelian cherries (*Cornus*, Cornaceae): tracking tertiary plant migration. Evolution. 2005;59(8):1685–700.
9. Olmstead RG, Kim KJ, Jansen RK, Wagstaff SJ. The phylogeny of the Asteridae Sensu late based on chloroplast *ndhF* gene sequences. Mol Phylogen Evol. 2000;16(1):96–112.
10. Chase MW, Soltis DE, Olmstead RG, Morgan D, Les DH, Mishler BD, et al. Phylogenetics of seed plants - an analysis of nucleotide-sequences from the plastid gene *rbcL*. Ann Mo Bot Gard. 1993;80(3):528–80.
11. Straub SCK, Parks M, Weitemier K, Fishbein M, Cronn RC, Liston A. Navigating the tip of the genomic iceberg: next-generation sequencing for plant systematics. Am J Bot. 2012;99(2):349–64.
12. Zeng LP, Zhang Q, Sun RR, Kong HZ, Zhang N, Ma H. Resolution of deep angiosperm phylogeny using conserved nuclear genes and estimates of early divergence times. Nat Commun. 2014;5:12.
13. Twyford AD, Ness RW. Strategies for complete plastid genome sequencing. Mol Ecol Resour. 2017;17(5):858–68.
14. Barrett CF, Baker WJ, Comer JR, Conran JG, Lahmeyer SC, Leebens-Mack JH, et al. Plastid genomes reveal support for deep phylogenetic relationships and extensive rate variation among palms and other commelinid monocots. New Phytol. 2016;209(2):855–70.
15. Birky CW. Uniparental inheritance of mitochondrial and chloroplast genes: mechanisms and evolution. Proc Natl Acad Sci U S A. 1995; 92(25):11331–8.
16. Tonti-Filippini J, Nevill PG, Dixon K, Small I. What can we do with 1000 plastid genomes? Plant J. 2017;90(4):808–18.
17. Jansen RK, Cai Z, Raubeson LA, Daniell H, Depamphilis CW, Leebens-Mack J, et al. Analysis of 81 genes from 64 plastid genomes resolves relationships in angiosperms and identifies genome-scale evolutionary patterns. Proc Natl Acad Sci U S A. 2007;104(49):19369–74.
18. Moore MJ, Bell CD, Soltis PS, Soltis DE. Using plastid genome-scale data to resolve enigmatic relationships among basal angiosperms. Proc Natl Acad Sci U S A. 2007;104(49):19363–8.
19. Ruhfel BR, Gitzendanner MA, Soltis PS, Soltis DE, Burleigh JG. From algae to angiosperms-inferring the phylogeny of green plants (Viridiplantae) from 360 plastid genomes. BMC Evol Biol. 2014;14:26.
20. Xi ZX, Ruhfel BR, Schaefer H, Amorim AM, Sugumaran M, Wurdack KJ, et al. Phylogenomics and a posteriori data partitioning resolve the cretaceous angiosperm radiation Malpighiales. Proc Natl Acad Sci U S A. 2012;109(43): 17519–24.
21. Ma PF, Zhang YX, Zeng CX, Guo ZH, Li DZ. Chloroplast phylogenomic analyses resolve deep-level relationships of an intractable bamboo tribe Arundinarieae (poaceae). Syst Biol. 2014;63(6):933–50.
22. Zhang SD, Jin JJ, Chen SY, Chase MW, Soltis DE, Li HT, et al. Diversification of Rosaceae since the late cretaceous based on plastid phylogenomics. New Phytol. 2017;214(3):1355–67.
23. Stull GW. Duno de Stefano R, Soltis DE, Soltis PS. Resolving basal lamiid phylogeny and the circumscription of Icacinaceae with a plastome-scale data set. Am J Bot. 2015;102(11):1794–813.
24. Zhang N, Wen J, Zimmer EA. Another look at the phylogenetic position of the grape order Vitales: chloroplast phylogenomics with an expanded sampling of key lineages. Mol Phylogen Evol. 2016;101:216–23.
25. Cai J, Ma PF, Li HT, Li DZ. Complete plastid genome sequencing of four *Tilia* species (malvaceae): a comparative analysis and phylogenetic implications. PLoS One. 2015;10(11):e0142705.
26. Reginato M, Neubig KM, Majure LC, Michelangeli FA. The first complete plastid genomes of Melastomataceae are highly structurally conserved. PeerJ. 2016;4:16.
27. Xu C, Dong W, Li W, Lu Y, Xie X, Jin X, et al. Comparative analysis of six *lagerstroemia* complete chloroplast genomes. Front Plant Sci. 2017;8:15
28. Zhong BJ, Deusch O, Goremykin VV, Penny D, Biggs PJ, Atherton RA, et al. Systematic error in seed plant phylogenomics. Genome Biol Evol. 2011;3:1340–8.
29. Goremykin VV, Nikiforova SV, Biggs PJ, Zhong BJ, Delange P, Martin W, et al. The evolutionary root of flowering plants. Syst Biol. 2013;62(1):50–61.
30. Goremykin VV, Nikiforova SV, Cavalieri D, Pindo M, Lockhart P. The root of flowering plants and total evidence. Syst Biol. 2015;64(5):879–91.
31. Sugiura M. The chloroplast genome. Plant Mol Biol. 1992;19:149–68.
32. Wicke S, Schneeweiss GM, dePamphilis CW, Müller KF, Quandt D. The evolution of the plastid chromosome in land plants: gene content, gene order, gene function. Plant Mol Biol. 2011;76(3):273–97.
33. Zhu A, Guo W, Gupta S, Fan W, Mower JP. Evolutionary dynamics of the plastid inverted repeat: the effects of expansion, contraction, and loss on substitution rates. New Phytol. 2015;209(4):1747–56.
34. Schwarz EN, Ruhlman TA, Sabir JSM, Hajrah NH, Alharbi NS, Al-Malki AL, et al. Plastid genome sequences of legumes reveal parallel inversions and multiple losses of *rps16* in papilionoids. J Syst Evol. 2015;53(5):458–68.
35. Hsu CY, Wu CS, Chaw SM. Birth of four chimeric plastid gene clusters in Japanese umbrella pine. Genome Biol Evol. 2016;8(6):1776–84.
36. Cusimano N, Wicke S. Massive intracellular gene transfer during plastid genome reduction in nongreen Orobanchaceae. New Phytol. 2016;210(2): 680–93.
37. Graham SW, Lam VKY, Merckx VSFT. Plastomes on the edge: the evolutionary breakdown of mycoheterotroph plastid genomes. New Phytol. 2017;214(1):48–55.
38. Yi TS, Jin GH, Wen J. Chloroplast capture and intra- and inter-continental biogeographic diversification in the Asian – new world disjunct plant genus Osmorhiza (Apiaceae). Mol Phylogen Evol 2015; 85(0):10–21.

39. Huang H, Shi C, Liu Y, Mao SY, Gao LZ. Thirteen *Camellia* chloroplast genome sequences determined by high-throughput sequencing: genome structure and phylogenetic relationships. BMC Evol Biol. 2014;14:151.

40. Knox EB. The dynamic history of plastid genomes in the Campanulaceae Sensu Lato is unique among angiosperms. Proc Natl Acad Sci U S A. 2014; 111(30):11097–102.

41. Logacheva MD, Schelkunov MI, Nuraliev MS, Samigullin TH, Penin AA. The plastid genome of mycoheterotrophic monocot *Petrosavia stellaris* exhibits both gene losses and multiple rearrangements. Genome Biol Evol. 2014;6(1):238–46.

42. Wu Z, Raven PH. Flora of China, vol. 8, 13, 14. Beijing: Science Press; Missouri Botanical Garden; 2005.

43. Doyle JJ, Doyle JL. A rapid DNA isolation procedure for small quantities of fresh leaf tissue. Phytochem Bul. 1987;19:11–5.

44. Yang JB, Li DZ, Li HT. Highly effective sequencing whole chloroplast genomes of angiosperms by nine novel universal primer pairs. Mol Ecol Resour. 2014;14(5):1024–31.

45. Zhang T, Zeng CX, Yang JB, Li HT, Li DZ. Fifteen novel universal primer pairs for sequencing whole chloroplast genomes and a primer pair for nuclear ribosomal DNAs. J Syst Evol. 2016;54(3):219–27.

46. Kearse M, Moir R, Wilson A, Stones-Havas S, Cheung M, Sturrock S, et al. Geneious basic: an integrated and extendable desktop software platform for the organization and analysis of sequence data. Bioinformatics. 2012;28(12):1647–9.

47. Wyman SK, Jansen RK, Boore JL. Automatic annotation of organellar genomes with DOGMA. Bioinformatics. 2004;20(17):3252–5.

48. Lohse M, Drechsel O, Bock R. OrganellarGenomeDRAW (OGDRAW): a tool for the easy generation of high-quality custom graphical maps of plastid and mitochondrial genomes. Curr Genet. 2007;52(5–6):267–74.

49. Darling AC, Mau B, Blattner FR, Perna NT. Mauve: multiple alignment of conserved genomic sequence with rearrangements. Genome Res. 2004; 14(7):1394–403.

50. Frazer KA, Pachter L, Poliakov A, Rubin EM, Dubchak I. VISTA: computational tools for comparative genomics. Nucleic Acids Res. 2004;32(Web Server issue):W273–9.

51. Katoh K, Standley DM. MAFFT multiple sequence alignment software version 7: improvements in performance and usability. Mol Biol Evol. 2013; 30(4):772–80.

52. Gielly L, Taberlet P. The use of chloroplast DNA to resolve plant phylogenies: noncoding versus *rbcL* sequences. Mol Biol Evol. 1994;11(5):769–77.

53. Yang Z. PAML 4: Phylogenetic analysis by maximum likelihood. Mol Biol Evol. 2007;24(8):1586–91.

54. Castresana J. Selection of conserved blocks from multiple alignments for their use in phylogenetic analysis. Mol Biol Evol. 2000;17(4):540–52.

55. Darriba D, Taboada GL, Doallo R, Posada D. jModelTest 2: more models, new heuristics and parallel computing. Nat Methods. 2012;9(8):772.

56. Stamatakis A. RAxML-VI-HPC: maximum likelihood-based phylogenetic analyses with thousands of taxa and mixed models. Bioinformatics. 2006; 22(21):2688–90.

57. Huelsenbeck JP, Ronquist F. MRBAYES: Bayesian inference of phylogenetic trees. Bioinformatics. 2001;17(8):754–5.

58. Lanfear R, Calcott B, Ho SYW, Guindon S. PartitionFinder: combined selection of partitioning schemes and substitution models for phylogenetic analyses. Mol Biol Evol. 2012;29(6):1695–701.

59. Kim JS, Kim JH. Comparative genome analysis and phylogenetic relationship of order Liliales insight from the complete plastid genome sequences of two Lilies (Lilium longiflorum and Alstroemeria aurea). PLoS One. 2013;8(6):11.

60. Ravi V, Khurana J, Tyagi A, Khurana P. An update on chloroplast genomes. Plant Syst Evol. 2008;271(1–2):101–22.

61. Kim KJ, Lee HL. Complete chloroplast genome sequences from Korean ginseng (Panax schinseng Nees) and comparative analysis of sequence evolution among 17 vascular plants. DNA Res. 2004;11(4):247–61.

62. Doyle JJ, Doyle JL, Palmer JD. Multiple independent losses of two genes and one intron from legume chloroplast genomes. Syst Bot. 1995;20(3):272–94.

63. Bausher MG, Singh ND, Lee SB, Jansen RK, Daniell H. The complete chloroplast genome sequence of *Citrus sinensis* (L.) Osbeck var 'Ridge Pineapple': organization and phylogenetic relationships to other angiosperms. BMC Plant Biol. 2006;6:11.

64. Gantt JS, Baldauf SL, Calie PJ, Weeden NF, Palmer JD. Transfer of *rpl22* to the nucleus greatly preceded its loss from the chloroplast and involved the gain of an intron. EMBO J. 1991;10(10):3073–8.

65. Jansen RK, Saski C, Lee SB, Hansen AK, Daniell H. Complete plastid genome sequences of three rosids (*Castanea, Prunus, Theobroma*): evidence for at least two independent transfers of *rpl22* to the nucleus. Mol Biol Evol. 2011; 28(1):835–47.

66. Steane DA. Complete nucleotide sequence of the chloroplast genome from the Tasmanian blue gum, *Eucalyptus globulus* (Myrtaceae). DNA Res. 2005; 12(3):215–20.

67. Raubeson LA, Peery R, Chumley TW, Dziubek C, Fourcade HM, Boore JL, et al. Comparative chloroplast genomics: analyses including new sequences from the angiosperms *Nuphar advena* and *Ranunculus macranthus*. BMC Genomics. 2007;8

68. Shi C, Liu Y, Huang H, Xia EH, Zhang HB, Gao LZ. Contradiction between plastid gene transcription and function due to complex posttranscriptional splicing: an exemplary study of *ycf15* function and evolution in angiosperms. PLoS One. 2013;8(3):e59620.

69. Talavera G, Castresana J. Improvement of phylogenies after removing divergent and ambiguously aligned blocks from protein sequence alignments. Syst Biol. 2007;56(4):564–77.

70. Som A. Causes, consequences and solutions of phylogenetic incongruence. Brief Bioinform. 2015;16(3):536–48.

71. Kainer D, Lanfear R. The effects of partitioning on phylogenetic inference. Mol Biol Evol. 2015;32(6):1611–27.

72. Xiang QY, Soltis DE, Soltis PS. Phylogenetic relationships of cornaceae and close relatives inferred from *matK* and *rbcL* sequences. Am J Bot. 1998;85(2):285–97.

73. Moore MJ, Soltis PS, Bell CD, Burleigh JG, Soltis DE. Phylogenetic analysis of 83 plastid genes further resolves the early diversification of eudicots. Proc Natl Acad Sci U S A. 2010;107(10):4623–8.

74. Huang DI, Hefer CA. Kolosova N, Douglas CJ, Cronk QC. Whole plastome sequencing reveals deep plastid divergence and cytonuclear discordance between closely related balsam poplars, *Populus balsamifera* and *P. trichocarpa* (Salicaceae). New Phytol. 2014;204(3):693–703.

11

Identification and characterization of *Prunus persica* miRNAs in response to UVB radiation in greenhouse through high-throughput sequencing

Shaoxuan Li[1,2], Zhanru Shao[3,4], Xiling Fu[1,2], Wei Xiao[1,2], Ling Li[1,2], Ming Chen[1,2], Mingyue Sun[1,2], Dongmei Li[1,2*] and Dongsheng Gao[1,2*]

Abstract

Background: MicroRNAs (miRNAs) are small non-coding RNAs that regulate gene expression of target mRNAs involved in plant growth, development, and abiotic stress. As one of the most important model plants, peach (*Prunus persica*) has high agricultural significance and nutritional values. It is well adapted to be cultivated in greenhouse in which some auxiliary conditions like temperature, humidity, and UVB etc. are needed to ensure the fruit quality. However, little is known about the genomic information of *P. persica* under UVB supplement. Transcriptome and expression profiling data for this species are therefore important resources to better understand the biological mechanism of seed development, formation and plant adaptation to environmental change. Using a high-throughput miRNA sequencing, followed by qRT-PCR tests and physiological properties determination, we identified the responsive-miRNAs under low-dose UVB treatment and described the expression pattern and putative function of related miRNAs and target genes in chlorophyll and carbohydrate metabolism.

Results: A total of 164 known peach miRNAs belonging to 59 miRNA families and 109 putative novel miRNAs were identified. Some of these miRNAs were highly conserved in at least four other plant species. In total, 1794 and 1983 target genes for known and novel miRNAs were predicted, respectively. The differential expression profiles of miRNAs between the control and UVB-supplement group showed that UVB-responsive miRNAs were mainly involved in carbohydrate metabolism and signal transduction. UVB supplement stimulated peach to synthesize more chlorophyll and sugars, which was verified by qRT-PCR tests of related target genes and metabolites' content measurement.

Conclusion: The high-throughput sequencing data provided the most comprehensive miRNAs resource available for peach study. Our results identified a series of differentially expressed miRNAs/target genes that were predicted to be low-dose UVB-responsive. The correlation between transcriptional profiles and metabolites contents in UVB supplement groups gave novel clues for the regulatory mechanism of miRNAs in *Prunus*. Low-dose UVB supplement could increase the chlorophyll and sugar (sorbitol) contents via miRNA-target genes and therefore improve the fruit quality in protected cultivation of peaches.

Keywords: *Prunus persica*, UVB, MicroRNA, High-throughput sequencing, Chlorophyll and carbohydrates

* Correspondence: dmli2002@163.com; dsgao@sdau.edu.cn
[1]College of Horticulture Science and Engineering, Shandong Agricultural University, Tai'an 271018, People's Republic of China
Full list of author information is available at the end of the article

Background

As an important environment signal, sunlight provides energy for the growth and development of plants [1], but its ultraviolet (UV) radiation part causes abiotic stress potentially influence the biological processes of plants. Since the late 1980s when awareness of stratospheric ozone layer depletion triggered concerns about the potentially harmful effects of increased UVB radiation, many studies have shown that UVB causes non-specific damage to DNA, proteins and lipids [2–4]. On the other hand, there is overwhelming evidence that other than substantially impeding plant growth, low-dose UVB is an environmental regulator affecting gene expression, cellular and metabolic activities, and growth and development [5–8]. Whether UVB radiation is a stressor or a regulator is determined by the fluence rate and exposure time [4]. Nevertheless, the regulatory mechanism of plants responding to the UVB-lack environment, for example in the greenhouse where the UVB radiation level is 30%–70% lower than outdoors, were rarely reported [9].

Most of the photomorphogenic responses to low-dose UVB are mediated by the photoreceptor UV RESISTANCE LOCUS8 (UVR8). Subsequent structural and functional characterization revealed that the UVR8 has a unique regulatory mechanism in photoreception [10–12]. After UVB treatment, UVR8 interacts with the E3 ubiquitin ligase (transducin/WD40 repeat-like superfamily protein) CONSTITUTIVELY PHOTOMORPHOGENIC1 (COP1), following the ubiquitination of the basic leucine-zipper (bZIP family) transcription factor ELONGATED HYPOCOTYL5 (HY5) which is primarily in the initiation of photo-morphogenesis [13–16].

MiRNAs, small endogenous non-coding RNAs approximately 21–24 nucleotides (nt) in length, play an important role in regulating gene expression at the post-transcriptional level [17–19]. A large number of miRNAs have been recently identified in plants via high-throughput sequencing, and numerous miRNAs have been entered into the miRBase 21. MiRNAs are involved in regulating growth, development, root initiation and development, hormone balance, floral morphogenesis and reproductive performance [20, 21]. Stress-regulated miRNAs in plants confer resistance to the extreme conditions, including UVB, drought, salt, cold and heat. In addition, the expression of miRNAs can alter the behavior of plants in response to both abiotic and biotic stresses [21, 22]. Previous reports have shown that miRNA induction was involved in regulating auxin signaling via miR160, miR167 and miR393 thus, becoming an important strategy for photomorphism in plants [23].

China is the largest producer of both outside-grown and inside-grown peaches and nectarines in the world. There are nearly 16,000 ha of protected peach and nectarines cultivation, 2.3% of the total area [24]. As a new agricultural form, protected production has been rapidly developed. Peach (*Prunus persica*), which has been cultivated for more than 4000 years, is one of the most important fruits in the world [25]. Peach has a small genome and it reaches reproductive maturity in a relatively short time. In 2010, the Genome Sequencing Project of the peach double haploid cultivar 'Lovell' was completed, which generated, 230 Mb genome sequence and 202 assembly scaffolds [26]. Therefore, peach is considered to be a useful forest model species for genetic and ecological research. Following these findings, several reports on the identification of miRNAs in different peach tissues have been published [27–29]. Meanwhile, peach has some unique biological features not commonly found in other model organisms, such as a 3–5 year juvenile period before blossom, the formation of fleshy fruit with a hardened endocarp and chilling-requirement dormancy mechanism [30–34]. Nectarine, because of its low-chill and nutritional value, is selected as one of the most important fruits in the protected cultivation industry which targeted early and high markets. In previous research, we found that the supplement of UVB radiation can improve the fruit quality and the ability to compete for C-assimilate [35, 36]. Considering the distinct environment especially the light condition in greenhouse, more deep studies related to the molecular and metabolic mechanism under UVB irradiance are needed.

In this study, we generated over 2 billion bases of high-quality RNA sequence with Illumina platform. In a single run, we identified 31,763,592 raw sequences including thousands of seed target and metabolism genes. Our results identified a series of differentially expressed miRNAs/target genes that were predicted to be low-dose UVB-responsive. The correlation between transcriptional profiles and metabolites' contents helped elucidate the regulatory mechanisms of peach under the UVB supplement. Our findings of correlation among miRNA, target genes and metabolites provided clues for breeding with high-quality fruit and other properties which are suitable for greenhouse cultivation.

Results

Analysis of miRNA sequences

Using Illumina sequencing, a total of 74,119,581 and 76,123,247 raw reads were obtained from control and UVB-treatment groups, respectively (Table 1). After discarding 3′ adapter deletions, insertion deletions, 5′ adapter contaminants, poly-A sequences and sequences less than 18 nt from the high-quality reads, 52,375,087 and 56,244,851 clean reads were used for further analysis. The proportions of clean reads were 70.79% and

Table 1 The quality control of the clean data

Sample	Num. of Reads	Raw Clean Reads %	Remove Adapter%	Insert Null %	N %	Too short %	Poly-A %	Too long %	Low quality %
CK1	23,619,871	76.50%	0.06%	2.18%	0.17%	13.86%	0.06%	7.04%	0.12%
CK2	25,100,163	68.65%	0.09%	1.27%	0.16%	24.63%	0.05%	5.05%	0.11%
CK3	25,399,547	67.22%	0.09%	1.34%	0.16%	29.78%	0.04%	1.25%	0.11%
T1	25,154,335	74.96%	0.07%	2.16%	0.17%	15.86%	0.06%	6.60%	0.11%
T2	26,168,148	75.67%	0.10%	0.65%	0.18%	14.36%	0.05%	8.88%	0.11%
T3	24,800,764	70.92%	0.11%	0.64%	0.16%	21.37%	0.05%	6.65%	0.11%

CK control groups without UVB treatment, *T* UVB treatment groups
N% means the percentage of loci that fail to distinguish specific bases

73.85% of the total reads obtained from the two libraries, respectively.

The small RNA (sRNA) reads were typically 18 to 30 nt in length (Fig. 1). Among these sequences, 21 nt sRNAs were the most abundant in the two libraries, accounting for 16.56% and 13.22% of the total reads, followed by 24 nt sRNAs, which accounted for 11.79% and 10.26% of the total reads, respectively. Furthermore, we observed that the number of less than 24 nt length sequences in the control libraries was more abundant than that in the treatment libraries (74.08% and 64.14%, respectively). Additionally, a large proportion of unique sequences (>85% in both libraries) were unclassified sRNAs, suggesting a broad existence of miRNAs in peach.

Known miRNAs in peach

Known miRNAs in peach affected by UVB were identified through homologous alignment analysis using the plant miRNA in miRBase 21. A total of 164 known miRNAs belonging to 59 families were obtained from the deep sequencing. The dominant miRNA families are shown in Additional file 1: Table S1, and most of these miRNAs were largely conserved in various plant species. The expression levels of a few miRNA families, such as miR166, miR1511 and miR398, were evidentially high in both libraries. Some conserved miRNAs were reported only in few species, such as miR3627 in five (*Vitis vinifera, Populus trichocarpa, Malus domestica, Solanum tuberosum, and P. persica*), miR1511 in three (*M. domestica, S. tuberosum, and P. persica*), while miR8133 only in *P. persica*.

MiRNAs have a broad range of expression levels, varying from several to millions of reads. Most of the conserved miRNAs were identified from the two libraries, and certain miRNAs were abundant in some samples but scarce or even lacking in other samples. For example, the expression of miR159 generated 323,238 and 128,812 reads in the CK and UVB libraries respectively. Moreover, the number of reads for different members of the same family varied widely. For example, the expression of miR7122a in both libraries generated 21,291 reads while only 417 reads of miR7122b in both libraries.

Novel miRNAs in peach

In the present study, 109 novel miRNAs from peach deep sequencing were identified (Additional file 2: Table S2). The most abundant miRNA was Pp04_27840-3p, with 33,630 and 42,292 reads in the UVB libraries and CK libraries, respectively. Many novel miRNAs from our database were conserved with miRNAs from other species to a certain degree (<50%), such as Pp02_15663-5p and Pp03–22,312-3p corresponding to miR172d and miR2950-5p in grape (*Vitis vinifera*), respectively.

Differential expression of miRNA

The comparison of the miRNA expression levels in CK and UVB groups showed that 164 known and 109 novel miRNAs were identified in the two libraries. The analysis of the differential expression of miRNAs in the UVB treatment and control libraries showed that 9 known and 6 novel differentially expressed miRNAs from the two libraries might play important roles in the UVB response(Fig. 2). In brief, 8 miRNAs were up-regulated, including Pp03–22,312-3p, Pp03–22,312-5p, Pp05–19,842-3p, Pp06–35,148-3p, Pp06–35,148-5p, miR397, miRNA171d-3p and miRNA3627-5p, and 7 miRNAs were down-regulated, including miRNA395d, miRNA395e, miRNA7122b-5p, miRNA399a, miRNA399b, miRNA8133-3p, and Pp05–28,899-3p in UVB treatment.

Target prediction and functional analysis

A large number of targets were predicted for most differentially expressed miRNAs. As for the known miRNAs, there were 1928 pathways that accounted for the largest percentage of the total targets. The results of Gene ontology classification and top 30 preferential KEGG pathway analysis revealed that target genes of these miRNAs were involved in various biological and biochemical processes in plant growth and development (Figs. 3 and 4) (Additional file 3: Table S3, Additional file 4: Table S4), such as porphyrin and chlorophyll metabolism (pper00860), pentose and glucuronate interconversions (pper00040), citrate cycle (TCA cycle, pper04712 and pper00020), circadian rhythm - plant and metabolic pathways (pper01100).

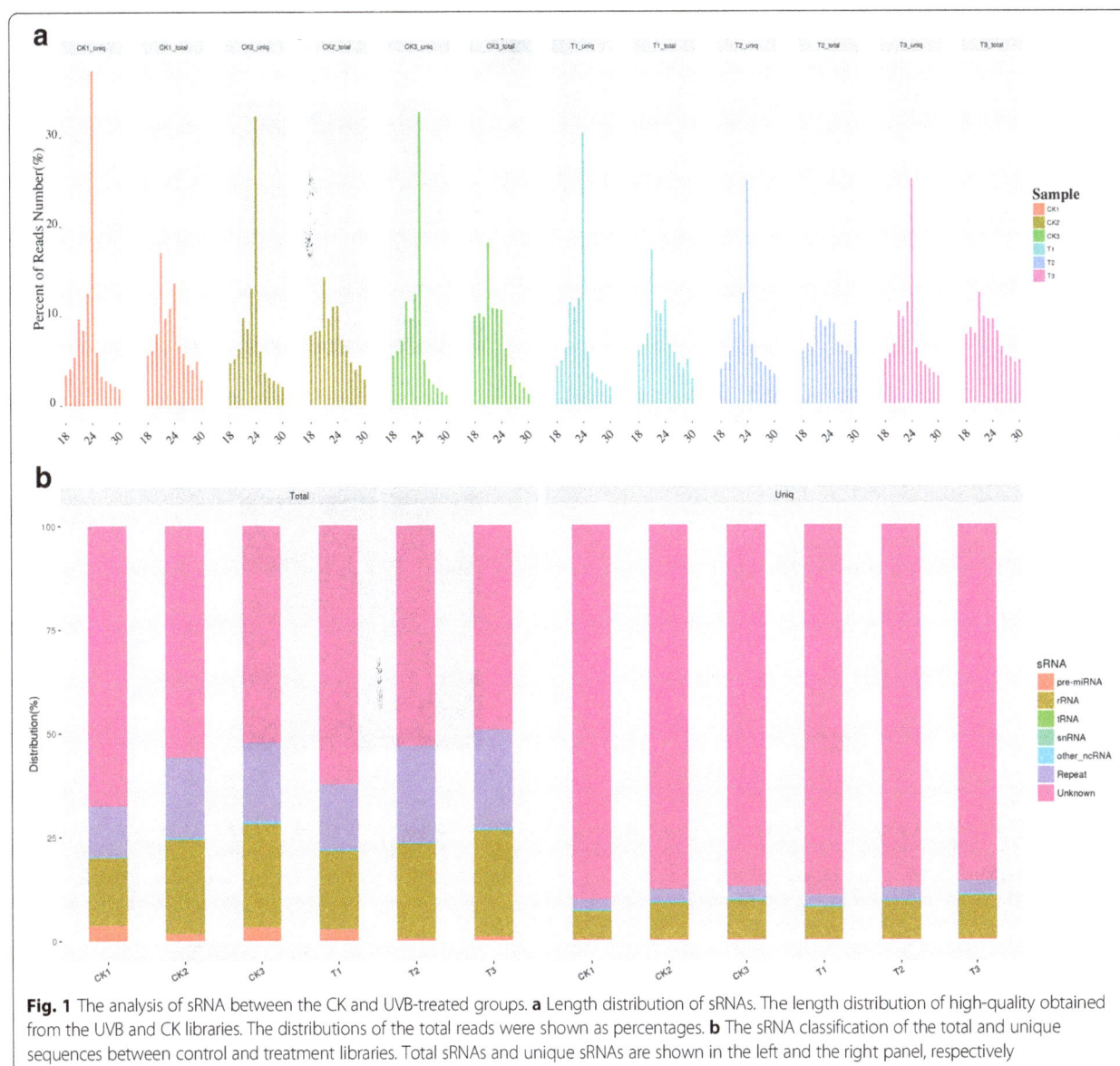

Fig. 1 The analysis of sRNA between the CK and UVB-treated groups. **a** Length distribution of sRNAs. The length distribution of high-quality obtained from the UVB and CK libraries. The distributions of the total reads were shown as percentages. **b** The sRNA classification of the total and unique sequences between control and treatment libraries. Total sRNAs and unique sRNAs are shown in the left and the right panel, respectively

Expression validation of UVB-responsive miRNA

Totally, we verified the expression of 31 miRNAs (Table 2) via qRT-PCR in which 15 miRNAs were retrieved in the database (Fig. 5). We selected miR5059 as reference miRNA [37] and 15 other miRNAs that related to UVB irradiance. The primers were listed in Additional file 5: Table S5. The qRT-PCR results showed the consistency with sequencing data except that miR397 levels under UVB treatment did not have significant difference with CK. The results showed some difference with the database. Many miRNAs were up-regulated other than down-regulation from high-throughput sequencing, such as miR398a-5p (6.67-fold), miR398a-3p (2.09-fold), miR6263 (5.21-fold), miR6260 (2.31-fold), and miR319a (1.75-fold). On the contrary, miR1511 (2.31-fold), miR171c (2.31-fold), and miR3627-3p (1.61-fold) were down-regulated

under UVB treatment (Table 2; Fig. 5). Some conserved UVB-responsive miRNAs were not remarkably expressed such as miR156a, miR160a, miR166a, miR393a, miR402a which were considered to be involved in plant acclimation and adaptation of biotic or abiotic stress.

Chlorophyll metabolism under UVB treatment

KEGG pathway analysis of miRNA target genes (Fig. 4) showed that genes involved in chlorophyll metabolism might be significantly regulated by UVB. We determined chlorophyll content and crucial miRNA/genes expression levels to give clues for UVB effects on chlorophyll metabolism. Spectrophotometry results showed that the content of chlorophyll was gradually increased during the whole development period (Fig. 6a). Its content reached up to 5.24 mg/g Fw in the mature stage of

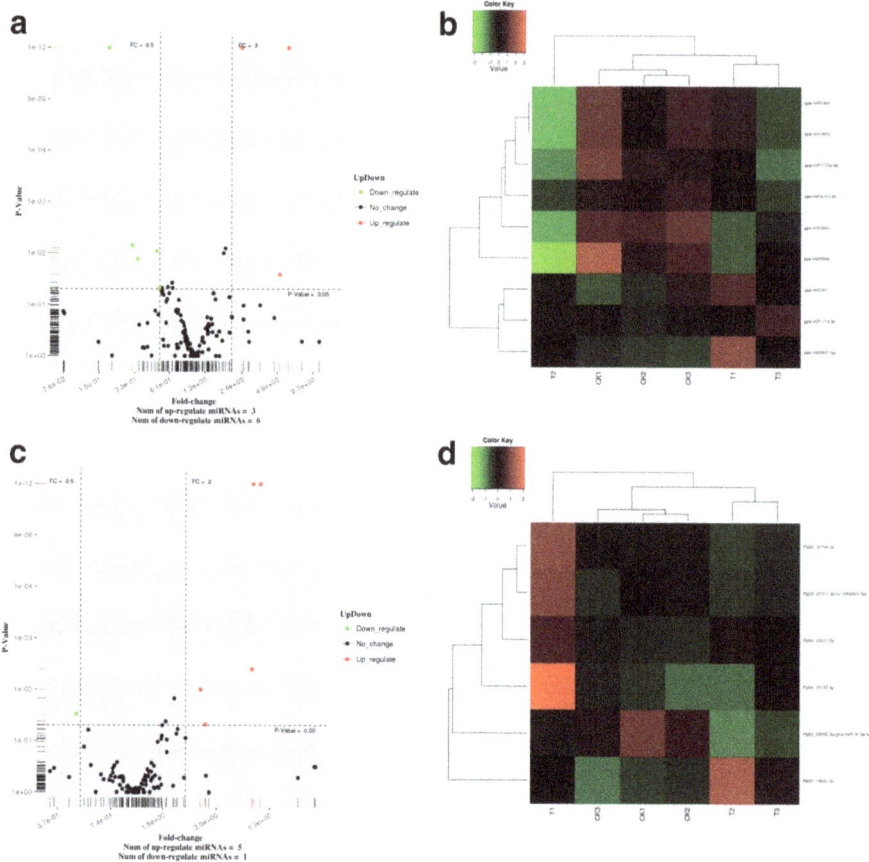

Fig. 2 The differentially expressed miRNAs between the CK and UVB-treated groups. **a** Scatter diagram of the differential read counts of known miRNAs. Each point in the figure represents a miRNA. **b** Heat map of differentially expressed known miRNAs between the control and UVB-treated groups. **c** Scatter diagram of the differential read counts of novel miRNAs. Each point in the figure represents a miRNA. **d** Heat map of differentially expressed novel miRNAs between the control and UVB-treated groups. Red points represent miRNAs showing a > 2-fold change of expression; green points represent miRNAs showing 1/2 < fold change ≤ 2; black points represent miRNAs showing a fold change ≤ 1/2

UVB-group, higher than 4.17 mg/g Fw of the control group. The expression levels of miR171c in UVB-group were lower than the untreated samples, which was 2.3-fold difference in the mature stage (Fig. 6b). Fig. 6c showed the different expression profiles of genes involved in chlorophyll metabolism. Scarecrow-like protein (*SCL*), the target gene of miR171c, was highly up-regulated during the developmental period. On the contrary, pchlide oxidoreductase C (*PORC*), the downstream gene of *SCL* was remarkably down-regulated. The others did not show significant difference between CK and UVB groups. Specifically, *SCL* was up-regulated by 2.2-fold after UVB treatment in fruit mature phase. *PORC* was down-regulated by 2.8-fold (Fig. 6c). The gene information and sequences of primers were listed in Additional file 6: Table S6.

Carbohydrate metabolism under UVB treatment

Carbohydrate metabolism was another pathway targeted by miRNAs predicted by KEGG analysis. The contents

of carbohydrates in leaves were significantly affected by UVB treatment (Fig. 7). Sorbitol, the main form of sugar, was up-regulated by 1.24-fold with UVB stimulation in the mature period (Fig. 7a), while the content of sucrose was dramatically decreased and did not show significant difference with/without UVB treatment (Fig. 7b). Fructose and glucose contents were increased with the developmental process, and the UVB-groups were higher than the control samples (Fig. 7c and d).

We screened 19 genes related with sugar metabolism from *P. persica* genome, and applied qRT-PCR assay to investigate the correlation between sugar synthesis and miRNA under UVB treatment (Fig. 8). The crucial genes involved in sucrose pathway did not show significant difference between UVB and control groups (Fig. 8a). However, genes in sorbitol metabolism were expressed distinctly after UVB stimulation (Fig. 8b). The expression levels of Sorbitol –3-orbitol –6-phosphate dehydrogenase (*S6PDH*), NADP dependent sorbitol dehydrogenase (*NADP-SDH*) and Sorbitol transporter (*SOT*) were

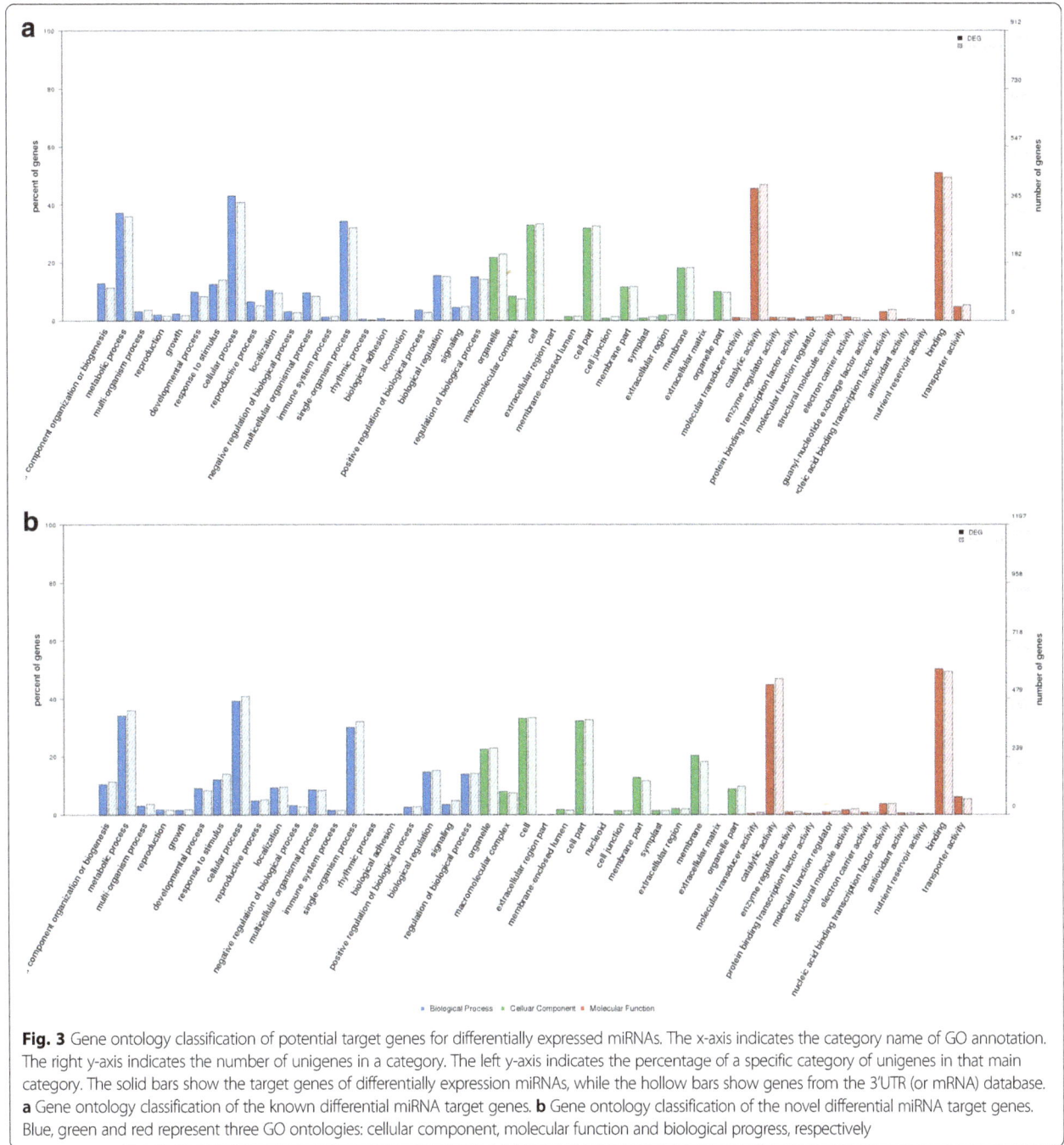

Fig. 3 Gene ontology classification of potential target genes for differentially expressed miRNAs. The x-axis indicates the category name of GO annotation. The right y-axis indicates the number of unigenes in a category. The left y-axis indicates the percentage of a specific category of unigenes in that main category. The solid bars show the target genes of differentially expression miRNAs, while the hollow bars show genes from the 3′UTR (or mRNA) database. **a** Gene ontology classification of the known differential miRNA target genes. **b** Gene ontology classification of the novel differential miRNA target genes. Blue, green and red represent three GO ontologies: cellular component, molecular function and biological progress, respectively

increased by 2.9-fold, 6.1-fold and 2.0-fold respectively under UVB treatment compared to the control groups. However, NAD dependent sorbitol dehydrogenase (*NAD-SDH*) was rarely expressed under UVB treatment. Moreover, the expression levels of Pyrophosphate–fructose 6-phosphate 1-phosphotransferase subunit alpha (*PFP-α*) and Phosphofructokinase (*PFK*) in hexose pathway were1.8-fold and 1.6-fold higher in UVB group than those in control group (Fig. 8c). Hexose carrier protein (*HEX6*) showed a 3.2-fold decrease after UVB stimulation (Fig. 8c).

These results indicated that UVB had a major influence on the interconversion and transportation of monosaccharides. The gene information and sequences of primers were listed in Additional file 6: Table S6.

Discussion

There are many problems in the protected cultivation of peach trees, such as the basic theory and growth/development pattern, the environmental intelligent control, the standardization system with high quality and

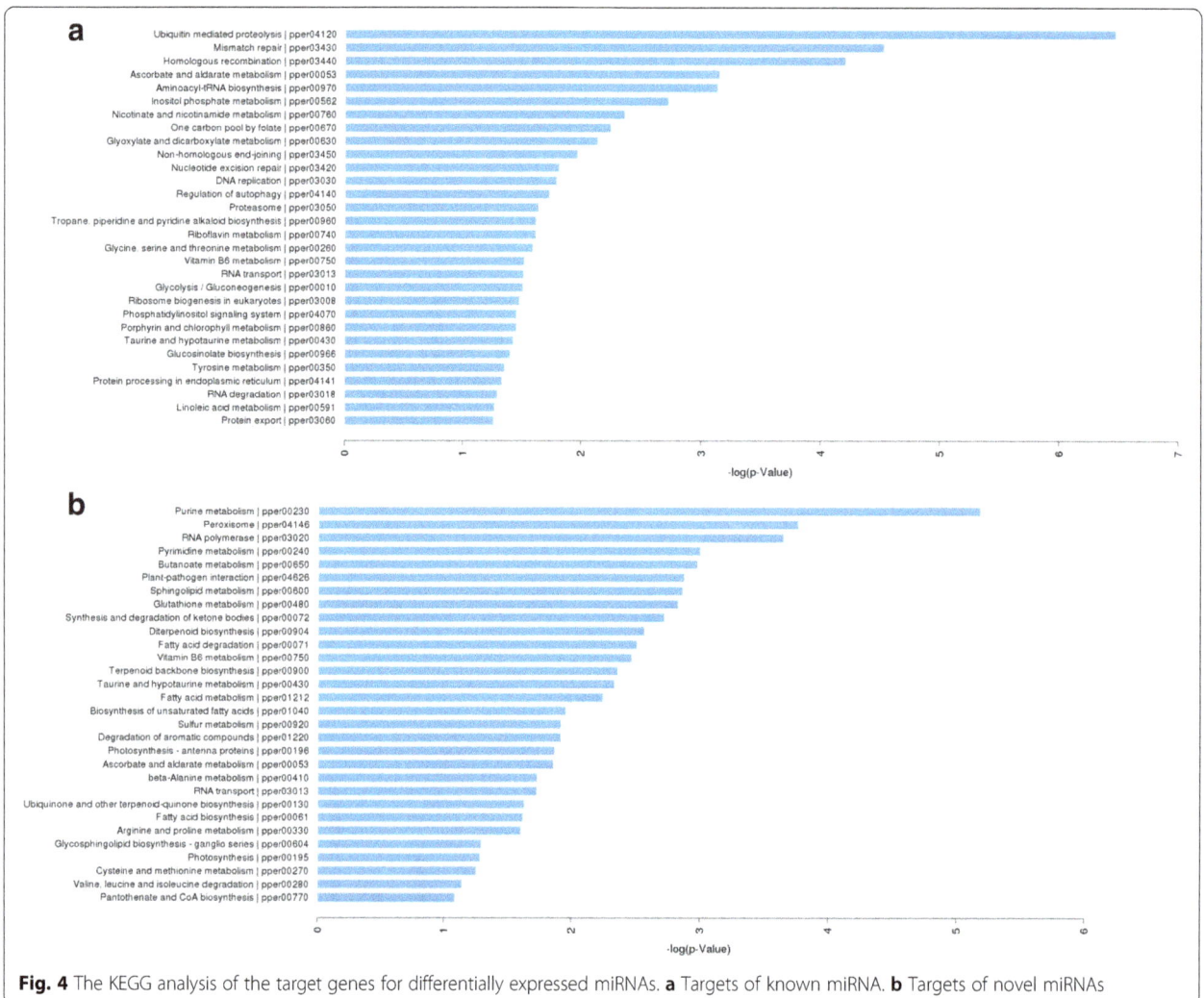

Fig. 4 The KEGG analysis of the target genes for differentially expressed miRNAs. **a** Targets of known miRNA. **b** Targets of novel miRNAs

efficiency etc. [38]. It is mainly ascribed to the weak light irradiance and quality, and the quality of fruits in the greenhouse is worse than that in the outdoor cultivation [39]. Our investigation on the regulatory mechanism of complementary UVB will provide significant effects for elucidating the growth and development pattern of *P. persica*, thus increasing the fruit quality.

High-dose UVB may produce phenols, ROS and cause damage to DNA, proteins, and membranes [40–42]. However, low-dose UVB may trigger early adaptation to environment and regulate plants seeding, development and growth such as the induction of alterations in antioxidant status, the regulation of phenylpropanoids, cinnamates, or flavonoids pathways, chlorophyll and pyridoxine biosynthesis pathways [43]. In our previous study, Chen et al. selected three UVB radiation levels and 1.44 Kj m^{-2} d^{-1} showed the most effective function on the improvement of total soluble solid, anthocyanin and the repression of total acid in peach cultivated in greenhouse [44]. Therefore the UVB radiation level (1.44

Kj m^{-2} d^{-1}) was applied in this study as the most suitable dose to regulate the growth and fruit quality of *P. persica* in the greenhouse environment.

MiRNAs play important roles in plant response and adaptation to environmental change. Thus, understanding the miRNA-mediated regulatory network of UVB supplement will lay the foundation for unraveling the complex molecular genetic mechanism of positive effects on fruit's agronomic traits improvement. A growing evidences suggested that miRNA-guided gene regulation could play a vital role in plant response to UVB radiation [42, 45, 46]. In the present study, a total of 4.02 M and 3.83 M unique sRNA sequences were obtained from the control and UVB-treatment libraries, respectively, suggesting adequate sequencing depth for further analysis. The majority of total sRNA reads ranged from 18 nt to 30 nt in length (Fig. 1), which was consistent with the typical size for Dicer-derived products [47]. The most abundant length is 21 nt followed by 24 nt which was consistent with previous studies in peach [48–51].

Table 2 The comparison of fold change (from sequencing libraries) and expression levels (via qRT-PCR) of typical miRNAs

	MiRNA	Source	Target	Annotation	Sequencing (Fold-change)	qRT-PCR (Fold-change)
Down-regulation	miR159	Ptc [27],Ath,Pte [22]	MYB33	leaf development	–	–
	Pp05–28,899-3p	Sequencing			0.47	–
	miR395d	Sequencing, Ath [22],Ptc [27]	APS,AST68	sulfate translocation and assimilation.	0.5	80.75
	miR7122b-5p	Sequencing	HOS		0.47	17.55
	miR8133-3p	Sequencing			0.32	13.93
	miR399a	Sequencing, Ath [22],Ptc [27]	UBC24		0.18	13.18
	miR395e	Sequencing, Ath [22],Ptc [27]	APS,AST68	sulfate translocation and assimilation.	0.5	2.89
	miR1511	Ppe [37]		unique	–	2.31
	miR171c	Ath [52]	SCL	chlorophyll synthesis	–	2.31
	miR399b	Sequencing			0.29	2.09
	miR3627-3p	Ppe [58]		TCA, EMP	–	1.61
	miR393a	Ath,Pte [22]	AFB2,TIR1, SCF	Antibacterial Resistance, abiotic stress tolerance	–	1.41
	miR5072	Ppe [37]		alternative reference	–	1.25
	miR166a	Ath,Pte [22]	HD-ZIP	abiotic stress tolerance	–	1.05
	miR156a	Ath,Pte [22]	SPL	abiotic stress tolerance	–	1.05
Reference	miR5059*	Ppe [37]		reference*	–	1
Up-regulation	miR397	Sequencing	Laccase		2.42	1.07
	miR160a	Ath,Pte [22]	ARF17	Leaf development	–	1.24
	miR402	Ppe [37]		abiotic stress tolerance	–	1.28
	miR319a	Ath [21]		flowing time	–	1.75
	Pp03–19,842-3p	Sequencing			5.23	2.04
	miR398a-3p	Ath,Pte [22]	CSD1,2	protection from oxidative stress	–	2.09
	miR6260	Ppe [37]		unique	–	2.31
	miR171d-3p	Sequencing			5.1	2.43
	Pp03–22,312-5p	Sequencing			–	3.77
	Pp03–22,312-3p	Sequencing			2.42	4.15
	miR6263	Ppe [37]		unique	4.67	5.21
	miR3627-5p	Sequencing			6.02	5.24
	Pp06–35,148-3p	Sequencing			2.57	6.09
	miR398a-5p	Ath,Pte [22]	CSD1,2	protection from oxidative stress	–	6.67
	Pp06–35,148-5p	Sequencing			4.75	11.99

Abbreviations: *Ath Arabidopsis thaliana, Ptc Populus trichocarpa, Pte Populus tremula, Ppe Prunus persica*

In our study, the expression levels of 2 known miRNAs were highly up-regulated (miR171d-3p, miR3627-5p) and 6 known miRNAs showed significant down-regulation (miR395d, miR395e, miR399a, miR399b, miR7122b-5p, miR8133-3p). These miRNAs were predicted to be involved in distinct metabolic pathways.

MiR171c was predicted to target *SCL6, SCL22* and *SCL27*, a family of transcription factors which were involved in the morphogenesis, proliferation of meristematic cells, polar organization and chlorophyll synthesis [52–54]. Further study found that its target gene tomato (*Solamum lycopersicum*) gras transcription factor gene (*SlGRAS24*) impacts multiple agronomical traits, such as plant height, flowering time, leaf architecture, lateral branch number, root length, fruit set and development, via regulating gibberellin and auxin

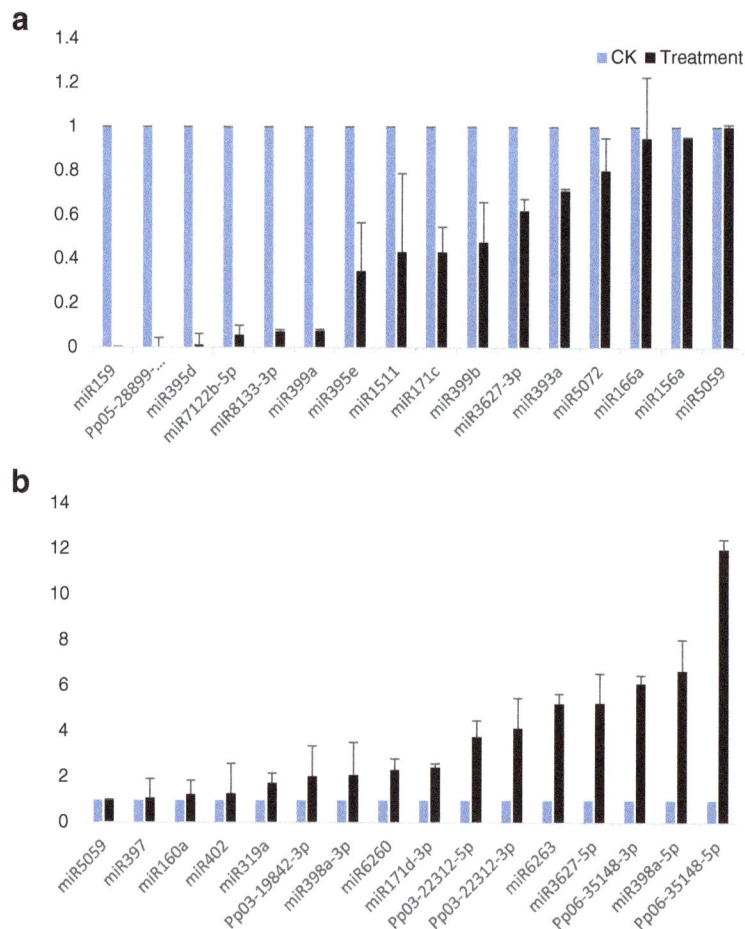

Fig. 5 Relative expression levels of the selective UVB-responsive miRNAs by qRT-PCR analysis. **a** Down-regulation. **b** Up-regulation

homeostasis [55, 56]. It is indicated that the UVB-responsive miRNA171 plays an important role in regulating the growth and development of plant. In the present study, we investigated the potential relation between miR171c and chlorophyll synthesis. The content of chlorophyll was gradually increased during the whole development period under UVB treatment, which was in accordance with the studies by Chen [44]. Although previous reports showed that miRNA171 was up-regulated under light [57], our results found that miRNA171c was less expressed after UVB stimulation. Further analysis on genes involved in miR171-SCL model showed that *SCL* and *PORC* were up-regulated and down-regulated respectively, which was in accordance with this model. *PORC* had positive effects as an upstream gene of chlorophyll synthetic pathway. However, our results showed an opposite regulation pattern between *PORC* and chlorophyll content. Interestingly, the expression levels of Chlorophyll a-b binding protein (*CBR*) and Red chlorophyll catabolite reductase (*PCCR*) (genes involved in chlorophyll degradation) had no change after UVB treatment, while the *CHLG* was up-regulated by 1.6-fold.

According to these results, we presumed that the increase of chlorophyll content under UVB might be not only related with miR171-SCL model but also regulated by non miR171-SCL pathway like the regulation of CHLG. Further verification is needed to illustrate the regulatory mechanism of chlorophyll metabolism under UVB.

A conserved miRNA family (miR3627) was reported only in five species (*Solanum tuberosum*, *Malus domestica*, *Populus trichocarpa*, *Citrus trifolia*, and *Prunus persica*), and was identified as a chilling responsive miRNA in *P. persica* [58]. In our study, miRNA3627-5p was up-regulated under UVB treatment, and we found an interesting physiological phenomenon that the germination rate of treatment group (89.2%) was 12% higher than the CK group (77.5%), which occurred 8 months after the termination of UVB treatment. It implied that UVB radiation may have a long-term and sustainable influence on plants via modification called "UVB memory" [59], and this phenomenon involved a series of miRNAs such as miR3627-5p. Bioinformatics analysis of miR3627-5p in our libraries showed that it had 415 target genes which referred to many pathways such as

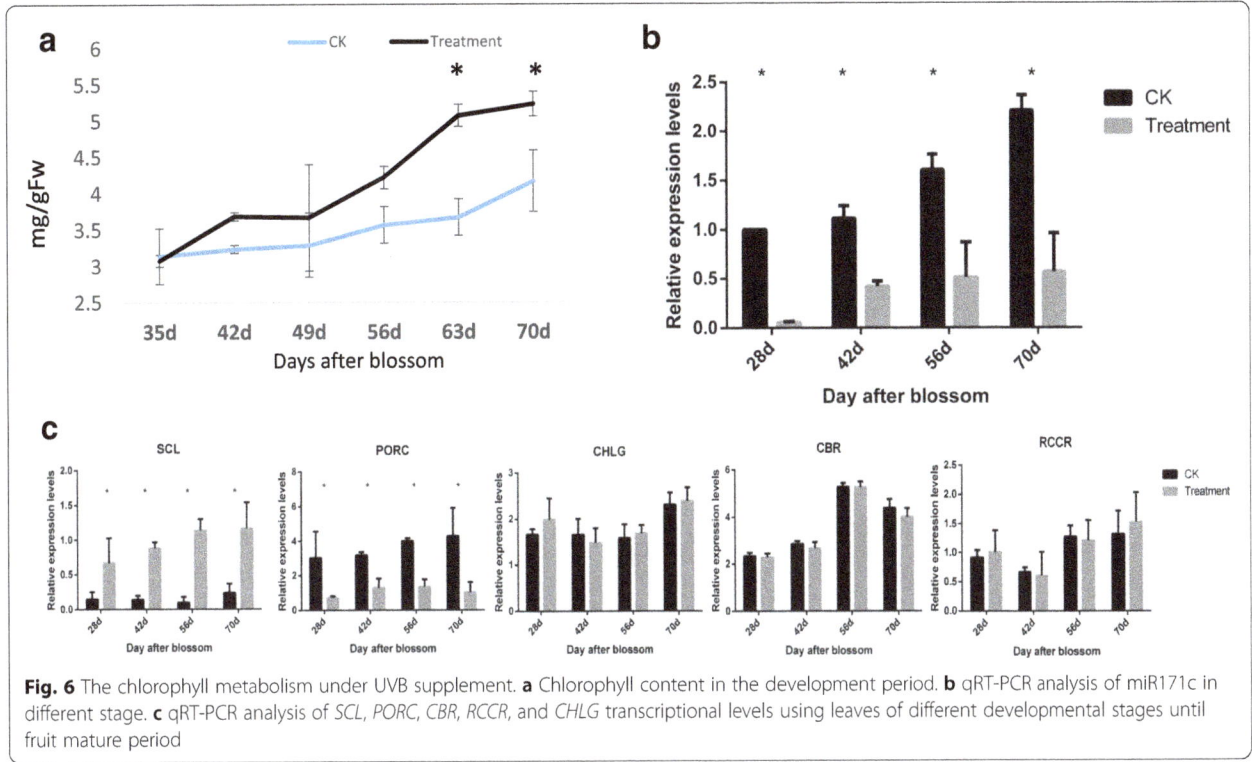

Fig. 6 The chlorophyll metabolism under UVB supplement. **a** Chlorophyll content in the development period. **b** qRT-PCR analysis of miR171c in different stage. **c** qRT-PCR analysis of *SCL*, *PORC*, *CBR*, *RCCR*, and *CHLG* transcriptional levels using leaves of different developmental stages until fruit mature period

metabolic, protein and amino acid metabolism, RNA transport and circadian rhythm. However, the molecular mechanism of miR3627-5p regulation in chilling still needs more experimental studies.

The target genes of miR3627-5p: ppa025787mg, ppa007623mg, ppa016917mg, ppa021860mg were involved in pentose and glucuronate interconversions, starch and sucrose metabolism. Besides, miR3627-5p

target gene ppa007934mg is involved in carbon metabolism and TCA cycle. The expression levels of these five target genes were all down-regulated (Additional file 7: Figure S1), which were consistent with the up-regulation of miR3627-5p. Also the predicted target genes of novel miRNA Pp03_22,312-5p (upregulated) were involved in sugar metabolism. This indicated that miRNA may participate in the formation of fruit sweetness, an essential

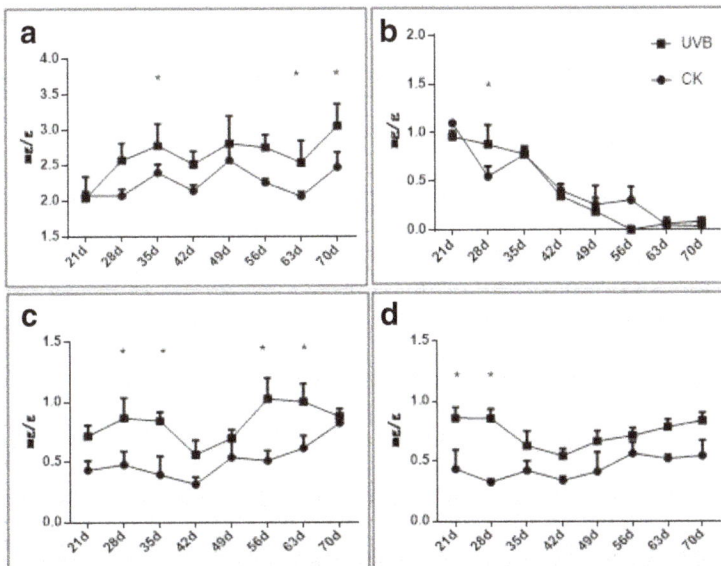

Fig. 7 Comparison of the contents of carbohydrates in leaves at mature stage in CK and UVB group. **a** Sorbitol. **b** Sucrose. **c** Glucose. **d** Fructose

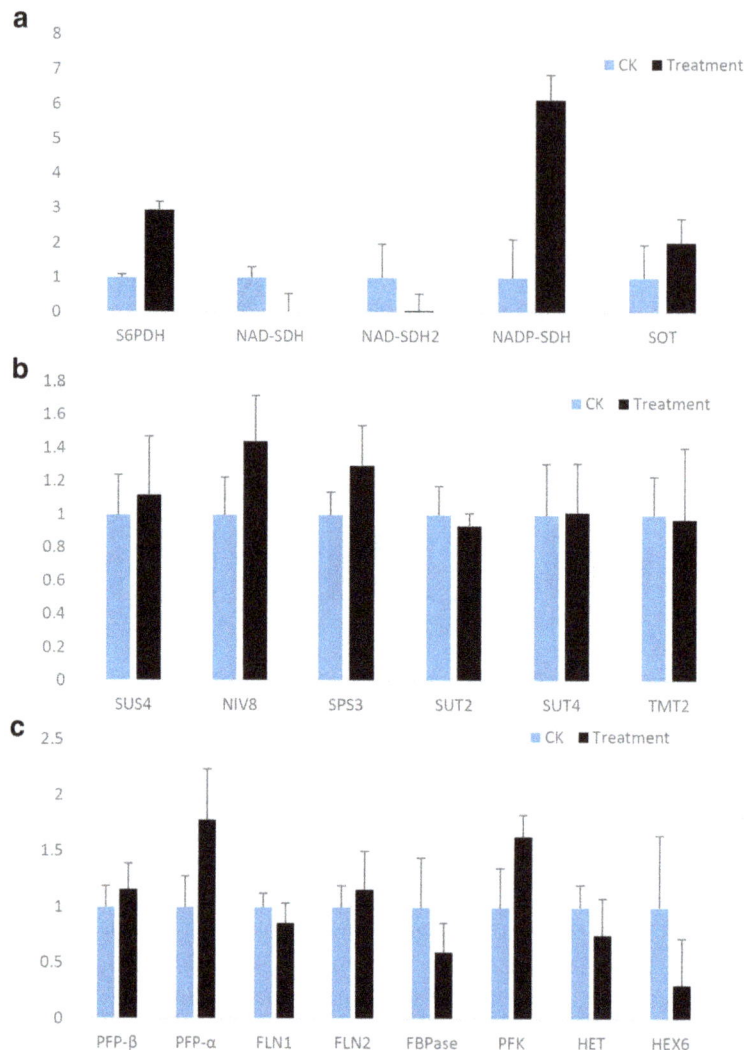

Fig. 8 qRT-PCR analysis of key genes which related to sugar metabolism. **a** *S6PDH, NAD-SDH, NAD-SDH2, NADP-SDH,* and *SOT* which are related to sorbitol metabolism. **b** *SUS4, NIV8, SPS3, SUT2, SUT4,* and *TMT2* which related to sucrose metabolism. **c** *PFP-β, PFP-α, FLN1, FLN2, FBpase, PFK, HET,* and *HEX6* which related to Hexose metabolism

characteristic of fruit quality which was in accordance with previous reports [60].

Different sugar components in leaves were analyzed by High Performance Liquid Chromatography (HPLC) and we found that the carbohydrate contents changed variously under UVB treatment (Fig. 7). Accordingly, we speculated some UVB-responsive miRNAs which targeted sugar-metabolizing pathway, such as miR3627-5p. Further qRT-PCR showed relative expression of the key genes in different sugar metabolic pathways (Fig. 8). Sorbitol is the main form of sugar in Rosaccae leaves and is transferred to fruits for the storage of carbohydrates in the mature stage. The content of sorbitol in our study was gradually increased with the developmental process, which was assistant with previous reports [61]. qPCR results showed that sorbitol-related genes

were distinctly expressed with UVB treatment. The relative expression levels of *S6PDH* was remarkably up-regulated in UVB-group, indicating the increase of *S6PDH* activities and therefore stimulating the synthesis of sorbitol. *NAD-SDH1* and *NAD-SDH2* were down-regulated while the levels of *NADP-SDH* were increased after UVB treatment. This suggested that the plants restricted *NAD-SDHs* expression to reduce sorbitol degradation while increased *NADP-SDH* transcripts for the synthesis of fructose from sorbitol under UVB treatment. *SOT* was up-regulated, responsible for more active transportation of sorbitol, which accorded with the increase of sorbitol contents.

The content of sucrose did not change obviously during the continuous growth period, which could be explained by the slight change of expression levels of genes involved in sucrose metabolism, such as Sucrose synthase (*SUS*), Sucrose transporter (*SUT*), Tonoplast monosaccharide transporter (*TMT*) etc. In hexose pathway, the crucial genes *PFP-a* and *PFK* were up-regulated while Fructose-1,6-bisphosphatase (*FBPase*) and *HEX6* were down-regulated, implying the UVB treatment had an influence on hexose metabolism.

In our study, the UVB treatment led to different expression patterns of genes related with sugar metabolism. Previous research have shown that the regulation was mainly mediated by miRNA. Our results were essentially in agreement with these reports, and more investigations on functional verification need to be performed.

Conclusions

In this study, we constructed 2 miRNA libraries for low-dose UVB radiation groups and control groups of *P. persica* in greenhouse. A total of 164 known and 109 novel miRNAs were identified. In brief, 8 miRNAs were highly up-regulated and 7 miRNAs were significantly down-regulated in the UVB treatment groups, which were mainly predicted to be targeted in signal transduction, carbohydrate metabolism and stress response etc. Combined with qRT-PCR tests and the measurement of the related metabolites, our results showed that low-dose UVB radiation could regulate the expression patterns of some miRNAs e.g. miR3627-5p and Pp_22,312, and cause the expression levels of genes in carbohydrate (sorbitol, fructose, and glucose etc.) and chlorophyll pathway, and therefore indirectly affect sugar contents and fruit quality (Fig. 9). Our study provided a comprehensive database of miRNA for *P. persica* and a theoretical basis for further investigations of the function of miRNA in regulating the biological features of peach in greenhouse.

Methods

Plant material and tissue collection

The experiment was carried out on 7-year-old peach trees (*Prunus persica* var. *nectarine* Zhongyou No.5) cultivated in experimental station of Shandong Agriculture university, Taian, China. The trees were treated with 1.44 Kj m^{-2} d^{-1} UVB radiation during the whole growing period. UVB was provided by the dedicated UV lamp of 40 W, 297 nm (Nanjing Kazhi), hanging at the position of 1.5 m above the plants. UVB-type single-channel UV irradiator (Beijing Normal University Photoelectric Instrument Factory) equipped with 297 nm probe was used to determine the UVB radiation dose of 1.44 Kj m^{-2} d^{-1}. The function leaves within the range of 80–120 cm below the lamp were selected. The on/off time of UV light was controlled through the electronic automatic control device, from 7 days after blossom to the fruit mature stage. UV light was kept on from 9 h30 to 10 h30 every day and stopped at cloudy, rainy and snowy days. Function leaves were sampled on the 7th day after blossom and every week after, until the mature period. These samples were washed with DEPC-treated H$_2$O, immediately frozen in liquid nitrogen and stored at –80 °C until use. The function leaves of fruit mature stage were selected for high-quality deep-sequencing.

Small RNA isolation and Illumina sequencing

Total RNA in peach was isolated using the mirVan miRNA Isolation kit (Ambion; Thermo Fisher Scientific, USA) and purified using the miRNeasy Mini kit (Qiagen, Germany), following the manufacturer's instructions. RNA was quantified using a spectrophotometer (NanoDrop, Thermo Fisher Scientific, USA). Purified RNA was frozen in liquid nitrogen and then stored at –80 °C until required.

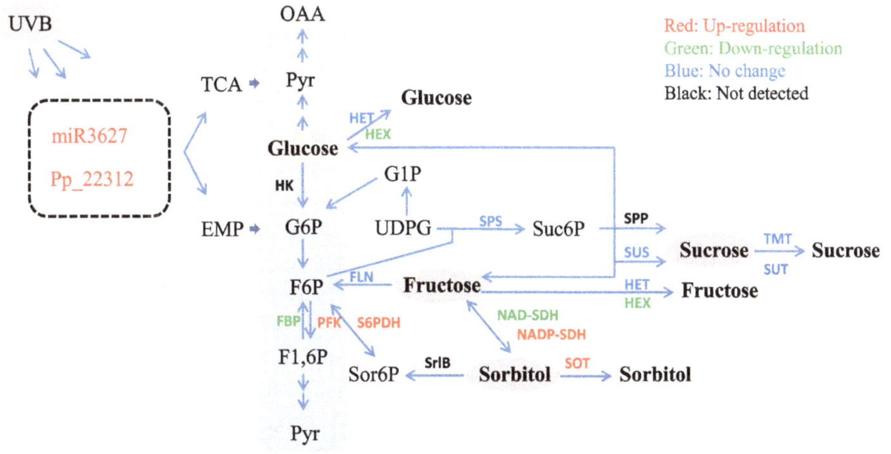

Fig. 9 The schematic of potential carbohydrate pathways regulated by miRNA under UVB radiation in greenhouse. Red: Genes were up-regulated transcriptional levels under UVB; Green: Down-regulation; Blue: No change; Black: Not detected by qRT-PCR

We constructed an RNA library using the NEBNext Ultra RNA Library Prep kit for Illumina (New England Biolabs, USA). There were four steps: Firstly, total RNA (approx. 1 µg) was spliced into shorter fragments (200–500 bp) in the NEBNext First Strand Synthesis Reaction Buffer and the fragments were used to produce the double-stranded cDNA; secondly, the cDNA was end-repaired and ligated with Illumina-specific adaptors; thirdly, we used 200 bp inserts from the library and selected suitable fragments for PCR amplification; last, we performed PCR using Phusion High-Fidelity DNA polymerase (New England Biolabs, USA) and purified the products with a QIAquick Nucleotide Removal kit (Qiagen, Germany). Then we sequenced the new RNA library on an Illumina HiSeq 2500 system using 2×150 base pairs paired-end sequencing.

Analysis of sequencing data

Clean reads were obtained from raw reads after removing low-quality and adapter reads. SOAP software was used for the mapping of clean reads to the peach genome. The non-coding RNAs, including rRNAs, scRNAs, snoRNAs, snRNAs, and tRNAs deposited in the NCBI GenBank database and Rfam (11.0) database, were removed. We also excluded the small RNAs corresponding to the exons and introns of mRNA and repeat sequences. The remaining sRNA sequences were aligned to the miRBase 21 database, with a maximum of two mismatches, to identify known miRNAs in *P. persica*. The obtained sequences were used to predict hairpin structures using the perpl program. The remaining unannotated sRNAs were used to predict novel miRNAs using Mireap software [62].

Target prediction of miRNAs and functional analysis

Target prediction of miRNAs followed rules referring to Allen et al. [63]: a. ≤two adjacent mismatches in the miRNA/target duplex; b. ≤ four mismatches between the sRNA and target gene; c. ≤ 2.5 mismatches at positions 1–12 of the 5′-miRNA/target duplex; d. no mismatches at 10–11 of the miRNA/target duplex; e. no adjacent mismatches at 2–12 of the miRNA/target duplex; f. the minimum free energy (MFE) of the miRNA/target duplex should be 75% of the MFE bound to the perfect complement. Target genes were searched using peach genome information. To better understand the roles of miRNAs in peach under low-dose UVB treatment, the potential target functions were annotated using the Gene Ontology and KEGG pathway database.

Differential expression analysis of miRNAs

The miRNA reads were used to analyze differential expression and determine significant differences between the control and treatment libraries. The frequency of miRNAs was normalized to one million to reduce potential errors before calculating the fold-change, P-value and ratio. Normalized expression = (actual miRNA counts/total counts of clean reads) × 1,000,000. Fold-change = log2 (miRNA normalized read counts in the treatment library/ in the control library).

A fold change larger than 1 or less than −1 and a P-value less than 0.01 suggested the highly significant difference in the miRNA expression between two libraries. When the fold-change was greater than 1 or less than −1 and the P-value was between 0.01 and 0.05, the expression of the miRNA was significantly different between the two libraries. The ratio of miRNA normalized read counts in treatment library/in control library was used to determine changes in the expression of an miRNA in the treatment samples compared with the control samples. When the ratio was more than 2, the miRNA was indicated as up-regulated, and when the ratio was less than 1/2, the miRNA was down-regulated [64].

Relative expression of miRNA and target genes

qRT-PCR was performed to verify the expression levels of identified miRNAs and targets genes using the IQ5 Quantitative Real-time PCR Detection System (Bio-Rad, California, USA) with the SYBR® PrimeScript™ miRNA RT-PCR Kit (TaKaRa, Dalian, China). The reactions were performed in a total volume of 25 µL containing 2.0 µL of diluted cDNA (100 ng/µL), 1 µL of each primer (10 µM), and 12.5 µL of SYBR Green premix Ex Taq II with the following reaction conditions: 95 °C for 30 s, followed by 40 cycles of 95 °C for 5 s and 60 °C for 20 s, then dissociation curve with 95 °C for 60 s, 55 °C for 30 s and 95 °C for 30 s. The reference genes for qRT-PCR of miRNA and target genes were miR5059 and beta-actin, respectively [65, 66]. Each sample was processed in triplicate. All validated primer sequences of miRNAs/target genes are listed in Table S5 and Table S6.

Metabolites analysis by HPLC

Four sugar components: sucrose, sorbitol, glucose and fructose of leaves, were analyzed according to the method of Karkacier et al. [67]. In a mortar pre-cooled in the −20 °C refrigerator, 100 mg of fresh leaves were ground with liquid nitrogen and extracted with 1 mL NANO pure water into a 2 ml tube. The tubes were then vigorously shaken for 15 s, sonicated for 15 min and centrifuged at 12,000 rpm for 15 min. The supernatant was sterilized by filtering through a 0.45 µm membrane filter and stored at −20 °C prior to sugar components' measurement using HPLC. The HPLC system was programmed to inject 50 µL crude extracts automatically. Online detection was performed using a Waters 410 differential refractometer detector and the data were analyzed by Oirigin75 software. The whole program used a MetaCarb 87 °C equipped with a guard column as the analytical column, and the deionized water as the mobile phase with a 0.5 mL min^{-1} flow rate. Glucose, sorbitol, fructose and sucrose purchased from company were used as standards [68].

Chlorophyll content

Function Leaves were taken every 7 days after flowering to measure chlorophyll content via spectrophotometry. The maximum UV absorption wavelength of chlorophyll a and chlorophyll b is 645 nm and 663 nm respectively. The total chlorophyll content was analyzed with the following formula [69]:

$$C_T = C_a + C_b = 20.29\ A_{645} + 8.05\ A_{663}.$$

Additional files

Additional file 1: Table S1. The summary of known miRNAs prediction and expression in control and UVB supplement libraries. (PDF 250 kb)

Additional file 2: Table S2. The summary of novel miRNAs prediction and expression in control and UVB supplement libraries. (PDF 162 kb)

Additional file 3: Table S3. Details of targets genes and their annotation, GO classification, and KEGG pathway for the known miRNAs. (PDF 2499 kb)

Additional file 4: Table S4. Details of targets genes and their annotation, GO classification, and KEGG pathway for the known miRNAs. (PDF 2823 kb)

Additional file 5: Table S5. Primers for qRT-PCR verification of miRNAs. (PDF 17 kb)

Additional file 6: Table S6. The names, Genbank IDs, involved pathways of verified target genes and their primers for qRT-PCR tests. (PDF 79 kb)

Additional file 7: Figure S1. qRT-PCR analysis of five target genes predicted for miR3627-5p. Beta-actin was the internal control. Each experiment was performed with three biological replicates. (PDF 66 kb)

Abbreviations

CBR: Chlorophyll a-b binding protein; CHLG: Chlorophyll synthase; COP1: Constitutively photomorphogenic 1; FBPase: Fructose-1,6-bisphosphatase, chloroplastic; FLN1: Fructokinase-like 1; FLN2: Fructokinase-like 2; HET: Hexose transporter; HEX6: Hexose carrier protein; HPLC: High Performance Liquid Chromatography; HY5: Elongated hypocotyl 5; NADP-SDH: NADP-sorbitol dehydrogenase; NAD-SDH: NAD-sorbitol dehydrogenase; NAD-SDH2: NAD-sorbitol dehydrogenase 2; NIV8: Neutral invertase 8; PFK: Phosphofructokinase; PFP-α: Pyrophosphate–fructose 6-phosphate 1-phosphotransferase subunit alpha; PFP-β: Pyrophosphate–fructose 6-phosphate 1-phosphotransferase subunit beta; PORC: Pchlide oxidoreductase C; RCCR: Red chlorophyll catabolite reductase; S6PDH: Sorbitol –6- phosphate dehydrogenase; SCL: Scarecrow-like protein; SlGRAS24: A tomato (*solanum lycopersicum*) GRAS transcription factor gene; SOT: Sorbitol transporter; SPS: Sucrose phosphate synthase; SUS4: Sucrose synthase 4; SUT2: Sucrose transporter 2; SUT4: Sucrose transporter 4; TMT2: Tonoplast monosaccharide transporter; UVB: Ultraviolet radiation B; UVR8: UV resistance locus 8

Acknowledgements

We thank WX in the lab for providing the trees and BBW and ZJZ for DNA extraction. We thank GW from the CapitalBio Technology for sequencing performed and bioinformatics support. We are grateful to all co-authors who participated in the studies mentioned in the text that were published by our groups. All authors declared no conflict of interest.

Funding

This work was supported by the National Natural Science Foundation of China [Grant number 31601706], Natural Science Foundation of Shandong Province [Grant number ZR2016CM09] and Science and Technology Innovation Team of Shandong Agriculture University-Facility Horticulture Advantages Team (SYL2017YSTD07). The funding institutions had no direct role in study design, sample collection, analysis, and interpretation of date, nor in manuscript writing. Annual reports were submitted to the funding institutions tracking the progress of the projects.

Authors' contributions

SXL and DML conceived and designed the research. SXL performed the experiments and wrote the manuscript. SXL and ZRS analyzed the data, and ZRS revised the intellectual content of this manuscript. XLF and LL contributed in the retrieval of genes and the analysis of qRT-PCR results. WX cultivated the plant materials and helped to analyze the physiological data. MC and MYS contributed in designing qRT-PCR primers. DML and DSG supervised the project as co-correspondence. All authors have read and approved the final manuscript.

Competing interests

The authors declare that they have no competing interests.

Author details

[1]College of Horticulture Science and Engineering, Shandong Agricultural University, Tai'an 271018, People's Republic of China. [2]State Key Laboratory of Crop Biology, Shandong Agricultural University, Tai'an 271018, People's Republic of China. [3]Key Laboratory of Experimental Marine Biology, Institute of Oceanology, Chinese Academy of Sciences, Qingdao 266071, People's Republic of China. [4]Laboratory for Marine Biology and Biotechnology, Qingdao National Laboratory for Marine Science and Technology, Qingdao 266237, People's Republic of China.

References

1. Kami C, Lorrain S, Hornitschek P, Fankhauser C. Light-regulated plant growth and development. Curr Top Dev Biol. 2010;91:29–66.
2. Lidon FJC, Teixeira M, Ramalho JC. Decay of the chloroplast pool of ascorbate switches on the oxidative burst in UV-B-irradiated rice. J Agron Crop Sci. 2012;198:130–44.
3. Pitzschke A, Forzani C, Hirt H. Reactive oxygen species signaling in plants. Antioxid Redox Signal. 2006;8:1757–64.
4. Gill SS, Tuteja N. Reactive oxygen species and antioxidant machinery in abiotic stress tolerance in crop plants. Plant Physiol Biochem. 2010;48:909–30.
5. Jenkins GI. Signal transduction in responses to UVB radiation. Annu Rev Plant Biol. 2009;60:407–31.
6. Brosche M, Strid A. Molecular events following perception of ultraviolet-B radiation by plants: UVB induced signal transduction pathways and changes in gene expression. Physiol Plant. 2003;117:1–10.

7. Ballare CL, Caldwell MM, Flint SD, Robinson SA, Bornman JF. Effects of solar ultraviolet radiation on terrestrial ecosystems. patterns, mechanisms, and interactions with climate change. Photochem Photobiol Sci. 2011;10:226–41.

8. Li FR, Peng SL, Chen BM, Hou YP. A meta-analysis of the responses of woody and herbaceous plants to elevated ultraviolet-B radiation. Acta Oecol. 2010;36:1–9.

9. Deckmyn G, Gaeyenberghs E, Ceulemans R. Reduced UV-B in greenhouses decreases white clover response to enhance CO_2. Environ Exp Bot. 2001;46(2):109–17.

10. Rizzini L, Favory JJ, Cloix C, Faggionato D, O'Hara A, Kaiserli E, Baumeister R, Schäfer E, Nagy F, Jenkins GI, et al. Perception of UV-B by the *Arabidopsis* UVR8 protein. Science. 2011;332(6025):103–6.

11. Christie JM, Arvai AS, Baxter KJ, Heilmann M, Pratt AJ. Plant UVR8 photoreceptor senses UVB by tryptophan-mediated disruption of cross-dimer salt bridges. Science. 2012;335:1492–6.

12. Wu D, Hu Q, Yan Z, Chen W, Yan C. Structural basis of ultraviolet-B perception by UVR8. Nature. 2012;484:214–9.

13. Heijde M, Ulm R. UVB photoreceptor-mediated signaling in plants. Trends Plant Sci. 2012;17:230–7.

14. Yi C, Deng XW. COP1-from plant photomorphogenesis to mammalian tumorigenesis. Trends Cell Biol. 2005;15:618–25.

15. Saijo Y, Sullivan JA, Wang H, Yang J, Shen Y. The COP1-SPA1 interaction defines a critical step in phytochrome A-mediated regulation of HY5 activity. Genes Dev. 2003;17:2642–7.

16. Osterlund MT, Hardtke CS, Wei N, Deng XW. Targeted destabilization of HY5 during light-regulated development of *Arabidopsis*. Nature. 2000;405:462–6.

17. Jones-Rhoades MW, Bartel DP, Bartel B. MicroRNAs and their regulatory roles in plants. Annu Rev Plant Biol. 2006;57:19–53.

18. Ramachandran V, Chen X. Small RNA metabolism in *Arabidopsis*. Trends Plant Sci. 2008;13:368–74.

19. Voinnet O. Origin, biogenesis, and activity of plant microRNAs. Cell. 2009; 136:669–87.

20. Chen X. MicroRNA biogenesis through Dicer-like 1 protein functions. Proc Natl Acad Sci U S A. 2004;101:12753–8.

21. Sunkar R, Li Y, Jagadeeswaran G. Functions of microRNAs in plant stress responses. Trends Plant Sci. 2012;17:196–203.

22. Sunkar R, Chinnusamy V, Zhu J, Zhu JK. Small RNAs as big players in plant abiotic stress responses and nutrient deprivation. Trends Plant Sci. 2007;12:301–9.

23. Pérez-Quintero AL, Quintero A, Urrego O, Vanegas P, López C. Bioinformatic identification of cassava miRNAs differentially expressed in response to infection by *Xanthomonas axonopodis pv. manihotis*. BMC Plant Biol. 2012;12:1–11.

24. Wang Z, Niu L. Peach industry status and recommendations. Fruit Growers' Friend (in Chinese). 2012;11:37–8.

25. Shulaev V, Korban SS, Sosinski B, Abbott AG, Aldwinckle HS, Folta KM, Lezzoni A, Main D, Arus P, Dandekar AM, et al. Multiple models for Rosaceae genomic. Plant Physiol. 2008;147(3):985–1003.

26. Verde I, Abbott AG, Scalabrin S, Jung S, Shu S, Marroni F, Zhebentyayeva T, Dettori MT, Grimwood J, Cattonaro F, et al. The high-quality draft genome of peach (*Prunus persica*) identifies unique patterns of genetic diversity, domestication and genome evolution. Nat Genet. 2013;45(5):487–94.

27. Zhu H, Xia R, Zhao BY, An YQ, Dardick DC, Callahan MA, Liu ZR. Unique expression, processing regulation, and regulatory network of peach (*prunus persica*) miRNAs. BMC Plant Biol. 2012;12(1):149.

28. Gao Z, Luo X, Shi T, Cai B, Zhang Z, Cheng ZM. Identification and validation of potential conserved microRNAs and their targets in peach (*Prunus persica*). Mol Cells. 2012;34:239–49.

29. Luo X, Gao Z, Shi T, Cheng Z, Zhang Z, Ni ZJ. Identification of miRNAs and their target genes in peach (*Prunus persica* L.) using high-throughput sequencing and degradome analysis. PLoS One. 2013;8(11):e79090.

30. Reig G, Alegre S, Gatius F, Iglesias I. Adaptability of peach cultivars [*Prunus persica* (L.) Batsch] to the climatic conditions of the Ebro Valley, with special focus on fruit quality. Sci Hortic-Amsterdam. 2015;190:149–60.

31. Byrne DH. Peach breeding trends: a worldwide perspective. Acta Hort. 2002;592:49–59.

32. Horvath DP, Anderson JV, Chao WS, Foley ME. Knowing when to grow: signals regulating bud dormancy. Trends Plant Sci. 2003;8(11):534–40.

33. Cirilli M, Bassi D, Ciacciulli A. Sugars in peach fruit: a breeding perspective. Hortic Res. 2016;2:15067.

34. Crisosto C, Costa G. The peach: botany, production, and uses. In: Layne DR, Bassi D, editors. . Cambridge: CAB International; 2008. p. 536–49.

35. Yu N, Li D, Tan Q, Zhang H, Gao D. Effect of UVB radiation on assimilate translocation and distribution in fruiting shoot of protected peach. Chin J Appl Environment Biol. 2013;19(1):157–63.

36. Yu N, Tan Q, Tan Y, Zhang H, Gao D. Effects of UVB radiation on ^{15}N urea absorption, utilization and distribution in fruiting shoot of peach under protected culture. Plant Nutrition Fertilizer Sci. 2012;18(2):491–8.

37. Luo X, Shi T, Sun H. Selection of suitable inner reference genes for normalization of microRNAs expression response to abiotic stresses by RT-qPCR in leaves, flowers and young stems of peach. Sci Hortic-Amsterdam. 2014;165(3):281–7.

38. Gao D. The current conditions and developing tendency of protected cultivation of fruit trees in China. Deciduous Fruits (in Chinese). 2016;48(1):1–4.

39. Li ZY, Gao DS, Qian S, Zhang JH, Li ZJ, Wang C. Effects of different light environments on the fruit quality of peach in greenhouse. J Anhui Agri Sci. 2009;37(21):9933V9934–63.

40. Brown BA, Jenkins GI. UV-B signaling pathways with different fluence-rate response profiles are distinguished in mature *Arabidopsis* leaf tissue by requirement for UVR8, HY5, and HYH. Plant Physiol. 2008;146:576–88.

41. Parul P, Samiksha S, Rachana S. Changing scenario in plant UV-B research: UV-B from a generic stressor to a specific regulator. J Photoch Photobio B. 2015;153:334–43.

42. Ulm R, Nagy F. Signalling and gene regulation in response to ultraviolet light. Curr Opin Plant Biol. 2005;8:477–82.

43. Hideg E, Jansen MA, Strid A. UV-B exposure, ROS, and stress: inseparable companions or loosely linked associates? Trends Plant Sci. 2013;18:107–15.

44. Chen XD. The effects of ultraviolet-B radiation intensity and different plasic film on development characteristics of peach flower and fruit in protected culture. Shandong Agriculture University, Horticulture and Engineering college: Master Thesis; 2009.

45. Hectors K, Prinsen E, De CW, Jansen MAK, Guisez Y. *Arabidopsis thaliana* plants acclimated to low dose rates of ultraviolet B radiation show specific changes in morphology and gene expression in the absence of stress symptoms. New Phytol. 2007;175:255–70.

46. Wang B, Sun Y, Song N. Identification of UVB-induced microRNAs in wheat. Genet Mol Res. 2013;12(4):4213–21.

47. Henderson IR, Zhang X, Lu C, Johnson L, Meyers BC, Green PJ. Dissecting *Arabidopsis thaliana* DICER function in small RNA processing, gene silencing and DNA methylation patterning. Nat Genet. 2006;38:721–5.

48. Chen L, Ren Y, Zhang Y, Xu J, Sun F, Zhang Z, Wang Y. Genome-wide identification and expression analysis of heat-responsive and novel microRNAs in *Populus tomentosa*. Gene. 2012;504:160–5.

49. Chen L, Zhang Y, Ren Y, Xu J, Zhang Z, Wang Y. Genome-wide identification of cold-responsive and new microRNAs in *Populus tomentosa* by high-throughput sequencing. Biochem Biophys Res Commun. 2012;417:892–6.

50. Li B, Duan H, Li J, Deng X, Yin W. Global identification of miRNAs and targets in *Populus euphratica* under salt stress. Plant Mol Biol. 2013;81:525–39.

51. Ren Y, Chen L, Zhang Y, Kang X, Zhang Z, Wang Y. Identification of novel and conserved *Populus tomentosa* microRNA as components of a response to water stress. Funct Integr Genomics. 2012;12:327–39.

52. Wang L, Mai YX, Zhang YC, Luo Q, Yang HQ. MicroRNA171c-targeted SCL6-II, SCL6-III, and SCL6-IV genes regulate shoot branching in *Arabidopsis*. Mol Plant. 2010;3:794–806.

53. Curaba J, Talbot M, Li Z, Helliwell C. Over-expression of microRNA171 affects phase transitions and floral meristem determinancy in barley. BMC Plant Biol. 2013;13:6.

54. Fan T, Li X, Yang W, Xia K, Ouyang J, Zhang M. Rice osa-mir171c mediates phase change from vegetative to reproductive development and shoot apical meristem maintenance by repressin four OsHAM transcription factors. PLoS One. 2015;10(5):e0125833.

55. Huang W, Peng S, Xian Z, Lin D, Hu G, Yang L, Ren M, Li Z. Overexpression of a tomato mir171 target gene *SIGRAS24* impacts multiple agronomical traits via regulating gibberellin and auxin homeostasis. Plant Biotechnol J. 2017;15(4):472–88.

56. Huang W, Xian Z, Kang X, Tang N, Li Z. Genome-wide identification, phylogeny and expression analysis of GRAS gene family in tomato. BMC Plant Biol. 2015;15:209.

57. Ma Z, Hu X, Cai W, Huang W, Xhou X, Luo Q, Yang H, Wang J, Huang J. *Arabidopsis* miR171-targeted scarecrow-like proteins bind to GT *cis*-elements and mediate gibberellin-regulated chlorophyll biosynthesis under light conditions. PLoS Genet. 2014;10(8):E1004519.

58. Barakat A, Sriram A, Park J, Zhebentyayeva T, Main D, Abbott A. Genome wide identification of chilling responsive microRNAs in *Prunus persica*. BMC Genomics. 2012;13:481.

59. Müller-Xing R, Xing Q, Goodrich J. Footprints of the sun: memory of UV and light stress in plants. Front Plant Sci. 2014;5:474.

60. Kroger M, Meister K, Kava R. Low-calorie sweeteners and other sugar substitutes: a review of the safety issues. Compr Rev Food Sci F. 2006;5(2):35–47.

61. Guo X, Li S, Liu G, Fu Z, Li S. Seasonal changes in carbohydrate content and reated enzyme activity in fruit and leaves of "Yanfengyihao" peach variety. J Fruit Science (in Chinese). 2004;21(3):196–200.

62. Ding D, Li W, Han M, Wang Y, Fu Z, Wang B, Tang J. Identification and characterization of maize microRNAs involved in developing ears. Plant Biol. 2013;16(1):9–15.

63. Allen E, Xie Z, Gustafson AM, Sung GH. Evolution of microRNA genes by inverted duplication of target gene sequences in *Arabidopsis thaliana*. Nat Genet. 2004;36(12):1282.

64. Robinson MD, McCarthy DJ, Smyth GK. EdgeR: a bioconductor package for differential expression analysis of digital gene expression data. Bioinformatics. 2010;26(1):139–40.

65. Luo X, Shi T, Sun H, Song J, Ni Z, Gao Z. Selection of suitable inner reference genes for normalisation of microRNA expression response to abiotic stresses by RT-qPCR in leaves, flowers and young stems of peach. Sci Hortic-Amsterdam. 2014;165(3):281–7.

66. Wang D, Gao Z, Du P, Xiao W, Tan Q, Chen X, Li L, Gao D. Expression of ABA metabolism-related genes suggests similarities and differences between seed dormancy and bud dormancy of peach (*Prunus persica*). Front Plant Sci. 2016;6:1248.

67. Karkacier M, Erbas M, Uslu MK, Aksu M. Comparison of different extraction and detection methods for sugars using amino-bonded phase HPLC. J Chromatogr Sci. 2003;41(6):331–43.

68. Sornkanok V, Zheng H, Peng Q, Jiang Q, Wang H, Fang T, Liao L, Wang L, He H, Han Y. Assessment of sugar components and genes involved in the regulation of sucrose accumulation in peach fruit. J Agric Food Chem. 2016;64:6723–9.

69. Shabala SN, Shabala SI, Martynenko AI, Babourina O, Newman IA. Salinity effect on bioelectric activity, growth, Na^+ accumulation and chlorophyll fluorescence of maize leaves: a comparative survey and prospects for screening. Aust J Plant Physiol. 1998;25:609–16.

12

Genomic signature of highland adaptation in fish: a case study in Tibetan Schizothoracinae species

Chao Tong[1,2,3*] iD, Fei Tian[1] and Kai Zhao[1*]

Abstract

Background: Genome-wide studies on highland adaptation mechanism in terrestrial animal have been widely reported with few available for aquatic animals. Tibetan Schizothoracinae species are ideal model systems to study speciation and adaptation of fish. The Schizothoracine fish, *Gymnocypris przewalskii ganzihonensis* had underwent the ecological niche shift from salt water to freshwater, and also experienced a recent split from *Gymnocypris przewalskii przewalskii*. In addition, *G. p. ganzihonensis* inhabited harsh aquatic environment including low temperature and hypoxia as well as other Schizothoracinae species, its genetic mechanism of highland adaptation have yet to be determined.

Results: Our study used comparative genomic analysis based on the transcriptomic data of *G. p. ganzihonensis* and other four fish genome datasets to investigate the genetic basis of highland adaptation in Schizothoracine fish. We found that Schizothoracine fish lineage on the terminal branch had an elevated dN/dS ratio than its ancestral branch. A total of 202 gene ontology (GO) categories involved into transport, energy metabolism and immune response had accelerated evolutionary rates than zebrafish. Interestingly, we also identified 162 genes showing signature of positive selection (PSG) involved into energy metabolism, transport and immune response in *G. p. ganzihonesis*. While, we failed to find any PSG related to hypoxia response as previous studies.

Conclusions: Comparative genomic analysis based on *G. p. ganzihonensis* transcriptome data revealed significant genomic signature of accelerated evolution ongoing within Tibetan Schizothoracinae species lineage. Molecular evolution analysis suggested that genes involved in energy metabolism, transport and immune response functions in Schizothoracine fish underwent positive selection, especially in innate immunity including toll-like receptor signaling pathway genes. Taken together, our result as a case study in Schizothoracinae species provides novel insights in understanding the aquatic animal adaptation to extreme environment on the Tibetan Plateau, and also provides valuable genomic resource for further functional verification studies.

Keywords: Comparative genomics, Schizothoracinae, Highland adaptation, Positive selection, Innate immunity

Background

It is of evolutionary interest to understand that how wildlife adapts to high altitude [1]. With an average elevation above 4000 m, the Tibetan Plateau (TP) is one of the earth's most significant continental-scale highlands [2] imposes an extremely inhospitable environment on most wildlife, including hypoxia, high ultraviolet radiation and low temperatures [3, 4]. Past research had indicated the adaptation of local wildlife to harsh living challenges. Recent studies employing genome-wide approaches on Tibetan terrestrial animal have primarily focused on response to hypoxia and energy metabolic pathways, including yak [4], Tibetan antelope [5], ground tit [2], Tibetan mastiff [6], Tibetan dog [7], Tibetan chicken [8]. Nevertheless, we know little about the mechanism of Tibetan aquatic animal adaptation to aquatic environment on the TP. Specifically, genetic mechanisms of adaptation in Schizothoracine fish have yet to be determined.

---

* Correspondence: tongchao1990@gmail.com; zhaokai@nwipb.cas.cn
[1]Key Laboratory of Adaptation and Evolution of Plateau Biota, Qinghai Key Laboratory of Animal Ecological Genomics, Laboratory of Plateau Fish Evolutionary and Functional Genomics, Northwest Institute of Plateau Biology, Chinese Academy of Sciences, Xining 810001, China
Full list of author information is available at the end of the article

Therefore, it may provide novel insights for understanding the mechanism of highland adaptation of Tibetan wildlife.

The Schizothoracinae is the largest and most diverse taxon of the TP ichthyofauna, which are distributed throughout the TP and its peripheral regions [9, 10]. Past research had revealed that Schizothoracinae species had well adapted to the harsh aquatic environment on the TP, including hypoxia, low temperature and even high salinity [9, 11–13], making them excellent models for investigating the genetic mechanism of aquatic animal adaptation to the extreme environment at high altitude. In addition, increasing studies focused on the speciation mechanism of the Schizothoracine fish and the uplift of the Tibetan Plateau [14–17]. A Schizothoracine fish, *Gymnocypris przewalskii ganzihonensis* is the only fish inhabiting the Ganzi River (Fig. 1a). Another Schizothoracine fish, *Gymnocypris przewalskii przewalskii* is also the only fish inhabiting the Lake Qinghai (the largest salt lake in China). Previous research had indicated that the Ganzi River once flowed into the Lake Qinghai before the Wei-Jin-Nanbei Dynasty (200 to 589 A.D) [18]. An additional survey had revealed that the Ganzi River had disconnected to the Lake Qinghai as offshore great sand dune movement and shrinking of lake shoreline, which resulted in *G. P. przewalskii*

colonized the freshwater habitat [18]. In addition, the taxonomists named this fish species as *G. p. ganzihonensis* based on morphological data [9], and also have been supported by mitochondrial evidences [17, 19–21]. Obviously, *G. p. ganzihonensis* had undergone the transition from salt water to freshwater, and this species also faced the challenges due to the low temperature and hypoxia environment in accord with other Schizothoracinae species. Therefore, it is an interesting issue to investigate the highland adaptation in fish species using *G. p. ganzihonesis* as a case study in Schizothoracinae.

Recent advances in sequencing technologies have offered the opportunity to map and quantify transcriptome in almost any species of interest that do not currently have a reference genome [22]. Noteworthy, most Schizothoracinae species are polyploidy, tetraploid, and even sixteen-ploid [9]. Transcriptome sequencing technology have been successfully applied in many polyploidy cases [23, 24], which is a rapid and effective approach to obtain massive protein-coding genes and molecular markers. This technology could facilitate investigations into the genetic basis of adaptations. Here we sequenced and generated the transcriptome of *G. p. ganzihonensis* as a case study in Tibetan Schizothoracinae species. We then performed comparative genomic analysis together with other previously available fish

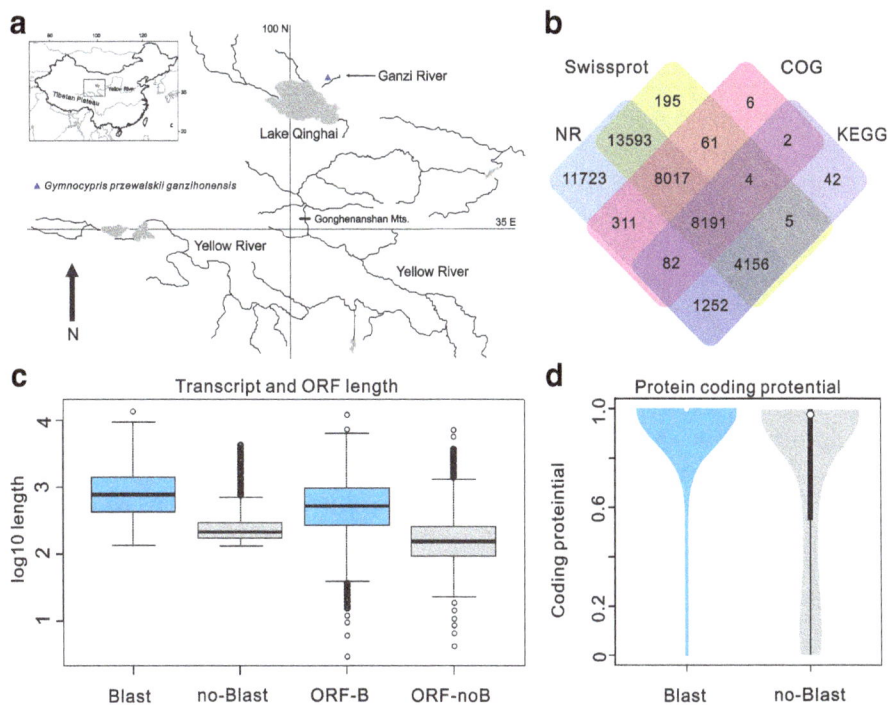

Fig. 1 Sampling site and annotation of *G. p. ganzihonensis* transcriptome. **a** The sampling map was created using the ArcGIS v10.1 (ESRI, CA, USA) and Adobe Illustrator CS5 (Adobe Systems Inc., San Francisco, CA). **b** Venn diagram shows shared and distinct genes under the annotations of NR, Swiss-Prot, COG and KEGG databases. Numbers indicating how many unigenes were annotated by each database. **c** Boxplot shows sequence characterization of the transcripts with and without detected homologs. **d** Protein coding potential were determined by CPAT and illustrated by boxplot

genomes to reveal the potential genetic mechanism of highland adaptation in fish.

Methods

Sample collection

Ten adult *G. p. ganzihonensis* samples were captured and identified from the Ganzi River using gill nets. Gender was determined for each specimen. Next, four individuals (2 male and 2 female) were selected and dissected after anesthesia with MS-222 (Solarbio, Beijing, China). Gill and kidney from each individual were collected respectively, and then immediately stored in liquid nitrogen at –80 °C.

RNA extraction and transcriptome sequencing

Total RNAs ($n = 8$, 4 gills and 4 kidney) of two tissues were isolated from each of four individuals using TRIzol (Invitrogen, Carlsbad, CA) according to the manufacturer's instructions. The quantity and quality of total RNA was verified by an Agilent 2100 bioanalyzer (Agilent Technologies, Palo Alto, CA) and gel electrophoresis. Approximately 10 μg of each RNA of same tissue from different individual were pooled for transcriptome library preparation (totally two independent libraries, gill and kidney), and sequenced on an Illumina HiSeq™ 2000 platform (parameters: 101-bp paired-end reads, 1 lane).

Transcriptome assembly and annotation

RNA-seq raw reads from each library were preprocessed to filter residual adapter sequences and low-quality reads ($Q < 20$), Then all clean reads were assembled using the Trinity v2.2.0 program (https://github.com/trinityrnaseq/trinityrnaseq/releases) with default parameters. Contigs from each sample's assembly were clustered by CD-HIT program [25] (percent identity: 80%; word size: 5) to generate a set of non-redundant unigenes, with a minimum overlap length of 200 bp. The assembled unigene sequences were aligned with a Blast-X search (cut-off E-value of 1×10^{-10}) in public NCBI non-redundant (NR), Swiss-Prot, Cluster of Orthologous Groups (COG) databases and Kyoto Encyclopedia of Genes and Genomes (KEGG) database. Gene ontology (GO) terms were obtained from NR hits using Blast2GO (version_3.2) [26] with default parameters. Next, the Getorf program in EMBOSS (version_6.4.0) [27] was applied to obtain the Open reading frames (ORFs) of *G. p. ganzihonensis* genes. The CPAT tool [28] was used to predict the protein-coding potential for the assembled unigenes, with previously downloaded zebrafish dataset (Zv9/danRer7) as the assembly database and 0.38 as the coding probability cutoff.

Orthologs identification

Orthologs between Schizothoracine fish (*G. p. ganzihonensis*) and zebrafish were identified using reciprocal BLAST best-hit method with an E value cutoff of 1×10^{-10} as used in prior investigations [29]. Then 1:1 orthologs between four fish genomes, including zebrafish (*Danio rerio*), fugu (*Takifugu rubripes*), medaka (*Oryzias latipes*), and spotted gar (*Lepisosteus oculatus*) were obtained from Ensembl server using BioMART (JCI_4.2.75) [30]. Noteworthy, only the longest transcript was considered if one gene had multiple transcripts. Each orthologous gene set was aligned using PRANK [31] (parameters: -f = fasta -F -codon -noxml -notree -nopost) and trimmed using GBlocks [31] (parameters: -t = c -b3 = 1 -b4 = 6 -b5 = n). Then we deleted all gaps and "N" from the alignments to lower the effect of ambiguous bases on the inference of positive selection. After deletion process, trimmed alignments shorter than 150 bp after removing sites with ambiguous data were discarded for subsequent analyses.

Molecular evolution analyses

The CODEML program in PAML 4.7a [32] with the free-ratio model (parameters: model = 1, NSsites = 0, fix_omega = 0, omega = 1) was run on each ortholog, a concatenation of all alignments of the orthologs, and 1000 concatenated alignments constructed from 150 randomly chosen orthologys, according to previous studies [29, 33–35]. The parameters of nonsynonymous (Ka or dN), synonymous (Ks or dS) and especially the substitution rate (ω = Ka/Ks or dN/dS) were used to meansure the lineage-specific evolutionary rates of above fish species. Based on GO term date which downloaded from BioMART (Ensembl, JCI_4.2.75), the orthologs were clustered into different functional GO terms and dN, dS and dN/dS ratio for each term was calculated, respectively. Finally, only GO categories with more than 20 orthologs were considered in this section analysis.

In addition, the CODEML program in PAML 4.7a with the branch-site model [36] (parameters: Null hypothesis: model = 2, NSsites = 2, fix_omega = 1, omega = 1) was used to identify positively selected genes (PSGs) in the Schizothoracine fish lineages, with other lineages being specified as the foreground branch. A LRT was constructed to compare a model that allows sites to be under positive selection ($\omega > 1$) on the foreground branch with the null model in which sites may evolve neutrally ($\omega = 1$) and under purifying selection ($\omega < 1$) with a posterior probability in excess of 0.95 based on the Bayes empirical Bayes (BEB) results [37]. Finally, the *P* values were computed based on rigorous Chi-square statistic adjusted by FDR method and genes with adjusted *P* value <0.05 were treated as candidates under positive selection.

Results

Transcriptome sequence analysis and assembly

Two pooled cDNA libraries derived from gill and kidney tissues of Schizothoracine fish, *G. p. ganzihonensis* were prepared and sequenced, totally generated 85,371,306 (gill) and 88,787,918 (kidney) raw 101-bp paired-end (PE) reads, respectively. After trimming adapters and removing low-quality reads, a total of 78,605,558 (gill) and 80,382,460 (kidney) clean reads were obtained from gill and kidney dataset, respectively. Finally, a total of 132,554 unigenes ranged from 201 to 16,310 bp, with an average length of 952 bp and an N50 of 1836 bp (Additional file 1: Table S1), the length distribution of all transcripts is shown in Additional file 2: Figure S1.

Functional annotation

To comprehensively annotate the transcriptome of G. p. ganzihonensis, all unigenes were queried against several public databases. A total of 94,321 (71.15%) sequences yielded at least one significant match to an existing gene model in Blast-X search (Fig. 1b, Additional file 3: Table S2). Statistics results of COG and GO classification of all annotated unigenes were shown in Additional file 4: Figure S2 and Additional file 5: Figure S3. Almost half (47.82%, $n = 63,404$) of homologs aligned to known proteins have identified between 80% and 100%. Due to fact of G. p. ganzihonensis is phylogenetically closer to zebrafish than some other fish species with complete genomic resources, it is not surprising that 81.72% ($n = 51,812$) of the best hits were similar with model organism zebrafish (Additional file 6: Table S3). Next, the assembly unigene dataset was divided into two subsets to characterize the sequence features in detail, including unigenes with and without protein homology in NR, namely "Blast" and "no-Blast" respectively. The "Blast" subset had significantly larger unigenes length and longer ORFs than the "no-Blast" subset with P value $<2.2 \times 10^{-5}$ in Wilcoxon rank sum test (Fig. 1c). In addition, further analysis of the potential for protein coding with CPAT tool showed a significantly lower protein-coding potential in the "no-Blast" subset with P value $<2.2 \times 10^{-6}$ (Fig. 1d).

Accelerated evolution of the Schizothoracine fish lineage

We identified the single-copy orthologs in Schizothoracine fish dataset and zebrafish, fugu, medaka, and spotted gar genome databases, resulting in a total of 6829 orthologs. Next, we used the species tree [38] in conjunction with a branch model constructed in PAML to determine dN, dS, and ω values across all 6829 orthologous genes. The result showed that the averaged ω value was significantly higher than other fish branches with P $< 2.2 \times 10-16$ in Wilcoxon rank sum test (Fig. 2a), implied that accelerated function evolution in Schizothoracine fish lineages. In addition, we analyzed the ω value for each branch for a concatenated alignment of all 6829 orthologs and 1000 concatenated alignments constructed from 150 randomly chosen orthologs. Intriguingly, using both comparison strategies, we found that Schizothoracine fish lineage exhibited a significantly

higher ω value than other four fish branches in our study ($P < 2.2 \times 10-16$) (Fig. 2b and c). Here showed a clear clue was that the Schizothoracine fish branch had an elevated ω value than its ancestral branch and trend to ongoing accelerated evolution under the extreme environment on the TP (Fig. 2a).

After a strict filtering analysis, we calculated the mean ω value for each GO category with at least 20 orthologs in Schizothoracine fish and zebrafish lineages, respectively. A total of 202 GO categories showing accelerated evolutionary rate ($P < 0.05$, binomial test) were detected in Schizothoracine fish and 113 in zebrafish (Fig. 3 and Additional file 7: Table S4), which also confirmed overall accelerated evolution in Schizothoracine fish lineage. We then focused on these accelerated categories potentially associated with highland adaptation. Interestingly, these GO categories were mainly involved into four functional groups. One group was mostly related to biotic and abiotic stress, such as "response to DNA damage stimulus" and "activation of immune response". As the fact is that G. p. ganzihonensis in fresh water is split from its ancestor G. P. przewalskii in salt water, one groups was related to transport function, such as "ion transport" and "lipid transport". The other two groups involved in energy metabolism and immune system, such as "regulation of lipid metabolic process" and "regulation of immune system process" (Fig. 3 and Additional file 7: Table S4).

Candidate genes under positive selection in Schizothoracine fish

To better understand the potential genes contributed to Schizothoracine fish adaptation to TP, we used branch-site model in PAML to identify candidate positively selected genes (PSGs) in Schizothoracine fish lineage. After applying strict filtering criteria, we totally identified 162 PSGs ($P < 0.05$) in G. p. ganzihonensis (Additional file 8: Table S5). Intriguingly, the PSGs were also had functions associated with three main groups. The first functional group were related to transport functions, including solute carrier family 12, member 1 (SLC12A1), solute carrier family 7, member 2 (SLC7A2), solute carrier family 38, member 4 (SLC38A4). The PSGs in second group were associated with energy metabolism, including NADH dehydrogenase 1 (ND1), ATPase family, AAA domain containing 2 (ATAD2), ADP-ribosylation factor 3 (ARL3). Innate immunity function group was the third one, such as toll-like receptor 3 (TLR3), interferon regulatory factor 8 (IRF8), interleukin 10 (IL10) and tumor necrosis factor receptor superfamily, member 1b (TNFRSF1b (Fig. 4a). Noteworthy, two infectious diseases "white spot" disease and saprolegniasis have been considered as the chief culprits and suffered high mortality rate of Schizothoracine fish

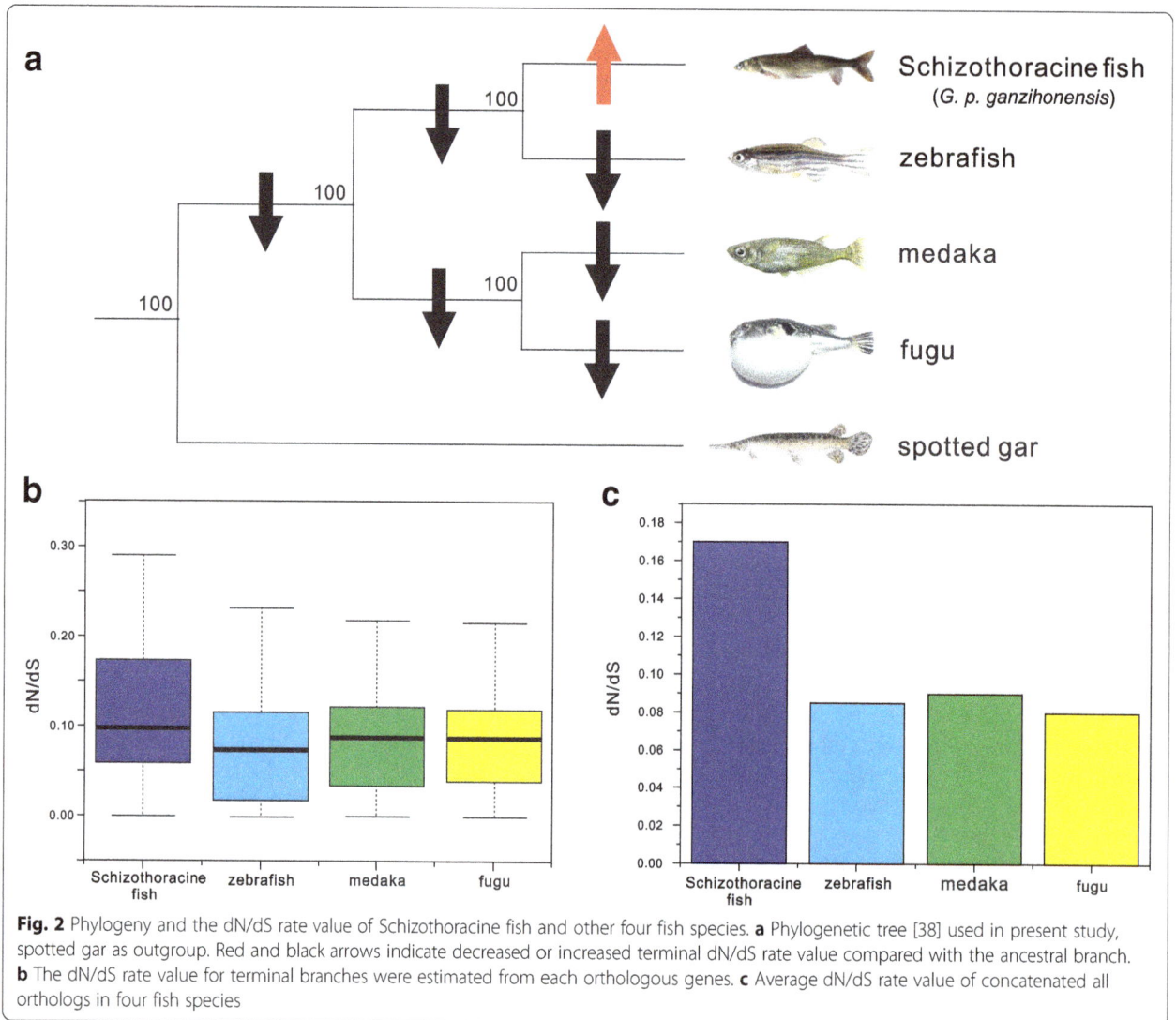

Fig. 2 Phylogeny and the dN/dS rate value of Schizothoracine fish and other four fish species. **a** Phylogenetic tree [38] used in present study, spotted gar as outgroup. Red and black arrows indicate decreased or increased terminal dN/dS rate value compared with the ancestral branch. **b** The dN/dS rate value for terminal branches were estimated from each orthologous genes. **c** Average dN/dS rate value of concatenated all orthologs in four fish species

when in aquaculture industry rather than native environment (Fig. 4b). This led us to hypothesize whether there is a link between weak immune ability and PSGs in innate immunity. In addition, ten of the candidate PSGs were identified and linked to energy metabolism, including ATP13a, ABCC2a, ATAD2 and MRPL45 (Additional file 8: Table S5). This finding also confirmed that GO category of energy metabolism in *G. p. ganzihonensis* lineage may undergo accelerated evolution.

Discussion

In evolutionary biology, comparative genomic analysis had been widely applied in understanding the genetic basis of organisms' speciation [39–41] and adaptation [2, 4, 8, 11, 29, 35]. Although whole genome sequencing data of nonmodel organisms have increasingly become available, most organisms still lack genomic resource. Transcriptome sequencing is an effective and accessible approach to initiate comparative genomic analysis on

nonmodel organisms, because it could also contain a large number of protein-coding genes likely enriched for targets of natural selection. In this study, we sequenced and annotated the transcriptome of the Schizothoracine fish, *G. p. ganzihonensis* [9, 18], and identified more than 6000 pairwise orthologs among five fish genomes. Then, we performed comparative genomic analysis on this Schizothoracine fish using its de novo assembly transcriptome dataset and other four fish genomes. Finally, this transcriptome resource could develop our understanding of genetic makeup of highland fishes and provide a foundation for further studies to identify candidate genes underlying adaptation to the Tibetan Plateau of Schizothoracine fishes.

How an organism adapts to environment change is an important issue in evolutionary biology [42]. Adaptive evolution may prefer to proceed at molecular level, expressed by an increase in ratio of nonsynonymous substitutions to synonymous substitutions [43]. Previous

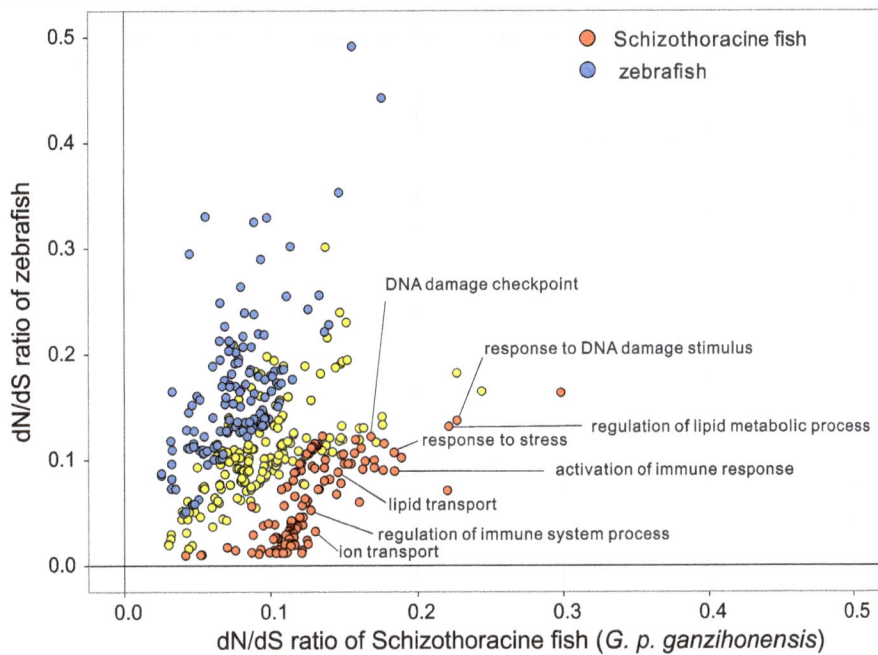

Fig. 3 Accelerated evolution within GO category. Scatter plot of mean dN/dS rate value for each GO category in Schizothoracine fish and zebrafish lineages. GO categories with significantly higher mean dN/dS rate value in Schizothoracine fish (red) and zebrafish (blue) are highlighted, respectively. Yellow points represent GO categories with higher but not statistically significant mean dN/dS rate value in both lineages

studies revealed that terrestrial organisms adapted to life at high altitude by gene family expansion, accelerated evolutionary rate and underwent positive selection on genes associated with specific function [2, 4, 7, 8]. Convergence is an independent evolution of similar physiological or morphological features in different species [44].

Its occurrence could support the hypothesis that specific ecological environment challenges can induce species to evolve in predictable and repeatable ways [45]. Our current analysis results suggested that Schizothoracine fish lineage trends to genome-wide accelerated evolution relative to other fish lineages. Past evidence indicated that accelerated

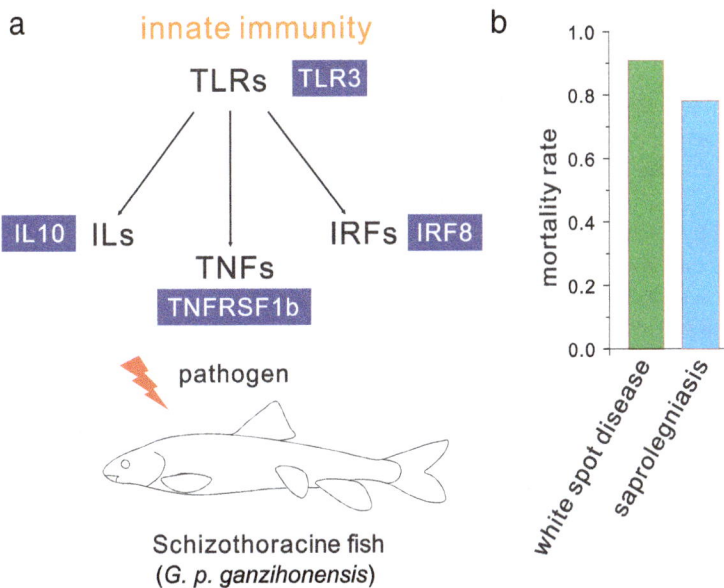

Fig. 4 Immune characterizations of Schizothoracine fish. **a** Schematic diagrams of innate immunity and Toll-like receptor (TLR) signing pathway. Four PSGs are highlighted in TLR pathway. The fish innate immune response to pathogens invasion, four major families are involved. TLR = toll-like receptor; IL = interleukin; IRF = interferon regulatory factor; TNF = tumor necrosis factor. **b** The morality rates of two culprit infectious diseases in Schizothoracine fish

evolution is usually driven by positive selection [32], we therefore speculated that Schizothoracine fish may adaptively speed up its evolutionary rate of genes for better adaptation to extreme environment of the TP. The relaxation of function constraint could possibly trigger accelerated evolution, which the hypothesis should need more cases based on population genomic analyses to support. Furthermore, compared with the ancestral branch, the terminal branch of Schizothoracine fish had underwent an elevated dN/dS ratio, implying that accelerated evolution only in the Schizothoracine fish lineage after diverged from zebrafish (also belonged to Cyprinidae). Previous studies had identified various adaptive processes that may be responsible for highland adaptation in terrestrial animal, including energy metabolism and hypoxia response [2, 4, 46]. Therefore it is not surprising that many GO categories related to energy metabolism and stress response in aquatic animal, Schizothoracine fish. A striking finding of the present study is that "transport function" genes may undergo accelerated evolution, this is consisted of the finding in recent study on the extremely alkaline environment adaptation mechanism in Amur ide, *Leuciscus waleckii* [47]. This finding implied that the adaptive evolution might play important role in this recent split Schizothoracine fish, *G. p. ganzihonesis* in transition of salt water to fresh water.

The functions of candidate PSGs were consistent with above identified functional groups of GO categories exhibiting accelerated evolution. Recent studies revealed the genetic basis of terrestrial animal adaptation to low oxygen and low temperature environment at high altitude [1, 2, 4, 7, 12, 13, 29]. In *G. p. ganzihonesis*, we failed to identify any PSG involved in hypoxia response. This may because the oxygen condition in Tibetan Plateau aquatic environment is different with ground. Previous evidence indicated that abundant and diverse of hydrophyte species in Ganzi River [18], these factors could have positive impacts on the dissolved oxygen content as the plant photosynthesis, which may help to explain the absence of PSGs related to response to hypoxia function. Low temperature is a typical feature of lake and river environment on the TP, which faced up this challenge for all aquatic animals. Accord to previous findings [2, 4], several candidate PSGs involved in energy metabolism were identified. For example, the ATP13a gene that encodes an accessory protein for ATP synthesis and decomposition, suggesting an important role in energy metabolism to adaptation to this low temperature water environment. In addition, SLC family play vital roles in transport function contribute to organism response to dynamic aquatic environment [48]. We also identified several SLC family members shown positive selection in Schizothoracine fish lineage, such as SLC12A1, SLC7A2, SLC38A4. This finding were similar to previous genome-wide study on the Amur ide in extremely alkaline environment [47], which indicated that adaptive evolution within genes involved in

transport function contribute to provide novel insight into adaptation to extreme aquatic environment on the Tibetan Plateau. Remarkably, here we provide another novel insight to understand the genetic mechanism of highland adaptation in Schizothoracine fish, the adaptive evolution of innate immunity. Recent evidence showed that Schizothoracine fish is susceptible to infectious diseases and triggered high mortality rate in non-native environment [49–52]. In addition, recent genome-wide study reveal that adaptive evolution of innate immunity contributed to fish well response to pathogen invasion [53]. Intriguingly, we identified significant positive selection signs in TLR3, IRF8, IL10 and TNFRSF1b involved in innate immunity. This finding is similar with our previous reports on the PSGs and neofunctionalization in TLR signaling pathway genes [49, 51, 52], suggesting that adaptive evolution of innate immunity may play important roles in Schizothoracine fish adaptation to high attitude aquatic life.

Conclusions
In summary, we have sequenced and annotated the first transcriptome of a recent split Schizothoracine fish, *G. p. ganzihonesis*. Comprehensive analyses of over 6000 orthologs among *G. p. ganzihonesis* and four fish genome databases identified evidence for sign of accelerated evolution in *G. p. ganzihonesis* lineage and only the terminal branch of *G. p. ganzihonesis* had an elevated dN/dS ratio than ancestral branches. Number of GO categories involved into energy metabolism, transport and immune response showed rapid evolution in compared with model zebrafish. Intriguingly, we found that many genes showing signature of positive selection in *G. p. ganzihonesis* lineage were enriched in functions associated with energy metabolism, and a novel finding is that PSGs also related to innate immunity, especially in toll-like receptor signaling pathway. While we failed to identify any PSG involved in hypoxia response as previous studies reported in terrestrial animal and several Tibetan fish species. This transcriptome dataset also provides a valuable resource for further functional verification study which will develop our understanding of ecological and evolutionary questions concerning fish species on the TP.

Notes
Chao Tong and Fei Tian contributed equally to this work.

Additional files

Additional file 1: Table S1. Summary of sequencing, assembly and analysis of *G. p. ganzihonensis* transcriptome. (DOC 36 kb)

Additional file 2: Figure S1. Length distribution of all transcripts. Transcripts of gill and kidney datasets are calculated respectively. Cumulative length of unigenes is also calculated. (PDF 138 kb)

Additional file 3: Table S2. Annotation of assembled unigenes in *G. p. ganzihonensis* transcriptome. (XLS 41915 kb)

Additional file 4: Figure S2. COG classification of assembled unigenes in *G. p. ganzihonensis* transcriptome. (PDF 22 kb)

Additional file 5: Figure S3. GO classification of assembled unigenes in *G. p. ganzihonensis* transcriptome. (PDF 21 kb)

Additional file 6: Table S3. Species distribution is calculated as a percentage of the total homologous sequences. (XLS 27 kb)

Additional file 7: Table S4. GO categories showing accelerated evolutionary rates within *G. p. ganzihonensis* and *Danio rerio*. (XLS 104 kb)

Additional file 8: Table S5. Positively selected genes (PSGs) identified in *G. p. ganzihonensis*. (XLS 64 kb)

Abbreviations
COG: Cluster of Orthologous Groups; FDR: false discovery ratio.; GO: Gene ontology; NR: non-redundant; ORFs: Open reading frames; PSG: Positively selected gene; TP: Tibetan Plateau

Acknowledgements
We thank Dr. Renyi Zhang, Yongtao Tang and Chenguang Feng in our group for critical comments on early manuscript. We also thank the editor and two anonymous reviewers for their valuable comments.

Funding
This work was supported by grants from the National Natural Science Foundation of China (30970341, 31572258 and 31700325), the Key Innovation Plan of Chinese Academy of Sciences (KSCX2-EW-N-004) and the Open Foundation from the Qinghai Key Laboratory of Tibetan Medicine Pharmacology and Safety Evaluation (2014-ZY-03).

Authors' contributions
CT and KZ conceived and designed the experiments. CT, FT and KZ collected the samples. CT analyzed the data and wrote the paper. All authors read and approved the final manuscript.

Competing interests
The authors declare that they have no competing interests.

Author details
[1]Key Laboratory of Adaptation and Evolution of Plateau Biota, Qinghai Key Laboratory of Animal Ecological Genomics, Laboratory of Plateau Fish Evolutionary and Functional Genomics, Northwest Institute of Plateau Biology, Chinese Academy of Sciences, Xining 810001, China. [2]University of Chinese Academy of Sciences, Beijing 100049, China. [3]Department of Biology, University of Pennsylvania, Philadelphia, PA 19104-6018, USA.

Reference
1. Bickler PE, Buck LT. Hypoxia tolerance in reptiles, amphibians, and fishes: life with variable oxygen availability. Annu Rev Physiol. 2007;69:145–70.
2. Qu Y, Zhao H, Han N, Zhou G, Song G, Gao B, Tian S, Zhang J, Zhang R, Meng X. Ground tit genome reveals avian adaptation to living at high altitudes in the Tibetan plateau. Nat Commun. 2013;4
3. An Z, John EK, Warren LP, Stephen CP. Evolution of Asian monsoons and phased uplift of the Himalaya–Tibetan plateau since late Miocene times. Nature. 2001;411(6833):62–6.
4. Qiu Q, Zhang G, Ma T, Qian W, Wang J, Ye Z, Cao C, Hu Q, Kim J, Larkin DM. The yak genome and adaptation to life at high altitude. Nat Genet. 2012;44(8):946–9.
5. Ge R, Cai Q, Shen Y, San A, Ma L, Zhang Y, Yi X, Chen Y, Yang L, Huang Y. Draft genome sequence of the Tibetan antelope. Nat Commun. 2013;4:1858.
6. Gou X, Wang Z, Li N, Qiu F, Xu Z, Yan D, Yang S, Jia J, Kong X, Wei Z. Whole-genome sequencing of six dog breeds from continuous altitudes reveals adaptation to high-altitude hypoxia. Genome Res. 2014;24(8):1308–15.
7. Wang G, Fan R, Zhai W, Liu F, Wang L, Zhong L, Wu H, Yang H, Wu S, Zhu C. Genetic convergence in the adaptation of dogs and humans to the high-altitude environment of the Tibetan plateau. Genome Biology and Evolution. 2014;6(8):2122–8.
8. Wang M-S, Li Y, Peng M-S, Zhong L, Wang Z-J, Li Q-Y, Tu X-L, Dong Y, Zhu C-L, Wang L. Genomic analyses reveal potential independent adaptation to high altitude in Tibetan chickens. Mol Biol Evol. 2015;32(7):1880–889.
9. Wu Y, Wu C. The fishes of the Qinghai-Xizang plateau: Sichuan Publishing House of Science & Technology; 1992.
10. Cao W, Chen Y, Wu Y, Zhu S. Origin and evolution of schizothoracine fishes in relation to the upheaval of the Qinghai-Tibetan plateau. Beijing: Science Press; 1981.
11. Zhang R, Ludwig A, Zhang C, Tong C, Li G, Tang Y, Peng Z, Zhao K. Local adaptation of *Gymnocypris przewalskii* (Cyprinidae) on the Tibetan plateau. Sci Rep. 2015;5:9780.
12. Guan L, Chi W, Xiao W, Chen L, He S. Analysis of hypoxia-inducible factor alpha polyploidization reveals adaptation to Tibetan plateau in the evolution of schizothoracine fish. BMC Evol Biol. 2014;14(1):192.
13. Xu Q, Zhang C, Zhang D, Jiang H, Peng S, Liu Y, Zhao K, Wang C, Chen L. Analysis of the erythropoietin of a Tibetan plateau schizothoracine fish (Gymnocypris Dobula) reveals enhanced cytoprotection function in hypoxic environments. BMC Evol Biol. 2016;16(1):1.
14. Duan Z, Zhao K, Peng Z, Li J, Diogo R, Zhao X, He S. Comparative phylogeography of the Yellow River schizothoracine fishes (Cyprinidae): Vicariance, expansion, and recent coalescence in response to the quaternary environmental upheaval in the Tibetan plateau. Mol Phylogenet Evol. 2009;53(3):1025–31.
15. Zhao K, Duan ZY, Peng ZG, Guo SC, Li JB, He SP, Zhao XQ. The youngest split in sympatric schizothoracine fish (Cyprinidae) is shaped by ecological adaptations in a Tibetan plateau glacier lake. Mol Ecol. 2009;18(17):3616–28.
16. Zhao K, Duan Z, Peng Z, Gan X, Zhang R, He S, Zhao X. Phylogeography of the endemic Gymnocypris chilianensis (Cyprinidae): sequential westward colonization followed by allopatric evolution in response to cyclical Pleistocene glaciations on the Tibetan plateau. Mol Phylogenet Evol. 2011;59(2):303–10.
17. Zhang R, Peng Z, Li G, Zhang C, Tang Y, Gan X, He S, Zhao K. Ongoing speciation in the Tibetan plateau *Gymnocypris* species complex. PLoS One. 2013;8(8):e71331.
18. Zhu S, Wu Y. Study of fish fauna in Qinghai Lake. Beijing: Science Press; 1975.
19. Qi D, Chao Y, Zhao L, Shen Z, Wang G. Complete mitochondrial genomes of two relatively closed species from Gymnocypris (Cypriniformes: Cyprinidae): genome characterization and phylogenetic considerations. Mitochondrial DNA. 2013;24(3):260–2.
20. Zhang R, Li G, Zhang C, Tang Y, Zhao K. Morphological differentiations of the gills of two *Gymnocypris przewalskii* subspecies in different habitats and their functional adaptations. Zool Res. 2013;34(4):387–91.
21. Tong C, Tang Y, Zhao K. The complete mitochondrial genome of Gymnocypris Przewalskii kelukehuensis (Teleostei: Cyprinidae). Conserv Genet Resour. 2017;9(3):443–5.
22. Wang Z, Gerstein M, Snyder M. RNA-Seq: a revolutionary tool for transcriptomics. Nat Rev Genet. 2009;10(1):57–63.
23. Bancroft I, Morgan C, Fraser F, Higgins J, Wells R, Clissold L, Baker D, Long Y, Meng J, Wang X. Dissecting the genome of the polyploid crop oilseed rape by transcriptome sequencing. Nat Biotechnol. 2011;29(8):762–6.
24. Dong S, Adams KL. Differential contributions to the transcriptome of duplicated genes in response to abiotic stresses in natural and synthetic polyploids. New Phytol. 2011;190(4):1045–57.
25. Li W, Godzik A. Cd-hit: a fast program for clustering and comparing large sets of protein or nucleotide sequences. Bioinformatics. 2006;22(13):1658–9.

26. Conesa A, Götz S, García-Gómez JM, Terol J, Talón M, Robles M. Blast2GO: a universal tool for annotation, visualization and analysis in functional genomics research. Bioinformatics. 2005;21(18):3674–6.

27. Rice P, Longden I, Bleasby A. EMBOSS: the European molecular biology open software suite. Trends Genet. 2000;16(6):276–7.

28. Wang L, Park HJ, Dasari S, Wang S, Kocher J-P, Li W. CPAT: coding-potential assessment tool using an alignment-free logistic regression model. Nucleic Acids Res. 2013;41(6):e74.

29. Yang L, Wang Y, Zhang Z, He S. Comprehensive transcriptome analysis reveals accelerated genic evolution in a Tibet fish, *Gymnodiptychus pachycheilus*. Genome Biology and Evolution. 2015;7(1):251–61.

30. Durinck S, Moreau Y, Kasprzyk A, Davis S, De Moor B, Brazma A, Huber W. BioMart and bioconductor: a powerful link between biological databases and microarray data analysis. Bioinformatics. 2005;21(16):3439–40.

31. Löytynoja A, Goldman N. An algorithm for progressive multiple alignment of sequences with insertions. Proc Natl Acad Sci U S A. 2005;102(30):10557–62.

32. Yang Z. PAML 4: phylogenetic analysis by maximum likelihood. Mol Biol Evol. 2007;24(8):1586–91.

33. Wang Y, Yang L, Zhou K, Zhang Y, Song Z, He S. Evidence for adaptation to the Tibetan plateau inferred from Tibetan loach transcriptomes. Genome biology and evolution. 2015;7(11):2970–82.

34. Yang Y, Wang L, Han J, Tang X, Ma M, Wang K, Zhang X, Ren Q, Chen Q, Qiu Q. Comparative transcriptomic analysis revealed adaptation mechanism of Phrynocephalus Erythrurus, the highest altitude lizard living in the Qinghai-Tibet plateau. BMC Evol Biol. 2015;15(1):101.

35. Tong C, Fei T, Zhang C, Zhao K. Comprehensive transcriptomic analysis of Tibetan Schizothoracinae fish Gymnocypris Przewalskii reveals how it adapts to a high altitude aquatic life. BMC Evol Biol. 2017;17(1):74.

36. Zhang J, Nielsen R, Yang Z. Evaluation of an improved branch-site likelihood method for detecting positive selection at the molecular level. Mol Biol Evol. 2005;22(12):2472–9.

37. Yang Z, Wong WS, Nielsen R. Bayes empirical Bayes inference of amino acid sites under positive selection. Mol Biol Evol. 2005;22(4):1107–18.

38. Near TJ, Eytan RI, Dornburg A, Kuhn KL, Moore JA, Davis MP, Wainwright PC, Friedman M, Smith WL. Resolution of ray-finned fish phylogeny and timing of diversification. Proc Natl Acad Sci. 2012;109(34):13698–703.

39. Fruciano C, Franchini P, Kovacova V, Elmer KR, Henning F, Meyer A. Genetic linkage of distinct adaptive traits in sympatrically speciating crater lake cichlid fish. Nat Commun. 2016;7

40. Elmer KR, Fan S, Kusche H, Spreitzer ML, Kautt AF, Franchini P, Meyer A. Parallel evolution of Nicaraguan crater lake cichlid fishes via non-parallel routes. Nat Commun. 2014;5

41. Brawand D, Wagner CE, Li YI, Malinsky M, Keller I, Fan S, Simakov O, Ng AY, Lim ZW, Bezault E. The genomic substrate for adaptive radiation in African cichlid fish. Nature. 2014;513(7518):375–81.

42. Smith NG, Eyre-Walker A. Adaptive protein evolution in drosophila. Nature. 2002;415(6875):1022–4.

43. Bakewell MA, Shi P, Zhang J. More genes underwent positive selection in chimpanzee evolution than in human evolution. Proc Natl Acad Sci. 2007;104(18):7489–94.

44. Feil R, Berger F. Convergent evolution of genomic imprinting in plants and mammals. Trends Genet. 2007;23(4):192–9.

45. Stern DL. The genetic causes of convergent evolution. Nat Rev Genet. 2013;14(11):751–64.

46. Beall CM, Decker MJ, Brittenham GM, Kushner I, Gebremedhin A, Strohl KP. An Ethiopian pattern of human adaptation to high-altitude hypoxia. Proc Natl Acad Sci. 2002;99(26):17215–8.

47. Xu J, Li J-T, Jiang Y, Peng W, Yao Z, Chen B, Jiang L, Feng J, Ji P, Liu G, et al. Genomic basis of adaptive evolution: the survival of Amur ide (*Leuciscus waleckii*) in an extremely alkaline environment. Mol Biol Evol. 2016;34(1):145–59.

48. Hediger MA, Romero MF, Peng JB, Rolfs A, Takanaga H, Bruford EA. The ABCs of solute carriers: physiological, pathological and therapeutic implications of human membrane transport proteinsIntroduction. Pflugers Arch - Eur J Physiol. 2004;447(5):465–8.

49. Tong C, Lin Y, Zhang C, Shi J, Qi H, Zhao K. Transcriptome-wide identification, molecular evolution and expression analysis of toll-like receptor family in a Tibet fish, *Gymnocypris przewalskii*. Fish and Shellfish Immunology. 2015;46(2):334–45.

50. Tong C, Zhang C, Zhang R, Zhao K. Transcriptome profiling analysis of naked carp (*Gymnocypris przewalskii*) provides insights into the immune-related genes in highland fish. Fish and Shellfish Immunology. 2015;46(2):366–77.

51. Tong C, Tian F, Tang Y, Feng C, Guan L, Zhang C, Zhao K. Positive Darwinian selection within *interferon regulatory factor* genes of *Gymnocypris przewalskii* (Cyprinidae) on the Tibetan plateau. Fish and Shellfish Immunology. 2016;50:34–42.

52. Tong C, Zhao K. Signature of adaptive evolution and functional divergence of TLR signaling pathway genes in Tibetan naked carp Gymnocypris Przewalskii. Fish and Shellfish Immunology. 2016;53:124.

53. Wu C, Zhang D, Kan M, Lv Z, Zhu A, Su Y, Zhou D, Zhang J, Zhang Z, Xu M. The draft genome of the large yellow croaker reveals well-developed innate immunity. Nat Commun. 2014;5

13

Pronounced strain-specific chemosensory receptor gene expression in the mouse vomeronasal organ

Kyle Duyck[1†], Vasha DuTell[1,3†], Limei Ma[1], Ariel Paulson[1] and C. Ron Yu[1,2*]

Abstract

Background: The chemosensory system plays an important role in orchestrating sexual behaviors in mammals. Pheromones trigger sexually dimorphic behaviors and different mouse strains exhibit differential responses to pheromone stimuli. It has been speculated that differential gene expression in the sensory organs that detect pheromones may underlie sexually-dimorphic and strain-specific responses to pheromone cues.

Results: We have performed transcriptome analyses of the mouse vomeronasal organ, a sensory organ recognizing pheromones and interspecies cues. We find little evidence of sexual dimorphism in gene expression except for *Xist,* an essential gene for X-linked gene inactivation. Variations in gene expression are found mainly among strains, with genes from immune response and chemosensory receptor classes dominating the list. Differentially expressed genes are concentrated in genomic hotspots enriched in these families of genes. Some chemosensory receptors show exclusive patterns of expression in different strains. We find high levels of single nucleotide polymorphism in chemosensory receptor pseudogenes, some of which lead to functionalized receptors. Moreover, we identify a number of differentially expressed long noncoding RNA species showing strong correlation or anti-correlation with chemoreceptor genes.

Conclusions: Our analyses provide little evidence supporting sexually dimorphic gene expression in the vomeronasal organ that may underlie dimorphic pheromone responses. In contrast, we find pronounced variations in the expression of immune response related genes, vomeronasal and G-protein coupled receptor genes among different mouse strains. These findings raised the possibility that diverse strains of mouse perceive pheromone cues differently and behavioral difference among strains in response to pheromone may first arise from differential detection of pheromones. On the other hand, sexually dimorphic responses to pheromones more likely originate from dimorphic neural circuits in the brain than from differential detection. Moreover, noncoding RNA may offer a potential regulatory mechanism controlling the differential expression patterns.

Keywords: Vomeronasal, Pheromone, Sexual dimorphic, Strain, Innate behavior, G-protein coupled receptor, Transcriptome

Background

In terrestrial animals, pheromones and olfactory cues mediate some key social behaviors [1–3]. Pheromones carry information about sex, reproductive status, genetic background, and individuality of the animals [1, 4]. In many vertebrate species, the vomeronasal organ (VNO) has evolved to specialize in detecting pheromone cues [5, 6]. The recent finding that the VNO responds to cues from other species expands its role in chemosensory perception [7]. In mice, the VNO expresses three major families of G protein coupled receptors: V1rs, V2rs, and formyl peptide receptors (FPRs) [8–13]. Additionally, some odorant and taste receptors are also detected in the VNO.

It has long been recognized that sexually dimorphic behaviors in male and female mice can be triggered by pheromone cues. For example, urine from mature female mice elicits sexual arousal in males, but suppresses sexual maturation and delays estrus cycle in

* Correspondence: cry@stowers.org

†Equal contributors

[1]Stowers Institute for Medical Research, 1000 East 50th Street, Kansas City, MO 64110, USA

[2]Department of Anatomy and Cell Biology, University of Kansas Medical Center, 3901 Rainbow Boulevard, Kansas City, KS 66160, USA

Full list of author information is available at the end of the article

females [14]. The origin of these sexually dimorphic behaviors may arise from brain circuitry that processes pheromone information, the differential recognition of pheromone signals by the sensory organs, or both. Previous studies have found moderate differences between male and female animals in the expression of a few genes in the VNO [15]. However, these studies have examined a single strain of mice, which may not be generalized to mice of different genetic backgrounds. True sexual dimorphism should be detected across different strains.

The patterns of activity in the mouse VNO can encode information about sex, genetic background and individuality of the carrier [16], as well as other species [7]. Several observations suggest that VNO is central in orchestrating innate behaviors. For example, some strains of mice exhibit the Bruce effect, when the presence of a stud male from a different strain causes a newly mated female to abort pregnancy [17]. Exhibition of the Bruce effect depends on not only the recognition of sex, but also strain information, by the VNO [18, 19]. Animals also display kinship recognition and respond stereotypically to cues from animals of different genetic backgrounds. Mice prefer sexual partners of a different genetic background [20, 21]. It is unknown whether kinship recognition and mating preferences directly arise from differential recognition of chemosensory cues mediated at the level of sensory organ.

The vomeronasal receptors are among the fastest evolving genes [22–35]. Comparison of receptor diversity among different species demonstrates highly divergent family members and receptor sequences [26, 28, 31–33, 36]. The diversity of receptor likely accommodates the variety of pheromone molecules. It is possible that co-evolution of pheromones and their receptors results in differential behavioral responses in various strains to influence mate choice, mating frequency and other reproductive behaviors. Differential expression of the receptors and associated proteins may also have a direct impact on how pheromones are recognized. In this study, we analyze VNO transcriptomes of both sexes from four inbred mouse strains. These analyses reveal a rich array of genes that are differentially expressed by the VNO with implications on how pheromones cues can be differentially recognized by different strains of mice.

Results

Lack of significant sexual dimorphism in VNO gene expression

We dissected VNO neuroepithelia from 6-week-old male and female animals of C57BL/6 (B6), 129Sv/J (129), SJL and SWR strains. The widely used B6 and 129 strains are derived from the Lathrop and Castle lineages, respectively [37]. In comparison, SJL and SWR lines descend from the Swiss lineage and are closely related to each other. We reason that sampling from these four

strains may provide information about strain and sex difference in VNO gene expression.

We extracted total RNA from individual VNO neuroepithelia and performed ribo-depletion to remove ribosomal RNA from the samples prior to library construction. Routine RNAseq was performed on HiSeq platform and high-quality reads were mapped to GRCm38 (mm10) mouse reference genome (Additional file 1: Figure S1). In total, we identified 44,957 genes as expressed by any of the samples. Principal components analysis (PCA) of the dataset indicated that the samples were well separated according to strains (Fig. 1a), with principal component 1 (PC1, 27.5% variance) separating B6 and 129 from each other and the Swiss strains, and PC2 (22.8% variance) separating 129 from both the B6 and Swiss strains. Within each strain, however, male and female samples were intermingled (Fig. 1a). Analyses of the first four PCs, which accounted for 72.9% of the variance, did not reveal an axis that separated the sexes. Only for PC5 and PC6 (4.11% and 3.1% variances, respectively) did we observe clear separation by sex for all samples (Fig. 1b). This result indicated that sex did not contribute significantly to the variance of gene expression in VNO, although some of the genes indeed showed sexual dimorphic expression.

Previous studies discovered limited sexual dimorphism in gene expression from the olfactory tissues of the B6 strain [15]. However, it was not clear whether the observed sexual dimorphism was also present in other strains as well. We reasoned that for a gene to be considered truly sexual dimorphic, the differential expression between male and female should be consistently observed across all strains. By comparing the male and female samples from all four strains, we found seven genes emerge as differentially expressed (DE) between the sexes with fold change (FC) greater than 2, or Log_2 fold change (LFC) greater than 1 ($p < 0.01$) (Fig. 1c). Among these were *Xist* (Fig. 1d), an X-linked non-coding RNA gene that plays an essential role in X-inactivation [38], and six Y chromosome genes: *Gm18665, Gm29650, Eif2s3y, Ddx3y, Kdm5d,* and *Uty (Kdm6c)*. When we examined these Y chromosome genes, we found that the expression levels of their X allele homologs were slightly lower in males than females (FC < 2; Fig. 1e). Moreover, apart from *Eif2s3*, the expression of the Y chromosome counterpart of the genes in males largely compensated the differences between male and female samples (Fig. 1e). These results suggested that the X-allele genes did not escape dosage-compensation in female VNO. After taking into account the expression of their Y chromosome counterparts, the functions of these genes were not sexually dimorphic. We found no other transcripts, including those related to chemosensory perception such as odorant receptor, vomeronasal receptor or pheromone binding

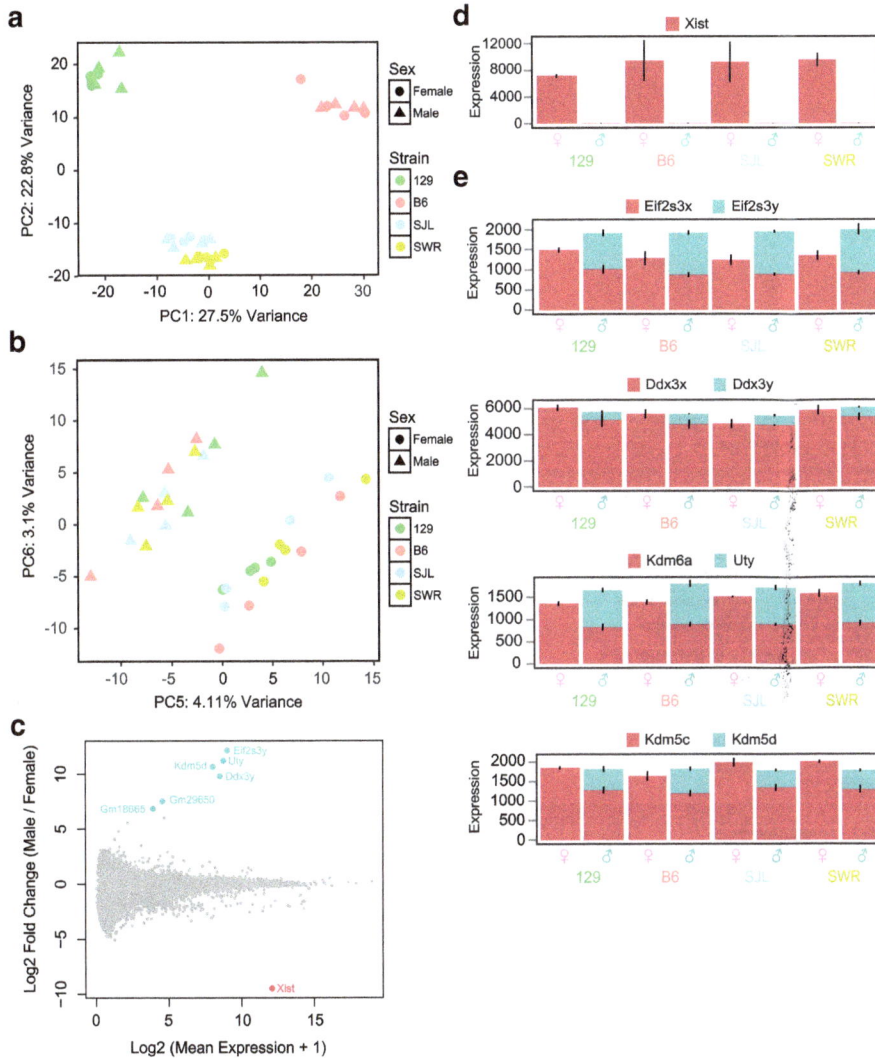

Fig. 1 Sex linked gene expression in the VNO. **a** and **b** Principal components analysis (PCA) of all expressed genes in VNO of B6, 129, SWR and SJL strains. **c** MA plot of gene expression in VNOs from male and female mice. Y-axis indicates the maximal value of fold change (FC) between male and female in log2 scale. Genes that exhibit significant DE are highlighted in color (weighted FC > 2; $p < 0.01$). **d** Bar plot of mean normalized expression of *Xist* in male and female mice. **e** Stacked bar plot of mean normalized value of Y chromosome linked genes and their X chromosome homologs. Genes expressed from the X and Y chromosomes are labeled red and teal, respectively. Error bars represent standard deviation of expression values

protein genes, to be differentially expressed in the VNO between the sexes. Thus, *Xist* was the only gene exhibiting sexually dimorphic expression in the VNO.

Differential gene expression among strains

We next examined whether gene expression in VNO was different among strains. Of the 44,957 genes expressed in the VNO, we identified a list of 5745 genes (12.8% of all expressed) that were DE among the strains with FC > 2, and false discovery rate (FDR) <0.05 (Fig. 2a). Of these DE genes, 1644 were annotated as either gene models (Gm) or Riken (Rik) genes. These putative genes constituted the largest subgroup (28.6%) of DE genes with no known

function. It was yet to be determined how these transcripts affected VNO functions.

For the remaining 4101 DE genes that had functional annotations, we performed a gene ontology (GO) analysis to investigate a possible enrichment of the GO terms in certain categories (Fig. 2b). This analysis indicated that G-protein coupled receptor (GPCR) activity and immune system related genes dominated the list. Enriched GO terms of the Biological Process category were related to the regulation of immune, stimulus, and inflammatory responses, as well as signaling (classic Fisher, $p < 1e-23$). In the Molecular Function category, GO terms were highly enriched for binding of calcium and glycosaminoglycan, activity of pheromone, transmembrane, and signal receptors, and

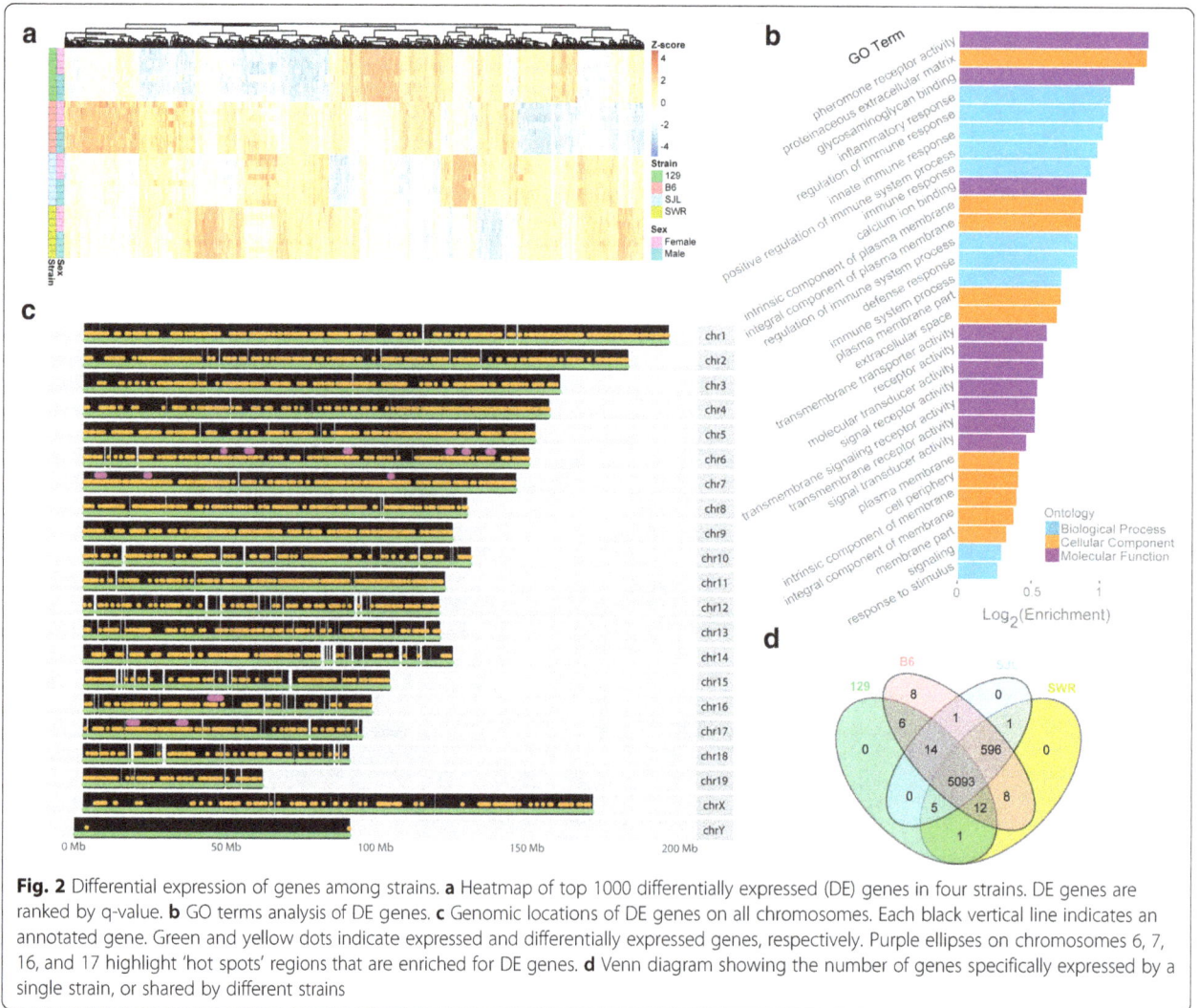

Fig. 2 Differential expression of genes among strains. **a** Heatmap of top 1000 differentially expressed (DE) genes in four strains. DE genes are ranked by q-value. **b** GO terms analysis of DE genes. **c** Genomic locations of DE genes on all chromosomes. Each black vertical line indicates an annotated gene. Green and yellow dots indicate expressed and differentially expressed genes, respectively. Purple ellipses on chromosomes 6, 7, 16, and 17 highlight 'hot spots' regions that are enriched for DE genes. **d** Venn diagram showing the number of genes specifically expressed by a single strain, or shared by different strains

transmembrane transporter activity (classic Fisher, $p < 1e\text{-}12$). GO terms in the Cellular Component category were enriched for the cellular periphery, plasma membrane, and extracellular space (classic Fisher, $p < 1e\text{-}25$).

Whereas differentially expressed genes were located throughout the entire genome, some chromosomal regions appeared to contain high numbers of DE genes. By applying a sliding window across all the expressed genes on each chromosome, we identified 12 "hot spots" – genomic regions in which there were a larger percentage of DE genes than random scattering would predict (Poisson test, FDR < 0.05). Interestingly, these clusters are enriched in genes from the chemoreceptor and immune system related gene families. We identified six hot spots on Chr. 6, three on Chr. 7, one on Chr. 16, and two on Chr. 17 (Fig. 2c, Table 1). Three of the six Chr. 6 hotspots, and two of the three Chr. 7 hotspots contained vomeronasal receptors, including vmn1r (Chr. 6) and vmn2r (Chr. 7). Of the 2 hotspots on Chr. 17, the largest one corresponded to a locus enriched in vmn2r genes.

The remaining hotspots largely contained immune system related genes. Three hotspots on Chr. 6 contained genes from the GIMAP, Clec, Klr families of genes. The hotspots on Chr. 7 and Chr. 16 contained Trim, and CD200/CD200 receptor genes, respectively. On Chr. 17, a 2.17 Mb hot spot was enriched with Butyrophilin-like and MHC class 1b, 2a, and 2b genes, with 20 of 28 of the MHC genes differentially expressed. The downstream end of the hotspot was enriched for MHC class 1b genes. This region was the most densely packed, with 35 expressed genes in a region less than 1 Mb in length, with almost half of them differentially expressed. In total, the hot spots covered 18.34 Mb and 575 expressed genes, 241 (41.9%) of which were DE. This percentage contrasts to the whole genome with an average of 12.8% DE genes.

Some differentially expressed genes were present in all strains but at different levels. Others were expressed exclusively in some strains but not others. 5093 (89%) of the DE genes were expressed by all

Table 1 Hot Spot of Differentially Expressed Genes

Chr	Start	End	Span (MB)	Expr. Genes	DE genes	% DE	Prominent Gene Families
6	48,448,229	48,754,210	0.31	25	11	44%	GIMAP
6	56,172,928	57,664,632	1.49	39	14	35.9%	Vmn1r (Clade C)
6	89,316,314	90,600,203	1.28	46	19	41.3%	Vmn1r (Clades A, B)
6	123,195,632	124,082,601	0.89	27	11	40.7%	Clec, Vmn2r (Clade B)
6	128,648,576	129,740,484	1.09	62	28	45.2%	Clec, Klr
6	136,506,167	138,079,916	1.57	30	12	40%	NA
7	7,171,330	9,389,264	2.22	67	32	47.8%	Vmn2r (Clade A4)
7	23,272,801	24,143,241	0.87	41	18	43.9	Vmn1r (Clade D)
7	104,140,623	104,601,779	0.46	32	13	40.6	Olfr, Trim
16	44,347,121	47,758,671	3.41	40	17	42.5	Cd200, Cd200r
17	17,830,352	20,405,756	2.58	44	19	43.2	Fpr, Vmn2r (Clade A8)
17	34,031,812	36,198,513	2.17	122	47	38.5	Btnl, MHC (Clades IIa, IIb, Ia, Ib)
Hot Spot Totals			18.34	575	241	41.9%	
Whole Genome				44,957	5745	12.8%	

four strains (Fig. 2d). The remaining 11% had no expression in the VNO of both sexes in at least one strain. Of these, eight genes were expressed solely in C57BL/6, and 627 genes were excluded in one strain.

Chemosensory receptor expressions in different strains

In our analyses, GPCRs (453 out of 5745) constituted a large group of DE genes (Fig. 3a), which included 114 V1r (Fig. 3b), 111 V2r (Fig. 3c), 141 olfactory receptors (Additional file 2: Figure S2), 4 formyl peptide receptors, and 2 taste receptor genes. Differentially expressed V1r genes were found in all clades (A - K) except L, which contains only one gene Vmn1r70 (Fig. 3b). The DE V2r genes were also found in all clades (A1- A5, A8, A9, B, C, D, and E) except for clade A6, which also contains only one gene, Vmn2r120 (Fig. 3c).

Interestingly, we observed a completely lack of expression of some chemosensory genes in one or more strains (Fig. 3d and Additional file 3:Figure S3). Some genes were expressed in a mutually exclusive fashion among the tested strains (Additional file 3: Figure S3). In the V1r family, for example, Vmn1r188 was expressed exclusively in B6, while Vmn1r76 was expressed in all strains except SWR. In the V2r family, Vmn2r-ps24 was expressed in all but the 129 strain. We also observed a similar scenario in the DE olfactory receptor genes. Olfr279 and Olfr116 were expressed in all but 129 mice. Overall, among the DE chemoreceptor genes, 12.3% (14/114) of the V1rs, 8.1% (9/111) of the V2rs, and 65.2% (92/141) of the ORs completely lacked expression in at least one strain. Some of the differentially expressed VRs show single nucleotide polymorphisms (SNPs) with both synonymous and non-synonymous changes. (Additional file 4: Figure S4).

The expression level of different VR genes varied widely. Some clades, such as V1r clade J, E and F, were expressed at higher levels than others (Fig. 3 and Additional file 5: Figure S5). Clade E and J members were shown to recognize female specific cues that identify the sex and the reproductive status of female mice (Fig. 3e) [39]. The function of the V1rf genes remained unknown.

FPRs are a family of chemosensory receptors expressed in the VNO implicated in the recognition of the health statuses of the animals [8, 9, 40]. Fpr-rs3 had the strongest expression among all FPR genes, which was about 3-fold higher than other FPR. It was also one of four FPR genes differentially expressed. In addition to differential expression, we also found SNPs in FPR genes specific to 129 strain mice (Additional file 6: Figure S6). SNPs in the coding regions of Fpr-rs3, Fpr-rs4 and Fpr-rs6 altered protein sequences. One synonymous SNP was found within the protein-coding region of Fpr3. The changes in both expression levels and coding sequences implied that the recognition of FRP ligands were likely to be different between 129 and the other strains.

Of the 141 olfactory receptors, only a few data points have more than 1 transcripts per million, indicating that their expression is either limited to an extremely small population of cells or is from leakage. Besides the classical chemosensory receptors, we identified 409 genes that were expressed in the VNO of at least one strain, and had GO terms related to GPCR activity or one of its children terms. Of the 409 expressed genes in this group, 138 were differentially expressed between the strains, however none was shown directly to be involved in VNO signaling (data not shown).

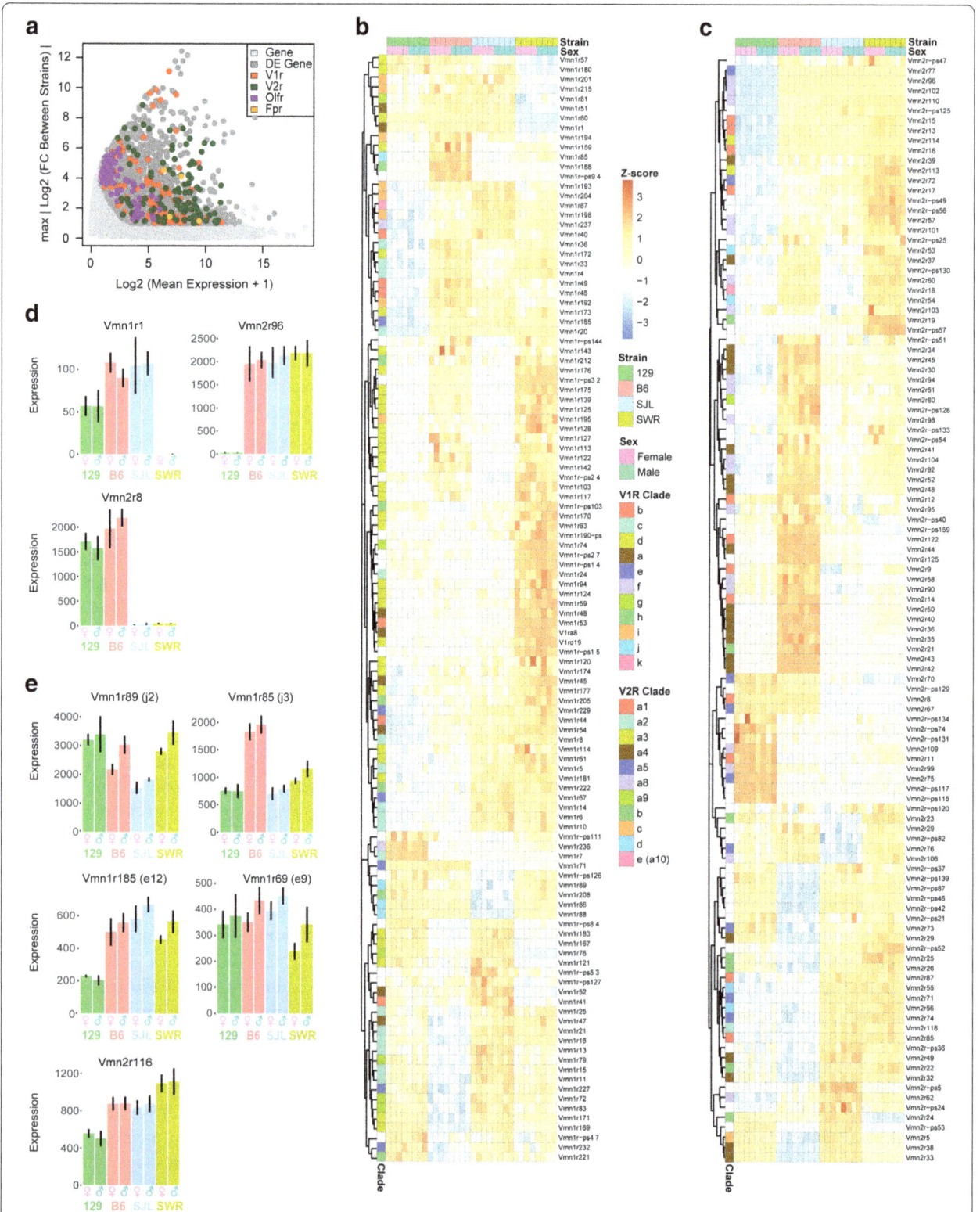

Fig. 3 Differential expression of chemosensory receptor genes. **a** MA plot highlighting DE chemosensory receptor genes. Y-axis indicates the maximal value of FC between any two strains in log2 scale. Chemosensory receptors genes are highlighted, including 114 V1r (red), 111 V2r (green), 141 Olfr (purple), and 4 Fpr (orange). **b-c** Heatmaps showing the DE chemoreceptor genes, including V1rs (B) and V2rs (C). Each clade is color-coded. **d** Example bar plots showing expression profiles for highly DE V1rs & V2rs across different strains. **e** Bar plots showing the expression levels of receptors identified as sex pheromone detecting receptors. Error bars represent standard deviation of expression values

VRs detecting sex pheromones

Only a handful of VRs have been assigned functions in pheromone signaling. This made it difficult to assess whether the differentially expressed receptors could affect pheromone-dependent behaviors. Previous studies have identified several receptors involved in sexually dimorphic behavior in mice [39, 41, 42]. We, therefore, specifically examined *Vmn1r69* (*V1re9*) and *Vmn1r185* (*V1re12*), two receptors known to respond to female sex-specific pheromone cues; *Vmn1r85* (*V1rj3*) and *Vmn1r89* (*V1rj2*), two receptors known to recognize estrus cues; and *Vmn2r116* (*V2rp5*), a receptor for the male-specific ESP-1 peptide (Fig. 3e) [39, 41, 42]. We found all four V1r genes in all strains suggesting the critical roles of these receptors in mating behavior. Three of these genes, *Vmn1r185* and *Vmn1r85*, *Vmn1r89*, were differentially expressed among the strains, with *Vmn1r185* expressed significantly less in the VNO of 129 strain mice, and *Vmn1r85* expressed at higher level in B6 mice than any other strains. Expression of *Vmn1r89* was slightly higher in male VNO of all strains, but the difference was not statistically significant. No genes exhibited preferential expression in the females.

We observe high levels of polymorphism in Vmn2r116 for 129 strain mice, although the difference in expression between strains is not significant given our stringent threshold of FC > 2 (Fig. 3e). There were six SNPs within the reading frame, five of which resulted in non-synonymous amino acid changes, including a Gly to Asp substitution within the predicted 7-TM domain. In contrast, no SNPs within the reading frames of *Vmn1r185* or *Vmn1r89* were detected. *Vmn1r69* contained only two SNPs, both found only in the Swiss mice, and only one of which resulted in a change in amino acid sequence. *Vmn1r85* contained no synonymous polymorphisms within the ORF.

Functionalized Pseudogenes

We identified a list of 504 DE genes that were annotated as pseudogenes in the reference genome. B6 had the lowest pseudogene expression (Fig. 4a). Many of these pseudogenes contained SNPs, some of which led them to encode functional proteins. Two Vmn1r pseudogenes, *Vmn1r-ps27*, and *Vmn1r-ps32*, as well as one Vmn2r pseudogene, *Vmn2r-ps53*, encoded functional receptors because of insertions that changed the reading frame and/or SNPs that removed stop codons. *Vmn1r-ps27* was expressed over 2-fold higher in SWR than in any other strain. It contained ten SNPs solely found in the SWR strain (both male and female samples) (Fig. 4b-c, Additional file 7: Figure S7). These SNPs resulted in an ORF over the entire gene length to encode a 329- amino acid protein that shared 84% protein identity (91% nucleic acid identity) with Vmn1r42 (Fig. 4c and Additional file 7: Figure S7). *Vmn1r-ps32*, which was expressed over 3-fold higher in SWR than in any other strain, contained a C insert 359 bp from the start codon that restored the reading frame such that the ORF encoded a 318-amino acid protein with 95% protein identity and 97% nucleic acid identity to *Vmn1r45*. We suspect that this phenomenon is more widespread than these two examples. However, due to the lack of complete reference genome for 129, SWR and SJL at the time of the study, we are not able to test whether all B6 psuedogenes listed in Fig. 4a have functional counterparts in the other three strains.

Immune system related genes

An interesting observation was that 2159 immune system related genes were found to be expressed in VNO epithelia and 591 of them showed differential expression among strains (Fig. 5a). It was not clear whether these genes simply reflected the genetic background of the mice or contributed to the VNO mediated pheromone response. The largest group included 32 MHC genes, whereas others included five fragment receptor (Fce/g), eight guanylate binding protein (Gbp), five interferon induced (Ifit), 13 interleukin (Il), 11 interleukin receptors, and eight Toll-like receptor (Tlr) family genes (Fig. 5a). Interestingly, five of the immune system related genes were polymorphic pseudogenes with protein coding sequences known to be intact in other individuals of the same species.

The class I MHC molecules present peptide antigens derived from intracellular proteins to elicit immune responses. The expression of these genes was expected to be strain specific. Of the DE MHC genes, two were of class 1a, five were class IIa, and three were class IIb molecules (Fig. 5b). H2-Bl, a polymorphic pseudogene was also found to be DE. A subset of the MHC class 1b genes, specifically those of the H2-Mv family (*H2-M1, H2-M9, H2M10.2–5*, and *H2-M11*) have been shown to be co-expressed with specific clades of Vmn2r genes, namely *V2ra1–5* and *V2rc* [43–45]. They have been suggested to be either co-receptors of the Vmn2r products or to facilitate their expression on the VNO neuron surface.

lncRNAs expression is correlated with chemoreceptors

Long non-coding (lnc) RNA have emerged as major regulators of gene expression in cell differentiation and development [46–48]. We found 446 lncRNA biotypes from the DE gene set (Fig. 6a). The majority of these DE genes were gene models or Riken transcripts with unknown functions. Two highly expressed lncRNAs showed differential expression among the strains: *Gm26870* and *Miat* (Fig. 6b). Both genes showed exclusive expression profiles with high expression level in some strains and virtually undetectable in others

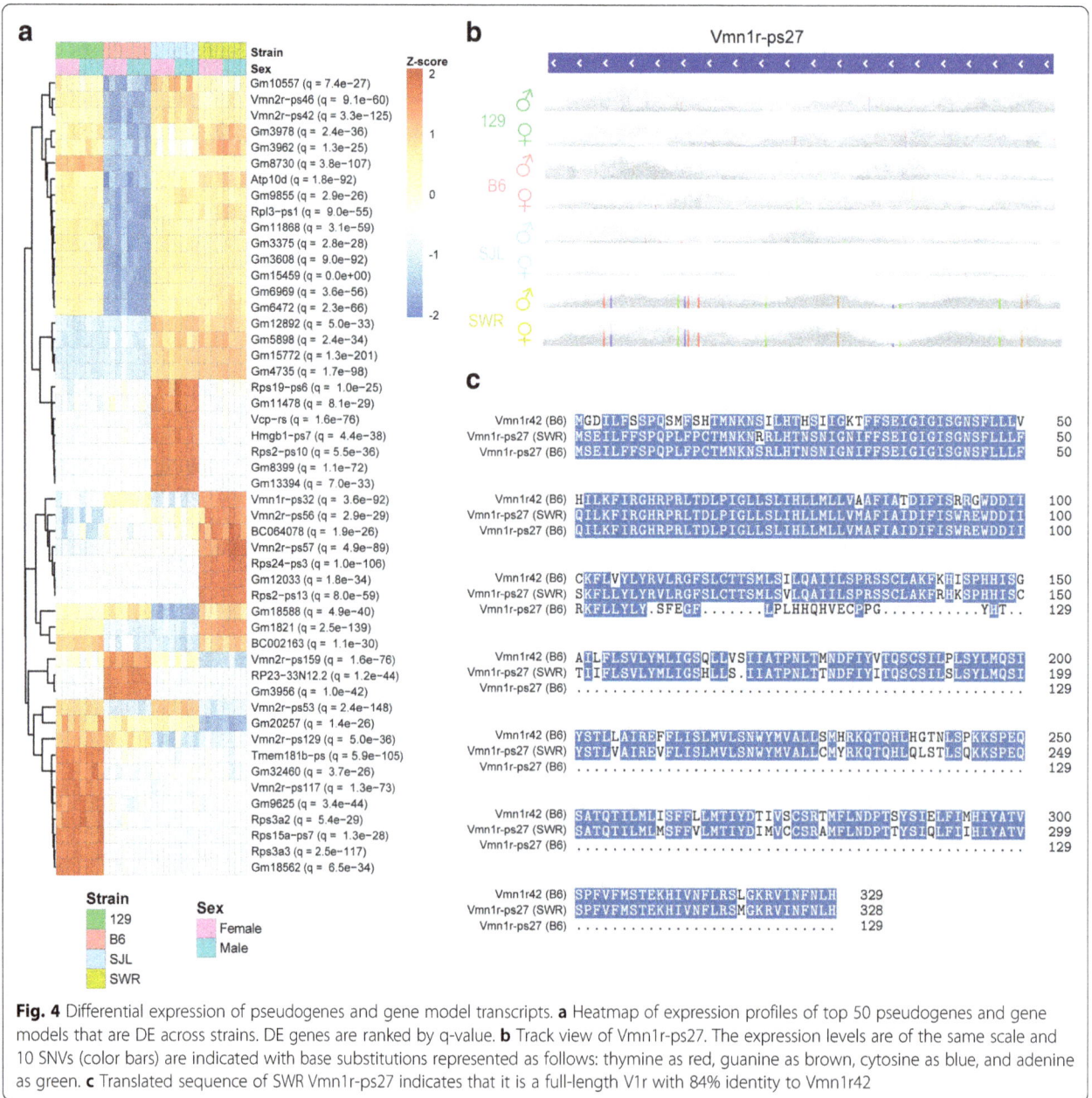

Fig. 4 Differential expression of pseudogenes and gene model transcripts. **a** Heatmap of expression profiles of top 50 pseudogenes and gene models that are DE across strains. DE genes are ranked by q-value. **b** Track view of Vmn1r-ps27. The expression levels are of the same scale and 10 SNVs (color bars) are indicated with base substitutions represented as follows: thymine as red, guanine as brown, cytosine as blue, and adenine as green. **c** Translated sequence of SWR Vmn1r-ps27 indicates that it is a full-length V1r with 84% identity to Vmn1r42

(Fig. 6c). *Miat* was expressed highly by 129, SJL, and SWR strains, but at low levels in B6. *Gm26870* was expressed in B6 and Swiss Strain, but was virtually absent in the 129 strain with only a few samples in SWR showing very low expression.

We examined whether there was a correlation between the differentially-expressed lncRNAs and chemosensory receptor genes. Upon cluster analysis, we found that one group of lncRNAs, including *Miat*, was negatively correlated with a number of chemosensory receptor genes, and a second group was positively correlated with the rest (Fig. 6d). This finding implied a possible link between some of these

lincRNAs and the differential expression of the chemosensory receptors.

Strain and sex specific expression of genes

Differentially expressed genes may be associated with specific combinations of sex and strain. These cases would be missed by our analyses when the data are aggregated in a phenomenon called the Simpson's paradox [49]. Therefore, we performed an analysis to identify genes that showed DE between males and females within individual strains. We identified 10 genes that were differentially expressed in this specific manner: *Ajuba* (SWR), *Vmn1r-ps47* (SJL and SWR), *Vmn2r9* and *Wnt7b*

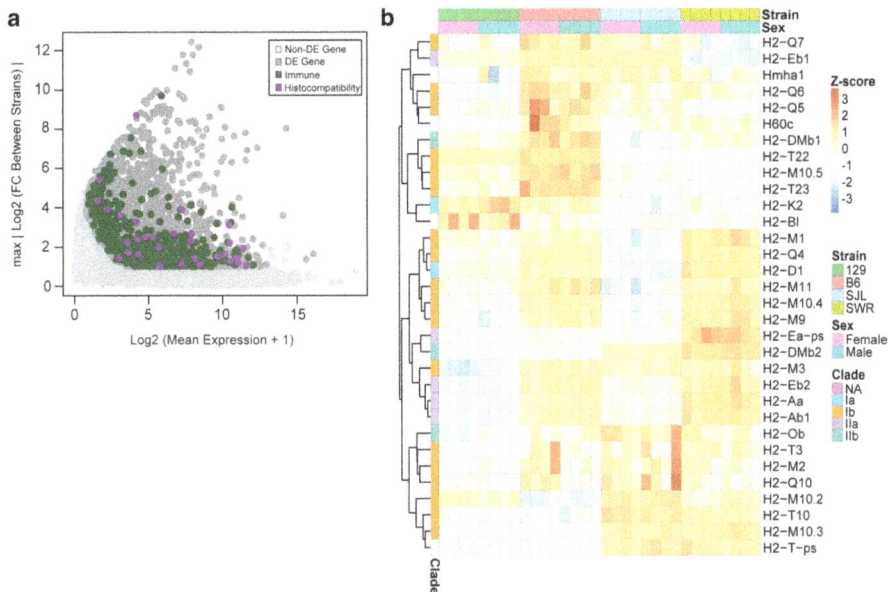

Fig. 5 Differential expression of immune system related genes. **a** MA plot highlighting DE immune system related genes. Y-axis indicates the maximal value of FC between any two strains in log2 scale. Immune system genes are highlighted, including immune response gene (green) and MHC (purple). **b** Heatmap of the expression profile of the MHC genes in all four strains

(SWR), 5 genes in B6 (*Batf, Gm4017, Gm25099, Rn18s-rs5* and *Ttc22) and Tspy-ps* (all 4 strains) (Fig. 7). Two of these genes, *Vmn2r9* and *Vmn1r-ps47*, encoded vomeronasal receptors and may mediate vomeronasal-based behaviors. No other gene has any known function in the VNO.

Phylogenetic inference of strain lineage

SJL and SWR strains originally diverged in 1920, with recombination occurring as late as 1932 [37]. The divergence between B6 and 129 mice occurred earlier, between 1903 and 1915 (Fig. 8a). Divergence of the strains can be reflected by nucleotide differences in the genes, as well as by differences in gene expression. Currently there is a lack of reference genomes that cover the strains we study here. Even though a rough reference genome exists for 129, close inspection of regions of the VR clusters indicate that they are thinly covered. In the absence of reference genomes, we built lineage relationships using gene expression level as traits and compared it to the breeding lineage map. Using genes with a normalized expression count above one, we generated a dendrogram of the strains. It revealed relationships among the strains that coincided well with the known lineage map, and suggested a closer relationship between 129 and the Swiss strains than with B6 (Fig. 8b, approximately unbiased p value au < 0.05). Similar phylogenetic relations were also established when all 5745 DE genes (Fig. 8c, au < 0.05), or 591 DE immune system related genes (Fig. 8d, au < 0.05) were used to generate the

dendrograms. In contrast, using the 453 differentially expressed GPCRs, most of which are vomeronasal receptors, the phylogenetic relation no longer respected the pattern suggested by other gene groups. In this case, B6 is still an outgroup from the other strains (129, SJL, and SWR; au < 0.05), but 129 and SJL are closer to one another (au < 0.05) than the Swiss strains (Fig. 8e, au < 0.14). Interestingly, the tree from 446 DE long non-coding RNA transcripts (Fig. 8f, au < 0.05) also did not conform to the other gene sets.

Discussion

Sensory neurons in the mammalian olfactory systems express the largest families of G-protein coupled receptors. Transcriptional regulation of these genes is highly coordinated to ensure each neuron expresses a unique set of genes. Through transcriptome analyses, we find that differentially expressed genes in the VNO are dominated by strain differences. A substantial number of GPCRs, as well as a chemosensory-related subclass of MHC family of genes, are differentially expressed among the strains. These genes are clustered in hotspot locations in the genome. A group of genes with unknown function, including many lncRNA genes and gene models, also show strain-specific expression. Intriguingly, our analyses reveal correlation and anti-correlation between lncRNAs and chemoreceptor genes, suggesting that they may be coordinately regulated. Importantly, we find that several chemoreceptors annotated as pseudogenes in the reference genome are expressed as functional genes due

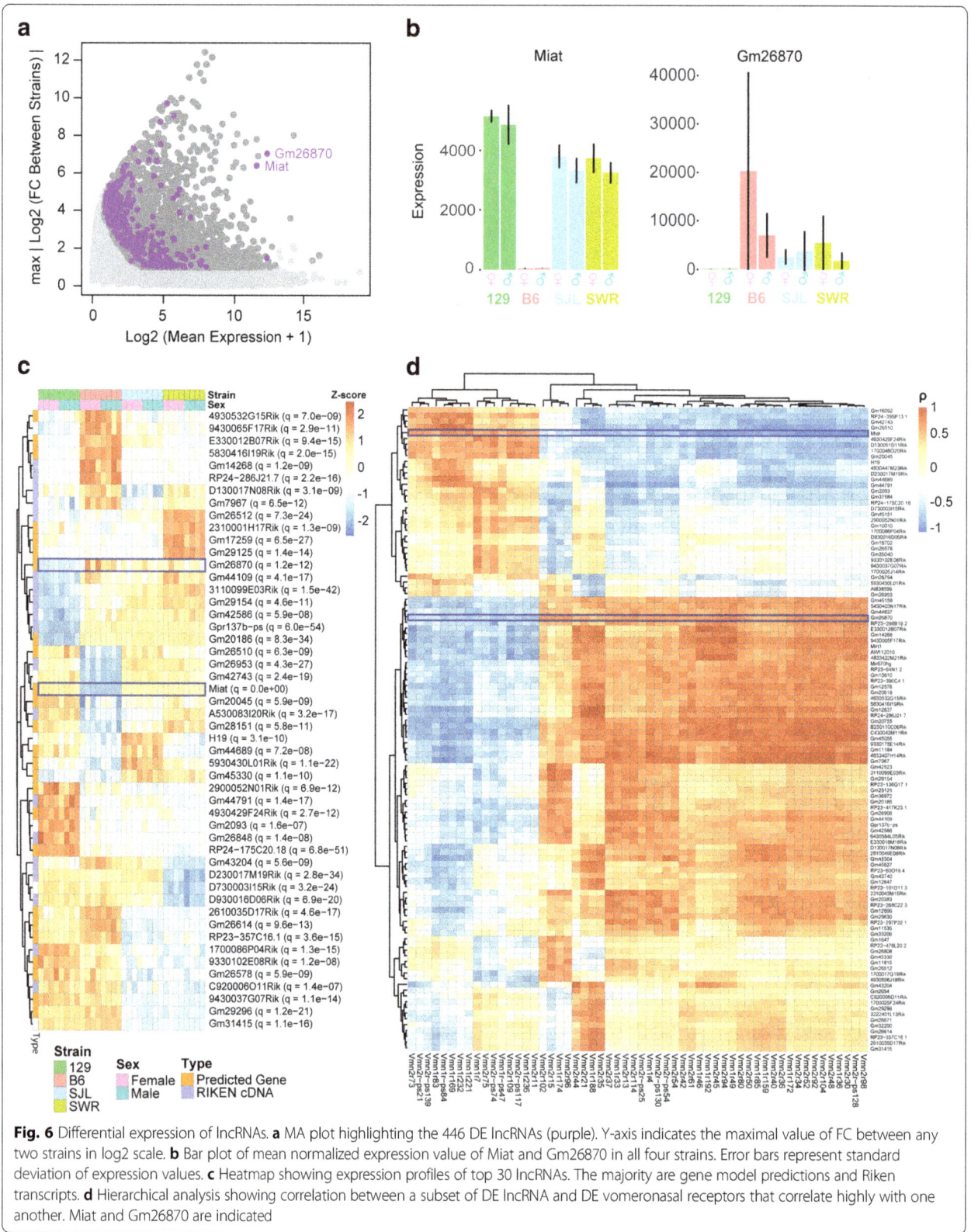

Fig. 6 Differential expression of lncRNAs. **a** MA plot highlighting the 446 DE lncRNAs (purple). Y-axis indicates the maximal value of FC between any two strains in log2 scale. **b** Bar plot of mean normalized expression value of Miat and Gm26870 in all four strains. Error bars represent standard deviation of expression values. **c** Heatmap showing expression profiles of top 30 lncRNAs. The majority are gene model predictions and Riken transcripts. **d** Hierarchical analysis showing correlation between a subset of DE lncRNA and DE vomeronasal receptors that correlate highly with one another. Miat and Gm26870 are indicated

to SNPs in non-B6 strains. These discoveries hint at important differences of VNO functions in detecting pheromones and inter-species chemosensory cues.

Sex specific gene expression

There is little evidence supporting sexually dimorphic gene expression in all strains. Except for *Xist* and Y

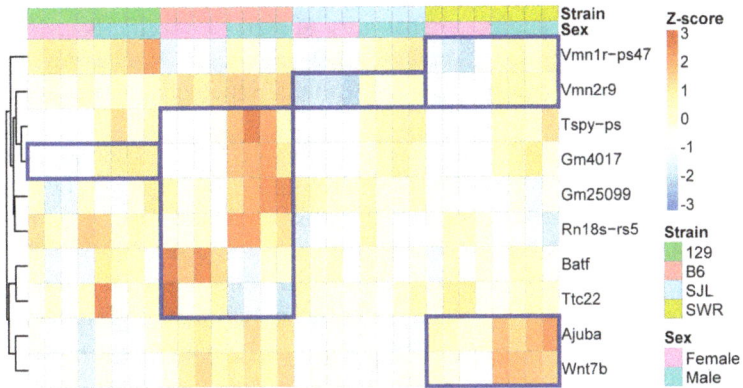

Fig. 7 Differential gene expression between male and female animals within same strains. Heatmap showing expression profiles of ten genes exhibiting sex-specific expression within strains. Boxes indicate the strains within which significant sexual dimorphic expressions are found

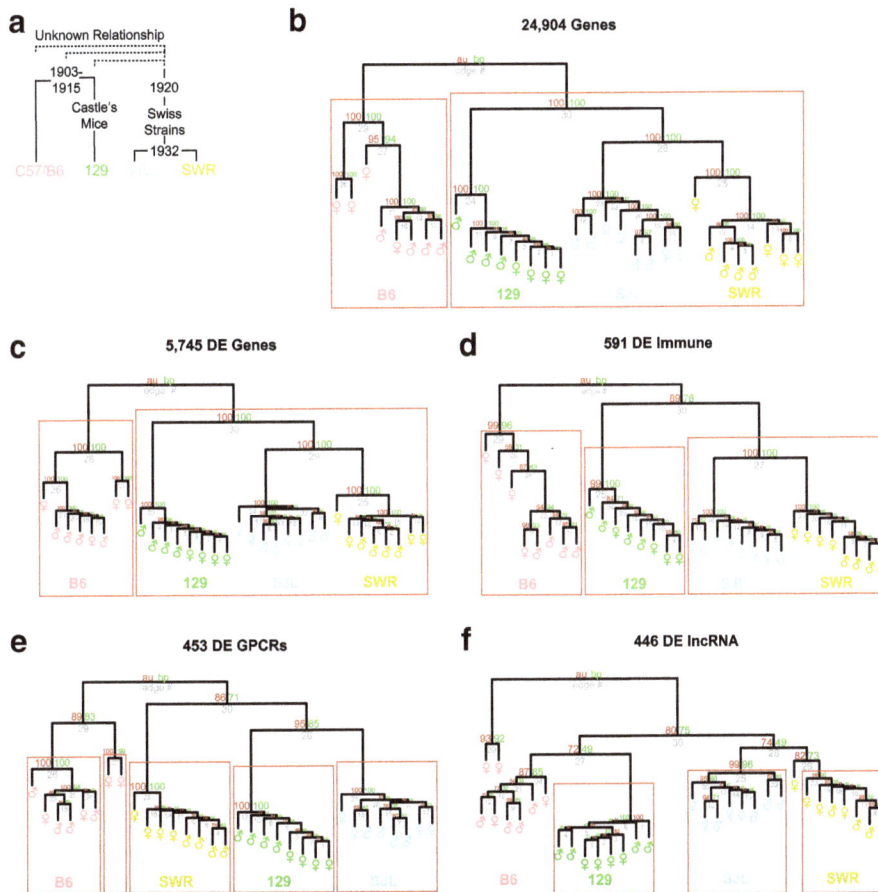

Fig. 8 Phylogenetic analyses of DE genes. **a** The genealogy of strains. **b-f** Phylogenetic dendrograms of the strains derived from the gene expression profiles of all expressed genes (**b**), all DE genes (**c**), lncRNA (**d**), immune system related genes (**e**), and GPCRs (**f**). In all cases except in (**e** and **f**), B6 is an outgroup to all the other strains. In (**f**), 129 is no longer an outgroup to the Swiss strains. Abbreviations: au: approximately unbiased *p* value; bp: bootstrap probability

chromosome genes, there are no other genes that can be considered as truly sexual dimorphic across strains. These X and Y chromosomal homologs have similar functions. *Eif2s3y* and *Ddx3y* are part of the translation initiation machinery [50]. Overexpression of *Eif2s3x* has been shown to substitute for loss of *Eif2s3y* [51] whereas the function of *Ddx3y* is thought to be replaceable by *Ddx3x* [52]. *Uty* is a putative histone demethylase, and *Utx* also a known histone demethylase [53]. *Kdm5d* and its X chromosome homolog *Kdm5c* (also known as *Jarid1d,c*) are both Lysine-specific demethylases and are functional homologs. The commonality between these genes pairs implies a dose-compensation mechanism that maintains the total product from X and Y chromosomes at constant levels in the VNO. When expressions from both sex chromosomes are considered, the overall expression levels of these genes no longer have significant difference.

We do not find any GPCRs, proteins associated with pheromone binding such as lipocalins or MHC proteins, as differentially expressed between the sexes. This observation is consistent with previous work suggesting no significant sexual dimorphism in the expression of receptors recognizing sex pheromones or in VNO response to urine stimulation [16, 39]. The absence of significant receptor differences suggests that sexually dimorphic behavioral responses are unlikely to originate from the VNO. They are more likely to be the result from the differential processing of pheromone cues in the brain circuitry [54].

Previous studies have implicated a few genes to be differentially expressed between the sexes [15]. While our study replicates these results, our data show that these differences are restricted to the strains examined. Products encoded by these genes may still contribute to differential function of the VNO in the B6 strain, but the differences cannot be generalized.

Strain-related differential gene expression

With respect to strains, the largest group of DE genes is related to immunological functions, including genes encoding MHC, cytokines and their receptors, as well as molecules involved in signaling pathways. It is well established that inbred strains of mice are distinguished by their haplotypes and all cells express MHC molecules. Therefore, it is not surprising that the MHC genes in the VNO also show strain differences as in other tissues. The differential expression of H2-Mv non-classic MHC molecules, on the other hand, may have implications in chemosensation. These genes have been shown to be coordinately expressed with V2r subfamilies V2ra1–5 and V2rc [43–45] and have been shown to allow for ultra-sensitive pheromone detection, possibly by influencing V2r surface expression [55]. Differential expression

in the H2-Mv genes may affect the affinity and sensitivity of V2rs to specific ligands, and their recognition by the animals. In this context, these DE H2-Mv genes could affect strain-specific recognition.

Strain-specific expression of chemosensory receptors

Despite the relatively recent lineage separation of different inbred lab strains, we find abundant examples of DE genes in the VNO. These differences include binary expression differences and modulated expression levels. In striking examples, we find SNPs that render some of the annotated pseudogenes functional in some strains, including both V1rs and V2rs.

The expression of particular sets of receptors may define the cue set each strain of mice can detect. Different clades of receptors appear to be tuned to specific sets of cues. We found these receptors exhibit differential expression among the strains. The expression of *Vmn1r85* is high in only B6 mice while *Vmn1r185* is high in all strains except 129 mice. Notably, the SJL is considered a 'challenging breeder' by Jackson Laboratories. It is possible that a reduced response to estrus cues may cause reduced mating in this strain.

Besides the V1rj and V1re clades, differentially expressed V1rs are found in all other clades except L, including eight members of V1ra, six of V1rb, and 17 of V1rc. Deletion of a genomic region encompassing both V1ra and V1rb genes results in decline in mating and aggression [56]. The V1rc receptors have been implicated in detecting cues present in female mice or predators. Differential expression of these V1rs may affect the recognition of environmental as well as species-specific cues.

The V2rs have long N-terminus domains and have been shown to recognize polypeptide pheromones. V2r-expressing cells respond robustly to MHC peptides and can also be activated by the MUPs [57, 58]. As these polypeptides may be specifically associated with strain and individuals, differentially expressed V2rs may lead to divergent recognition of strain information and trigger biased responses.

The expression of *Fpr-3* shows strain differences. Formyl peptides are present in the mitochondria of bacteria and are released when bacteria die. The presence of formyl peptides triggers chemotaxis of immune cells in response to infection. FPR expression in the VNO is thought to allow the animals to detect the health status of other animals [9]. The differential expression *Fpr-3* may bias this recognition.

Taken together, differential expression of the VRs may lead to the recognition of a particular set of cues in one strain but not the other. It is worth noting, however, that even though the VRs are highly specific in their ligand recognition, there is certain redundancy in how pheromones are recognized. For example, *Vmn1r85* (*V1rj3*) and *Vmn1r89* (*V1rj2*) receptors are activated by sulfated estrogens, but they display different sensitivities to the ligands [39, 59]. Given that many of the differentially expressed VRs have paralogs in the genome, differential expression of the VRs may reduce or enhance the sensitivity to certain pheromones, rather than create a situation in which a pheromone is recognized by one strain but not the other.

lncRNAs

Both genetic and epigenetic mechanisms may contribute to differential gene expression among the strains. We find strong anti-correlation between the expression of *Miat* and *Gm26870*, two lncRNAs, which along with others show strong positive and negative correlations with chemoreceptor expression. lncRNAs are expressed highly in the nervous system [60] and are known to control gene expression by directly regulating gene-specific transcription and splicing, as well as epigenetic modifications [46, 61]. *Miat*, also known as *RNCR2* or *Gamufu* [62, 63], is one of the most strongly DE lncRNAs among the strains. It is known to regulate cell specification in the developing retina [62]. While the functional roles played by *Miat* and the other lncRNAs in regulating VNO gene expression are not clear, the strong correlations among the transcripts raised the possibility that they may be coordinately regulating differential gene expression among the strains.

Implication in strain evolution

Although differences in gene expression are not equivalent to genetic differences at the nucleotide level, they are nonetheless important traits that can provide information about evolutionary divergence among the mouse strains. The phylogenetic relationship inferred from the expression of GPCR genes does not conform to those by other genes, nor to that of the genealogy. Several VNO receptor genes marked as pseudogenes in the reference B6 genome are functional in other strains. Moreover, we find SNPs that result in synonymous and missense changes in protein coding in many V1r and V2r genes. These observations, together with the observation that several hotspots of DE genes are enriched in VNO receptors, suggest that the VNO receptors genes and their expression may have followed a different evolutionary path from the rest of the genome. These differentially expressed chemosensory receptors may enable different strains of mice to sense social cues emitted by conspecific animals, react to the health status of another animals, or respond to heterospecific signals including predators in distinct manners. The differential detection of social cues may therefore underlie

some of the strain-specific behavior differences observed in mice.

Conclusions

Transcriptome analyses provide little support of sexual dimorphism in gene expression in the VNO. In contrast, there are profound variations in the expression of immune response related genes, vomeronasal and G-protein coupled receptor genes among different strains of mice. These differentially expressed genes are concentrated in hotspots on the genome, indicating rapid evolution of genes involved in pheromone detection. These findings suggest it is likely that diverse strains of mouse perceive pheromone cues differently. Behavioral difference among strains in response to pheromone may thus first arise from differential detection of pheromones by the vomeronasal organ. On the other hand, sexually dimorphic responses to pheromones more likely originate from dimorphic neural circuits in the brain than from differential detection.

Methods

RNA library preparation & sequencing

All strains of animals were purchased from Jackson laboratory. Mice are maintained in Lab Animal Services Facility at Stowers Institute with a 14:10 light cycle, and provided with food and water ad libitum. Experimental protocols were approved by the Institutional Animal Care and Use Committee at Stowers Institute and in compliance with the NIH Guide for Care and Use of Animals. Total RNA was isolated from VNO epithelia of individual mouse using TRIzol solution (Thermo Fisher Scientific) followed by spin-column (Zymo Research) purification. Ribodepletion was performed using Ribo-Zero Gold rRNA Removal kit (Illumina) to remove rRNA from the sample prior to library preparation. Sequencing libraries were generated using TruSeq Stranded Total RNA Kit (Illumina) and sequenced as 125 bp paired-end stranded reads on Illumina Hi-Seq 2500 platform. Preliminary analysis including basecalling was performed using HiSeq Control Software (v2.2.58) with fastq files generated using bcl2fastq. FastQC [64] reports were generated for each sample to ensure sequencing quality. Trim Galore was used with default parameters to trim reads with leftover adapter sequence and low quality scores [65].

Sequence alignment

GRCm38 (mm10) mouse reference genome was used to align the reads with STAR aligner version 2.5.2b (Dobin, et al. 2013). Ensembl reference annotation version 87 [66] was used to define gene models for mapping quantification. Uniquely mapped reads for each gene model were produced using STAR parameter "–quantMode GeneCounts" and raw stranded counts were extracted from the fourth column of the output matching the orientation produced by the True-seq stranded preparation protocol used. All the options chosen are equivalent to the HTSeq command "htseq-count option -s reverse".

Differential expression analyses

Differential Expression analysis was performed using the R package DESeq2 [67]. Under the assumption of negative binomial distribution, we normalized the data for technical variation in sequencing depth among each sample. Each gene was then fit to a generalized linear model and dispersion coefficients were tested using cooks distance for independent filtering of high variance genes. For genes that passed independent filtering, Log_2 fold changes (LFC) between groups and their standard errors were used in a Wald test for differential expression. Genes were considered differentially expressed if any of groups passed independent filtering and had a FC > 2 and FDR < 0.05.

Additional downstream analyses

For PCA analysis, we used DESeq2 internal methods to calculate and plot principal components using all expressed genes instead of the default top 500 varying genes. Gene expression heatmaps were created with the R package pheatmap using regularized log transformed normalized counts from DESeq2 [68]. GO analyses on the groups of DE genes were performed in R using topGO [69] and based on GO annotations from BiomaRt [70]. To identify hot spots that contained a high percentage of DE genes, we used rollapply from the zoo package (https://cran.r-project.org/web/packages/zoo/index.html) to create sliding windows of 25 expressed genes and slid the window across each chromosome separately to calculate the probability of observing DE genes that exceed random chance. Within the sliding window, we performed the Poisson test using the function ppois to compare the percentage of DE genes within the window with the percentage of DE genes in the entire genome. Data were visualized using GenomicRanges [71] and ggbio [72].

Tracks for SNP identification and visualization were created using Integrative Genomics Viewer [73]. Identified SNPs were incorporated into the reference sequences from Ensembl, and translated to proteins using the ExPASY online translate tool [74], aligned using ClustalW [75], and visualized using MView [76] through the EMBL-EBI online web services [77]. Homologous sequences were identified using NCBI's Blastn and Blastp [78]. Correlation analysis dendrograms were created in R by running PVclust [79] using the 'average' method for clustering and a custom spearman implementation for calculating distance, parallelized with 10,000 bootstraps.

Additional files

Additional file 1: Figure S1. Distribution of uniquely mapped, multi-mapped and unmapped reads among the samples presented as total reads (A) and percentage of reads (B). (PDF 464 kb)

Additional file 2: Figure S2. Differentially expressed chemosensory genes other than V1r and V2r families. (PDF 138 kb)

Additional file 3: Figure S3. Strain-specific expression of vomeronasal receptors. Data is displayed as expressed (blue) or not expressed (white) to highlight the exclusive patterns of expression for some of the genes. (PDF 87 kb)

Additional file 4: Figure S4. High degree of polymorphism and differential expression of VR genes among strains. (**A-F**) Example track files illustrating the mapping of reads to individual VR genes. Each track is a superposition of four individual samples with SNPs highlighted as vertical lines with substitutions represented as follows: thymine as red, guanine as brown, cytosine as blue, and adenine as green. (PDF 4541 kb)

Additional file 5: Figure S5. Differences in expression level among different clades of VRs. (**A**) Expression of all V1r clades are represented in all strains. Clade J receptor genes are more highly expressed than receptors of other clades. (**B**) Expression of all V2r clades are represented in all strains. (PDF 558 kb)

Additional file 6: Figure S6. Polymorphism and differential expression of FPR genes. A) Fpr-rs3. B) Fpr3. (PDF 2426 kb)

Additional file 7: Figure S7. Sequence comparison of functionalized pseudogene. Alignment of *Vmn1r-ps27* from SWR with *Vmn1r-ps27* and *Vmn1r42* from B6. (PDF 814 kb)

Abbreviations
DE: Differentially expressed; FC: Fold change; FDR: False discovery rate; FPR: Formyl peptide receptor; GO: Gene ontology; GPCR: G-protein coupled receptor; LFC: Log_2 fold change; MHC: Major histocompatibility complex; ORF: Open reading frame; PC: Principal component; PCA: Principal component analysis; VNO: Vomeronasal organ; VR: Vomeronasal receptor

Acknowledgements
We thank members of Molecular Biology, Bioinformatics, Lab Animal Services at Stowers Institute for their assistance on this project.

Funding
This work is supported by funding from NIH (DC008003) and the Stowers Institute.

Authors' contributions
CRY and LM designed the experiments and supervised the research. LM performed the experiments. KD and VD performed bioinformatics analyses. APA participated in the analyses and provided consultation. VD, LM and CRY wrote the paper. All authors have read and approved the manuscript.

Ethics approval
All experimental protocols concerning animal use were approved by the Institutional Animal Care and Use Committee at Stowers Institute and in compliance with the NIH Guide for Care and Use of Animals.

Competing interests
The authors declare no competing financial interests.

Author details

[1]Stowers Institute for Medical Research, 1000 East 50th Street, Kansas City, MO 64110, USA. [2]Department of Anatomy and Cell Biology, University of Kansas Medical Center, 3901 Rainbow Boulevard, Kansas City, KS 66160, USA. [3]Redwood Center for Theoretical Neuroscience, University of California, 567 Evans Hall, Berkeley 94720, USA.

References

1. Wyatt TD. Pheromones and animal behaviour : communication by smell and taste. Cambridge, UK. New York: Cambridge University Press; 2003.

2. Powers JB, Winans SS. Vomeronasal organ: critical role in mediating sexual behavior of the male hamster. Science. 1975;187(4180):961–3.

3. Baum MJ, Cherry JA. Processing by the main olfactory system of chemosignals that facilitate mammalian reproduction. Horm Behav. 2015;68:53–64.

4. Karlson P, Lüscher M: 'Pheromones': a new term for a class of biologically active substances. 1959.

5. Halpern M. The organization and function of the vomeronasal system. Annu Rev Neurosci. 1987;10:325–62.

6. Tirindelli R, Dibattista M, Pifferi S, Menini A. From pheromones to behavior. Physiol Rev. 2009;89(3):921–56.

7. Isogai Y, Si S, Pont-Lezica L, Tan T, Kapoor V, Murthy VN, Dulac C. Molecular organization of vomeronasal chemoreception. Nature. 2011;478(7368):241–5.

8. Liberles SD, Horowitz LF, Kuang D, Contos JJ, Wilson KL, Siltberg-Liberles J, Liberles DA, Buck LB. Formyl peptide receptors are candidate chemosensory receptors in the vomeronasal organ. Proc Natl Acad Sci U S A. 2009;106(24):9842–7.

9. Riviere S, Challet L, Fluegge D, Spehr M, Rodriguez I. Formyl peptide receptor-like proteins are a novel family of vomeronasal chemosensors. Nature. 2009;459(7246):574–7.

10. Dulac C, Axel R. A novel family of genes encoding putative pheromone receptors in mammals. Cell. 1995;83(2):195–206.

11. Herrada G, Dulac C. A novel family of putative pheromone receptors in mammals with a topographically organized and sexually dimorphic distribution. Cell. 1997;90(4):763–73.

12. Ryba NJ, Tirindelli R. A new multigene family of putative pheromone receptors. Neuron. 1997;19(2):371–9.

13. Matsunami H, Buck LB. A multigene family encoding a diverse array of putative pheromone receptors in mammals. Cell. 1997;90(4):775–84.

14. Vandenbergh JG. Pheromones and reproduction in mammals. New York: Academic Press; 1983.

15. Ibarra-Soria X, Levitin MO, Saraiva LR, Logan DW. The olfactory transcriptomes of mice. PLoS Genet. 2014;10(9):e1004593.

16. He J, Ma L, Kim S, Nakai J, CR Y. Encoding gender and individual information in the mouse vomeronasal organ. Science. 2008;320(5875):535–8.

17. Bruce HM. An exteroceptive block to pregnancy in the mouse. Nature. 1959;184:105.

18. Halpern M, Martinez-Marcos A. Structure and function of the vomeronasal system: an update. Prog Neurobiol. 2003;70(3):245–318.

19. Kelliher KR, Spehr M, Li XH, Zufall F, Leinders-Zufall T. Pheromonal recognition memory induced by TRPC2-independent vomeronasal sensing. Eur J Neurosci. 2006;23(12):3385–90.

20. Singer AG, Beauchamp GK, Yamazaki K. Volatile signals of the major histocompatibility complex in male mouse urine. Proc Natl Acad Sci U S A. 1997;94:2210–4.

21. Bruce HM. Continued suppression of pituitary luteotrophic activity and fertility in the female mouse. J Reprod Fertil. 1962;4:313–8.

22. Wynn EH, Sanchez-Andrade G, Carss KJ, Logan DW. Genomic variation in the vomeronasal receptor gene repertoires of inbred mice. BMC Genomics. 2012;13:415.

23. Zhang J, Webb DM. Evolutionary deterioration of the vomeronasal pheromone transduction pathway in catarrhine primates. Proc Natl Acad Sci U S A. 2003;100(14):8337–41.

24. Grus WE, Zhang J. Rapid turnover and species-specificity of vomeronasal pheromone receptor genes in mice and rats. Gene. 2004;340(2):303–12.

25. Yang H, Shi P, Zhang YP, Zhang J. Composition and evolution of the V2r vomeronasal receptor gene repertoire in mice and rats. Genomics. 2005; 86(3):306–15.

26. Shi P, Bielawski JP, Yang H, Zhang YP. Adaptive diversification of vomeronasal receptor 1 genes in rodents. J Mol Evol. 2005;60(5):566–76.

27. Shi P, Zhang J. Comparative genomic analysis identifies an evolutionary shift of vomeronasal receptor gene repertoires in the vertebrate transition from water to land. Genome Res. 2007;17(2):166–74.

28. Grus WE, Zhang J. Origin of the genetic components of the vomeronasal system in the common ancestor of all extant vertebrates. Mol Biol Evol. 2009;26(2):407–19.

29. Grus WE, Zhang J. Origin and evolution of the vertebrate vomeronasal system viewed through system-specific genes. BioEssays. 2006;28(7):709–18.

30. Grus WE, Shi P, Zhang J. Largest vertebrate vomeronasal type 1 receptor gene repertoire in the semiaquatic platypus. Mol Biol Evol. 2007;24(10):2153–7.

31. Young JM, Trask BJ. V2R gene families degenerated in primates, dog and cow, but expanded in opossum. Trends Genet. 2007;23(5):212–5.

32. Young JM, Massa HF, Hsu L, Trask BJ. Extreme variability among mammalian V1R gene families. Genome Res. 2010;20(1):10–8.

33. Young JM, Kambere M, Trask BJ, Lane RP. Divergent V1R repertoires in five species: amplification in rodents, decimation in primates, and a surprisingly small repertoire in dogs. Genome Res. 2005;15(2):231–40.

34. Lane RP, Young J, Newman T, Trask BJ. Species specificity in rodent pheromone receptor repertoires. Genome Res. 2004;14(4):603–8.

35. Lane RP, Cutforth T, Axel R, Hood L, Trask BJ. Sequence analysis of mouse vomeronasal receptor gene clusters reveals common promoter motifs and a history of recent expansion. Proc Natl Acad Sci U S A. 2002;99(1):291–6.

36. Grus WE, Shi P, Zhang YP, Zhang J. Dramatic variation of the vomeronasal pheromone receptor gene repertoire among five orders of placental and marsupial mammals. Proc Natl Acad Sci U S A. 2005;102(16):5767–72.

37. Beck JA, Lloyd S, Hafezparast M, Lennon-Pierce M, Eppig JT, Festing MF, Fisher EM. Genealogies of mouse inbred strains. Nat Genet. 2000;24(1):23–5.

38. Heard E, Clerc P, Avner P. X-chromosome inactivation in mammals. Annu Rev Genet. 1997;31(1):571–610.

39. Haga-Yamanaka S, Ma L, He J, Qiu Q, Lavis LD, Looger LL, CR Y. Integrated action of pheromone signals in promoting courtship behavior in male mice. elife. 2014;3:e03025.

40. Dietschi Q, Tuberosa J, Rosingh L, Loichot G, Ruedi M, Carleton A, Rodriguez I. Evolution of immune chemoreceptors into sensors of the outside world. Proc Natl Acad Sci U S A. 2017;

41. Haga S, Hattori T, Sato T, Sato K, Matsuda S, Kobayakawa R, Sakano H, Yoshihara Y, Kikusui T, Touhara K. The male mouse pheromone ESP1 enhances female sexual receptive behaviour through a specific vomeronasal receptor. Nature. 2010;466(7302):118–22.

42. Kimoto H, Sato K, Nodari F, Haga S, Holy TE, Touhara K. Sex- and strain-specific expression and vomeronasal activity of mouse ESP family peptides. Curr Biol. 2007;17(21):1879–84.

43. Ishii T, Hirota J, Mombaerts P. Combinatorial coexpression of neural and immune multigene families in mouse vomeronasal sensory neurons. Curr Biol. 2003;13(5):394–400.

44. Loconto J, Papes F, Chang E, Stowers L, Jones EP, Takada T, Kumanovics A, Fischer Lindahl K, Dulac C. Functional expression of murine V2R pheromone receptors involves selective association with the M10 and M1 families of MHC class Ib molecules. Cell. 2003;112(5): 607–18.

45. Silvotti L, Moiani A, Gatti R, Tirindelli R. Combinatorial co-expression of pheromone receptors, V2Rs. J Neurochem. 2007;103(5):1753–63.

46. Bruce HM. Pheromones. Br Med Bull. 1970;26:10–3.

47. Bruce HM. Pheromones and behavior in mice. Acta Neurol Psychiatr Belg. 1969;69:529–38.

48. Vance KW, Ponting CP. Transcriptional regulatory functions of nuclear long noncoding RNAs. Trends Genet. 2014;30(8):348–55.

49. Blyth CR. On Simpson's paradox and the sure-thing principle. J Am Stat Assoc. 1972;67(338):364–6.

50. Mazeyrat S, Saut N, Grigoriev V, Mahadevaiah SK, Ojarikre OA, Rattigan A, Bishop C, Eicher EM, Mitchell MJ, Burgoyne PS. A Y-encoded subunit of the translation initiation factor Eif2 is essential for mouse spermatogenesis. Nat Genet. 2001;29(1):49–53.

51. Xu J, Watkins R, Arnold AP. Sexually dimorphic expression of the X-linked gene Eif2s3x mRNA but not protein in mouse brain. Gene Expr Patterns. 2006;6(2):146–55.

52. Vong QP, Li Y, Lau YF, Dym M, Rennert OM, Chan WY. Structural characterization and expression studies of Dby and its homologs in the mouse. J Androl. 2006;27(5):653–61.

53. Xu J, Deng X, Disteche CM. Sex-specific expression of the X-linked histone demethylase gene Jarid1c in brain. PLoS One. 2008;3(7):e2553.

54. MV W, Manoli DS, Fraser EJ, Coats JK, Tollkuhn J, Honda S, Harada N, Shah NM. Estrogen masculinizes neural pathways and sex-specific behaviors. Cell. 2009;139(1):61–72.

55. Leinders-Zufall T, Ishii T, Chamero P, Hendrix P, Oboti L, Schmid A, Kircher S, Pyrski M, Akiyoshi S, Khan M, et al. A family of nonclassical class I MHC genes contributes to ultrasensitive chemodetection by mouse vomeronasal sensory neurons. J Neurosci. 2014;34(15):5121–33.

56. Del Punta K, Leinders-Zufall T, Rodriguez I, Jukam D, Wysocki CJ, Ogawa S, Zufall F, Mombaerts P. Deficient pheromone responses in mice lacking a cluster of vomeronasal receptor genes. Nature. 2002; 419(6902):70–4.

57. Chamero P, Marton TF, Logan DW, Flanagan K, Cruz JR, Saghatelian A, Cravatt BF, Stowers L. Identification of protein pheromones that promote aggressive behaviour. Nature. 2007;450(7171):899–902.

58. Leinders-Zufall T, Lane AP, Puche AC, Ma W, Novotny MV, Shipley MT, Zufall F. Ultrasensitive pheromone detection by mammalian vomeronasal neurons. Nature. 2000;405(6788):792–6.

59. Haga-Yamanaka S, Ma L, CR Y. Tuning properties and dynamic range of type 1 vomeronasal receptors. Front Neurosci. 2015;9:244.

60. Derrien T, Johnson R, Bussotti G, Tanzer A, Djebali S, Tilgner H, Guernec G, Martin D, Merkel A, Knowles DG, et al. The GENCODE v7 catalog of human long noncoding RNAs: analysis of their gene structure, evolution, and expression. Genome Res. 2012;22(9):1775–89.

61. Bergmann JH, Spector DL. Long non-coding RNAs: modulators of nuclear structure and function. Curr Opin Cell Biol. 2014;26:10–8.

62. Rapicavoli NA, Poth EM, Blackshaw S. The long noncoding RNA RNCR2 directs mouse retinal cell specification. BMC Dev Biol. 2010;10:49.

63. Drickamer LC, Assmann SM. Acceleration and delay of puberty in female housemice: methods of delivery of the urinary stimulus. Dev Psychobiol. 1981;14:487–97.

64. Andrews S: FastQC: a quality control tool for high throughput sequence data. In.; 2010.

65. Krueger F: Trim Galore. A wrapper tool around Cutadapt and FastQC to consistently apply quality and adapter trimming to FastQ files 2015.

66. Flicek P, Amode MR, Barrell D, Beal K, Billis K, Brent S, Carvalho-Silva D, Clapham P, Coates G, Fitzgerald S, et al. Ensembl 2014. Nucleic Acids Res. 2014;42(Database issue):D749–55.

67. Love MI, Huber W, Anders S. Moderated estimation of fold change and dispersion for RNA-seq data with DESeq2. Genome Biol. 2014; 15(12):550.

68. Kolde R: pheatmap: Pretty Heatmaps. R package version 1.0. 8. 2015. In.; 2015.

69. Zeileis A, Grothendieck G, Ryan JA, Andrews F, & Zeileis MA : Package 'zoo'; 2015.

70. Kerkhoven R, van Enckevort FH, Boekhorst J, Molenaar D, Siezen RJ. Visualization for genomics: the microbial genome viewer. Bioinformatics. 2004;20(11):1812–4.

71. Lawrence M, Huber W, Pages H, Aboyoun P, Carlson M, Gentleman R, Morgan MT, Carey VJ. Software for computing and annotating genomic ranges. PLoS Comput Biol. 2013;9(8):e1003118.

72. Yin T, Cook D, Lawrence M. ggbio: an R package for extending the grammar of graphics for genomic data. Genome Biol. 2012;13(8):R77.

73. Robinson JT, Thorvaldsdottir H, Winckler W, Guttman M, Lander ES, Getz G, Mesirov JP. Integrative genomics viewer. Nat Biotechnol. 2011;29(1):24–6.

74. Sawkins MC, Farmer AD, Hoisington D, Sullivan J, Tolopko A, Jiang Z, Ribaut JM. Comparative map and trait viewer (CMTV): an integrated bioinformatic tool to construct consensus maps and compare QTL and functional genomics data across genomes and experiments. Plant Mol Biol. 2004;56(3):465–80.

75. Faith JJ, Olson AJ, Gardner TS, Sachidanandam R. Lightweight genome viewer: portable software for browsing genomics data in its chromosomal context. BMC Bioinformatics. 2007;8:344.

76. Brown NP, Leroy C, Sander C. MView: a web-compatible database search or multiple alignment viewer. Bioinformatics. 1998;14(4):380–1.

77. Li W, Cowley A, Uludag M, Gur T, McWilliam H, Squizzato S, Park YM, Buso N, Lopez R. The EMBL-EBI bioinformatics web and programmatic tools framework. Nucleic Acids Res. 2015;

78. Coletta A, Molter C, Duque R, Steenhoff D, Taminau J, de Schaetzen V, Meganck S, Lazar C, Venet D, Detours V, et al. InSilico DB genomic datasets hub: an efficient starting point for analyzing genome-wide studies in GenePattern, integrative genomics viewer, and R/bioconductor. Genome Biol. 2012;13(11):R104.

79. Hummel H, Karlson P. Hexanoic acid as constituent of the trail pheromone of the termite Zootermopsis Nevadensis Hagen. Hoppe-Seyler's Zeitschrift fur physiologische Chemie. 1968;349(5):725–7.

14

Framework for reanalysis of publicly available Affymetrix® GeneChip® data sets based on functional regions of interest

Ernur Saka[1], Benjamin J. Harrison[2,3], Kirk West[4], Jeffrey C. Petruska[2] and Eric C. Rouchka[1*]

Abstract

Background: Since the introduction of microarrays in 1995, researchers world-wide have used both commercial and custom-designed microarrays for understanding differential expression of transcribed genes. Public databases such as ArrayExpress and the Gene Expression Omnibus (GEO) have made millions of samples readily available. One main drawback to microarray data analysis involves the selection of probes to represent a specific transcript of interest, particularly in light of the fact that transcript-specific knowledge (notably alternative splicing) is dynamic in nature.

Results: We therefore developed a framework for reannotating and reassigning probe groups for Affymetrix® GeneChip® technology based on functional regions of interest. This framework addresses three issues of Affymetrix® GeneChip® data analyses: removing nonspecific probes, updating probe target mapping based on the latest genome knowledge and grouping probes into gene, transcript and region-based (UTR, individual exon, CDS) probe sets. Updated gene and transcript probe sets provide more specific analysis results based on current genomic and transcriptomic knowledge. The framework selects unique probes, aligns them to gene annotations and generates a custom Chip Description File (CDF). The analysis reveals only 87% of the Affymetrix® GeneChip® HG-U133 Plus 2 probes uniquely align to the current hg38 human assembly without mismatches. We also tested new mappings on the publicly available data series using rat and human data from GSE48611 and GSE72551 obtained from GEO, and illustrate that functional grouping allows for the subtle detection of regions of interest likely to have phenotypical consequences.

Conclusion: Through reanalysis of the publicly available data series GSE48611 and GSE72551, we profiled the contribution of UTR and CDS regions to the gene expression levels globally. The comparison between region and gene based results indicated that the detected expressed genes by gene-based and region-based CDFs show high consistency and regions based results allows us to detection of changes in transcript formation.

Keywords: Custom CDF, Affymetrix®, Probe group, Microarrays, Probe set, Probeset

Background

A DNA microarray (DNA chip or biochip) is a technology used to identify and measure the expression level of specific mRNA molecules in order to ascertain transcriptional profiles in response to differing conditions. The most commonly used microarray is the Affymetrix® GeneChip® family of arrays. Each GeneChip® consists of a silicon chip with fixed locations called cells, spots or features [1]. Each spot contains millions of identical 25 base oligonucleotides (probes) which are selected to be complementary to various transcript regions of a gene [2]. In order to determine transcript expression, which directly infers gene expression, groups of 11-20 probes matching the same gene/transcript are arranged in a probe set. Given a particular Affymetrix® GeneChip® platform, the design of the probes is fixed based on earlier genome assemblies and annotation available at that time. Since the design of the first Affymetrix®

* Correspondence: eric.rouchka@louisville.edu
[1]Department of Computer Engineering and Computer Science, University of Louisville, Louisville, KY, USA
Full list of author information is available at the end of the article

GeneChip®, rapid progress has been made in genome sequencing resulting in more accurate databases of annotated coding and non-coding genes.

The significant differences between old and new genome assemblies and annotations make it necessary to update probe-gene targeting according to current knowledge to get more accurate interpretations from experimental results. Affymetrix® does attempt to provide compatibility between genomic changes by updating links between probe sets and their corresponding genes/transcripts via NetAffx™ [3]. Table 1 shows release dates of source databases used by Affymetrix® for both the incorporated version and the most recently available version. In all cases, there is at least a two year difference between the incorporated and most recent release dates which can lead to inconsistent interpretation.

In addition, updating links between probe sets and their corresponding genes/transcripts does not provide a solution for problems caused by individual probes such as single nucleotide polymorphisms (SNPs) [4, 5], probes that target genes other than the designated gene of a probe set, and probes that no longer align to a genomic location. For example in the Affymetrix® GeneChip® HG-133 Plus 2 array, a total of 40,680 probes out of 603,158 (excluding quality control probes) do not have a perfect match to the most recent human genome assembly (hg38).

Even though the design of the probes is fixed, the methods with which the resulting experiments can be analyzed are dynamic in nature due to the ability to annotate and arrange probes into uniquely defined groupings. This is particularly important since there are publicly available repositories of microarray datasets, such as NCBI's Gene Expression Omnibus (GEO) [6] which contains 1,802,922 different samples as of 5/18/2016 that can be reanalyzed computationally based on current knowledge without the need for new biological experiments. As a case in point, each of the four most commonly used species have samples that have been analyzed using the original CDFs (Table 2).

Several research groups have reassigned probes into new probe sets by creating their own custom Chip Description Files (CDF) [7–13], which are specially formatted files used to store the layout information for an Affymetrix® GeneChip® array. Given a CDF, the intensity values of probes located in the CEL file can be extracted and summarized as a defined probe set to detect the expression level of genes or transcripts.

These approaches have a similar workflow of mapping probes but differ in terms of the groupings of probe sets, including: data source used, the selected target level (gene or transcript), whether to create probe sets from scratch or redesign the existing groups and sharing probes between probe sets.

In terms of annotations used, most approaches have mapped the probe sequences to the transcripts obtained from one or more databases such as GenBank, NCBI RefSeq and Ensembl. Unlike other approaches, Harbig et al. [14] mapped to the target sequences of probes obtained from Affymetrix® rather than the actual mRNA sequences themselves, where the target sequence is an exemplar region of a specific transcript ≤600 bases in length. After mapping, they grouped probes to unique transcripts or genes based on the mapping results. Some approaches update the original probe set groups by removing select probes and changing the link between probe set and gene/transcript. The most comprehensive study for probe annotation remapping was achieved by Dai et al. (brainarray CDFs) [8]. Rather than focusing on one reference database or combining multiple sources to create one custom CDF, they mapped probes to different annotation databases and created a specific custom CDF for each database.

Although the inherent effects of using dated probe gene mapping designs to analyze microarray data sets might seem obvious, the overwhelming majority of experimental results have only been analyzed using the original CDFs designed by Affymetrix®. For example as of May 2016, GEO has 120,920 samples which were analyzed via the original Affymetrix® CDFs for the

Table 1 Release dates of databases used by NetAffx v35 annotations and current database versions

GEO Platform		Organism		UniGene		
				NetAffx		Current
GPL570		Homo sapiens		Mar-10		Nov-12
GPL1261		Mus musculus		Jan-10		Jul-12
GPL1355		Rattus norvegicus		Mar-10		Nov-12
GPL198		Arabidopsis thaliana		May-09		Jul-12
Databases Common to All Four GEO Platforms						
	Ensembl	RefSeq		GenBank	Entrez Gene	Mirbase
NetAffx	Aug-14	Jul-14		Jun-14	May-14	Jul-12
Current	Mar-16	Mar-16		Apr-16	May-16	Jun-14

Table 2 Top Affymetrix® in situ oligonucleotide arrays found in GEO

GEO Platform	Title	Number of Probes (PM)	Number of Probe Sets	Number of Samples
GPL570	Human Genome U133 Plus 2.0 Array	604,258	54,675	120,920
GPL1261	Mouse Genome 430 2.0 Array	496,468	45,101	48,087
GPL1355	Rat Genome 230 2.0 Array	342,410	31,099	18,912
GPL198	Arabidopsis ATH1 Genome Array	251,078	22,810	12,624

HG-U133 Plus 2 array (Table 2). On the other hand only 6403 samples were analyzed using custom CDFs, mostly produced by brainarray (Table 3). Given that fewer than 5% of all samples in GEO have been analyzed by alternative CDFs, an opportunity exists to reanalyze existing datasets according to updated transcript knowledge or functional regions of interest.

While microarrays have been successfully utilized for understanding differential expression at the gene or probe set level, less attention has been given to the potential analysis at the individual exon, alternative transcript, and untranslated region (UTR) level. While the selection bias of probes on the 3′ ends of genes for earlier iterations of Affymetrix® GeneChip® designs presents limitations on the completeness of transcript information, more recent designs allow for a more complete coverage of exons and exon junctions. However, information concerning individual exons can still be extracted from earlier GeneChip® designs, particularly in the 3′ UTR regions that have been shown to play important roles in cancer [15–17], development [18–22], and localization in the nervous system [23–27]. In fact, over 40% of genes have been shown to generate multiple mRNAs with variable 3′ UTR lengths [28]. These 3′ UTRs harbor binding sites for molecules including microRNAs (miRNAs) and RNA-binding proteins. Thus, mRNA isoforms with lengthened 3′ UTRs have increased numbers of sites for these cis-interacting factors. The diversity of 3′ UTRs is predominantly regulated by alternative polyadenylation (APA),

which employs alternative mRNA cleavage sites that lie progressively distal to the stop codon. APA-driven mRNA diversity is required for normal physiology, and misregulation of this process is associated with diverse disease states [29]. We therefore have developed a framework for analysis of Affymetrix® GeneChip® data by regrouping probes into probe sets based on Ensembl annotations at the gene, transcript, individual exon, and UTR levels in order to detect changes in gene expression that may occur within specific regions of the transcript.

Methods

We developed an Affymetrix® GeneChip® probe remapping protocol at the level of genes, transcripts, untranslated regions (UTRs), coding sequences (CDS) and individual exons based on the latest genome (hg38, mm10, rn6) and Ensembl annotations (ENS-85) for human, mouse, and rat. The protocol takes annotations in a General/Gene Transfer Format (GTF) [30] file, generates a custom CDF where probes are grouped into probe sets based on region (UTR, CDS, individual exon), transcript or gene level. Here, we define individual exons as coding exons within protein coding genes, or all exons within structural RNAs (such as miRNA and lncRNA). In effect, the individual exons refer to all non-UTR portions of exons. Figure 1 shows the flow chart of annotation and grouping of probes based on the region of a gene. It is composed of three main steps: mapping probes to the genome, annotation of probes, and assignment of probes to probe sets based on annotations.

Mapping of perfect match probes to a genome

PM probe sequences, which can be obtained from the Affymetrix® Netaffx™ web site, are aligned to the indexed genome using Bowtie version 1.0.1 [31] with the parameters -v 0 and −m 1, requiring that probes align to a single genomic location with 100% identity, thereby reducing cross-hybridization effects. Note that Bowtie version 1 is best at aligning shorter sequences (25-50 bp) as found with microarray probes while the most recent versions of Bowtie are optimized for long sequence reads (>50 bp). Mismatch (MM) probes are not considered in the mapping step, although they could theoretically map uniquely to genomic regions. Rather, the MM probes are set aside and are included with their corresponding PM probe during the final CDF construction step once the PM probes have been assigned to a probe set. During this analysis, only probes perfectly matching to a region are considered. Therefore, probes crossing splice junctions will be discarded.

Table 3 Alternative CDFs for the top Affymetrix® in situ oligonucleotide arrays found in GEO

GEO Platform	Number of Alternative CDFs	Number and Percent of Samples Using Alternative CDFs
GPL570	54	6403 (5.0%)
GPL1261	36	1984 (4.0%)
GPL1355	12	460 (2.4%)
GPL198	9	642 (4.8%)

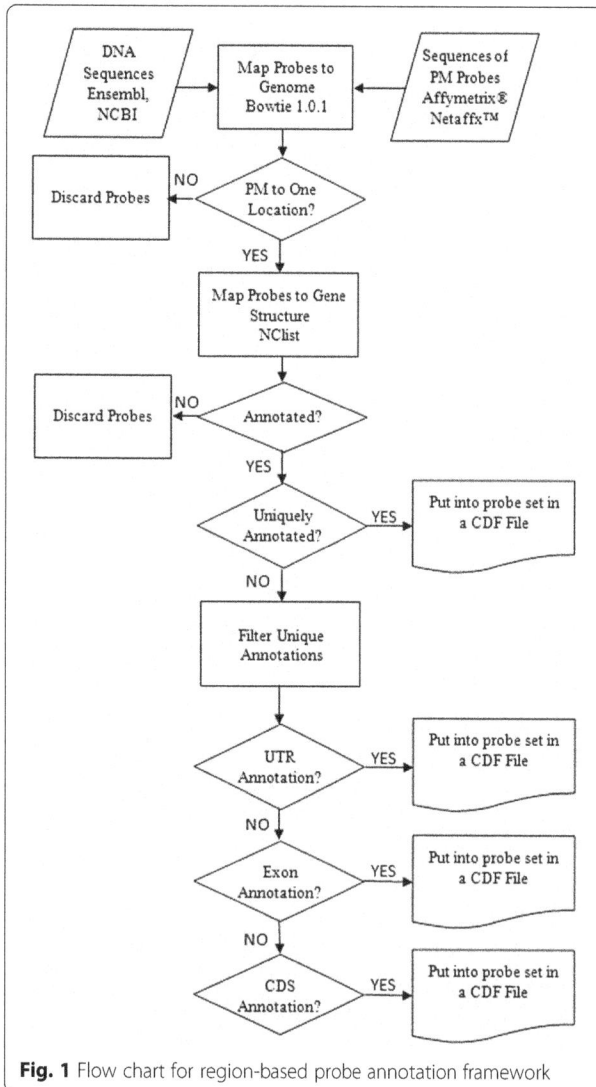

Fig. 1 Flow chart for region-based probe annotation framework

Annotation of perfect match probes via nested containment list (NCList)

Probes are annotated based on the overlap between probes and genomic intervals by the following steps.

I. GTF [30] files for the mouse, rat, and human genome were obtained from the Ensembl ftp server [32]. Each GTF is a tab-delimited text file used to represent gene structure information, including the start and end positions of a gene together with chromosome location. Each structure is tagged with a feature which can be gene, transcript, exon, start_codon, stop_codon, CDS or UTR. Ensembl GTFs were used since the annotations are determined by an automated system based on experimentally verified data combined from multiple databases such as RefSeq, EMBL and UniProtKB. It also contains manual curation for selected species.

II. A nested containment list (NCList) [33] was created for each chromosome from intervals (start and end points) of gene structures. The intervals of the NCList were selected based on the target of the probe sets. When the probe sets were constructed based on regions of a gene, we used UTR, individual exon and CDS intervals. For gene/transcript targeted probe sets, we used gene/transcript intervals.

III. Probe intervals were searched in the NCList and annotated according to the overlapping results. Probes were split based on the matched chromosome. Each probe group interval was searched in the same chromosome's NCList. When an overlap was found, the probe was annotated with the list node. Only complete overlaps were accepted; both the low and high ends of the interval have to be included in the list node. The probes which did not overlap the nodes were discarded. As a result, probes partially overlapping UTRs, individual exons, and CDS regions will not be included at the region and gene level, but will be present at the transcript level.

IV. A probe's start and end points may overlap multiple gene structures. It may overlap with the UTR and exon region of the same gene or with multiple genes or transcripts. In order to remove cross hybridization and ensure probes uniquely map to a single region, gene or transcript, we choose one of the annotations for each probe and remove the remaining matches. The rule for assigning these probes occurs with the following priority (I) 5′ and 3′ UTRs; (II) exons; (III) CDS. Thus, although UTR regions technically occur within exons, the more specific UTR assignment will be used. When the annotation was based on gene or transcript the first obtained annotation was selected.

V. Probes with the same annotation were grouped together to form a probe set. Figure 2 shows the grouping of probes for three types of CDFs. These CDFs are:
  - Region-based CDF: Probe sets are designed to target a specific region of a gene and consist of probes which map to the same region (UTR, individual exon, CDS) of a gene. In Fig. 2, green probes were mapped to the UTR region of Gene_1; therefore, those probes cluster together to form the Gene_1 UTR region probe set. Based on the same logic, blue colored probes form the probe set for Gene_1 exon and pink colored probes form the probe set for Gene_1 CDS.
  - Gene-based CDF: Probe sets are designed to target genes and consist of probes which map to the same gene. In Fig. 2, green, blue and pink colored probes, which mapped to Gene_1, cluster together to form the Gene_1 probe set.

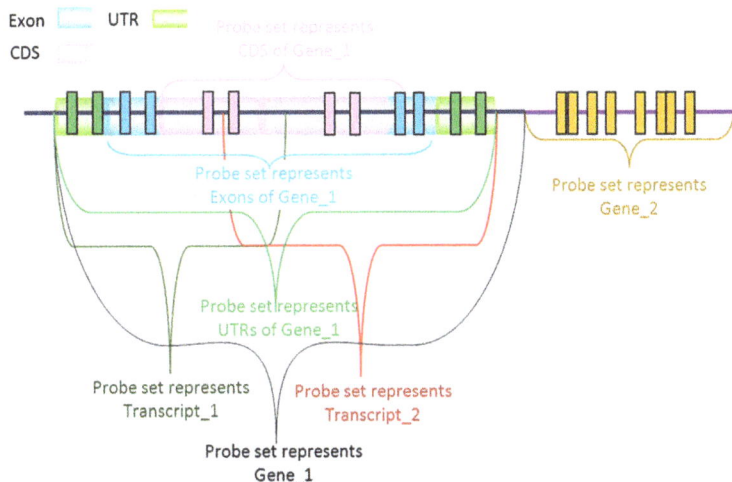

Fig. 2 Creating probe sets for different types of custom CDF based on probe mapping to gene regions

- Transcript-based CDF: Probes that map to same transcript of a gene compose a probe set. In Fig. 2, the orange and red arrow show the start and end positions of Transcript_1 and Transcript_2. The probes mapped to the Transcript_1 (two greens, two blue and two pink) cluster together to form the probe set for Transcript_1.

VI. Probe sets were saved into binary and ASCII format CDF files. The CDF files were created via the affxparser [34] Bioconductor package. In addition to the probes specific for a particular gene, Affymetrix® GeneChips® contain a number of different control probes such as probes that are added during sample preparation, providing evidence that assay was performed properly. We added those probe sets to our CDFs without any change. R CDF libraries were created via the makecdfenv [35] R Bioconductor package. The custom CDFs for three species (rat, mouse, and human) can be obtained from bioinformatics.louisville.edu/RegionCDFDesc.html

Probe set naming

Since GTF files obtained from Ensembl were used, Ensembl gene ids were employed to distinguish different genes and Ensembl transcript ids were used to distinguish different transcripts. When the generated CDF was based on regions of genes, the region was suffixed to the Ensembl gene id. Table 4 shows example probe set names taken from custom CDFs for the Affymetrix® GeneChip® HG-133 Plus 2.

We applied our framework to the three most widely used GeneChips®: HG-U133 Plus 2, Rat 230 2.0 and Mouse 430 2.0 (summarized in Tables 5 and 6). We also examined the effect of probe reannotation over the differentially expressed genes. Three types of CDFs were created for every selected organism. Our results discussed here are restricted to the analysis of the HG-U133 Plus 2 and Rat Genome 230 2.0 GeneChip® for brevity. After CDF creation, we reanalyzed the publicly available data series GSE48611 [36] and GSE72551 [24] from GEO via our custom CDFs.

Results
Custom CDF generation
Probes mapping to the genome

Using the bowtie parameters as discussed in the methods section, we were able to identify probes that uniquely map with 100% identity for each of the respective genomes. As a result, 87% PM probes of the HG-U133 Plus 2, 84% PM probes of the Rat 230 2.0 and 86% PM probes of the Mouse 430 2.0 were uniquely mapped to the genome and were used in the subsequent steps (Table 7).

Probe annotations and probe sets

To annotate probes, we mapped uniquely aligned probes to gene regions using the most recent Ensembl genome and GTF file for each respective organism. We used the specific regions based on the custom CDF type (gene,

Table 4 Custom CDF naming examples

CDF Type	Probe Set Name
Region-based	ENSG00000001036_exon_-
	ENSG00000001084_UTR_-
	ENSG00000001167_CDS_+
Gene-based	ENSG00000001461
Transcript-based	ENST00000489806

Table 5 Summary of probes used for gene and transcript based custom CDFs

	Homo sapiens		Rattus norvegicus		Mus musculus	
	Gene	Transcript	Gene	Transcript	Gene	Transcript
Number of Probes Used	414,701	504,419	162,356	205,671	323,917	395,884
Number of Probe Sets Constructed	22,651	26,096	13,150	14,466	19,282	20,980
Average Number of Probes Per Probe Set	18	18	12	14	16	18

transcript or region-based). Consequently we produced three types of custom CDFs (Tables 5 and 6).

The human gene based CDF has 22,651 custom designed probe sets composed from 414,701 probes and 62 original control probe sets. 442,025 annotations were identified between genes and the probes. 27,324 annotations were filtered after shared probes were removed. In order to validate our probe set annotations, we compared the original CDF probe sets with the custom CDF. A total of 21,585 annotated genes were shared between the two CDFs, with 3068 unique to the original CDF, and 1066 unique to our custom CDF. In order to determine why some genes were not covered in our CDF, we examined those unique to the original CDF. First we obtained the probe sets which represent these genes in the original CDF, yielding 2781 probe sets. We retrieved both the PM and MM probe sequences for each of these. We observed that for 667 probe sets, every probe was removed during probe mapping to the genome due to either non-unique mappings or mapping rates less than 100%. 30,150 probes from the remaining 2114 probe sets were not used in our CDF since they either did not map to the genome or they were MM probes. 14,028 probes were used in our newly constructed probe sets which target different genes than the original assignment by Affymetrix® and 2656 probes were not aligned to gene structures and not annotated. As a result, the differences between the original CDF and our method occurs because of probes removed during genome alignment, probes that no longer map to gene structure or probes that map to gene structures different from the original annotation.

For the rat 230 2.0 GeneChip®, the restriction of three probes per probeset yields 12,534 uniquely identified Ensembl genes at the gene level. We determined that for this specific GeneChip®, reorganization of the Affymetrix® probes into mRNA region-specific probesets provides 4024 unique Ensembl gene identifiers with

probesets in both the 3′ UTR and CDS. Using this subset of probesets, differential expression of the CDS can then be compared to the 3′ UTR.

Analysis with custom CDFs

We reanalyzed the publicly available data series GSE72551 and GSE48611. Both of these studies involve the nervous system, where differences in 3′ UTRs are likely to have phenotypic effects on transcript localization. The GSE72551 data series examines gene expression changes associated with collateral sprouting and includes 5 naïve controls, 7 replicates at day 7 post-surgery and 7 replicates at day 14 post-surgery. The GSE48611 data series examines Down syndrome gene expression monitoring. This data set includes mRNA samples from the isogenic trisomy of chromosome 21 (Ts21) and control pluripotent stem cells (iPSCs) (DS1, DS4, and DS2U) between passages 24 and 48 and from day 30 neurons. Three biological replicates were present for each condition. Prior to analysis, we removed probe sets with two or fewer probes from the custom CDFs in order to achieve more accurate results for target expression levels. Robust Multiarray Averaging (RMA) normalization [37] was used for preprocessing. A p-value cutoff of 0.05 was used as the threshold for all experiments.

In the GSE72551 data series, differentially expressed genes (DEGs) were determined for two pairwise comparisons: naïve vs. both 7 and 14 days using region and gene based custom CDFs. We also reanalyzed the data using the brainarray Ensembl CDF version 20. Figure 3 shows a Venn diagram representing the number of differentially expressed genes using region, gene and brainarray custom CDFs for both cases.

Further examination of the 7 day versus naïve ENSEMBL genes found to be differentially expressed in either the gene-based or region-based CDF shows high concordance, with 975 ENSEMBL genes

Table 6 Summary of probes used for region based custom CDFs

	Homo sapiens	Rattus norvegicus	Mus musculus
Number of Probes Aligned to Genome	822,681	321,905	637,942
Number of Probes Used	414,701	162,356	323,917
Number of Probe Sets Constructed	33,916	19,839	28,963
Average Number of Probes Per Probe Set	12	8	11

Table 7 Number of mapped probes for custom CDF construction

GeneChip®	Number of PM Probes	Number of PM Probes Mapped Uniquely	Number of PM Probes Mapped to Multiple Locations	Number of PM Probes Not Aligned
Human Genome U133 Plus 2.0 Array	603,158	525,985	36,493	40,680
Rat Genome 230 2.0 Array	341,459	288,319	26,027	27,113
Mouse Genome 430 2.0 Array	495,374	427,758	28,444	39,173

determined to be differentially expressed using both CDFs (Fig. 3a). Examination of the *p*-values shows a significant correlation between both the gene and the 3′ UTR region ($r = 0.439$; $p = 1.480E\text{-}58$) as well as between the gene and the exon region ($r = 0.101$; $p = 0.001$). The higher correlation with the 3′ UTR region is to be expected, due to a higher abundance of probes designed in these regions.

One hundred sixty genes are found to be differentially expressed using the gene-based approach only. Three genes are omitted completely from the region-based CDF. Further examination of the remaining 157 genes measured using both the gene-based and region-based

a 7 versus naïve

b 14 versus naïve

Fig. 3 Number of common and different differentially expressed genes using our custom region and gene-based CDFs compared to brain array custom CDFs. **a** Day 7 versus naïve. **b** Day 14 versus naïve

methods shows that 122 of these (78%) have a gene-based *p*-value >0.03, and 80 (50%) have a gene-based *p*-value >0.04, indicating the detected differences are just below the cutoff level. Analysis of the region-based *p*-values show that 120 of these (77%) have a region-based *p*-value <0.10, and 146 (94%) have a region-based *p*-value <0.20, putting these genes just above the significance threshold.

An additional 423 genes are found to be differentially expressed using the region-based approach only, with 203 from the 3′ UTR only, 10 from the 5′ UTR only, 206 from the exon only, and 4 from both the 3′ UTR and exon. Unlike the DEGs uniquely found in the gene-based approach, those genes found to be differentially expressed in the region-based approach typically have a much higher *p*-value in gene-based analysis, with only 31% having a *p*-value between 0.05 and 0.10. This supports our reasoning that separating into functional regions allows detection of subtle changes in transcript formation that may have a larger functional impact of those transcripts which has been further validated by experimental work showing differential expression of the 3′ UTR of the *CAMKIV* gene plays a role in localization [23].

In order to determine why some genes were only detected by the brainarray CDF, we examined the probe sequences of those genes that are brainarray specific. Of these, 39 were excluded from our CDFs since they aligned to multiple locations in the rn6 genome. An additional ten of these probes did not match to known Ensembl gene structures and were thus removed. Eighteen of these probes were excluded because the probe set contained fewer than three probes. An additional 40 of the brainarray probes were used in our CDFs, but with annotations differing from brainarray due to changes in annotation information.

In the GSE48611 data series, DEGs were determined for two pairwise comparisons: isogenic Ts21 vs. control iPSCs for both DS1 and DS4. We reanalyzed the data using region, gene and the original Affymetrix® supplied CDF obtained from the Affymetrix® Netaffx™ web site. For DS1, our gene-based CDF identified an additional 194 DEGs not found using the original CDF and 616 DEGs identified by both methods. For DS2, our gene base CDF identified an additional 331 DEGs found using our method only and 337 DEGs identified by both methods (Table 8).

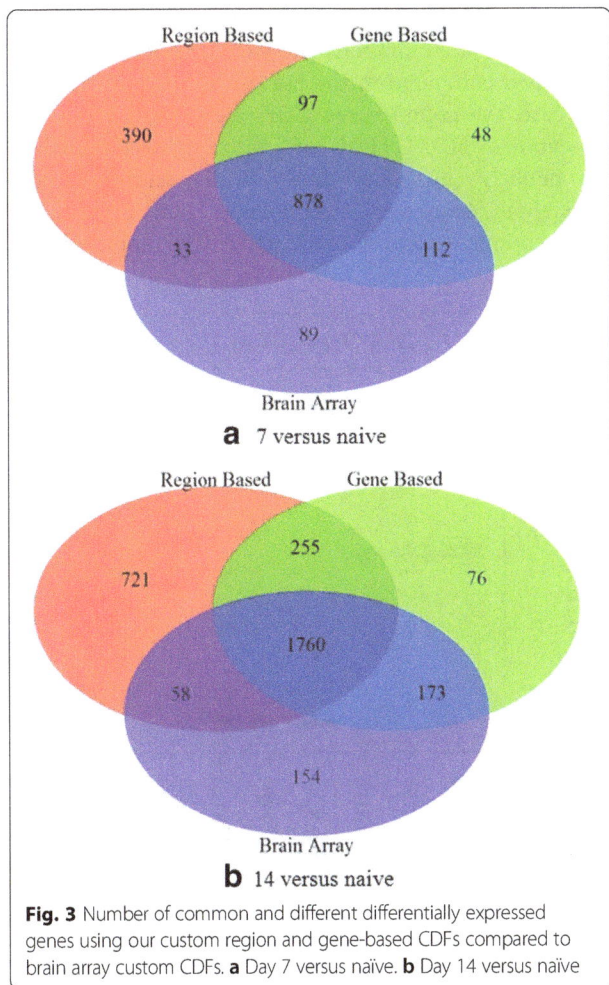

Table 8 DEGs detected by our gene based CDF and GPL570

Cases	Our Gene Based CDF	GPL570	Common
DS1 versus Ts21	810	2421	616
DS2 versus Ts21	668	1840	337

Discussion

One of the limitations of microarray technologies is the design of probes based on available sequence and annotation data at the time of design. Based on our analysis, the percentage of uniquely mapping probes varies from 84% (rat) to 87% (human), indicating that changing knowledge about the genome itself plays a role in probe utilization. In terms of annotation, the rat genome is known to have more incomplete information when compared to mouse and human, which is reflected in the fact that only 47% of the rat probes lie in region-based locales (exons and UTRs) compared to 65% for mouse, and 69% for human. Since this can potentially lead to a small number of probes in each annotated region (and thus increased false positive rates), we have further required at least three probes be present in each probe set for our analysis. Both unrestricted (1 or more) and restricted (3 or more) probe groupings are available as CDFs.

To further illustrate the importance of region-based CDFs, using the subset of 4024 genes with probesets in both the CDS and 3′ UTR regions, we were able to identify 203 differential expression events at the 3′ UTR level that do not show differential expression within the CDS. In addition, these events are not detected using the standard Affymetrix® CDF. Further analysis of these 203 genes yields some genes of particular interest. For instance, the 3′ UTR of *GRIK4* (Glutamate Ionotropic Receptor Kainate Type Subunit 4) was up-regulated (*p*-value 0.0450) while the CDS was not significantly regulated (FC = 1.07; p-value 0.4525), suggesting the 3′ UTR of this gene was lengthened (Fig. 4). *GRIK4* regulates kainite-receptor signaling and neuroplasticity [38] and its missregulation is associated with neurological diseases including Alzheimer's [39], bipolar disorder [40], and others. Interestingly, a deletion variant specific to the 3′ UTR of *GRIK4* is protective of bipolar disorder [40]. Alongside our observation, this suggests that regulation of this plasticity-associated gene occurs though its 3′ UTR. We also observed that the 3′ UTR of *VEGFA* (vascular endothelial growth factor-A) was downregulated (−1.17 FC; $p = 0.0102$) and expression of its CDS was unchanged (1.01 FC; $p = 0.8334$) (Fig. 5). The 3′UTR of *VEGFA*, a potent neuromodulator, undergoes a well-described binary switch to regulate its expression [41]. Our observations suggest the *VEGFA* 3′ UTR undergoes an additional layer of regulation by shortening during collateral sprouting.

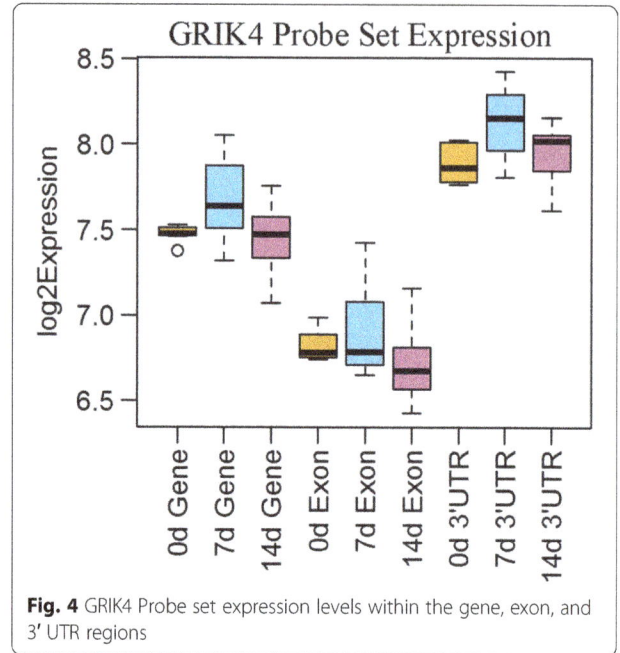

Fig. 4 GRIK4 Probe set expression levels within the gene, exon, and 3′ UTR regions

As our analysis with the GSE48611 and GSE72551 datasets show, reanalysis of publicly available datasets using updated annotations can yield additional information when compared to the use of the original CDFs. In our case, the region-based CDFs allow for a better understanding of 3′ UTR dynamics through the reanalysis of publicly available data. While current high-throughput sequencing technologies may allow for a more complete picture, this custom CDF approach will

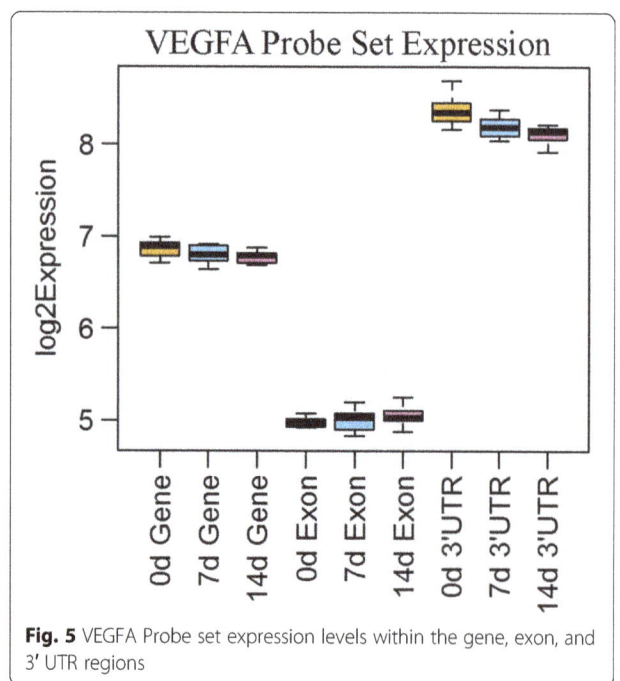

Fig. 5 VEGFA Probe set expression levels within the gene, exon, and 3′ UTR regions

Framework for reanalysis of publicly available Affymetrix® GeneChip® data sets based on functional...

171

allow for deeper insight with only minimal computational cost, taking advantage of the high volume of publicly available GeneChip® data.

Conclusions

We proposed a framework for reannotating and reassigning probe groups for Affymetrix® GeneChip® technology based on functional regions of interest. Our work differs from others in that we annotated probes in UTR and exon levels in addition to gene and transcript (isoform) levels. We illustrated how this framework affects the detection of differentially expressed genes, particularly when focusing on functional regions of interest. Removing probes that no longer align to the genome without mismatches or align to multiple locations can help to reduce false-positive differential expression, as can removal of probes in regions overlapping multiple genes.

The main motivation of our work was profiling the contribution of UTR and exon regions to the gene expression levels globally. Our results indicate that features differentially expressed in either the gene-based or region-based CDF show high concordance and separating out into functional regions allows for the detection of subtle changes in transcript formation.

Acknowledgements

The authors wish to thank members of the Kentucky Biomedical Research Infrastructure Network Bioinformatics Core, the University of Louisville Bioinformatics Journal Club, and members of the University of Louisville Bioinformatics and Biomedical Computing Laboratory for helpful insight, project review, and suggestions.

Funding

Publication charges provided by National Institutes of Health grant P20GM103436. Research support provided by NIH grants P20GM103436 and R01NS094741. The contents of this manuscript are solely the responsibility of the authors and do not represent the official views of NIH.

About this supplement

This article has been published as part of *BMC Genomics* Volume 18 Supplement 10, 2017: Selected articles from the 6th IEEE International Conference on Computational Advances in Bio and Medical Sciences (ICCABS): genomics. The full contents of the supplement are available online at https://bmcgenomics.biomedcentral.com/articles/supplements/volume-18-supplement-10.

Authors' contributions

ES was responsible for code preparation, development of the project, and manuscript preparation. ECR and JCP developed the overall project goals. ECR supervised the overall project, provided the necessary lab space and computational resources for project completion, and led development of the manuscript. JCP and BJH provided test data, analyzed results, and reviewed the manuscript. KW performed testing of UTR analysis of microarrays and reviewed the manuscript. All authors have read and approved the final manuscript.

Competing interests

The authors declare that they have no competing interests.

Author details

[1]Department of Computer Engineering and Computer Science, University of Louisville, Louisville, KY, USA. [2]Department of Anatomical Sciences and Neurobiology, School of Medicine University of Louisville, Louisville, KY, USA. [3]Department of Biological Sciences, University of New England, Biddeford, ME, USA. [4]Department of Biochemistry and Molecular Biology University of Arkansas for Medical Science, Little Rock, AR, USA.

References

1. Causton HC, Quackenbush J, Brazma, A: Microarray gene expression data analysis: a beginner's guide. Malden, MA: Wiley-Blackwell; 2009.
2. Knudsen S. Guide to analysis of DNA microarray data. 2nd ed. Hoboken, NJ: Wiley-Liss; 2004.
3. Liu G, Loraine AE, Shigeta R, Cline M, Cheng J, Valmeekam V, Sun S, Kulp D, Siani-Rose MA. NetAffx: Affymetrix probesets and annotations. Nucleic Acids Res. 2003;31(1):82–6.
4. Flight RM, Eteleeb AM, Rouchka EC. Affymetrix® mismatch (MM) probes: useful after all. In: 2012 ASE/IEEE international conference on BioMedical Computing (BioMedCom). Washington: IEEE Computer Society; 2012. pp. 6-13.
5. Rouchka EC, Phatak AW, Singh AV. Effect of single nucleotide polymorphisms on Affymetrix match-mismatch probe pairs. Bioinformation. 2008;2(9):405–11.
6. Barrett T, Wilhite SE, Ledoux P, Evangelista C, Kim IF, Tomashevsky M, Marshall KA, Phillippy KH, Sherman PM, Holko M, et al. NCBI GEO: archive for functional genomics data sets—update. Nucleic Acids Res. 2013; 41(Database issue):D991–5.
7. Chalifa-Caspi V, Yanai I, Ophir R, Rosen N, Shmoish M, Benjamin-Rodrig H, Shklar M, Stein TI, Shmueli O, Safran M, et al. GeneAnnot: comprehensive two-way linking between oligonucleotide array probesets and GeneCards genes. Bioinformatics. 2004;20(9):1457–8.
8. Dai M, Wang P, Boyd AD, Kostov G, Athey B, Jones EG, Bunney WE, Myers RM, Speed TP, Akil H, et al. Evolving gene/transcript definitions significantly alter the interpretation of GeneChip data. Nucleic Acids Res. 2005;33(20):e175.
9. Gautier L, Moller M, Friis-Hansen L, Knudsen S. Alternative mapping of probes to genes for Affymetrix chips. BMC Bioinformatics. 2004;5:111.
10. Liu H, Zeeberg BR, Qu G, Koru AG, Ferrucci A, Kahn A, Ryan MC, Nuhanovic A, Munson PJ, Reinhold WC, et al. AffyProbeMiner: a web resource for computing or retrieving accurately redefined Affymetrix probe sets. Bioinformatics. 2007;23(18):2385–90.
11. Lu J, Lee JC, Salit ML, Cam MC. Transcript-based redefinition of grouped oligonucleotide probe sets using AceView: high-resolution annotation for microarrays. BMC Bioinformatics. 2007;8:108.
12. Risueno A, Fontanillo C, Dinger ME, De Las RJ. GATExplorer: genomic and transcriptomic explorer; mapping expression probes to gene loci, transcripts, exons and ncRNAs. BMC Bioinformatics. 2010;11:221.
13. Yin J, McLoughlin S, Jeffery IB, Glaviano A, Kennedy B, Higgins DG. Integrating multiple genome annotation databases improves the interpretation of microarray gene expression data. BMC Genomics. 2010;11:50.
14. Harbig J, Sprinkle R, Enkemann SA. A sequence-based identification of the genes detected by probesets on the Affymetrix U133 plus 2.0 array. Nucleic Acids Res. 2005;33(3):e31.
15. Akman HB, Oyken M, Tuncer T, Can T, Erson-Bensan AE. 3'UTR shortening and EGF signaling: implications for breast cancer. Hum Mol Genet. 2015; 24(24):6910–20.
16. Fu Y, Sun Y, Li Y, Li J, Rao X, Chen C, Xu A. Differential genome-wide profiling of tandem 3' UTRs among human breast cancer and normal cells by high-throughput sequencing. Genome Res. 2011;21(5):741–7.
17. Wang L, Hu X, Wang P, Shao ZM. The 3'UTR signature defines a highly metastatic subgroup of triple-negative breast cancer. Oncotarget. 2016;
18. Hilgers V, Perry MW, Hendrix D, Stark A, Levine M, Haley B. Neural-specific elongation of 3' UTRs during drosophila development. Proc Natl Acad Sci. 2011;108(38):15864–9.

19. Ji Z, Lee JY, Pan Z, Jiang B, Tian B. Progressive lengthening of 3′ untranslated regions of mRNAs by alternative polyadenylation during mouse embryonic development. Proc Natl Acad Sci. 2009;106(17):7028–33.

20. Kuersten S, Goodwin EB. The power of the 3[prime] UTR: translational control and development. Nat Rev Genet. 2003;4(8):626–37.

21. Revil T, Gaffney D, Dias C, Majewski J, Jerome-Majewska LA. Alternative splicing is frequent during early embryonic development in mouse. BMC Genomics. 2010;11:399.

22. Thomsen S, Azzam G, Kaschula R, Williams LS, Alonso CR. Developmental RNA processing of 3′UTRs in Hox mRNAs as a context-dependent mechanism modulating visibility to microRNAs. Development. 2010;137(17):2951–60.

23. Harrison BJ, Flight RM, Gomes C, Venkat G, Ellis SR, Sankar U, Twiss JL, Rouchka EC, Petruska JC. IB4-binding sensory neurons in the adult rat express a novel 3′ UTR-extended isoform of CaMK4 that is associated with its localization to axons. J Comp Neurol. 2014;522(2):308–36.

24. Harrison BJ, Venkat G, Hutson T, Rau KK, Bunge MB, Mendell LM, Gage FH, Johnson RD, Hill C, Rouchka EC, et al. Transcriptional changes in sensory ganglia associated with primary afferent axon collateral sprouting in spared dermatome model. Genom Data. 2015;6:249–52.

25. Jansen RP. mRNA localization: message on the move. Nat Rev Mol Cell Biol. 2001;2(4):247–56.

26. Prakash N, Fehr S, Mohr E, Richter D. Dendritic localization of rat vasopressin mRNA: ultrastructural analysis and mapping of targeting elements. Eur J Neurosci. 1997;9(3):523–32.

27. Willis DE, Xu M, Donnelly CJ, Tep C, Kendall M, Erenstheyn M, English AW, Schanen NC, Kirn-Safran CB, Yoon SO, et al. Axonal localization of Transgene mRNA in mature PNS and CNS neurons. J Neurosci. 2011;31(41):14481–7.

28. Derti A, Garrett-Engele P, Macisaac KD, Stevens RC, Sriram S, Chen R, Rohl CA, Johnson JM, Babak T. A quantitative atlas of polyadenylation in five mammals. Genome Res. 2012;22(6):1173–83.

29. Curinha A, Oliveira Braz S, Pereira-Castro I, Cruz A, Moreira A. Implications of polyadenylation in health and disease. Nucleus. 2014;5(6):508–19.

30. The Brent Lab: GTF2.2: A Gene Annotation Format. http://mblab.wustl.edu/GTF22.html. Accessed 20 Sep 2016.

31. Langmead B, Trapnell C, Pop M, Salzberg SL. Ultrafast and memory-efficient alignment of short DNA sequences to the human genome. Genome Biol. 2009;10(3):R25.

32. Cunningham F, Amode MR, Barrell D, Beal K, Billis K, Brent S, Carvalho-Silva D, Clapham P, Coates G, Fitzgerald S, et al. Ensembl 2015. Nucleic Acids Res. 2015;43(D1):D662–9.

33. Alekseyenko AV, Lee CJ. Nested containment list (NCList): a new algorithm for accelerating interval query of genome alignment and interval databases. Bioinformatics. 2007;23(11):1386–93.

34. Bengtsson H, Bullard J, Hanson K: Affxparser: Affymetrix file parsing SDK. R package version 1.40.0. 2015.

35. Irizarry RA, Gautier L, Huber W, Bolstad B: makecdfenv: CDF Environment Maker. R package version 1.44.0. 2006.

36. Weick JP, Held DL, Bonadurer GF 3rd, Doers ME, Liu Y, Maguire C, Clark A, Knackert JA, Molinarolo K, Musser M, et al. Deficits in human trisomy 21 iPSCs and neurons. Proc Natl Acad Sci U S A. 2013;110(24):9962–7.

37. Irizarry RA, Hobbs B, Collin F, Beazer-Barclay YD, Antonellis KJ, Scherf U, Speed TP. Exploration, normalization, and summaries of high density oligonucleotide array probe level data. Biostatistics. 2003;4(2):249–64.

38. Fernandes HB, Catches JS, Petralia RS, Copits BA, Xu J, Russell TA, Swanson GT, Contractor A. High-affinity kainate receptor subunits are necessary for ionotropic but not metabotropic signaling. Neuron. 2009; 63(6):818–29.

39. Jacob CP, Koutsilieri E, Bartl J, Neuen-Jacob E, Arzberger T, Zander N, Ravid R, Roggendorf W, Riederer P, Grunblatt E. Alterations in expression of glutamatergic transporters and receptors in sporadic Alzheimer's disease. J Alzheimers Dis. 2007;11(1):97–116.

40. Pickard BS, Knight HM, Hamilton RS, Soares DC, Walker R, Boyd JK, Machell J, Maclean A, McGhee KA, Condie A, et al. A common variant in the 3′UTR of the GRIK4 glutamate receptor gene affects transcript abundance and protects against bipolar disorder. Proc Natl Acad Sci U S A. 2008;105(39): 14940–5.

41. Ray PS, Jia J, Yao P, Majumder M, Hatzoglou M, Fox PL. A stress-responsive RNA switch regulates VEGFA expression. Nature. 2009;457(7231):915–9.

Genome-wide identification of miRNAs and lncRNAs in *Cajanus cajan*

Chandran Nithin[1†], Amal Thomas[1,3†], Jolly Basak[2*] and Ranjit Prasad Bahadur[1*] (iD)

Abstract

Background: Non-coding RNAs (ncRNAs) are important players in the post transcriptional regulation of gene expression (PTGR). On one hand, microRNAs (miRNAs) are an abundant class of small ncRNAs (~22nt long) that negatively regulate gene expression at the levels of messenger RNAs stability and translation inhibition, on the other hand, long ncRNAs (lncRNAs) are a large and diverse class of transcribed non-protein coding RNA molecules (> 200nt) that play both up-regulatory as well as down-regulatory roles at the transcriptional level. *Cajanus cajan*, a leguminosae pulse crop grown in tropical and subtropical areas of the world, is a source of high value protein to vegetarians or very poor populations globally. Hence, genome-wide identification of miRNAs and lncRNAs in *C. cajan* is extremely important to understand their role in PTGR with a possible implication to generate improve variety of crops.

Results: We have identified 616 mature miRNAs in *C. cajan* belonging to 118 families, of which 578 are novel and not reported in MirBase21. A total of 1373 target sequences were identified for 180 miRNAs. Of these, 298 targets were characterized at the protein level. Besides, we have also predicted 3919 lncRNAs. Additionally, we have identified 87 of the predicted lncRNAs to be targeted by 66 miRNAs.

Conclusions: miRNA and lncRNAs in plants are known to control a variety of traits including yield, quality and stress tolerance. Owing to its agricultural importance and medicinal value, the identified miRNA, lncRNA and their targets in *C. cajan* may be useful for genome editing to improve better quality crop. A thorough understanding of ncRNA-based cellular regulatory networks will aid in the improvement of *C. cajan* agricultural traits.

Keywords: miRNA, lncRNA, *Cajanus Cajan*, SSR signature, genome-wide analysis

Background

Cajanus cajan is a major source of protein for the poor communities of many tropical and subtropical regions of the world [1]. The high protein and carbohydrate contents make it not only important to the human diet, but also suitable as high protein feed and fodder ingredient to livestock [2]. With its greater tolerance to heat, drought, and low soil fertility, *C. cajan* is a valuable component of low external input agricultural farming systems where the farmers have scarcity of resources [3–6]. *C. cajan* is a good source of sulphur containing amino acids, crude fibre, iron, sulphur, calcium, potassium, manganese and water soluble vitamins especially thiamine, riboflavin and niacin [7, 8]. In addition to these, several flavonoids, isoflavonoids, tannins and protein fractions have been isolated from the different parts of *C. cajan* and their medicinal uses have been established [9].

The ncRNAs are a wide class of non-coding RNAs that are transcribed but not translated and play a major role in post-transcriptional gene regulations. Based on their length, ncRNAs are generally classified into small non-coding RNAs (sncRNAs) and long non-coding RNAs (lncRNAs). MicroRNAs (miRNA) are an abundant class of sncRNAs (~22nt long), which negatively regulate gene expression at the levels of messenger RNAs (mRNAs) stability and translation inhibition. In addition to this, the miRNAs are also known to interact with lncRNAs as well as competing endogenous RNAs (ceRNAs) that de-repress the gene expression. Identification of the various miRNAs and their targets is important in understanding the dynamics of gene regulation and in designing new breeds of crops with higher productivity and better disease

* Correspondence: jolly.basak@visva-bharati.ac.in;
r.bahadur@hijli.iitkgp.ernet.in
†Equal contributors
²Department of Biotechnology, Visva-Bharati, Santiniketan, India
¹Computational Structural Biology Lab, Department of Biotechnology, Indian Institute of Technology Kharagpur, Kharagpur, India
Full list of author information is available at the end of the article

resistance. In spite of having immense importance, there are only few studies have ventured into identifying the miRNAs in *C. cajan* [10]. Additionally, miRNAs of *C. cajan* are still missing in miRBase 21 [11].

The miRNAs are known to have sequence conservation and are grouped into various miRNAs families in miR-Base. The presence of orthologs and paralogs among miRNA sequences allows the identification of miRNAs by using computational methods starting from the sequence similarity. The mere presence of a sequence match on a genome does not imply that the identified region is a miRNA. miRNA precursor sequences (pre-miRs) are known to have features distinct from other small RNA. The mapping of known miRNAs to the genome followed by extraction and analysis of pre-miRs is an effective strategy in miRNA discovery [10, 12–17]. Various sequence based information as well as structural attributes of the pre-miRs can be useful to establish whether a given match is a miRNA sequence or not. To begin with, the miRNA precursors have a distinct range in which the nucleotide composition falls [18]. The pre-miRNAs also have a distinct pattern in the free energy of folding [19]. The minimal folding energy index (MFEI), which is the free energy associated with folding, normalized per GC content per hundred nucleotides, is used as a parameter in predicting miRNAs. The miRNAs are also shown to have distinct region in the probability distributions of RNA folding measures, namely, normalized Shannon entropy (NQ), normalized base pairing propensity (Npb) and normalized base pairing distance (ND) [16]. Simple sequence repeats (SSRs) are one to six nucleotides long repeat sequences present in the pre-miRs [20], and can be used as a parameter to efficiently predict miRNAs [16].

lncRNAs are a large and diverse class of transcribed non-protein coding RNA molecules with a length of more than 200 nucleotides. The evidence for regulatory role of lncRNAs in important biological processes was first identified during the 1980s from genetic analyses of the *Drosophila* bithorax complex [21]. Compared to the protein coding mRNAs, lncRNAs have certain specific properties, namely, shorter length, lower abundance, restriction to particular tissues or cells and less frequent conservation between species [22]. The lncRNA biogenesis is very similar to protein coding mRNAs but some lncRNAs are transcribed by RNA polymerase III [23]. The lncRNAs also have the post-transcriptional modifications like 5' capping, splicing and polyadenylation [24]. While most of the lncRNAs are localized within the nucleus, there are a few exceptions that perform functions in the cytosol [25, 26]. The origin of lncRNAs can range from intronic, exonic, intergenic, intragenic, promoter regions, 3'- and 5'- UTRs and enhancer sequences. The transcription of lncRNAs can happen either in sense or in antisense directions [27]. They play both down regulatory as well as up regulatory roles at the transcriptional level. The lncRNAs originating from protein coding loci competes for the RNA polymerase II and other initiation factors or cause the premature termination of elongation complex [28]. The lncRNAs can enhance the accessibility of target site to RNA polymerase and thereby upregulate the gene expression [29]. Some lncRNAs bind to the promoter DNA of target gene, forming a RNA-dsDNA triplex that prevents the preinitiation complex from accessing the target gene promoter [30]. There are also lncRNAs which are reported to regulate the gene expression by inhibiting the RNA polymerase activities or by controlling the subcellular localization of transcription factors [31–33]. In addition to the transcriptional regulation, lncRNAs also play a role in post-transcriptional modulations of mRNA processing. They play role in pre-mRNA alternate splicing, transport, translation and degradation [34]. The lncRNAs can also cause the degradation of target mRNA through the formation of a double stranded RNA duplex, which is processed into endo-siRNAs [35].

In this study, we have identified 616 miRNAs in *C. cajan*, of which 578 are novel and not reported in MirBase21. Besides, we have also predicted 3919 lncRNAs. Additionally, the protein coding genes targeted by many of the miRNAs are identified in this study, facilitating a functional annotation to the predicted miRNAs. Moreover, we have identified the lncRNAs that are targeted by miRNAs. These findings will significantly contribute to the present knowledge of ncRNAs in *C. cajan*, and will enhance our understanding for genome editing and improving the crop varieties in plants.

Methods

Dataset collection and preparation

The dataset of known miRNAs and pre-miRs was downloaded from miRBase 21 [36], which consists of 4800 mature and 8480 pre-miRs belonging to 73 species of Viridiplantae. Besides, we have also downloaded the draft genome sequence of *C. cajan* [37]. The coding DNA sequences composed of 21,434 transcriptome assembly contigs, ccTAv2.0, was downloaded from Legume Information system [38]. The protein sequences of Viridiplantae was curated from NCBI [39]. The UniProt proteome, UP000075243, with 47,180 entries was downloaded along with the UniProt-GOA annotation data [40, 41]. The SWISS-PROT database [42] was downloaded for running the BLAST [43] search.

Prediction of miRNAs

The dataset of known miRNAs was BLAST searched against the genome of *C. cajan*. The BLAST hits with zero to three mismatches with the known miRNAs were selected, and were further used for analysis in the prediction pipeline. The upstream and downstream nucleotides from the BLAST hit was extracted following Nithin, et al. [16], and the protein coding sequences were removed by performing BLASTX with the protein sequences of

Viridiplantae. The sequences were selected based on the cut-off value for each of the following parameters: MFEI, NQ, ND, Npb and SSRs [16]. The MFEI value for a sequence of length L was calculated using the adjusted MFE (AMFE), which represents the MFE for 100 nucleotides.

$$MFEI = \frac{AMFE}{(G+C)\%} \quad and \quad AMFE = -\frac{MFE}{L} \times 100$$

The genRNAstats program [19] was used to calculate the NQ, ND and Npb for all known pre-miRs of Viridiplantae. Npb is the measure of total number of base pairs present in the RNA secondary structure per length of the sequence, and the value can range from 0.0 (no base-pairs) to 0.5 (L/2 base-pairs) [44]. The base-pairing probability distribution (BPPD) per base in a sequence were measured using NQ [45], while the base-pair distance for all the pair of structures were measured using ND [46]. Both the parameters ND and NQ were calculated from the MaCaskill base pair probability p_{ij} between the two bases, i and j:

$$where \quad p_{ij} = \sum_{S_\alpha \in S(s)} P(S_\alpha)\delta_{ij}$$

$$P(S_\alpha) = \frac{e^{\frac{-E_\alpha}{RT}}}{\sum_{S_\alpha \in S(s)} e^{\frac{-E_\alpha}{RT}}}$$

$$and \quad \delta_{ij} = \begin{cases} 1, & x_i \text{ pairs } x_j \\ 0, & \text{otherwise} \end{cases}$$

$$NQ = -\frac{1}{L}\sum_{i<j} p_{ij}\log_2\left(p_{ij}\right) \quad and \quad ND = \frac{1}{L}\sum_{i<j} p_{ij}\left(1-p_{ij}\right)$$

The signature SSRs for different miRNA families at window size of three were taken from Nithin, et al. [16]. The conserved SSR signatures were normalized per 100 nucleotides (R). The pipeline followed in the prediction of pre-miRs is depicted in Fig. 1.

Prediction of lncRNAs

The coding DNA sequences (CDS) of C. cajan were used as the starting point for the prediction of lncRNAs. The CDS with length greater than 200 nucleotides [47] were retained and the ORFs were computed using the EMBOSS getorf standalone [48]. The ORFs with length less than 120 amino acids were retained for further analysis [47]. The coding potential for the sequences were checked by two different algorithms: Coding Potential Calculator (CPC), developed on support vector machine [49], and Coding Potential Assessment Tool (CPAT), which is an alignment-free algorithm [50]. Based on CPC score (S), sequences were classified into non-coding (S ≤ -0.5), neutral (-0.5 < S

< 1.0) and coding (S ≥ 1.0) [49]. The sequences classified as neutral were further checked by CPAT. Sequences having CPAT score < 0.2 were classified as ncRNAs [51]. The sequences were further searched using BLASTX [52] against the SWISS-PROT database [42] with an e-value cut-off of 0.001. Sequences with more than 40 % identity were removed, and the remaining sequences were selected as lncRNAs. The pipeline followed for the prediction is represented as a flowchart in Fig. 2.

Prediction of miRNA targets

The targets for mature miRNAs were predicted using psRNATarget server [53] by submitting the mature miRNAs as query and the CDS sequences of C. cajan as subject. To reduce the number of false predictions, the maximum expectation threshold was set to a stringent value of 2.0. The cut-off length of nucleotides for complementarity scoring, hspsize [54], was set as the length of the mature miRNAs. The maximum energy of unpairing (UPE) the target site was set as 25 kcal [54]. The flanking length around the target site was selected as 17 nucleotides upstream and 13 nucleotides downstream [55]. Due to the variable length of the mature miRNAs, the sequence range of the central mismatch was adjusted as described by Nithin, et al. [16]. To predict the function of the target sequences, the sequences were mapped to the UniProt proteome. The miRNAs, targeting the lncRNAs, were predicted by submitting the mature miRNAs as query and the lncRNAs of C. cajan as subject following the same pipeline. The interaction networks of miRNAs with target mRNAs were constructed using Cystoscape [56].

Results and Discussion

Identification of miRNAs

In this study, we have identified 616 miRNAs from the genome of C. cajan by following the prediction pipeline explained in the 'Materials and Methods' section. The method has been computationally validated by predicting the miRNAs of model plants *Arabidopsis thaliana* and *Glycine max*. In both the cases, we obtained high specificity and sensitivity [16]. Moreover, we experimentally validated 97 miRNAs, predicted using the pipeline, in a small RNA library prepared from *P. vulgaris* cv. Anupam [16]. Both *G. max* and *P. vulgaris* are members of Fabaceae family and are phylogenetically closely related species of C. cajan. Hence, we believe that our hypothesis for prediction of miRNAs and their targets will also holds for C. cajan.

The known miRNAs of Viridiplantae were BLAST searched against the genome of C. cajan with an e-value cut-off of 1000, allowing zero to three mismatches. The mismatches permitted during the BLAST search allows the identification of miRNAs, which are identical to known miRNAs but novel to the plant species that are not reported in the miRBase. From the BLAST search, a total of 1831779306

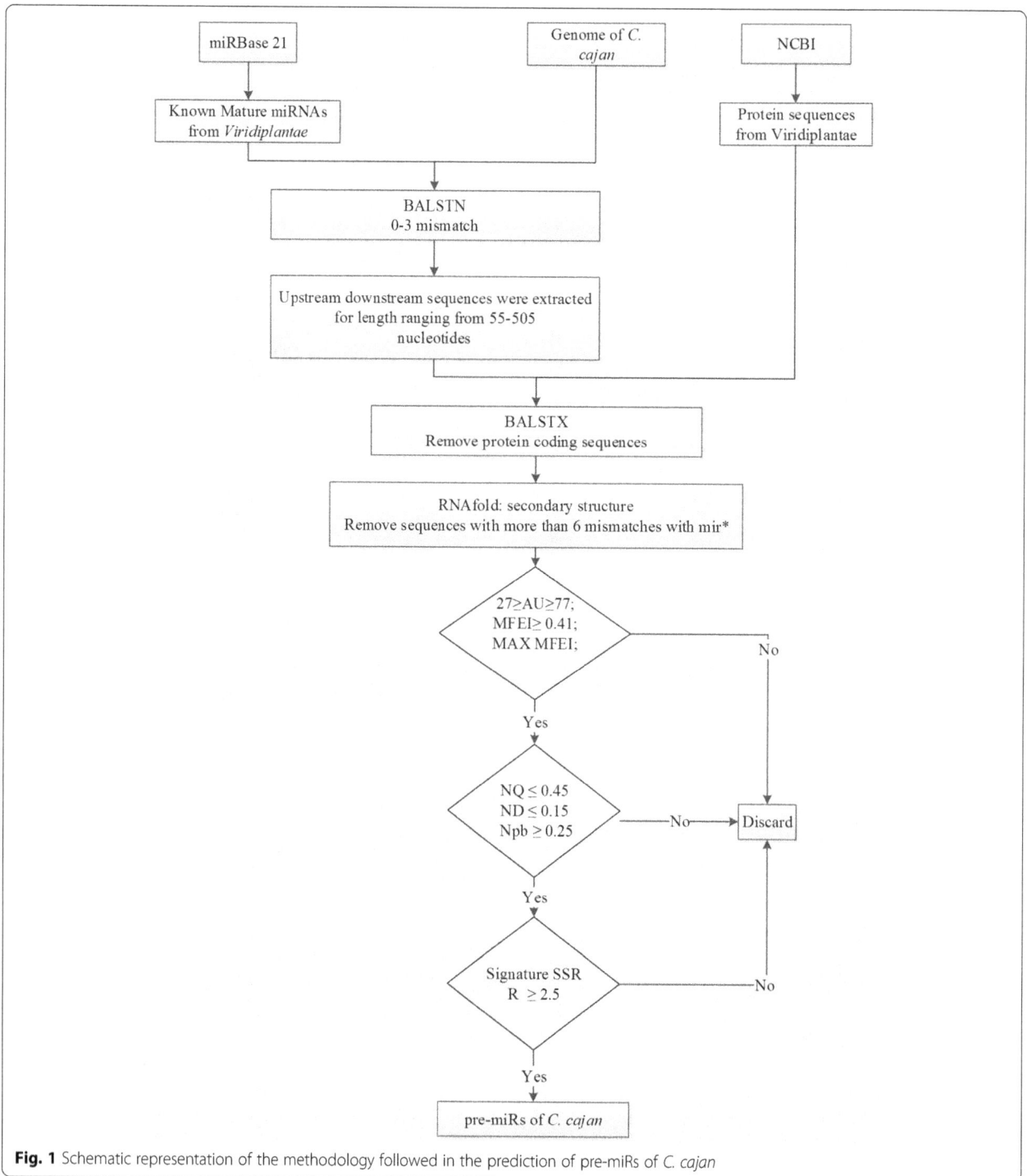

Fig. 1 Schematic representation of the methodology followed in the prediction of pre-miRs of *C. cajan*

sequences, that do not code for proteins were extracted with all possible lengths. In case of multiple sequences resulting from a single BLAST hit fulfilling the criteria, the one with the maximum MFEI and the maximum R was retained. A total of 616 miRNAs belonging to 341 miRNA families were identified by the prediction pipeline (Additional file 1: Table S1). A previous study by Kompelli, et al. [10] had identified only 142 miRNAs in *C. cajan*. This lower

number of predicted miRNA may be due to the fact that they have used a smaller search space in identifying miR-NAs. Of the miRNAs identified in this study, 578 are novel with respect to both plant miRNAs available in miRBase 21 as well as those identified by Kompelli, et al. [10].

The length of mature miRNAs of *C. cajan* varies from 15 to 24 nucleotides (nt) with an average of 20 nt (s.d. is ±1.4).

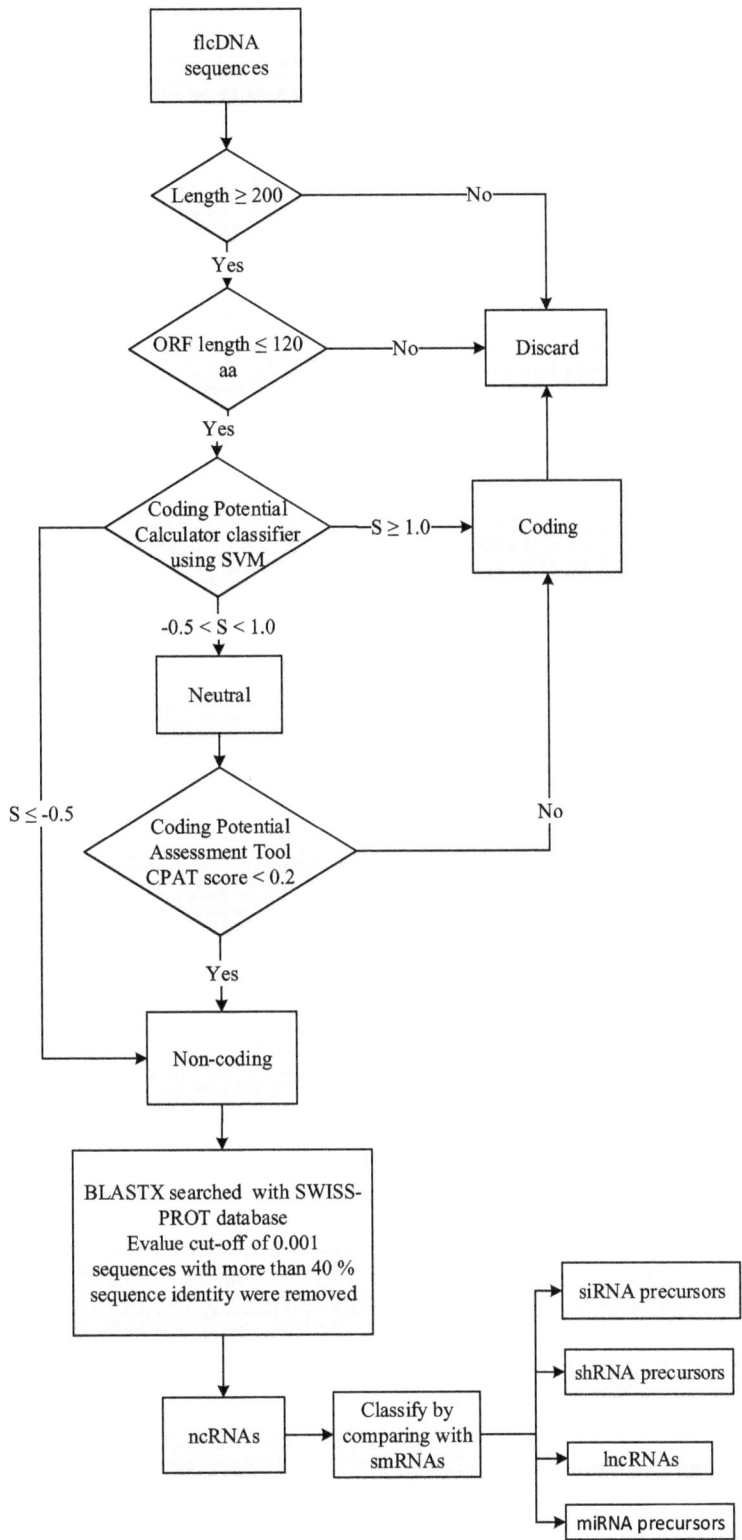

Fig. 2 Schematic representation of the methodology followed in the prediction of lncRNAs of *C. cajan*

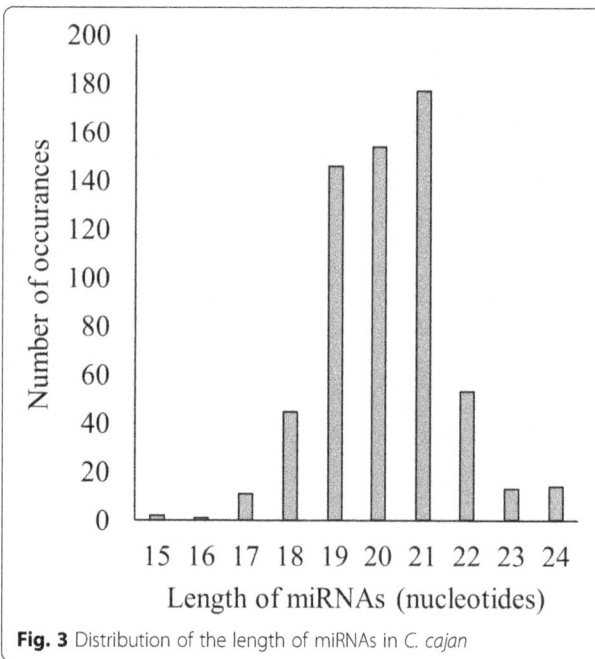

Fig. 3 Distribution of the length of miRNAs in *C. cajan*

Majority of them (93 %) falls within the range of 18 to 22 nt. Figure 3 shows the distribution of length of miRNAs identified in this study. MiRBase 21 has classified the 4800 plant miRNAs into 2290 families. The 616 miRNAs identified in this study belongs to 341 different families (Table 1). The distribution of miRNAs across various families is highly heterogeneous. Majority (85 %) of the families have only either one or two member(s). The highest number of members is observed in the miR171 family followed by miR477, miR169 and miR167 with 14, 10, 9 and 8 members, respectively. In the remaining 49 families, the number of miRNA varies from two to seven (Fig. 4). This distribution is in agreement with the diversity observed in other plant species [57]. Figure 5 shows the distribution of different miRNAs across the 11 chromosomes of *C. cajan*.

The SSR signatures in various miRNA families of the kingdom Viridiplantae, the family Fabaceae and the species *C. cajan* are presented in Table 2, while their relative distributions are shown in Fig. 6. We observe only 45 signature SSRs present in the miRNA families of *C. cajan*. The highest frequency is observed for AAU in Viridiplantae, Fabaceae and *C. cajan*. A total of 19 SSRs are absent in miRNA families of *C. cajan* while 11 of them are present only in one family. The signatures GUG and CCG are absent in other Fabaceae species while they are present in *C. cajan*. The former signature is present in miR1171 while the latter is present in miR2102 and miR5075 families.

miRNA targets on coding sequences

The mature miRNAs play a major role in the regulation of gene expression either by inhibiting translation or by degrading coding mRNAs [58, 59]. The number of targets for an miRNA may range from one to hundreds [60]. However, many mRNA targets in plants contain single miRNA-complementary site, which perfectly complement with the corresponding miRNAs and cleave the target [61]. We have used the psRNATarget server for the prediction of miRNA targets. Due to the absence of *C. cajan* target candidates in the psRNATarget server, the CDS sequences of *C. cajan* were used as target candidates. For 259 miRNAs, belonging to 180 families, 1373 target sequences were predicted. In order to characterise the targets, BLASTX was used with the predicted target sequences as query and the entire protein sequences of Viridiplantae as subject. Using 80% sequence identity cut-off, 298 targets for 122 miRNAs were characterised (Additional file 2: Table S2).

In majority of the cases, the predicted targets in this study were in accordance with the already published reports in other plant species. Wu, et al. [62] have showed that miR156 and miR172 families work in coordination to regulate the transition from the juvenile to the adult phase of plants. miR156 targets squamosa promoter binding protein-Like (SPL) transcription factor (TF) gene family to control the transition from the vegetative phase to the floral phase in *Arabidopsis*, rice and maize [63–68]. The cca-miR156b also targets SPL and is in agreement with the observation found in the literature. Members of the miR164 family target the NAC family of TF genes in *A. thaliana*, *Picea abies* and *Vitis vinifera* [69–73]. The NAC family of TFs play a major role in regulation of the boundary domain around developing primordia at the shoot apical and floral meristems [74]. cca-miR164e also targets NAC domain proteins. Scarecrow-like transcription factor is already an established target for miR171 family in *Arabidopsis* [75] and *Oryza sativa* [76]. Similar results were obtained in our study where cca-miR171b was predicted to bind Scarecrow-like (SCL) TF. SCL TFs are known to negatively regulate chlorophyll biosynthesis by suppressing the expression of the key gene PROTOCHLOROPHYLLIDE OXIDOREDUCTASE (POR). The miR172 family control plant development by regulating the trichome growth in *Arabidopsis* [62]. It is already established that MYB transcription factors are the negative controllers of the trichome growth [77]. The cca-miR172b family targets the MYB transcription factor mRNAs, and by cleaving these transcription factors they positively control the trichome growth. miR172 functions in regulating the transitions between developmental stages and in specifying floral organ identity. During flower development, miRNA172 represses the expression of APETALA2 (AP2) [78]. This regulation is crucial for the proper development of the reproductive organs and for the timely termination of floral stem cells [79]. The cca-miR172c targets floral homeotic protein AP2. The cca-miR397a targets laccase (LAC) enzymes, and is in agreement with established targets of miR397 family in *A. thaliana*, *Populus trichocarpa* and *O. sativa*. In rice, it is reported that the miR397 overexpression leads to greater

Table 1 Distribution of miRNAs across various miRNA families.

Number of members	miRNA families	Number of miRNA families
1	miR403, miR417, miR444, miR476, miR478, miR535, miR771, miR774, miR781, miR816, miR825, miR827, miR831, miR835, miR838, miR854, miR857, miR900, miR952, miR1025, miR1039, miR1061, miR1087, miR1088, miR1097, miR1128, miR1130, miR1153, miR1171, miR1217, miR1426, miR1428, miR1430, miR1439, miR1446, miR1507, miR1510, miR1518, miR1520, miR1535, miR1852, miR1854, miR1916, miR1917, miR2055, miR2079, miR2086, miR2090, miR2101, miR2102, miR2105, miR2108, miR2118, miR2119, miR2199, miR2275, miR2600, miR2604, miR2608, miR2611, miR2642, miR2646, miR2657, miR2671, miR2866, miR2871, miR2878, miR2905, miR2912, miR2920, miR2923, miR2928, miR3433, miR3436, miR3438, miR3441, miR3444, miR3447, miR3512, miR3515, miR3522, miR3626, miR3627, miR3629, miR3630, miR3631, miR3704, miR3712, miR3950, miR3979, miR4223, miR4237, miR4238, miR4244, miR4249, miR4340, miR4414, miR5039, miR5055, miR5075, miR5139, miR5140, miR5171, miR5174, miR5183, miR5201, miR5205, miR5219, miR5234, miR5237, miR5248, miR5253, miR5261, miR5264, miR5265, miR5285, miR5288, miR5291, miR5292, miR5368, miR5372, miR5373, miR5374, miR5379, miR5382, miR5521, miR5523, miR5532, miR5555, miR5558, miR5561, miR5668, miR5672, miR5716, miR5722, miR5745, miR5757, miR5770, miR5773, miR5775, miR5778, miR5828, miR5837, miR6034, miR6111, miR6140, miR6148, miR6173, miR6182, miR6191, miR6196, miR6230, miR6231, miR6271, miR6291, miR6443, miR6449, miR6457, miR6459, miR6462, miR6466, miR6476, miR6478, miR6483, miR6485, miR7124, miR7125, miR7127, miR7484, miR7488, miR7508, miR7516, miR7532, miR7534, miR7540, miR7545, miR7696, miR7728, miR7736, miR7741, miR7745, miR7753, miR7757, miR7767, miR7812, miR7814, miR7816, miR7817, miR7982, miR8007, miR8014, miR8030, miR8035, miR8044, miR8047, miR8049, miR8123	197
2	miR158, miR160, miR161, miR162, miR164, miR168, miR394, miR397, miR398, miR408, miR414, miR419, miR530, miR837, miR846, miR862, miR868, miR1023, miR1027, miR1030, miR1046, miR1051, miR1134, miR1320, miR1508, miR1511, miR1512, miR1514, miR1525, miR1527, miR1533, miR1534, miR2089, miR2093, miR2595, miR2606, miR2607, miR2628, miR2630, miR2641, miR2650, miR2655, miR2665, miR2673, miR2868, miR3434, miR3711, miR3951, miR4233, miR4246, miR4248, miR4371, miR4376, miR4413, miR4415, miR5040, miR5041, miR5054, miR5057, miR5142, miR5240, miR5255, miR5256, miR5257, miR5260, miR5281, miR5369, miR5512, miR5559, miR5565, miR5712, miR5721, miR5741, miR6135, miR6169, miR6202, miR6218, miR6232, miR6281, miR6299, miR6300, miR6464, miR7535, miR7543, miR7742, miR7776, miR7822, miR7823, miR7834, miR8051, miR8140	91
3	miR390, miR395, miR400, miR437, miR828, miR829, miR859, miR860, miR902, miR1044, miR1438, miR1521, miR1863, miR2111, miR2676, miR2873, miR2931, miR4245, miR5163, miR6025, miR6470, miR7699, miR8011, miR8041	24
4	miR159, miR172, miR393, miR399, miR1516, miR2592, miR3513, miR5031, miR7701, miR8005, miR8040	11
5	miR166, miR821, miR1078, miR1435, miR1522, miR5185, miR6288	7
6	miR319, miR396, miR1530, miR5568	4
7	miR156, miR482, miR845	3
8	miR167	1
9	miR169	1
10	miR477	1
14	miR1710	1

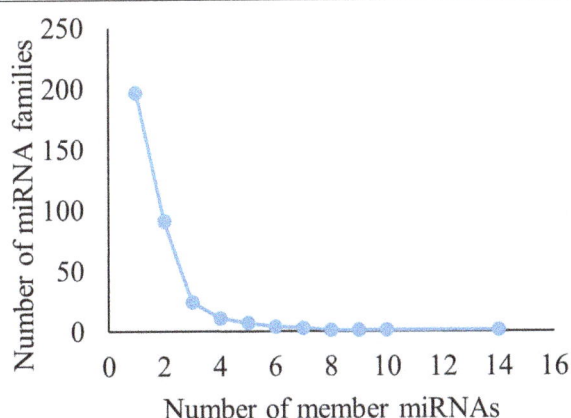

Fig. 4 Frequency distribution of miRNAs across the miRNA families in C. cajan

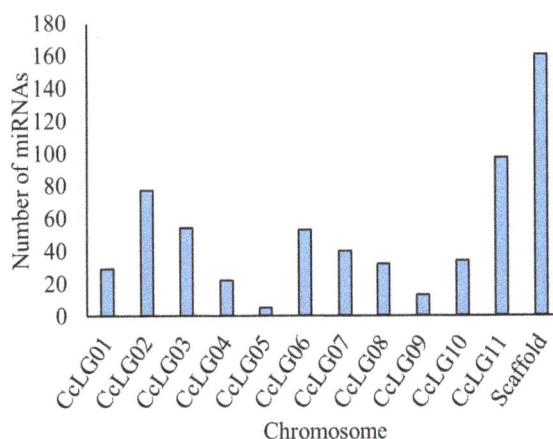

Fig. 5 Distribution of miRNAs across different chromosomes of C. cajan

Table 2 Distribution of SSR signatures in various miRNA families of Viridiplantae, Fabaceae and *C. cajan*

	A			C			G			U			
	V[a]	F[b]	C[c]	V[a]	F[b]	C[c]	V[a]	F[b]	C[c]	V[a]	F[b]	C[c]	
A	4.92	4.44	1.36	1.32	1.82	0.09	1.48	2.02	0.68	2.96	1.41	0.85	A
	1.43	1.41	0.17	0.63	0.61	0.09	1.06	0.40	0.17	2.38	3.03	0.60	C
	3.07	4.04	0.77	0.37	0.20	0.00	0.69	0.20	0.09	3.91	2.83	1.36	G
	7.45	9.70	2.39	0.79	1.01	0.09	0.79	0.61	0.09	8.77	10.71	4.43	U
C	1.22	1.01	0.51	0.79	0.61	0.00	0.32	0.20	0.00	0.37	0.00	0.00	A
	0.26	0.20	0.00	0.05	0.00	0.00	0.69	0.00	0.00	0.79	0.20	0.00	C
	0.37	0.20	0.00	0.63	0.00	0.17	0.74	0.40	0.09	0.63	0.40	0.17	G
	1.90	2.22	0.51	0.32	0.40	0.00	0.37	0.81	0.00	2.11	2.22	0.60	U
G	1.48	2.42	0.34	0.69	0.40	0.17	0.79	0.81	0.09	0.32	0.00	0.00	A
	0.16	0.40	0.00	0.74	0.20	0.00	0.90	0.20	0.00	0.26	0.20	0.00	C
	0.58	0.20	0.26	0.69	0.20	0.00	0.21	0.00	0.00	0.16	0.00	0.09	G
	1.59	1.62	0.51	0.58	0.81	0.17	0.42	0.40	0.09	0.95	1.41	0.26	U
U	2.01	1.82	1.02	1.80	2.63	0.26	2.70	3.03	1.28	2.17	3.23	0.60	A
	0.21	0.00	0.00	0.85	0.61	0.09	1.48	0.40	0.34	2.27	3.03	0.51	C
	0.58	0.61	0.17	0.58	0.81	0.09	1.53	1.62	0.26	6.18	6.46	2.22	G
	2.59	2.83	0.94	2.17	1.82	1.02	2.48	1.62	1.02	6.29	6.87	2.05	U

V[a]- The percentage of miRNA families belonging to Viridiplantae with a particular signature SSR
F[b]- The percentage of miRNA families belonging to Fabaceae with a particular signature SSR. The data for V[a] and F[b] are taken from our previous study [16]
C[c]- The percentage of miRNA families belonging to *C. cajan* with a particular signature SSR. There are 118 miRNA families to which *C. cajan* miRNAs belong

number of branches, increased number of grains per main panicle, increased grain size and substantially enhanced grain yield. In case of *A. thaliana*, overexpression of miR397b causes a reduction in lignification of vascular and interfascicular tissue as well as an increase in inflorescence shoots number and seed size.

The UniProt proteome, UP000075243, was used to map the target mRNAs and retrieve the corresponding UniProt

protein identifiers. Of the 1373 targets, 1312 were mapped to the proteome. The visualization of these targets is provided as an interaction network between the miRNA and the corresponding UniProt entry in Fig. 7(a). The network consists of 1525 nodes. The number of targets ranges from 1 to 111 mRNAs (Fig. 7(b)). The highest number of targets is observed in cca-miR8123a. The characterized targets for cca-miR8123a includes mRNAs coding for ribosomal

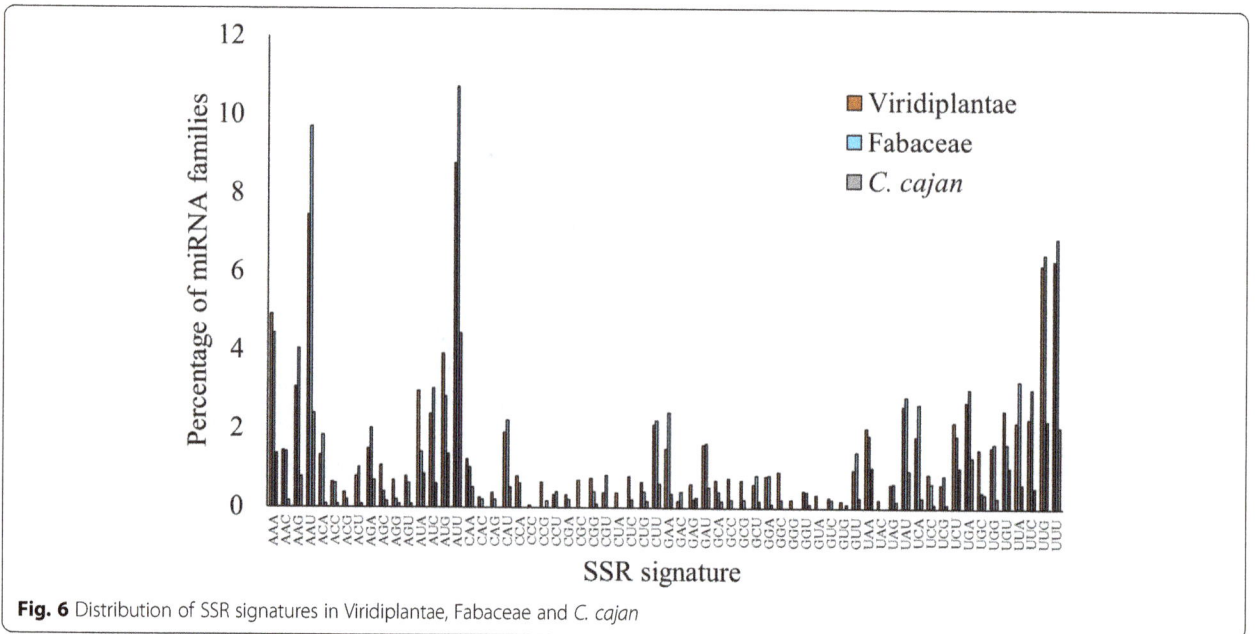

Fig. 6 Distribution of SSR signatures in Viridiplantae, Fabaceae and *C. cajan*

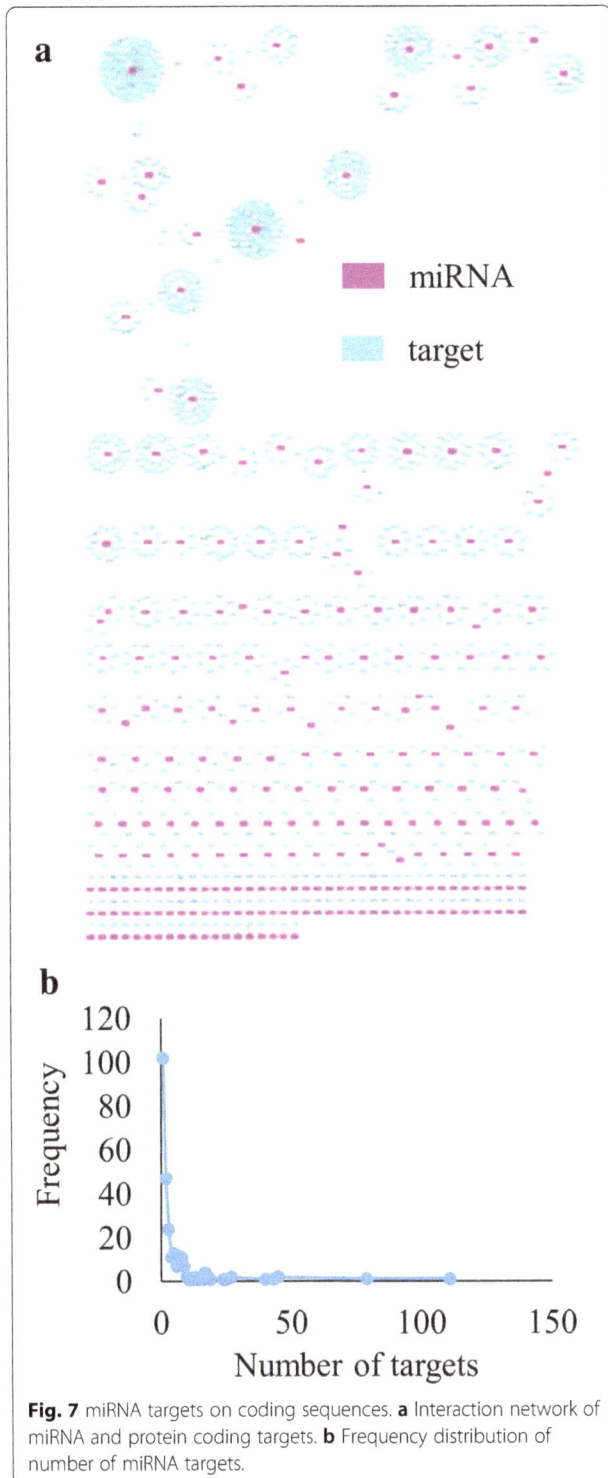

Fig. 7 miRNA targets on coding sequences. **a** Interaction network of miRNA and protein coding targets. **b** Frequency distribution of number of miRNA targets.

proteins, hydrolases, kinases and CLIP-associated proteins. There are 102 miRNAs which target only one mRNA. For example, cca-miR171k targets the mRNA, which codes for F-box protein.

The GO annotations for the targets were taken from UniProt-GOA. The biological processes, molecular functions and cellular components of the targets are shown in Fig. 8. Under Biological process, majority of the targets (60 %) are involved in the metabolic and cellular process (Fig. 8(a)). Around 10 % of the targets are responsible for the response to stimuli, while 8.5 % are involved in the regulation of biological process. The remaining (21.5 %) are involved in a plethora of processes including reproduction, development, component organization, localization and other cellular processes. The molecular functions performed by the targets cover almost all aspects of plant metabolism (Fig. 8(b)). Majority of the targets perform functions in binding (52.5 %) and catalytic activity (37.5 %). The remaining 10 % functions in nutrient reservation, transportation, signal reception and transduction, transportation and regulatory activities. The proteins coded by miRNA targets localize in different cellular components (Fig. 8(c)). A large number of proteins localize in membrane and membrane parts (43.7 %), protoplasm (24.5 %), cell organelles (19.0 %), macromolecular complexes (6.4 %) and extracellular region (5.3 %). The remaining (1.1 %) are localized in microtubules, virion parts and other regions in cell.

Prediction of lncRNAs

The full length cDNA sequences of *C. cajan* were used as the starting point for predicting the lncRNAs. The sequences longer than 200 nucleotides and does not have an ORF coding for more than 120 residues were only selected as the input for prediction pipeline. The coding potential of these sequences were used as a measure to remove the potential protein coding sequences and to retain the non-coding sequences. A total of 3919 lncRNAs were predicted by this pipeline.

lncRNAs have emerged as important regulators of gene expression in a variety of biological processes in multiple species. lncRNAs are increasingly recognized as functional regulatory components in eukaryotic gene regulation. In plants, they are transcribed by different RNA polymerases and show diverse structural features. Recent studies have showed that the lncRNAs play a major role in growth and cell differentiation [80], phosphate homeostasis [81], chromatin modification [82, 83] and protein re-localization [84, 85]. Three major mechanisms of action are mainly proposed for the functioning of lncRNAs: decoys, scaffolds and guides [86]. lncRNAs act as decoys that prohibit the access of regulatory proteins to DNA. They also act as adaptors to bring two or more proteins into discrete complexes and guides in localizing specific proteins [87]. The miRNA target

proteins, chaperones, kinases, transporters, receptors, signal transducers, ubiquitination proteins and spliceosomal RNAs (Additional file 2 Table S2). Another major node in the interaction network is cca-miR902a with 79 targets. The targets include mRNAs coding for RNA polymerases, kinases, U-box proteins, methyltransferase, retrotransposon

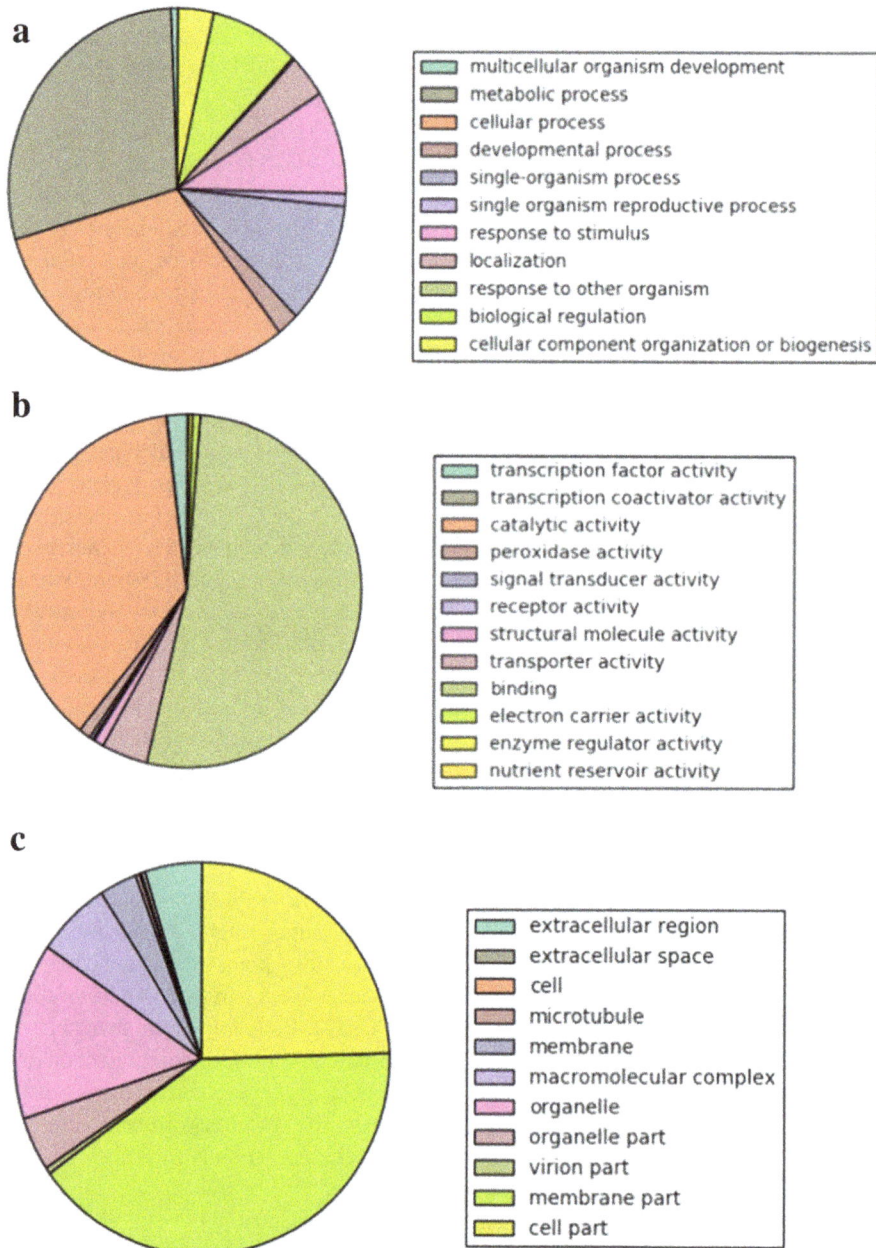

Fig. 8 Targets of miRNA distributed among three different GO terms: (**a**) Biological processes, (**b**) Molecular functions and (**c**) Cellular components

mimicking by lncRNA can be exemplified with *Induced by Phosphate Starvation1* (*IPS1*) lncRNA, which has a stretch of 23 conserved nucleotides that is partially complementary to miR399. The *IPS1* acts as a non-cleavable target mimic for miR399 in *Medicago truncatula* [88], rice [89] and *Arabidopsis* [90, 91]. Chromatin remodelling is demonstrated by the action of two classes of lncRNAs identified in the regulation of *FLC* (*Flowering Locus C*) expression. *FLC* is a floral repressor, which is repressed during the process of vernalization and it is mediated by polycomb repressive complex *PRC2*, which

is a repressive chromatin modifier. Two classes of lncRNAs – cold induced antisense intragenic RNA (*COOLAIR*) and cold assisted intronic non-coding RNA (*COLDAIR*) are involved in this process of stable silencing of FLC [82, 92–94]. The transcription of *COOLAIR* is repressed in warm temperatures by stabilization of a RNA-DNA hybrid structure (R-loop) in its promoter region [95]. The COLDAIR is involved in the enrichment of H3K27me3 by direct interaction with CURLY LEAF (CLF), which is a component of *PRC2*, thereby repressing *FLC* [82].

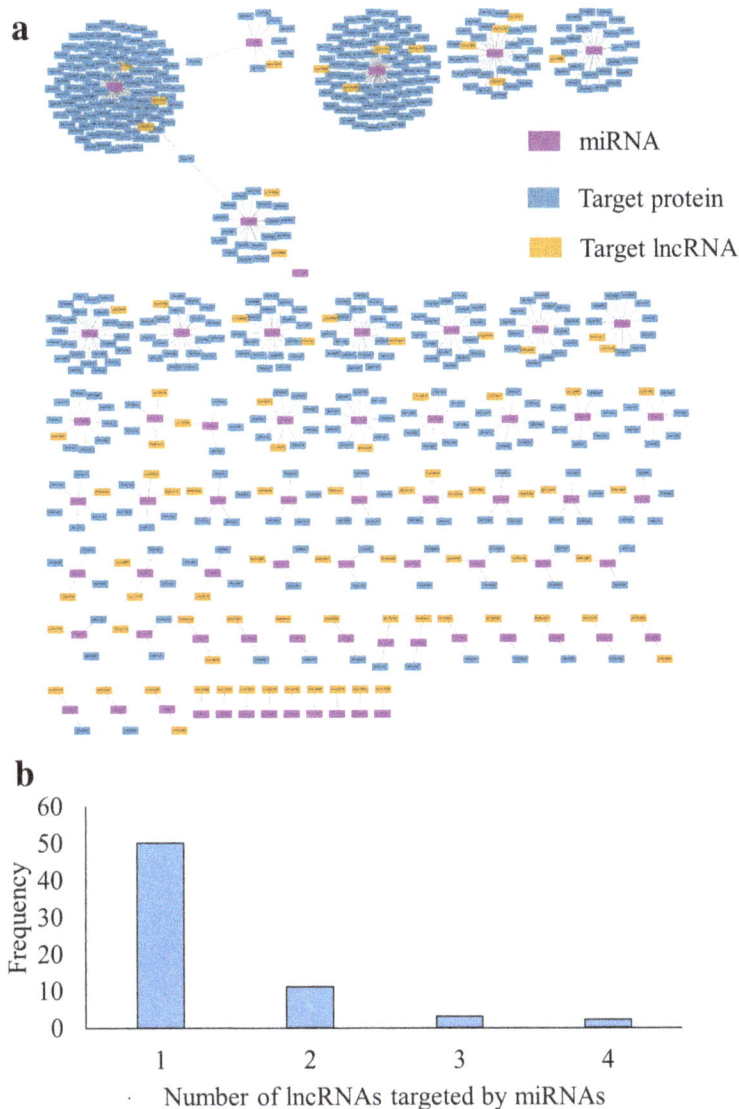

Fig. 9 miRNA targets on lncRNAs. **a** Interaction network of miRNA and their coding and non-coding targets. (**b**) Frequency distribution of number of miRNA targets on lncRNAs

Prediction of miRNA targets on lncRNAs

In this study, we have identified both the miRNAs and lncRNAs belonging to *C. cajan*. In order to study the direct targeting of lncRNAs by miRNAs, we have identified the targets of lncRNAs on miRNAs using psRNATarget server. A total of 66 miRNAs were identified to target 87 lncRNAs. The details of miRNAs, targeting lncRNAs, are available in the Additional file 3: Table S3. The interaction network of miRNAs that target lncRNAs is shown in Fig. 9(a). The network consists of 665 nodes. The number of lncRNAs targeted by a single miRNA varies from one to four, with a majority of them (76 %) targeting only one (Fig. 9(b)). cca-miR3979a and cca-miR902a targets four lncRNAs. These

miRNAs also have relatively higher number of protein targets, 26 and 79 respectively. cca-miR8123a, which has the highest number of proteins targets, has three lncRNAs as targets. cca-miR1527a and cca-miR403a has three target lncRNAs, however, both of them target two proteins each.

Conclusion

In the present study, we have identified the miRNAs from the genome of *C. cajan* and their corresponding targets. A total of 616 miRNAs belonging to 341 different families were identified. Of the identified miRNAs, 578 are novel that are not reported in the MiRBase 21. We have also identified 1379 targets for 259 miRNAs, of which 298 were characterized at protein level. Moreover, we have identified 3919 lncRNAs of *C. cajan*, 87 of which are found to be targeted by 66 miRNAs. It is well known that ncRNAs and their target mimics has the potential to be used for crop improvement programmes as proper management of them can generate crop cultivars with improved agronomic traits leading to increased yield and high nutritional value. Thorough understanding of interaction of miRNAs and their targets can provide valuable insight into molecular pathways controlling plant stress responses. Accordingly, our findings will enhance the knowledge of ncRNAs in economically important pulse crop *C. cajan* and their role in PTGR, will contribute in genome editing and thereby development of better crop varieties.

Abbreviations

PTGR: post transcriptional regulation of gene expression; ncRNAs: non-coding RNAs; sncRNAs: small non-coding RNAs; lncRNAs: long non-coding RNAs; miRNAs: microRNAs; ceRNAs: competing endogenous RNAs; pre-miRs: precursor miRNAs; MFEI: Minimal Folding Free Energy Index; NQ: normalised Shannon entropy; Npb: normalized base-pairing propensity; ND: normalized base-pair distance; SSR: simple sequence repeat; CDS: coding DNA sequences; ORF: open reading frame

Acknowledgements

CN is thankful to the IIT Kharagpur for research fellowship. RPB and JB acknowledge DBT, India.

Funding

None.

Authors' contributions

RPB and JB conceived the study and participated in its design and coordination. CN and AT performed the study. CN, AT, JB and RPB wrote the manuscript.

Competing interests

The authors declare that they have no competing interests.

Author details

[1]Computational Structural Biology Lab, Department of Biotechnology, Indian Institute of Technology Kharagpur, Kharagpur, India. [2]Department of Biotechnology, Visva-Bharati, Santiniketan, India. [3]Present address: Molecular and Computational Biology, Department of Biological Sciences, University of Southern California, Los Angeles, California, USA.

References

1. Singh U, Eggum BO. Factors affecting the protein quality of pigeonpea (Cajanus cajan L). Plant Foods for Human Nutrition. 1984;34(4):273–83.
2. Sharma S, Agarwal N, Verma P: Pigeon pea (Cajanus cajan L.): A Hidden Treasure of Regime Nutrition. Journal of Functional And Environmental Botany 2011, 1(2):91-101.
3. Nene Y, Hall SD, Sheila V: The pigeonpea: CAB International; 1990.
4. Flower D, Ludlow M. Variation among accessions of pigeonpea (Cajanus cajan) in osmotic adjustment and dehydration tolerance of leaves. Field Crops Res. 1987;17(3):229–43.
5. Subbarao GV, Chauhan YS, Johansen C. Patterns of osmotic adjustment in pigeonpea - its importance as a mechanism of drought resistance. Eur J Agron. 2000;12(3-4):239–49.
6. Sinha SK: Food legumes: distribution, adaptability and biology of yield: FAO; 1977.
7. Saxena K, Kumar R, Rao P. Pigeonpea nutrition and its improvement. J Crop Prod. 2002;5(1-2):227–60.
8. Saxena KB, Kumar RV, Sultana R. Quality nutrition through pigeonpea—a review. Health. 2010;2(11):1335.
9. Pal D, Mishra P, Sachan N, Ghosh A: Biological activities and medicinal properties of Cajanus cajan (L) Millsp, vol. 2; 2011.
10. Kompelli SK, Kompelli VSP, Enjala C, Suravajhala P. Genome-wide identification of miRNAs in pigeonpea (Cajanus cajan L). Aust J Crop Sci. 2015;9(3):215–22.
11. Kozomara A, Griffiths-Jones S. miRBase: annotating high confidence microRNAs using deep sequencing data. Nucleic Acids Res. 2014;42(D1):D68–73.
12. Chi X, Yang Q, Chen X, Wang J, Pan L, Chen M, Yang Z, He Y, Liang X, Yu S. Identification and characterization of microRNAs from peanut (Arachis hypogaea L.) by high-throughput sequencing. PloS one. 2011;6(11):e27530.
13. Guo N, Ye W, Yan Q, Huang J, Wu Y, Shen D, Gai J, Dou D, Xing H. Computational identification of novel microRNAs and targets in Glycine max. Mol Biol Rep. 2014;41(8):4965–75.
14. Hu J, Sun L, Ding Y. Identification of conserved microRNAs and their targets in chickpea (Cicer arietinum L). Plant Signal Behav. 2013;8(4):e23604.
15. Hu J, Zhang H, Ding Y. Identification of conserved microRNAs and their targets in the model legume Lotus japonicus. J Biotechnol. 2013;164(4):520–4.
16. Nithin C, Patwa N, Thomas A, Bahadur RP, Basak J. Computational prediction of miRNAs and their targets in Phaseolus vulgaris using simple sequence repeat signatures. BMC Plant Biol. 2015;15:140.
17. Zhu J, Li W, Yang W, Qi L, Han S. Identification of microRNAs in Caragana intermedia by high-throughput sequencing and expression analysis of 12 microRNAs and their targets under salt stress. Plant Cell Rep. 2013;32(9): 1339–49.
18. Zhang BH, Pan XP, Cox SB, Cobb GP, Anderson TA. Evidence that miRNAs are different from other RNAs. Cell Mol Life Sci. 2006;63(2):246–54.
19. Ng Kwang Loong S, Mishra SK. Unique folding of precursor microRNAs: quantitative evidence and implications for de novo identification. RNA. 2007;13(2):170–87.
20. Chen M, Tan Z, Jiang J, Li M, Chen H, Shen G, Yu R. Similar distribution of simple sequence repeats in diverse completed Human Immunodeficiency Virus Type 1 genomes. FEBS Lett. 2009;583(17):2959–63.
21. Lipshitz H, Peattie D, Hogness D. Novel transcripts from the Ultrabithorax domain of the bithorax complex. Genes Dev. 1987;1(3):307–22.
22. Derrien T, Johnson R, Bussotti G, Tanzer A, Djebali S, Tilgner H, Guernec G, Martin D, Merkel A, Knowles DG, et al. The GENCODE v7 catalog of human long noncoding RNAs: Analysis of their gene structure, evolution, and expression. Genome Res. 2012;22(9):1775–89.
23. Dieci G, Fiorino G, Castelnuovo M, Teichmann M, Pagano A. The expanding RNA polymerase III transcriptome. Trends Genet. 2007;23(12):614–22.
24. The FANTOM Consortium, Carninci P, Kasukawa T, Katayama S, Gough J, Frith MC, Maeda N, Oyama R, Ravasi T, Lenhard B, et al. The Transcriptional Landscape of the Mammalian Genome. Science. 2005;309(5740):1559–63.
25. Louro R, El-Jundi T, Nakaya HI, Reis EM, Verjovski-Almeida S. Conserved tissue expression signatures of intronic noncoding RNAs transcribed from human and mouse loci. Genomics. 2008;92(1):18–25.
26. Mercer TR, Dinger ME, Sunkin SM, Mehler MF, Mattick JS. Specific expression of long noncoding RNAs in the mouse brain. Proc Natl Acad Sci USA. 2008; 105(2):716–21.
27. Nie L, HJ W, Hsu JM, Chang SS, Labaff AM, Li CW, Wang Y, Hsu JL, Hung MC. Long non-coding RNAs: versatile master regulators of gene expression and crucial players in cancer. Am J Transl Res. 2012;4(2):127–50.

28. Mazo A, Hodgson JW, Petruk S, Sedkov Y, Brock HW. Transcriptional interference: an unexpected layer of complexity in gene regulation. J Cell Sci. 2007;120(16):2755–61.

29. Hirota K, Miyoshi T, Kugou K, Hoffman CS, Shibata T, Ohta K. Stepwise chromatin remodelling by a cascade of transcription initiation of non-coding RNAs. Nature. 2008;456(7218):130–4.

30. Martianov I, Ramadass A, Serra Barros A, Chow N, Akoulitchev A. Repression of the human dihydrofolate reductase gene by a non-coding interfering transcript. Nature. 2007;445(7128):666–70.

31. Mariner PD, Walters RD, Espinoza CA, Drullinger LF, Wagner SD, Kugel JF, Goodrich JA. Human Alu RNA Is a Modular Transacting Repressor of mRNA Transcription during Heat Shock. Mol Cell. 2008;29(4):499–509.

32. Nguyen VT, Kiss T, Michels AA, Bensaude O. 7SK small nuclear RNA binds to and inhibits the activity of CDK9/cyclin T complexes. Nature. 2001;414(6861): 322–5.

33. Willingham AT, Orth AP, Batalov S, Peters EC, Wen BG, Aza-Blanc P, Hogenesch JB, Schultz PG. A Strategy for Probing the Function of Noncoding RNAs Finds a Repressor of NFAT. Science. 2005;309(5740):1570–3.

34. Beltran M, Puig I, Peña C, García JM, Álvarez AB, Peña R, Bonilla F, de Herreros AG. A natural antisense transcript regulates Zeb2/Sip1 gene expression during Snail1-induced epithelial–mesenchymal transition. Genes Dev. 2008;22(6):756–69.

35. Golden DE, Gerbasi VR, Sontheimer EJ. An Inside Job for siRNAs. Mol Cell. 2008;31(3):309–12.

36. Kozomara A. Griffiths-Jones S: miRBase: annotating high confidence microRNAs using deep sequencing data. Nucleic Acids Res. 2014; 42(Database issue):D68–73.

37. Varshney RK, Chen WB, Li YP, Bharti AK, Saxena RK, Schlueter JA, Donoghue MTA, Azam S, Fan GY, Whaley AM, et al. Draft genome sequence of pigeonpea (Cajanus cajan), an orphan legume crop of resource-poor farmers. Nat Biotechnol. 2012;30(1):83–U128.

38. Schmutz J, McClean PE, Mamidi S, GA W, Cannon SB, Grimwood J, Jenkins J, Shu S, Song Q, Chavarro C, et al. A reference genome for common bean and genome-wide analysis of dual domestications. Nat Genet. 2014;46(7):707–13.

39. NCBI. Resource Coordinators: Database resources of the National Center for Biotechnology Information. Nucleic Acids Res. 2016;44(D1):D7–D19.

40. The UniProt Consortium. UniProt: the universal protein knowledgebase. Nucleic Acids Res. 2017;45(D1):D158–69.

41. Huntley RP, Sawford T, Mutowo-Meullenet P, Shypitsyna A, Bonilla C, Martin MJ, O'Donovan C. The GOA database: Gene Ontology annotation updates for 2015. Nucleic Acids Res. 2015;43(D1):D1057–63.

42. O'Donovan C, Martin MJ, Gattiker A, Gasteiger E, Bairoch A, Apweiler R. High-quality protein knowledge resource: SWISS-PROT and TrEMBL. Briefings Bioinf. 2002;3(3):275–84.

43. Camacho C, Coulouris G, Avagyan V, Ma N, Papadopoulos J, Bealer K, Madden TL. BLAST+: architecture and applications. BMC Bioinformatics. 2009;10(1):421.

44. Schultes EA, Hraber PT, LaBean TH. Estimating the contributions of selection and self-organization in RNA secondary structure. J Mol Evol. 1999;49(1):76–83.

45. Huynen M, Gutell R, Konings D. Assessing the reliability of RNA folding using statistical mechanics. J Mol Biol. 1997;267(5):1104–12.

46. Moulton V, Zuker M, Steel M, Pointon R, Penny D. Metrics on RNA secondary structures. J Comput Biol. 2000;7(1-2):277–92.

47. Boerner S, McGinnis KM. Computational Identification and Functional Predictions of Long Noncoding RNA in Zea mays. PloS one. 2012;7(8):e43047.

48. Rice P, Longden I, Bleasby A. EMBOSS: The European Molecular Biology Open Software Suite. Trends Genet. 2000;16(6):276–7.

49. Kong L, Zhang Y, Ye Z-Q, Liu X-Q, Zhao S-Q, Wei L, Gao G. CPC: assess the protein-coding potential of transcripts using sequence features and support vector machine. Nucleic Acids Res. 2007;35(suppl 2):W345–9.

50. Wang L, Park HJ, Dasari S, Wang S, Kocher J-P, Li W. CPAT: Coding-Potential Assessment Tool using an alignment-free logistic regression model. Nucleic Acids Res. 2013;41(6):e74.

51. Weikard R, Hadlich F, Kuehn C. Identification of novel transcripts and noncoding RNAs in bovine skin by deep next generation sequencing. BMC Genomics. 2013;14(1):789.

52. Altschul SF, Gish W, Miller W, Myers EW, Lipman DJ. Basic local alignment search tool. J Mol Biol. 1990;215(3):403–10.

53. Dai X, Zhao PX. psRNATarget: a plant small RNA target analysis server. Nucleic Acids Res. 2011;39(Web Server issue):W155–9.

54. Zhang Y: miRU: an automated plant miRNA target prediction server. Nucleic Acids Res 2005, 33(Web Server issue):W701-704.

55. Kertesz M, Iovino N, Unnerstall U, Gaul U, Segal E. The role of site accessibility in microRNA target recognition. Nat Genet. 2007;39(10):1278–84.

56. Shannon P, Markiel A, Ozier O, Baliga NS, Wang JT, Ramage D, Amin N, Schwikowski B, Ideker T. Cytoscape: a software environment for integrated models of biomolecular interaction networks. Genome Res. 2003;13(11):2498–504.

57. Zhang B, Pan X, Cannon CH, Cobb GP, Anderson TA. Conservation and divergence of plant microRNA genes. Plant J. 2006;46(2):243–59.

58. Carrington JC, Ambros V. Role of microRNAs in plant and animal development. Science. 2003;301(5631):336–8.

59. Djuranovic S, Nahvi A, Green R. A Parsimonious Model for Gene Regulation by miRNAs. Science. 2011;331(6017):550–3.

60. Brennecke J, Stark A, Russell RB, Cohen SM. Principles of MicroRNA–Target Recognition. PLoS Biol. 2005;3(3):e85.

61. Kidner CA, Martienssen RA. The developmental role of microRNA in plants. Curr Opin Plant Biol. 2005;8(1):38–44.

62. Wu G, Park MY, Conway SR, Wang JW, Weigel D, Poethig RS. The sequential action of miR156 and miR172 regulates developmental timing in Arabidopsis. Cell. 2009;138(4):750–9.

63. Chuck G, Cigan AM, Saeteurn K, Hake S. The heterochronic maize mutant Corngrass1 results from overexpression of a tandem microRNA. Nat Genet. 2007;39(4):544–9.

64. Gandikota M, Birkenbihl RP, Höhmann S, Cardon GH, Saedler H, Huijser P. The miRNA156/157 recognition element in the 3' UTR of the Arabidopsis SBP box gene SPL3 prevents early flowering by translational inhibition in seedlings. Plant J. 2007;49(4):683–93.

65. Jiao Y, Wang Y, Xue D, Wang J, Yan M, Liu G, Dong G, Zeng D, Lu Z, Zhu X, et al. Regulation of OsSPL14 by OsmiR156 defines ideal plant architecture in rice. Nat Genet. 2010;42(6):541–4.

66. Miura K, Ikeda M, Matsubara A, Song X-J, Ito M, Asano K, Matsuoka M, Kitano H, Ashikari M. OsSPL14 promotes panicle branching and higher grain productivity in rice. Nat Genet. 2010;42(6):545–9.

67. Yang L, Conway SR, Poethig RS. Vegetative phase change is mediated by a leaf-derived signal that represses the transcription of miR156. Development. 2011;138(2):245–9.

68. Yamaguchi A, M-F W, Yang L, Wu G, Poethig RS, Wagner D. The MicroRNA-Regulated SBP-Box Transcription Factor SPL3 Is a Direct Upstream Activator of LEAFY, FRUITFULL, and APETALA1. Dev Cell. 2009;17(2):268–78.

69. Larsson E, Sundström JF, Sitbon F, von Arnold S. Expression of PaNAC01, a Picea abies CUP-SHAPED COTYLEDON orthologue, is regulated by polar auxin transport and associated with differentiation of the shoot apical meristem and formation of separated cotyledons. Ann Bot. 2012;110(4):923–34.

70. Raman S, Greb T, Peaucelle A, Blein T, Laufs P, Theres K. Interplay of miR164, CUP-SHAPED COTYLEDON genes and LATERAL SUPPRESSOR controls axillary meristem formation in Arabidopsis thaliana. Plant J. 2008;55(1):65–76.

71. Kim JH, Woo HR, Kim J, Lim PO, Lee IC, Choi SH, Hwang D, Nam HG. Trifurcate Feed-Forward Regulation of Age-Dependent Cell Death Involving miR164 in Arabidopsis. Science. 2009;323(5917):1053–7.

72. Laufs P, Peaucelle A, Morin H, Traas J. MicroRNA regulation of the CUC genes is required for boundary size control in Arabidopsis meristems. Development. 2004;131(17):4311–22.

73. Sun G. MicroRNAs and their diverse functions in plants. Plant Mol Biol. 2011; 80(1):17–36.

74. Aida M, Ishida T, Tasaka M. Shoot apical meristem and cotyledon formation during Arabidopsis embryogenesis: interaction among the CUP-SHAPED COTYLEDON and SHOOT MERISTEMLESS genes. Development. 1999;126(8):1563–70.

75. Sunkar R, Zhu JK. Novel and stress-regulated microRNAs and other small RNAs from Arabidopsis. Plant Cell. 2004;16(8):2001–19.

76. Zhou L, Liu Y, Liu Z, Kong D, Duan M, Luo L. Genome-wide identification and analysis of drought-responsive microRNAs in Oryza sativa. J Exp Bot. 2010;61(15): 4157–68.

77. Dubos C, Stracke R, Grotewold E, Weisshaar B, Martin C, Lepiniec L. MYB transcription factors in Arabidopsis. Trends Plant Sci. 2010;15(10):573–81.

78. Chen X. A microRNA as a translational repressor of APETALA2 in Arabidopsis flower development. Science. 2004;303(5666):2022–5.

79. Zhu QH, Helliwell CA. Regulation of flowering time and floral patterning by miR172. J Exp Bot. 2011;62(2):487–95.

80. Ben Amor B, Wirth S, Merchan F, Laporte P, d'Aubenton-Carafa Y, Hirsch J, Maizel A, Mallory A, Lucas A, Deragon JM, et al. Novel long non-protein coding RNAs involved in Arabidopsis differentiation and stress responses. Genome Res. 2009;19(1):57–69.

81. Franco-Zorrilla JM, Valli A, Todesco M, Mateos I, Puga MI, Rubio-Somoza I, Leyva A, Weigel D, Garcia JA, Paz-Ares J. Target mimicry provides a new mechanism for regulation of microRNA activity. Nat Genet. 2007;39(8):1033–7.

82. Heo JB, Sung S. Vernalization-Mediated Epigenetic Silencing by a Long Intronic Noncoding RNA. Science. 2011;331(6013):76–9.

83. He Y. Noncoding RNA-Mediated Chromatin Silencing (RmCS) in Plants. Mol Biol. 2013;2:e106.

84. Zhu Q-H, Wang M-B. Molecular Functions of Long Non-Coding RNAs in Plants. Genes. 2012;3(1):176.

85. Campalans A, Kondorosi A, Crespi M. Enod40, a short open reading frame-containing mRNA, induces cytoplasmic localization of a nuclear RNA binding protein in Medicago truncatula. Plant Cell. 2004;16(4):1047–59.

86. Wang Kevin C, Chang Howard Y. Molecular Mechanisms of Long Noncoding RNAs. Mol Cell. 2011;43(6):904–14.

87. Rinn JL, Chang HY. Genome regulation by long noncoding RNAs. Annu Rev Biochem. 2012;81:145–66.

88. Burleigh SH, Harrison MJ. A novel gene whose expression in Medicago truncatula roots is suppressed in response to colonization by vesicular-arbuscular mycorrhizal (VAM) fungi and to phosphate nutrition. Plant Mol Biol. 1997;34(2):199–208.

89. Wasaki J, Yonetani R, Shinano T, Kai M, Osaki M. Expression of the OsPI1 gene, cloned from rice roots using cDNA microarray, rapidly responds to phosphorus status. New Phytol. 2003;158(2):239–48.

90. Burleigh SH, Harrison MJ. The Down-Regulation of Mt4-Like Genes by Phosphate Fertilization Occurs Systemically and Involves Phosphate Translocation to the Shoots. Plant Physiol. 1999;119(1):241–8.

91. Martin AC, del Pozo JC, Iglesias J, Rubio V, Solano R, de La Pena A, Leyva A, Paz-Ares J. Influence of cytokinins on the expression of phosphate starvation responsive genes in Arabidopsis. Plant J. 2000;24(5):559–67.

92. Swiezewski S, Liu F, Magusin A, Dean C. Cold-induced silencing by long antisense transcripts of an Arabidopsis Polycomb target. Nature. 2009;462(7274):799–802.

93. Helliwell CA, Robertson M, Finnegan EJ, Buzas DM, Dennis ES. Vernalization-Repression of Arabidopsis FLC Requires Promoter Sequences but Not Antisense Transcripts. PloS one. 2011;6(6):e21513.

94. Heo JB, Sung S. Encoding memory of winter by noncoding RNAs. Epigenetics. 2011;6(5):544–7.

95. Sun Q, Csorba T, Skourti-Stathaki K, Proudfoot NJ, Dean C. R-Loop Stabilization Represses Antisense Transcription at the Arabidopsis FLC Locus. Science. 2013;340(6132):619–21.

16

An improved Bayesian network method for reconstructing gene regulatory network based on candidate auto selection

Linlin Xing[1], Maozu Guo[2*], Xiaoyan Liu[1], Chunyu Wang[1], Lei Wang[3] and Yin Zhang[3]

Abstract

Background: The reconstruction of gene regulatory network (GRN) from gene expression data can discover regulatory relationships among genes and gain deep insights into the complicated regulation mechanism of life. However, it is still a great challenge in systems biology and bioinformatics. During the past years, numerous computational approaches have been developed for this goal, and Bayesian network (BN) methods draw most of attention among these methods because of its inherent probability characteristics. However, Bayesian network methods are time consuming and cannot handle large-scale networks due to their high computational complexity, while the mutual information-based methods are highly effective but directionless and have a high false-positive rate.

Results: To solve these problems, we propose a Candidate Auto Selection algorithm (CAS) based on mutual information and breakpoint detection to restrict the search space in order to accelerate the learning process of Bayesian network. First, the proposed CAS algorithm automatically selects the neighbor candidates of each node before searching the best structure of GRN. Then based on CAS algorithm, we propose a globally optimal greedy search method (CAS + G), which focuses on finding the highest rated network structure, and a local learning method (CAS + L), which focuses on faster learning the structure with little loss of quality.

Conclusion: Results show that the proposed CAS algorithm can effectively reduce the search space of Bayesian networks through identifying the neighbor candidates of each node. In our experiments, the CAS + G method outperforms the state-of-the-art method on simulation data for inferring GRNs, and the CAS + L method is significantly faster than the state-of-the-art method with little loss of accuracy. Hence, the CAS based methods effectively decrease the computational complexity of Bayesian network and are more suitable for GRN inference.

Keywords: Gene regulatory networks, Bayesian network, Candidate auto selection, Breakpoint detection, Search space reduction

Background

Life activities are regulated through complex interconnections of genes and their products [1]. These interactions between genes form so-called gene regulatory networks (GRNs) in living cells. Inferring Gene regulatory networks (GRNs), also known as reverse engineering, is a critical problem in computational biology [2–4]. The advent of high throughput technologies has provided such an opportunity to biologists and bioinformatics researchers so that they can collect large amount of omics data that can quantify the activities of genes or their products. GRNs that constructed from gene expression data reflect the interactions of the regulatory elements in biological systems, such as genes and proteins [5–7], and the structure of GRN reveals the inner complex mechanism in adaptability to the environment and the growth and development of

* Correspondence: guomaozu@bucea.edu.cn
[2]School of Electrical and Information Engineering, Beijing University of Civil Engineering and Architecture, Beijing, China
Full list of author information is available at the end of the article

organisms [1, 8]. So enthusiasm for inferring GRN has continued unabated for years.

The availability of transcriptome data have been immensely improved by high throughput technologies such as DNA microarrays in recent years. This has led to the fast development of computational approaches for the reconstruction of GRN [9]. In computing complexity aspect, there are also various degrees of flexibility for modeling GRNs that range from complex differential equation method [10] to simple methods based on correlation coefficients [11]. Each model has its own special feature: pairwise or systematic, linearity or nonlinearity, etc.

The pairwise methods, which are relatively simple way, compute the correlation coefficients of genes and then set different threshold to construct GRNs [12, 13]. Commonly used methods to calculate correlations include Pearson Correlation Coefficient (PCC), mutual information (MI) [14], Granger causality [15, 16], etc. Most pairwise methods are low complexity, fast computing speed and adapting to large data set. Nevertheless, most of pairwise methods cannot identify the directions of regulatory interactions and cannot identify casual connections on system level. In addition, pairwise methods suffer from false positive/negative problems due to the simplicity of model and uncertainty of parameters.

Rather than the pairwise method, systematic approaches try to model the GRN from a holistic perspective. There are mainly three types of mathematical model in systematic approaches: Boolean network method [17–20], Bayesian Network method [21–24] and differential equation method [25–27]. These systematic methods can provide the researchers a deeper understanding of the regulatory mechanism at network level and can also identify the directions of regulations in the network. However, the problem of computational complexity makes them difficult to handle large-scale networks. With gradually increasing of computing complexity, the data size they can process rapidly goes down. Boolean network, which was first introduced by Kauffman [28], uses a set of Boolean variables and Boolean functions to describe gene-gene interactions. Probabilistic Boolean network, first introduced by Shmulevich et al., is a stochastic extension of Boolean network that integrates rule-based dependencies between variables [29]. Obviously, these crude simplifications of genes and their interactions cannot reflect the genetic reality. Differential equation method uses a set of differential equations to directly describe dynamic changes of the mRNA content in a precise manner. Obviously, the differential equation method can capture more details about the regulation relationships, but we could not bear such a high degree of computational complexity in most cases. Bayesian Network methods is in the middle of all the methods in complexity and

scale. Bayesian Network is a probabilistic graphical model and tries to find a directed acyclic graph (DAG) that fits the expression data reasonably. Among all the models, BN is always a concern, because of its inherent probabilistic nature. In this paper, we focus on developing systematic methods based on Bayesian network to construct GRNs with higher accuracy and better scalability.

Yet, the Bayesian network method has some limitations. The reconstruction of GRN based on Bayesian network is NP-hard with respect to the number of genes, so the exact network structure can be learned only for relatively small datasets [30]. For large-scale networks, some variants of heuristic approaches are applied [31]. Due to the decomposable scoring function and some reasonable assumptions, the score-search framework for learning network structures are efficient [32]. However, heuristic methods do not guarantee the globally optimal network structure. Furthermore, most of the time is wasted on examining unreasonable candidates due to the scale and sparsity of biological networks. Hence, many researchers have devoted themselves to accelerating the learning process through reducing the search space [33]. Sparse candidate [21], maximum number of parents limitation [34] (also called maxP technique) and Max-Min Hill-Climbing (MMHC) [35] are typical methods for speeding up the structure learning.

The sparse candidate method is an iterative optimal algorithm by combining two steps: restricting the parents of each variable to a small subset of candidates and then searching for a network structure that satisfies these constraints. The learned network will improve the quality of candidates in the next iteration. The optimal GRN structure is learned by taking this iteration strategy. The maxP technique further simplify this idea by directly limiting the maximum indegree of each node. In the learning process, the parents of a node no longer increase until the indegree threshold is reached. The MMHC algorithm uses a more reasonable way to learn the candidates. It learns the neighbors of each node by using a local neighbor discovery algorithm called Max-Min Parents and Children (MMPC) [35].

Commonly, an important tuning parameter k is needed to indicate the size of candidate set or the maximum number of parents in sparse candidate or maxP algorithms. However, we are still far away from understanding the complex regulation mechanism of biological networks, so that we know rather little about GRNs to guide the selection of parameter k, even for model organisms [36, 37]. According to statistics in Table 1, degree distributions of known biological networks have obvious differences. This critical problem illustrates that the estimation of degree distribution using known networks is unreasonable. Therefore, no

Table 1 indegree distribution of different network

–	% of total gene	
Indegree	E.coli	Yeast
1	5.1757	0.698
2	37.4441	35.6902
3	22.9393	24.454
4	13.8658	14.5913
>4	8.754	7.9261

reliable estimates exist for parameter k. The arbitrary selection of parameter k ignores the complexity of biological networks and compulsorily cut the search space in a rigid manner. How about using unified correlation threshold to select candidates? Figure 1 shows the MI distribution of each node in alarm network. As we can see in Fig. 1, the distribution varies significantly at different node. Hence, the selection of candidates is also unreasonable through a unified correlation threshold, e.g. MI.

MMHC's heuristic heavily depends on the accuracy of conditional independence estimation. In MMPC step, the number of samples grows exponentially to the size of the conditioning set for accurately estimating the conditional independence. It is impossible to obtain so much data in a biological sense. So you can see that the situation, typically known as "large p, small n" problem, greatly limits the use of MMPC method for biological network reconstruction. Moreover, users also need to set a p value as the threshold of independence. Hence, the MMPC algorithm does not work well on small dataset, and leads to a huge amount of false positives.

Based on the analysis above, we can draw the following conclusions: a) Screening of candidates using global parameters is unreasonable and unrealistic. These methods cannot reduce searching time for examining unreasonable candidates and cannot solve the dependency

on tuning parameter. In addition, we know little to guide the selection of the tuning parameters. b) Overly complex model is not suitable for GRN inference due to the limitation of biological data.

As is known to all, MI draws much attention in biological data analysis because of its ability to measure the non-linear relationships, which are common in biology. Previous studies has discussed and compared the advantages and power of using MI in measuring non-linear relations [38, 39]. And according to Frenzel's work [40], MI can differentiate between direct and indirect interactions to some extent. Thus far, many researchers have used the MI measure in biological data analysis and network reconstruction, and get some achievements [41–43]. Some pairwise methods, such as CMI, also use an improved MI method as measurements of independence [44].

In this work, we proposed a novel candidate selection algorithm based on mutual information and the concept of breakpoint detection, named CAS (candidate auto selection). The CAS algorithm is designed to reduce the search space of structure learning and get rid of an unwanted dependency on tuning parameters. Firstly, the CAS algorithm utilizes the capability of MI to measure the non-linear regulatory interactions. Then, the candidate selection problem is formalized as a hypothesis test problem by using breakpoint detection. More importantly, this algorithm is a polynomial-time approach and do not depend on turning parameters. Further, based on the CAS algorithm, a globally optimal greedy algorithm (CAS + G) and a local learning algorithm (CAS + L) are also proposed for reconstruction of GRN. The proposed CAS + G algorithm aims at finding the optimal BN structure from data in the restricted search space. Meanwhile, the CAS + L algorithm which learning the structure in a local way pays more attention to the learning

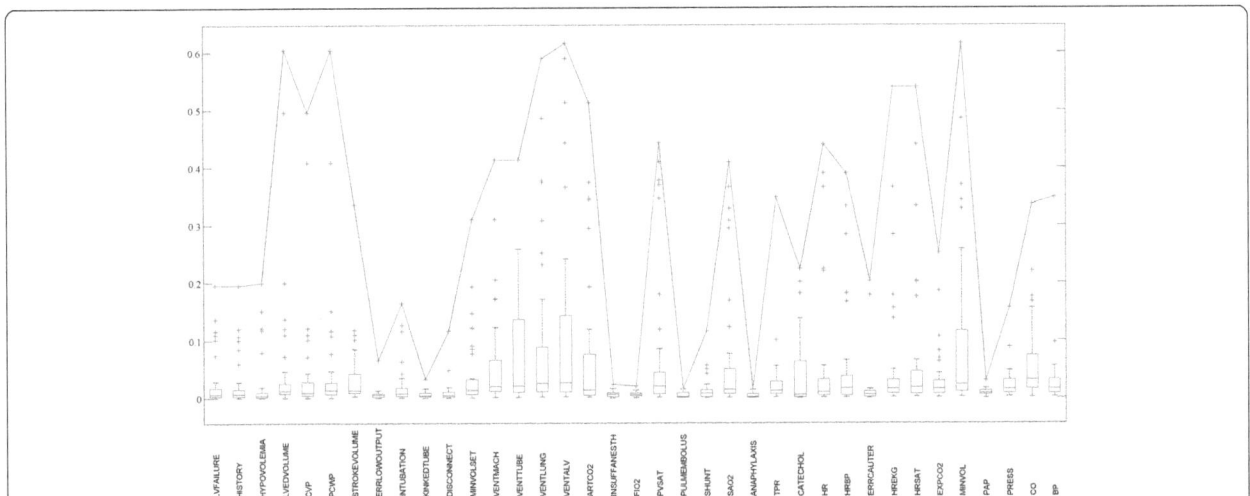

Fig. 1 Mutual information distribution of each node in alarm network

rate at little expense of quality. To evaluate the proposed methods, they are compared with a state-of-the-art method on different datasets.

Methods

This section consists of three parts: a) Candidate auto selection algorithm based on mutual information and breakpoint detection. b) Local learning algorithm (CAS + L) for reconstruction of the GRN. c) Globally-optimal greedy algorithm (CAS + G) for reconstruction of GRN. The overall diagram of aforementioned methods is shown in Fig. 2, the process on the left side is CAS + G and the process on the right side is CAS + L.

A. Candidate auto selection (CAS) algorithm based on mutual information and breakpoint detection

Usually, MI is used as a metric of the correlation between two variables. Here we choose MI as a correlation measure of genes, which has been widely used to construct GRN from gene expression data due to its capability of capturing the non-linear relationships between genes as mentioned above. However, without considering the other variables, MI tends to overestimate the regulation strengths between genes (i.e., false positive problem). High MI value indicates that there may be a close relationship between the variables (genes) X and Y, while low MI value implies their independence.

MI of two discrete variables X and Y is defined as in (1):

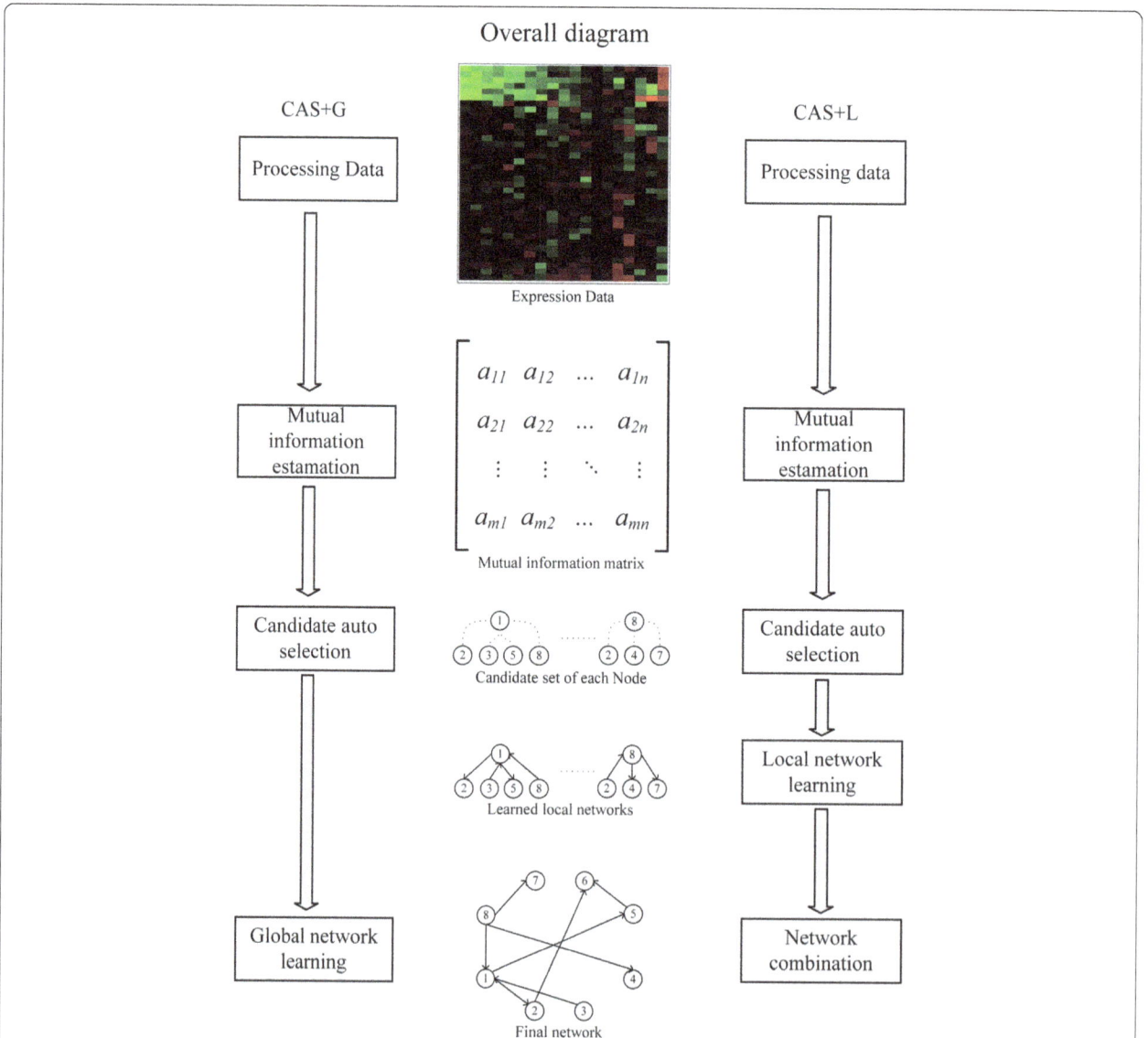

Fig. 2 Overall diagram of our method. 1) process the expressing data, 2)estimate mutual information of each pair of genes, 3) construct candidate set for all nodes using CAS algorithm 4) learn sub-networks on each node for CAS + L (right column), 5) combine all sub-networks into the final network for CAS + L(right column), or directly learn the final network for CAS + G(left column)

$$MI(X, Y) = -\sum_{x \in X, y \in Y} p(x, y) \log \frac{p(x, y)}{p(x)p(y)} = H(X) + H(Y) - H(X, Y)$$
$$(1)$$

where $p(x, y)$ is the joint probability of X and Y under specific value x, y, and $p(x), p(y)$ are the marginal probability; $H(X), H(Y)$ are entropies of X,Y; $H(X, Y)$ are joint entropy of X and Y.

Obviously, in a real case of inferring GRN, node Z or its descendants are always not in evidence. Hence, we show all the possible relationships of node X and Y in the case of no other observations, as shown in Fig. 3: Node X is closely related to Y in the case of direct connection and common cause, the correlation will decrease with distance in the case of indirect connection, and Node X and Y are independent with each other in the case of common effect (also known as V-structure). In biological networks, the correlations between a gene and its regulators or targets are closer than that between this gene and irrelevant genes. That is, the correlations between a gene and the genes in common effect branch is different corresponding to Fig. 3.

With the above analysis and mentioned research [32, 44], we can conclude that the MI distribution of highly correlated nodes is different from the distribution of uncorrelated nodes. According to this conclusion, a mutual information and breakpoint detection based method is proposed in this paper aiming at exactly selecting the candidates. This method achieves the goal of reducing search space by cutting off all the unrelated nodes in common effect branch.

To describe the whole method, we start by calculating the potential neighbors of one target node. The process of calculating the potential neighbors of one target node is as

follows. The input consists of the data D on the node set V of size n and a target node $g_i \in V$. The output is candidate set C_i for node g_i, which consists of the potential neighbors of g_i. In the first step, all the MIs between node g_i and other nodes are computed and stored in vector $\mathcal{X}$. These MIs are summarized based on ascending order of values. At this time, Vector $\mathcal{X}$ of size n-1 stores all the MIs between node g_i and other nodes in ascending order. Then suppose there is a position (breakpoint) in vector $\mathcal{X}$ that divides the nodes into two parts: related nodes and unrelated nodes. The task of breakpoint detection can be formalized as a hypothesis-testing problem, which can be solved using maximum likelihood method. So this breakpoint can be found by constructing a statistic and locating the maximum. The nodes on the left side of the breakpoint, whose MI value is smaller than the MI value of the breakpoint, are identified as unrelated and discard. The nodes on the right side of the breakpoint with bigger MI values are added to candidate set C_i. Then nodes in set V are processed one by one.

Now, the key question is how to construct the statistics and do hypothesis testing. As mentioned above, given node g_i's MI vector $\mathcal{X}$, the goal is to determine whether a significant breakpoint exists or not. The null hypothesis and the alternative hypothesis are stated as follows:

H0: Null hypothesis – no breakpoint exists.

H1: alternative hypothesis– one significant breakpoint exists.

Then we construct statistic Q to decide whether or not the null hypothesis should be rejected in hypothesis testing. The incoming data is the MI vector $\mathcal{X} = \{x'_1, x'_2, ..., x'_m\}$, $m = n-1$. Based on the prior analysis, we suppose that the MI of node g_i with related nodes and node g_i with unrelated nodes are coming from different distributions. For a model with a breakpoint at $k \in [1, m]$, the maximum likelihood is defined as in (2)

$$ML(k) = \log\Big(p(\mathcal{X}_{1:k}, \theta_1)\Big) + \log\Big(p(\mathcal{X}_{k+1:n}, \theta_2)\Big) \quad (2)$$

Where $p(X | \theta)$ in (2) is the probability density function, θ_1, θ_2 is the corresponding parameters of each distribution. Now the testing statistic Q can be defined as (3):

$$Q = 2[ML(k) - \log(p(\mathcal{X}_{1:n}|\theta))] \quad (3)$$

$p(X_{1:n} | \theta)$ in (3) is the probability of the null hypothesis.

Usually, we need to choose a constant value c as the threshold. If $Q > c$, then we reject the null hypothesis. At this point, all the position k satisfying $Q > c$ are breakpoints which dividing the vector into two different parts. Obviously, here c is underdetermined. However, for the goal of finding the best candidates, the maximum of Q can be considered as the criterion to determine the best breakpoint, as in (4).

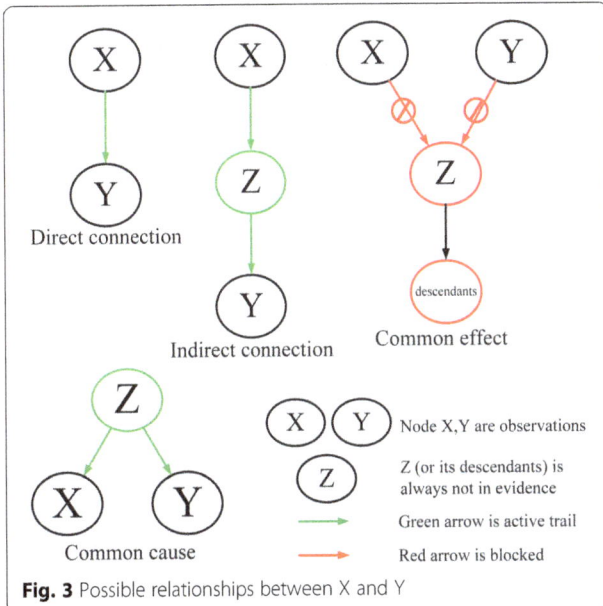

Fig. 3 Possible relationships between X and Y

$$k = argmax(Q) \qquad (4)$$

The nodes on the left side of the breakpoint k are unrelated nodes in common effect branch. Then all the nodes corresponding to MI on the right side of the breakpoint k are identified as the candidates.

At this time, let us solve the problem of how to calculate the probability. That is to say, how to make reasonable assumptions about the MI distributions of two types of nodes. If two nodes (g_i and g_j) are independent in the case of common effect, that is, knowing g_i does not give any information about g_j and vice versa, so their mutual information is zero by definition. However, due to the limit of sample size, the computed MI of these nodes are statistical noise, which can be modeled by a normal distribution. In other words, the MI distribution of unrelated nodes is a normal distribution with parameter $\theta_1 = (\mu_1, \sigma_1)$. Here, μ_1 is the mean or expectation of the distribution. The parameter σ_1 is its standard deviation. As we know, the normal distribution is often used to represent real-valued random variables whose distributions are not known. Therefore, it is also a reasonable assumption that the MI of related nodes is also normally distributed. Obviously, it has a different parameter $\theta_2 = (\mu_2, \sigma_2)$.

The full algorithm is as follows:

Algorithm 1 The Candidate auto selection algorithm (CAS)

Input: Node set $V = \{g_1, g_2, \cdots, g_n\}$

   Discrete Expression data D on nodes set V

Output: candidate Set of each node $\{C_i \mid i=1:n\}$

For $i=1:n$

   $N = V / g_i$

   //calculate and store mutual information
   For each node u in N

   $\mathcal{X} \leftarrow MI(g_i, u)$.

   End for
   $\mathcal{X} \leftarrow$ Sort ($\mathcal{X}$) in ascending order
   // Find position k that maximize Q
   //Q is defined as in(3)

   $k = argmax(Q)$

   $C_i = \{$nodes in the right side of breakpoint $k\}$

End for

In algorithm CAS, the while loop is executed n times. The time requirement inside the loop is dominated by the procedure of finding breakpoint (the position k) to maximize the statistic Q. For finding the breakpoint of a specified node, we must go through all the possible locations of the MI vector. In this process, the posterior probability is calculated at most n times for each possible position. So the time complexity is $O(n^2)$. Thus, the total time requirement of the algorithm is $O(n^3)$.

B. Local learning algorithm CAS + L for reconstruction of GRN

Local learning is a common idea in network structure learning problem. Based on the proposed CAS algorithm above, we can obtain the neighbors of each node exactly. Thus, local learning is a good solution for inferring the network structure in this case. Hence, we present a local learning method based on the CAS algorithm for inferring GRNs. To describe the whole idea of this local learning algorithm, we start with learning the local structure of a specified node. In the first step, we compute the candidate set $C(v)$ of specified node v by using CAS algorithm. Then construct potential edge set $E(v)$ of node v: $E(v) = \{(v, u), (u, v); u \in C(v)\}$. At last, a typical score-search framework is applied to finding the best local structure $G(v)$ of node v on the search space defined by $E(v)$. The high-level pseudocode of the full algorithm is as follows:

Algorithm 2 CAS+L

Input: Node set N dataset D on node set V

Output: DAG G

For node v in V

   Compute $C(v)$ by using CAS algorithm

   Construct $E_v = \{(v,u),(u,v) \mid u \in C(v)\}$

   $G(v) \leftarrow$ GreedySearch($E(v), D$)

End for

$G \leftarrow$ Combine($\{G(v) \mid v \in V\}$)

To combine all the subgraphs in a simple way, we accept the edge one by one, which do not introduce cycles. By analyzing this algorithm, we can find that this algorithm can be parallelized easily.

Given a node set V of size n, suppose that k candidates are selected for a node v. That means the algorithm should only check k possible candidates on one iteration. To learn the local structure of node v, $2k$ candidate edges should be examined, so the search space complexity is $O(2^{2k})$. The total time for learning the structure is $O(n \times 2^{2k})$. That is, the boundary of time is restricted by the size of candidate set. As we know, structure sparsity is one of the important properties of biological networks. At this point, we can suppose that the mean number of candidates is far smaller than n, e.g. $k = \log(n)$, then the space complexity is $O(n \times 2^{2 \log n})$.

C. Globally-optimal greedy algorithm (CAS + G) for reconstruction of GRN

The local learning algorithm is a quick solution, but does not guarantee a global optimum. For getting a global optimal solution, a global optimal algorithm based on the CAS method (CAS + G) is also proposed in this paper. The difference between the local method and the global method is that the CAS + G algorithm learning the structure as a whole in the restricted search space generated by CAS rather than learning substructures of each node.

Algorithm 3 CAS+G

Input: Node set V dataset D on node set

Output: DAG G V

Candidate Edge Set $E = \varnothing$

For node v in V

Compute C_v by using CAS algorithm

Construct $E_v = \{(v, u), (u, v) | u \in C_v\}$

$E = E \cup E(v)$

End For

$G = GreedySearch(E, D)$

The greedy search procedure adds the highest rated edge at each iteration with the hope of finding a global optimal GRN structure. Moreover, the algorithm only needs to examine the limited edge set at each iteration. Let us suppose that the average size of candidate set is k. The search space complexity is $O(2^{nk})$. At this point, we also suppose that the mean number of candidates is far smaller than n according to the sparsity of the structure, e.g. $k = n/4$ is a small constant c, then the space complexity is $O(2^{n \times n/4})$.

Result and discussion

Used networks and datasets for evaluation

In order to evaluate the CAS algorithm presented in this article, experiments are carried out on two types of networks: a. known Bayesian networks, b. Dream challenge networks. Known Bayesian networks are constructed by experts, and all the parameters are known. These networks are insurance network (27 nodes, 52 arcs), alarm network (37 nodes, 46 arcs), Barley network (48 nodes, 84 arcs), Hailfinder network (56 nodes, 66 arcs). Five datasets of different sizes (n = 50, 100, 200, 500, 1000) are sampled from each network. Ten 100-gene networks are collected from DREAM3 and DREAM4 challenge. The corresponding simulation data (210 samples) for these in-silico networks are generated by GNW software [45] and then converted into discrete data by K-Means discretization algorithm.

Results of CAS algorithm

In this section, the effectiveness of the CAS algorithm is illustrated and compared with MMPC algorithm. The aim of the CAS and MMPC algorithm is to identify the neighbors of each node in as few candidates as possible. So, in classification point, the true positive (TP) is defined as neighbor nodes correctly identified as candidates, and the recall in this context is defined as the

number of true positives divided by the total number of the neighbors, as in (5).

$$recall = \frac{TP}{TP + FN} \tag{5}$$

False negative (FN) are neighbor nodes incorrectly identified as unrelated nodes. Therefore, TP + FN equals to the number of true neighbors. In the following test, the recall and the average number of candidates are the most important indicators.

We studied the influence of the network size, network type and the dataset size to validate the effectiveness of the CAS method on the aforementioned networks. The alarm network is analyzed in detail.

We mark the identification result of four representative nodes on the true network to illustrate the detail result of CAS algorithm. As shown in Fig. 4, triangles with four colors indicate four different nodes in alarm network: red, black, blue and green corresponding to LVEDVOLUME, FIO2, ERRCAUTER, PVSAT, respectively. LVEDVLOUME (indicated by red triangles) has two parent nodes and two sibling nodes. Figure 4 shows that the CAS algorithm identified 9 candidates, and these 9 candidates completely cover all 4 true neighbors. Meanwhile, the candidates except VENTLUNG are closely related to LVEDVOLUME as mentioned in Fig. 3. For node FIO2, the candidates identified by CAS contains its child node. In addition, the common effect branch of FIO2 (the parents of PVSAT) is correctly cut off by CAS algorithm. For node ERRCAUTER, when HREKG and HRSAT are not given, HR and ERRCAUTER are independent in the common effect case. Hence, as mentioned in the method section, its child nodes (HREKG, HRSAT) are correctly identified by the CAS algorithm and the nodes in the common effect branch (all ancestors and descendants of node HR) are discarded. At last, we analyze another representative node PVSAT, which has a complicated patrilineal family. In this situation, the CAS algorithm also identified its parents and child correctly, and discard the nodes in common effect branch. We can see that the candidate set cover more ancestors rather than descendants. This phenomenon illustrates that node PVSAT is more affected by its ancestor and is not closely related to its descendants.

By analyzing the results of CAS algorithm on all nodes of alarm network, we can conclude that the CAS algorithm can effectively discard the unrelated nodes in common effect branch and can identify most of the true candidates correctly in most cases. In addition, the identification rates increase simultaneously with the amount of data. Nevertheless, there are still a few nodes that cannot be identified correctly and these situations

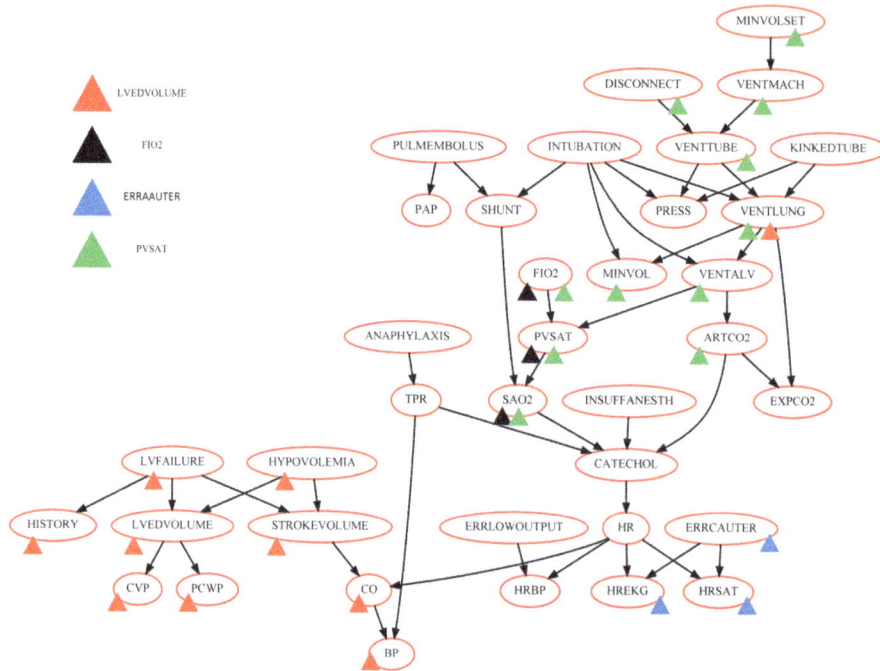

Fig. 4 Identification result of CAS algorithm on alarm network. Red triangle indicates the candidates of LVEDVOLUME identified by CAS algorithm. Likewise, the black triangle and blue triangle green triangle correspond to the candidates of FIO2, ERRCAUTER, PVAST, respectively

are not improving with the amount of data in our experiment. Further analysis indicates that the generated data cannot reflect the real probability distribution correctly due to its very small conditional probability in the conditional probability table. Hence, the generated data could not provide enough power to distinguish its neighbors and other nodes, such as INSUFFANESTH. For these nodes, more data is needed to confirm its correlation with others.

Figure 5 summarizes the trends of recall and the average number of candidates of CAS and MMPC on different networks when sample size gets larger. Based on the analysis of all these charts, the average recall of CAS (green line) increases with sample size, while the average number of candidates (blue bar) decreases. This situation illustrates that sufficient samples can improve the performance of CAS algorithm. Nevertheless, there are different situations in MMPC method. The average number of candidates decreases as samples increase in numbers (red bar), but the recall also declines to different extents (purple line).

As we can see from Fig. 5, when the sample size is small, the MMPC algorithm takes nearly all nodes as neighbors. This means that MMPC algorithm breaks down because of the lack of samples. Especially, it is more obvious in Barley network. That is mainly because multivariate variables in Barley network, so that

the data is not enough for accurately estimating the conditional associations. In insurance network, the decrease of average recall is mainly because of the undue screening. Hence, the MMPC cannot provide adequate performance when the data is insufficient. However, the CAS algorithm has better filtering ability in most cases. For example, the CAS algorithm gives out less candidates than MMPC algorithm for nearly the same recall rate in alarm network and Hailfinder network, when the data size is 50. Even in barley network, the CAS algorithm still shows better performance than MMPC with the increase of samples. We can draw a conclusion that the CAS algorithm is more effective on small data sets. Thus, it is more suitable for GRN inference which is a typical "large p, small n" problem.

Moreover, to assess the effectiveness further, the evaluations are also carried out using the simulation data from DREAM challenge. Although the complexity of the biological networks and the imperfection of simulation data will impose some performance penalties, it remains effective to a much larger degree, which will be validated in the learning phase. The CAS algorithm can effectively reduce the search space through cutting off the common effect branch. Based on the above analysis, one can draw a conclusion that the CAS algorithm outperforms the MMPC algorithm, especially on small data sets.

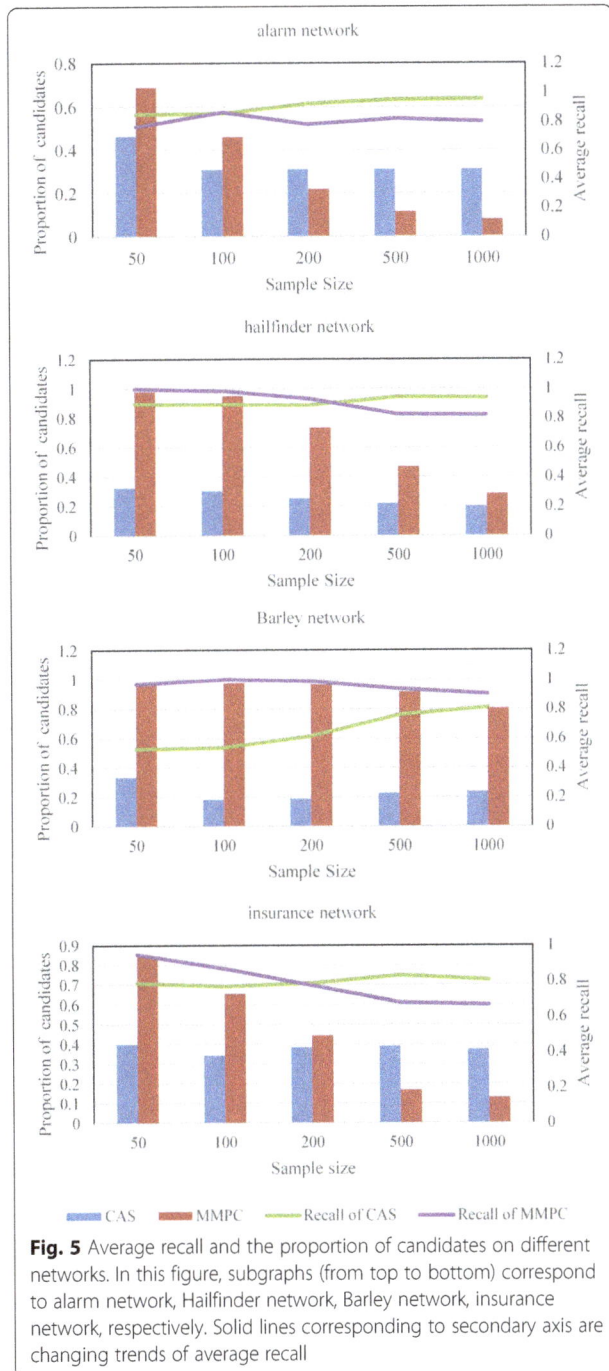

Fig. 5 Average recall and the proportion of candidates on different networks. In this figure, subgraphs (from top to bottom) correspond to alarm network, Hailfinder network, Barley network, insurance network, respectively. Solid lines corresponding to secondary axis are changing trends of average recall

Evaluation of the CAS + G and CAS + L algorithm for inferring GRNs

The performance of structure learning was evaluated using the following measures: TP, FP, TN, FN, Precision, Accuracy, Recall, Specificity and F-score. Experimental comparisons of CAS based methods against MMHC on various sample sizes are carried out. A local learning algorithm named MMHC + L using MMPC as the candidate selection algorithm is also in comparison. For purpose of comparison, we choose original greedy search

as benchmark method, which applies the score-search framework to find the optimal structure. The leading reason for the choice of greedy search is that there are no limits on the search space with the hope of finding a global optimum. Hence, the performance improvement is the ratio of the performance of a particular algorithm to the performance of the benchmark. Here BDeu score [46] is selected from other well-known scores. We examine all the networks described above, but only some are analyzed in detail limited by the space. Firstly, the alarm network and insurance network are analyzed and discussed.

Figures 6 and 7 show the comparison result on the alarm and insurance network in different sample size. The first five charts corresponding to five different sample sizes show comparison results under aforementioned measures. The sixth subgraph shows the runtime result with different sample sizes. When sample size is 50, the behavior of MMHC is nearly the same as greedy search according to the candidate selection results in Fig. 5. As you can see by comparing subgraphs 1 to 5, the CAS + G algorithm outperforms the MMHC algorithm with the increase of sample size, mainly because ample data can improve the performance. Especially in insurance network, the CAS based methods outperform MMPC based methods in all sample sized. It is not just about the learning phase, but also about the quality of candidate selection, which generates a more exact search space. When the sample size is larger to more accurately estimating the conditional independence, the performance of MMHC algorithm gets better quickly. Nevertheless, when the sample size is 1000, the recall of the MMPC algorithm is much lower than CAS according to Fig. 5. That is, MMPC overly reduces the search space and rejects many true neighbors. Finally, the proposed global optimal algorithm CAS + G is superior to the benchmark and better than MMHC when the sample size is 1000.

The sixth subgraph is the runtime results of different sample sizes. Firstly, we can see that the proposed CAS-based learning methods run more stable than MMPC-based methods and greedy search by taking into the number of candidates. In particular, CAS + L algorithm has an obvious advantage on runtime in both networks, especially in insurance network. The CAS algorithm does not need to enumerate all the conditional sets, so it is much faster than MMPC algorithm. Hence, the total time of CAS + L algorithm is less than the other methods. For the CAS + L algorithm, the main advantage is its speed without losing much of quality. Nevertheless, much more false positives are found because of fluctuations in the data and the simple combination phase. It may be solved by replacing the simple combination method with a heuristic phase.

In general, sufficient samples provide a performance boost because of the improvement of CAS algorithm

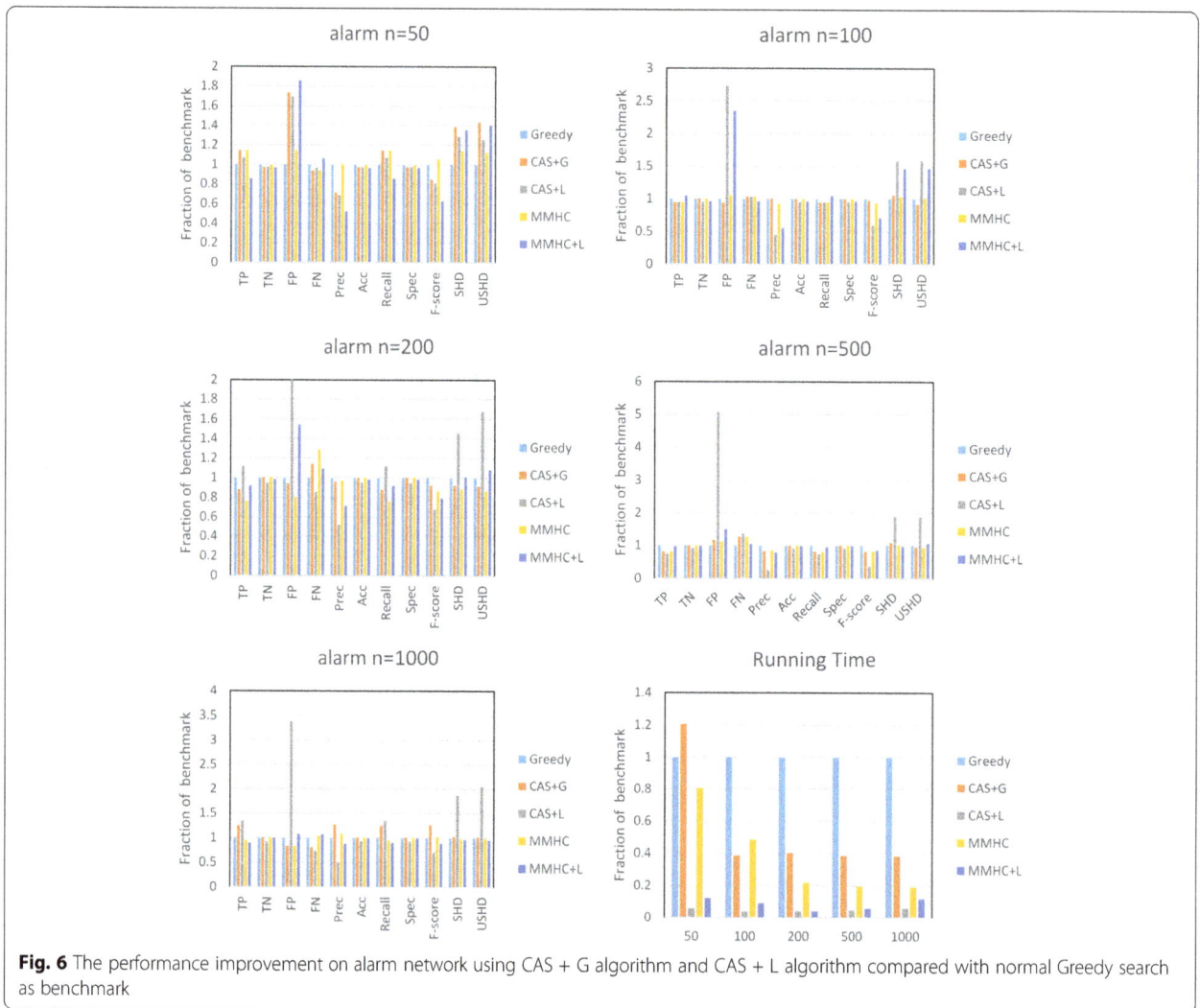

Fig. 6 The performance improvement on alarm network using CAS + G algorithm and CAS + L algorithm compared with normal Greedy search as benchmark

and the score-search phase. There are still swings due to their heuristic nature, which easily trap into a non-optimal "local maximum". According to the above analysis, the CAS algorithm is more suitable for small datasets.

Knowing the complexity of biological networks and the "large p, small n" situation, evaluation of the proposed methods should be made on biological datasets. Therefore, we carried out experiments on the simulation data of DREAM3 and DREAM4 challenge.

Figure 8 shows the comparison results of five 100-gene networks from DREAM3 challenge. As can be seen from Fig. 8, the CAS + G algorithm identified more true positives than MMHC method. Experimental results showed that CAS based method compared favorably against other approaches in the F-score and recall. As previously mentioned, the MMPC algorithm cannot accurately estimate the conditional association due to the lack of samples. Hence, the performance of MMHC algorithm is similar with Greedy Search. From the results of Yeast2 network, we can see that both algorithms

trapped into local optimum due to the lack of samples. However, the performance of the CAS + G and CAS + L algorithms are better than MMHC algorithm, mainly because of the conciseness of CAS algorithm, which reduces data dependencies of the algorithms.

Figure 9 shows the comparison results of five 100-gene networks from DREAM4 challenge. As can be seen from Fig. 9, the performances are very similar to DREAM 3 networks overall. Experimental results showed that the CAS-based approaches outperform other approaches in the F-score and recall and identify more true positives than MMHC method on network 1, 3, 4. What we find once again here is that the performance of MMHC algorithm is similar to Greedy Search due to the limitation of sample size. However, the CAS + G algorithm identified less true positives on network 2 and 5. In addition, we even find the CAS + G algorithm failed on network 2. According to the analysis, we found that network 2 and 5 are more complex than others. This resulted in more performance degradation than the MMHC algorithm,

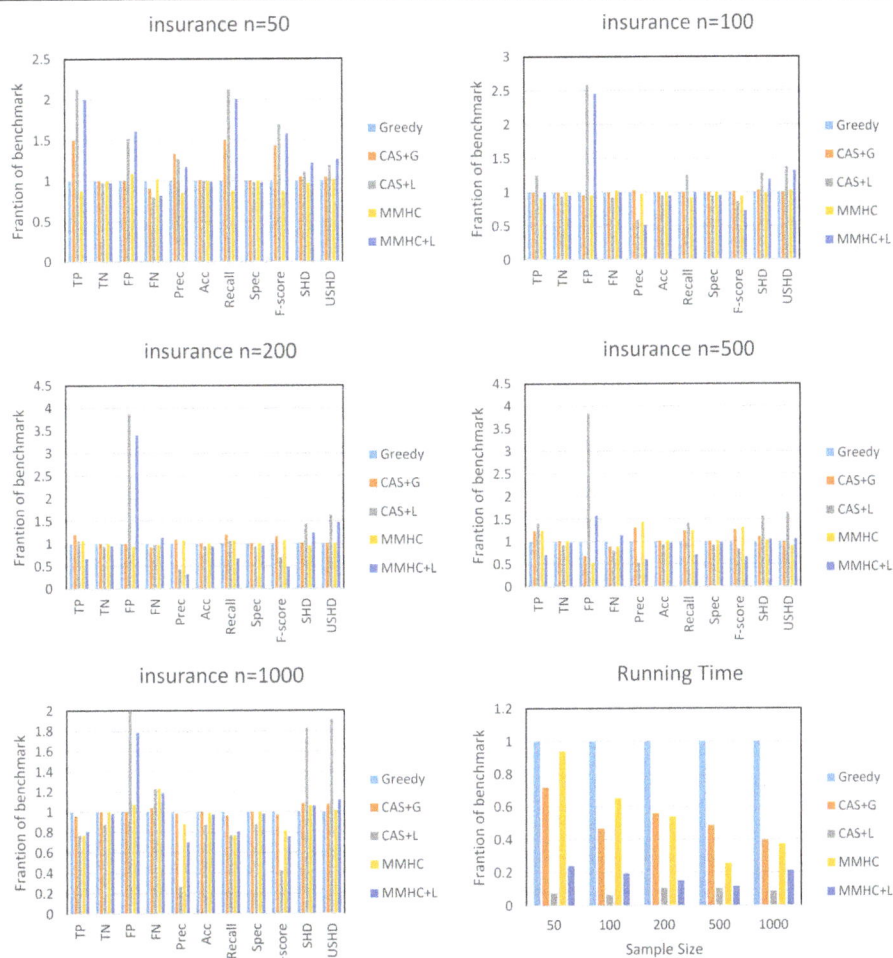

Fig. 7 The performance improvement on insurance network using CAS + G algorithm and CAS + L algorithm compared with normal Greedy search as benchmark

which behaves exactly like Greedy search. Yet even so, MMHC algorithm has little advantage depending on SHD metric. Form the sixth subgraph, we find that the proposed CAS-based methods have an obvious advantage on runtime. Especially the CAS + L algorithm—it takes advantage of both the reduced search space and the locality. Nevertheless, much more false positives are found due to noise in the data and the oversimplified combination phase. We can draw a conclusion that the CAS + G algorithm has equal or better performance than the MMHC algorithm and greedy search for GRN reconstruction, but runs faster. Meanwhile, CAS + L algorithm runs significantly faster than all others.

The CAS algorithm has several advantages because of its conciseness and locality. First, there is no turning parameter to be determined. Erroneous estimation of turning parameter affects not only efficiency but also the quality of the reconstruction. This is very important for unknown situations of biological networks. The second one is its weak data dependencies,

which makes it more suitable for GRN reconstruction. However, there are still limitations to consider, especially the assumption on the distribution of related nodes. The selection of Gaussian distribution is mostly based on experience and lack of a detailed study. In the future, we will study to make a more reasonable assumption. Another extension is to remove false positive candidates by using biological prior knowledge. In brief, we believe that the proposed CAS algorithm can identify the candidates effectively and improve the search process effectively. Furthermore, the global and local learning algorithms outperform the other methods on simulation data. It is clearly that the CAS based algorithms are very suitable for GRN reconstruction, which is always eager for data.

Conclusion

In this work, we first proposed a novel method CAS to select candidates of each node automatically. This algorithm is designed to reduce the search space by

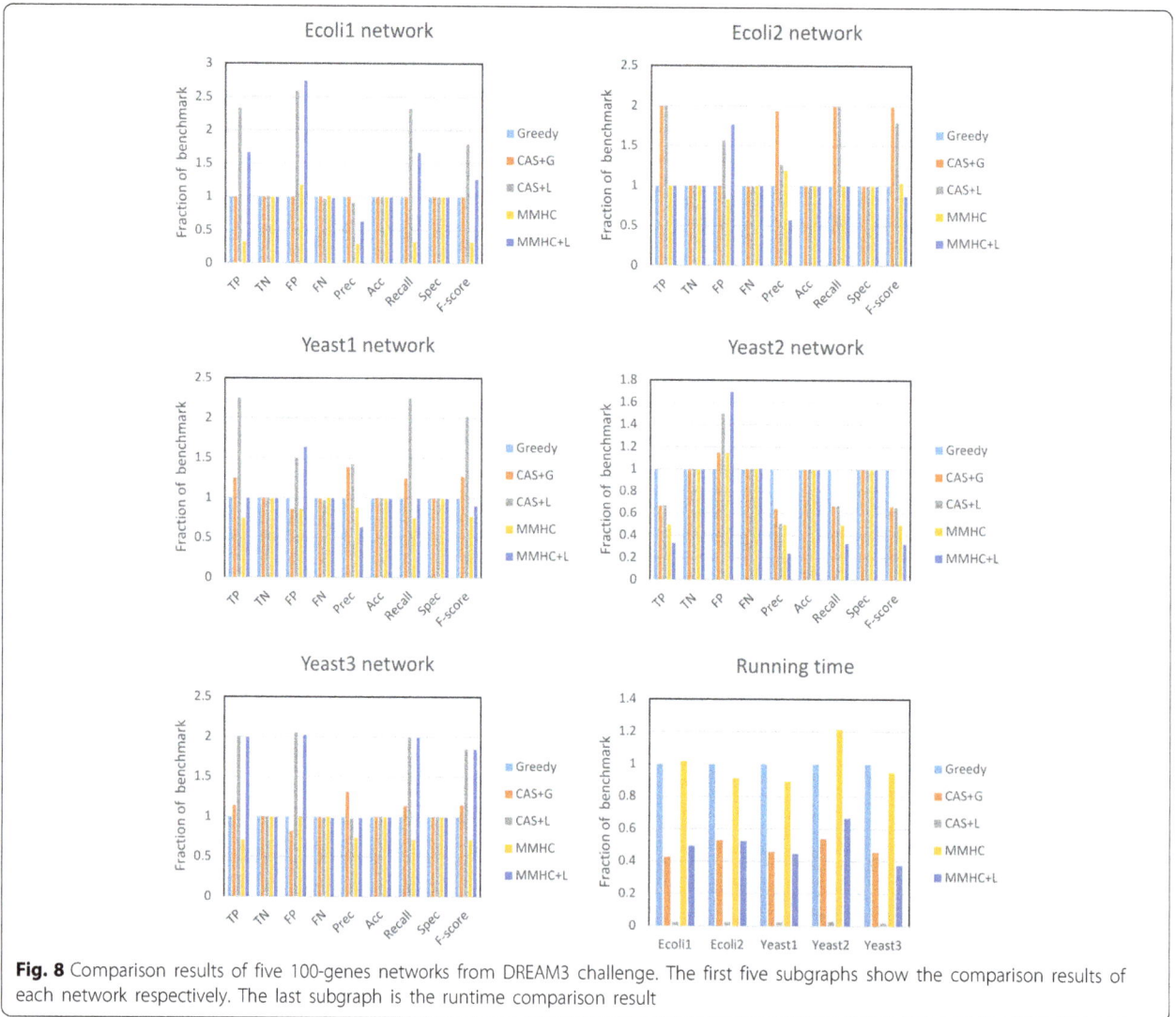

Fig. 8 Comparison results of five 100-genes networks from DREAM3 challenge. The first five subgraphs show the comparison results of each network respectively. The last subgraph is the runtime comparison result

restricting the neighbors of each node to a small candidate set. Firstly, MIs between nodes are calculated to reflect the independence. It is reasonable to assume that the distribution of the MIs of two types of nodes are different. That is to say, there is a breakpoint in the MI vector of each node to distinguish related nodes and unrelated nodes. Then, the breakpoint is located by hypothesis testing. So far, the candidates of each node are obtained. In the later learning phase, these candidates exactly restrict the search space. Hence, based on CAS algorithm, we propose a global optimal method (CAS + G), which focuses on finding the high-scoring network structure, and a local learning method (CAS + L), which focus on faster learning the structure with small loss of quality. At last, we validate the proposed algorithms on through experiments. Firstly, they are verified on known Bayesian networks. Then, the proposed methods are migrated to simulated biological data.

In candidate selection phase, the CAS algorithm correctly identifies the candidates of each node in polynomial time by discarding all nodes in common effect branch. The algorithm achieves relatively high performance on known Bayesian network datasets. However, it degrades on simulated data. Actually, it is not complicated to understand by considering the complexity of biological systems and the limitations of simulated data. The results show that CAS algorithm outperforms the MMPC algorithm. Especially, the CAS algorithm shows better performance on limited samples.

In structure learning phase, evaluations of CAS + G algorithm and CAS + L algorithm are carried out. The comparisons results show that the CAS + G method can learn the optimal structure and can avoid the local optimum to some extent benefited from the exactly restricted search space. Meanwhile, the CAS + L algorithm has obvious superiority in speed compared with other methods. Therefore, the proposed algorithms are

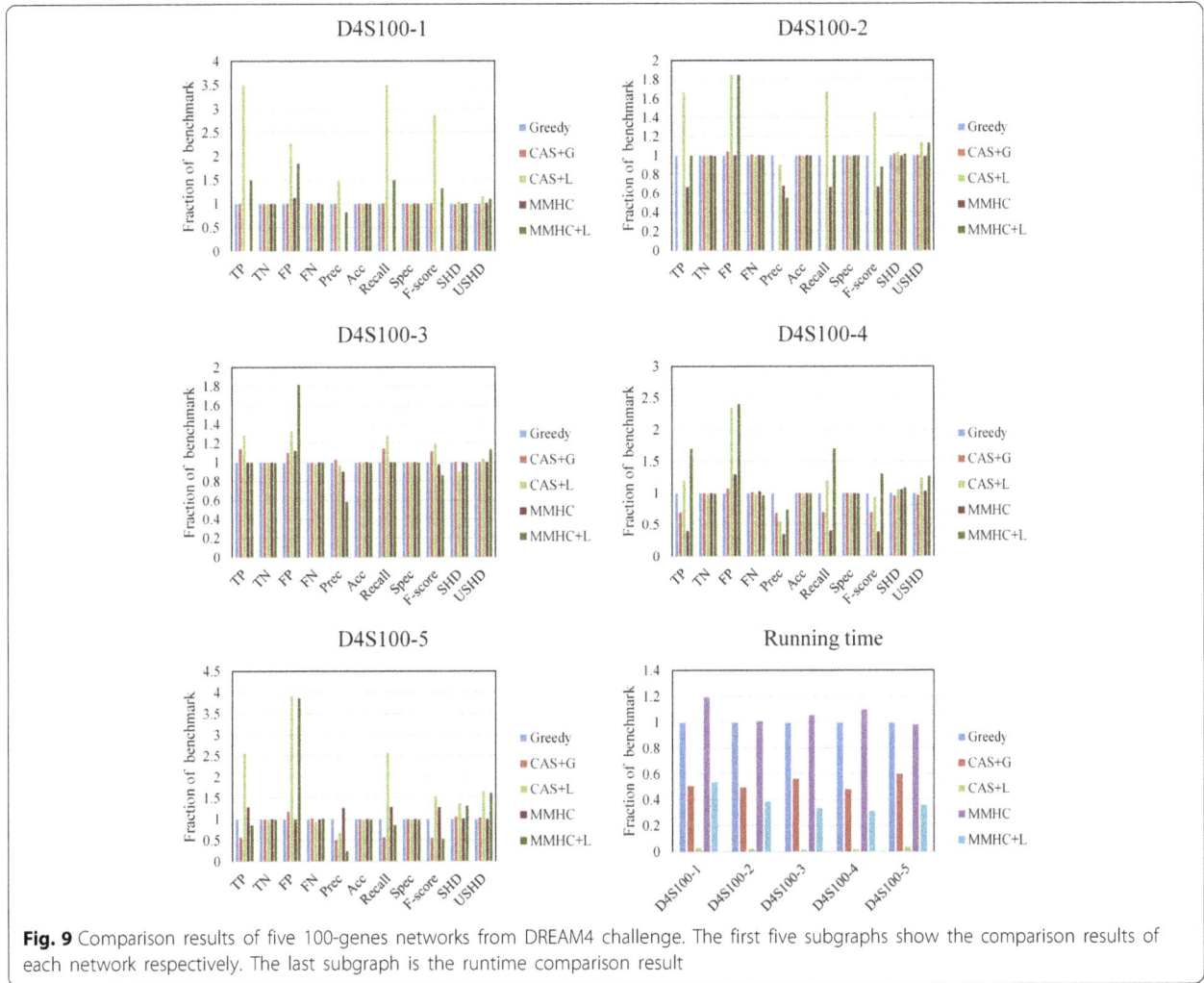

Fig. 9 Comparison results of five 100-genes networks from DREAM4 challenge. The first five subgraphs show the comparison results of each network respectively. The last subgraph is the runtime comparison result

effective and more suitable for GRN inference than MMHC algorithm.

Finally, the CAS algorithm is practical to reduce the search space, especially for limited samples, and provides enough flexibility to be extended in other fields. In the future, we would like to study the CAS algorithm in Markov Blanket view and consider a parallel implementation of the proposed algorithms.

Acknowledgements
We thank the members of the Natural Computing group for thoughtful discussions.

Funding
This work was supported by the Natural Science Foundation of China (Grant No. 61571163, 61532014, 61671189, and 61402132), and the National Key Research and Development Plan Task of China (Grant No. 2016YFC0901902). Grant No. 61571163 funded the publication costs.

About this supplement
This article has been published as part of *BMC Genomics* Volume 18 Supplement 9, 2017: Selected articles from the IEEE BIBM International Conference on Bioinformatics & Biomedicine (BIBM) 2016: genomics. The full contents of the supplement are available online at https://bmcgenomics.biomedcentral.com/articles/supplements/volume-18-supplement-9.

Authors' contributions
LX, MG, XL, CW, LW, YZ conceived and planed the research. LX designed and carried out the comparative study, wrote the code, and drafted the manuscript. LX, MG, XL and CW reviewed and edited the manuscript. All authors read and approved the manuscript.

Ethics approval and consent to participate
Not applicable.

Competing interests
All authors declare that they have no competing interests.

Author details
[1]School of Computer Science and Technology, Harbin Institute of Technology, Harbin, China. [2]School of Electrical and Information Engineering, Beijing University of Civil Engineering and Architecture, Beijing, China. [3]Institute of Health Service and Medical Information, Academy of Military Medical Sciences, Beijing, China.

References

1. Davidson EH, Rast JP, Oliveri P, Ransick A, Calestani C, Yuh CH, Minokawa T, Amore G, Hinman V, Arenas-Mena C, et al. A genomic regulatory network for development. Science. 2002;295(5560):1669–78.

2. D'Haeseleer P, Liang S, Somogyi R. Genetic network inference: from co-expression clustering to reverse engineering. Bioinformatics. 2000;16(8):707–26.

3. De Jong H. Modeling and simulation of genetic regulatory systems: a literature review. J Comput Biol. 2002;9(1):67–103.

4. Basso K, Margolin AA, Stolovitzky G, Klein U, Dalla-Favera R, Califano A. Reverse engineering of regulatory networks in human B cells. Nat Genet. 2005;37(4):382–90.

5. Lee TI, Rinaldi NJ, Robert F, Odom DT, Bar-Joseph Z, Gerber GK, Hannett NM, Harbison CT, Thompson CM, Simon I, et al. Transcriptional regulatory networks in Saccharomyces Cerevisiae. Science. 2002;298(5594):799–804.

6. Qin S, Ma F, Chen L. Gene regulatory networks by transcription factors and microRNAs in breast cancer. Bioinformatics. 2015;31(1):76–83.

7. Long TA, Rady SM, Benfey PN. Systems approaches to identifying gene regulatory networks in plants. *Ann Rev Cell Dev Biol*. 2008;24:81–103.

8. Hecker M, Lambeck S, Toepfer S, van Someren E, Guthke R. Gene regulatory network inference: data integration in dynamic models—a. Biosystems. 2009;96:86–103.

9. Kærn M, William J, Blake A, Collins JJ. The engineering of gene regulatory networks. Ann Rev Biomed Eng. 2003;5(1):179–206.

10. Chen T, He HL, Church GM. Modeling gene expression with differential equations. Pac Symp Biocomput. 1999;4:29–40.

11. Bansal M, Della Gatta G, di Bernardo D: Inference of gene regulatory networks and compound mode of action from time course gene expression profiles. Bioinformatics 2006, 22(7):815-822.

12. Zhang B, Horvath S. A general framework for weighted gene co-expression network analysis. Stat Appl Genet Mol Biol. 2005;4:1–45. doi:10.2202/1544-6115.1128.

13. Langfelder P, Horvath S. WGCNA: an R package for weighted correlation network analysis. Bmc Bioinformatics. 2008;9:559.

14. Song L, Langfelder P, Horvath S. Comparison of co-expression measures: mutual information, correlation, and model based indices. Bmc Bioinformatics. 2012;13(1):328.

15. Luo Q, Liu X, Yi D. Reconstructing gene networks from microarray time-series data via granger causality. In: Zhou J, editor. Complex Sciences. Berlin: Springer; 2009. pp. 196-209.

16. Siyal MY, Furqan MS, Monir SMG. Granger causality: Comparative analysis of implementations for Gene Regulatory Networks. In: Control Automation Robotics & Vision (ICARCV), 2014 13th International Conference on. Marina Bay Sands: IEEE; 2014. pp. 793-8.

17. Li P, Zhang C, Perkins EJ, Gong P, Deng Y. Comparison of probabilistic Boolean network and dynamic Bayesian network approaches for inferring gene regulatory networks. Bmc Bioinformatics. 2007;8(Suppl 7):S13.

18. Shmulevich I, Dougherty ER. Probabilistic Boolean networks: the modeling and control of gene regulatory networks. Philadelphia: Society for Industrial and Applied Mathematics; 2010. doi:10.1137/1.9780898717631

19. Higa CH, Andrade TP, Hashimoto RF. Growing seed genes from time series data and thresholded Boolean networks with perturbation. IEEE/ACM Trans Comput Biol Bioinform. 2013;10(1):37–49.

20. Yi Ming Z. Boolean networks with multi-expressions and parameters. IEEE/ACM Trans Comput Biol Bioinform. 2013;99(PrePrints):1–1.

21. Friedman N, Nachman I, Peér D. Learning bayesian network structure from massive datasets: the 'sparse candidate' algorithm. In: Proceedings of the Fifteenth conference on Uncertainty in artificial intelligence(UAI); 1999. Stockholm: Morgan Kaufmann Publishers Inc.; 1999. pp. 206-15.

22. Friedman N, Linial M, Nachman I, Pe'er D. Using Bayesian networks to analyze expression data. J Comput Biol. 2000;7(3–4):601–20.

23. Murphy K, Mian S. Modelling gene expression data using dynamic bayesian networks. Berkeley: Tech rep, Computer Science Division, University of California; 1999.

24. Friedman N, Koller D. Being Bayesian about network structure. A Bayesian approach to structure discovery in Bayesian networks. Mach Learn. 2003;50(1–2):95–125.

25. Chen T, He HL, Church GM. Modeling gene expression with differential equations. In: Pacific symposium on biocomputing(PSB): 1999; Mauna Lani, Hawaii: World Scientific Publishing; 1999:29–40.

26. Li Z, Li P, Krishnan A, Liu J. Large-scale dynamic gene regulatory network inference combining differential equation models with local dynamic bayesian network analysis. Bioinformatics. 2011;27(19):2686–91.

27. Henriques D, Rocha M, Saez-Rodriguez J, Banga JR. Reverse engineering of logic-based differential equation models using a mixed-integer dynamic optimization approach. Bioinformatics. 2015;31(18):2999–3007.

28. Kauffman SA. Metabolic stability and epigenesis in randomly constructed genetic nets. J Theor Biol. 1969;22(3):437–67.

29. Shmulevich I, Dougherty ER, Kim S, Zhang W. Probabilistic Boolean networks: a rule-based uncertainty model for gene regulatory networks. Bioinformatics. 2002;18(2):261–74.

30. Koivisto M, Sood K. Exact Bayesian structure discovery in Bayesian networks. J Mach Learn Res. 2004;5(5):549–73.

31. Zou M, Conzen S. A new dynamic Bayesian network (DBN) approach for identifying gene regulatory networks from time course microarray data. Bioinformatics. 2005;21(1):71–9.

32. De Campos LM. A scoring function for learning Bayesian networks based on mutual information and conditional independence tests. The Journal of Machine Learning Research. 2006;7:2149–87.

33. Parviainen P, Koivisto M. Exact structure discovery in Bayesian networks with less space. In: Proceedings of the twenty-fifth conference on uncertainty in artificial intelligence. Montreal: AUAI Press; 2009. p. 436–43.

34. Nair A, Chetty M, Wangikar PP. Improving gene regulatory network inference using network topology information. Mol BioSyst. 2015;11(9):2449–63.

35. Tsamardinos I, Brown LE, Aliferis CF. The max-min hill-climbing Bayesian network structure learning algorithm. Mach Learn. 2006;65(1):31–78.

36. Rottger R, Ruckert U, Taubert J, Baumbach J. How little do we actually know? On the size of gene regulatory networks. Comput Biol Bioinformat, IEEE/ACM Transact. 2012;9(5):1293–300.

37. Teichmann SA, Babu MM. Gene regulatory network growth by duplication. Nat Genet. 2004;36(5):492–6.

38. Wang YX, Waterman MS, Huang H. Gene coexpression measures in large heterogeneous samples using count statistics. Proc Natl Acad Sci U S A. 2014;111(46):16371–6.

39. Steuer R, Kurths J, Daub CO, Weise J, Selbig J. The mutual information: detecting and evaluating dependencies between variables. Bioinformatics. 2002;18:S231–40.

40. Frenzel S, Pompe B. Partial mutual information for coupling analysis of multivariate time series. Phys Rev Lett. 2007;99(20):204101.

41. Butte A, Kohane I. Mutual information relevance networks: functional genomic clustering using pairwise entropy measurements. Pac Symp Biocomputing. 2000;5:415–26.

42. Margolin AA, Nemenman I, Basso K, Wiggins C, Stolovitzky G, Dalla Favera R, Califano A. ARACNE: an algorithm for the reconstruction of gene regulatory networks in a mammalian cellular context. Bmc Bioinformatics. 2006;7(Suppl 1):S7.

43. Altay G, Emmert-Streib F. Revealing differences in gene network inference algorithms on the network level by ensemble methods. Bioinformatics. 2010;26(14):1738–44.

44. Zhang X, Zhao X-M, He K, Lu L, Cao Y, Liu J, Hao J-K, Liu Z-P, Chen L. Inferring gene regulatory networks from gene expression data by path consistency algorithm based on conditional mutual information. Bioinformatics. 2012;28(1):98–104.

45. Schaffter T, Marbach D, Floreano D. GeneNetWeaver: in silico benchmark generation and performance profiling of network inference methods. Bioinformatics. 2011;27(16):2263–70.

46. Heckerman D, Geiger D, Chickering DM. Learning Bayesian networks - the combination of knowledge and statistical-data. Mach Learn. 1995;20(3):197–243.

17

Whole genome analysis of CRISPR Cas9 sgRNA off-target homologies via an efficient computational algorithm

Hong Zhou[1*], Michael Zhou[2], Daisy Li[2], Joseph Manthey[1], Ekaterina Lioutikova[1], Hong Wang[3] and Xiao Zeng[4]

Abstract

Background: The beauty and power of the genome editing mechanism, CRISPR Cas9 endonuclease system, lies in the fact that it is RNA-programmable such that Cas9 can be guided to any genomic loci complementary to a 20-nt RNA, single guide RNA (sgRNA), to cleave double stranded DNA, allowing the introduction of wanted mutations. Unfortunately, it has been reported repeatedly that the sgRNA can also guide Cas9 to off-target sites where the DNA sequence is homologous to sgRNA.

Results: Using human genome and Streptococcus pyogenes Cas9 (SpCas9) as an example, this article mathematically analyzed the probabilities of off-target homologies of sgRNAs and discovered that for large genome size such as human genome, potential off-target homologies are inevitable for sgRNA selection. A highly efficient computationl algorithm was developed for whole genome sgRNA design and off-target homology searches. By means of a dynamically constructed sequence-indexed database and a simplified sequence alignment method, this algorithm achieves very high efficiency while guaranteeing the identification of all existing potential off-target homologies. Via this algorithm, 1,876,775 sgRNAs were designed for the 19,153 human mRNA genes and only two sgRNAs were found to be free of off-target homology.

Conclusions: By means of the novel and efficient sgRNA homology search algorithm introduced in this article, genome wide sgRNA design and off-target analysis were conducted and the results confirmed the mathematical analysis that for a sgRNA sequence, it is almost impossible to escape potential off-target homologies. Future innovations on the CRISPR Cas9 gene editing technology need to focus on how to eliminate the Cas9 off-target activity.

Keywords: sgRNA, Off-target homology, Crispr, Cas9, Computational algorithm, Genome wide

Background

Derived from the microbial clustered, regularly interspaced, short palindromic repeats (CRISPR) system, the Cas9 endonuclease has become an effective and reliable tool for genome editing in eukaryotes [1–6]. The magnificence of the working mechanism of Cas9 is that it can be guided by a 20-base sgRNA, immediately upstream the short DNA motif of Cas9, the so called protospacer adjacent motif (PAM), to almost any genome loci where the DNA sequence is

complementary to the sgRNA [1–4]. The PAM sequence is absolutely required for Cas9 to function and depends on the species of Cas9. For SpCas9, the most used Cas9 species, the PAM sequence is NGG, where N can be either A, C, G, or T. The very first step in making use of the sgRNA-Cas9 system for genome editing is to locate a primary PAM within the target region. Immediately upstream the PAM, the 20 bases of DNA sequence is the guide RNA sequence. Though they can be on either the sense or antisense strand, the PAM and sgRNA sequences must be on the same DNA strand.

Certain rules regarding the design of active sgRNAs have been proposed [6, 7]. As the gene editing

* Correspondence: hzhou@usj.edu
[1]Department of Mathematical Science, University of Saint Joseph, 1678 Asylum Avenue, West Hartford, CT 06117, USA
Full list of author information is available at the end of the article

mechanism of sgRNA-Cas9 is to generate indels via DNA repairing mechanisms, it is not difficult to understand that for mRNA genes, the target site should better be inside the gene coding sequence and be near the start codon. Another design rule is the GC content. It was found that higher sgRNA GC content could result in higher Cas9 activities [8]. In addition, the design of sgRNA should avoid certain sequences, for example, polyT [7].

One of the most important design rules is to avoid potential Cas9 off-target activity. Unfortunately, a significant number of experiments discovered undesired off-target cleavages by Cas9 at off-target genome sites where the DNA sequences are homologous to the 20-base sgRNA, though with one or more mismatches [7–16]. Considering the large size of some genomes, for example human, mouse and rat genomes, avoiding off-target Cas9 activities immediately becomes the most critical challenge in the application of the sgRNA-Cas9 technology. Systematic research has revealed sequence features governing sgRNA off-target interaction. However, the possible off-target Cas9 cleavages remain a defect and a challenge in sgRNA-Cas9 applications.

The large number of off-target studies of the sgRNA-Cas9 system has led to significant discoveries. Jinek et al. was the first to identify a seed sequence that is less tolerant to mismatches for sgRNA-Cas9 activity [1]. The definition of the seed sequence is generally considered to be the 12 bases on the 3′ end of sgRNA sequence, immediately upstream PAM [1, 10–12]. Mali et al. found that sgRNA-Cas9 system can tolerate one to three target mismatches, and two mismatches inside the seed sequence can eliminate off-target activity [11]. Based on their data, Fu et al. concluded that off-target activity can be observed with up to five mismatches when the concentrations of both sgRNA and Cas9 are relatively high [9]. Hsu et al. discovered that off-target activity depends on the number and positions of the mismatches between sgRNA and target DNA sequence [10]. Lin et al. systematically studied the sgRNA-Cas9 off-target activities when there are indels between target DNA and sgRNA sequences [13]. Their results showed that sgRNAs with low GC content have less tolerance to mismatches. They also found, that a bulge in sgRNA or DNA preserves less Cas9 activity, a result later confirmed by Doench et al. [7].

Making the off-target activity of sgRNA-Cas9 system even more complicated, it has been observed that secondary PAM sequences, in addition to the NGG motifs, can render Cas9 activity [3, 7, 17, 18]. Though these secondary PAMs are far less effective compared to the NGG PAMs, they must be taken into consideration for off-target searches [3, 7]. For SpCas9, the secondary PAMs include NAG, NCG, and NGA [3, 7].

The complexity of the Cas9-sgRNA off-target interaction and the large size of human genome led us to wonder the probability that a given sgRNA sequence has at least one off-target homology. Theoretically, will it be possible to apply the Cas9-sgRNA system without any potential off-target homologies that may introduce unwanted genome editing? In this article, we analyze this question from a mathematical perspective, and then present a very efficient algorithm for sgRNA off-target homology search. This algorithm can complete a whole genome sgRNA design and off-target search in about 40 h under a default setting, an efficiency that cannot be achieved by other available sgRNA software. Via this algorithm, we searched the off-target homologies for all sgRNAs designed for all human mRNA genes. The computational results confirmed our mathematical analysis.

Methods

The human genome was the sequence source used in this study. As SpCas9 is the most widely used CRISPR-Cas9 system, this study focuses on the mathematical and computational analysis of sgRNA-SpCas9 system. Human mRNA refseq sequence was downloaded from NCBI as the source for sgRNA sequence design. The off-target site search for designed sgRNA sequences were conducted on human chromosome sequences hs_ref_GRCh38.p2 which were also downloaded from NCBI. Computational programs were implemented in Java and executed on a 2016 Dell Precision 7510 laptop computer with Intel(R) Core(TM) i7-6820HQ CPU @ 2.7 GHz and 64.00 GB RAM.

Mathematical analysis

One crucial assumption made in this mathematical analysis is that the nucleotides A, C, G, T appear randomly at any single location. As there are repeated sequences in human genome, treating the human genome as a purely random combination of A, C, G, T must be regarded as a simplifying assumption. Furthermore, we also assume that human genome has exactly three billion 23-base regions for sgRNA off-target search on one DNA strand. Since the sgRNA can be designed on both the sense and antisense strands, the off-target homologies must be searched on both DNA strands. Thus, the total length of human genome contains six billion 23-base regions. For off-target homology search, we then make the following assumptions:

1. All off-target homologies must have a primary NGG PAM or a secondary PAM immediately downstream the sgRNA binding location.

2. All off-target homologies can have up to four base mismatches within a given sgRNA sequence. If there are at least five base mismatches, the DNA sequence in study is not considered an off-target homology. The reason for defining four instead of five base mismatches as the cut-off is because we have found only one active off-target homology with five base mismatches in the literature, and the off-target activity in that case could be eliminated by lowering both the Cas9 and sgRNA concentrations [9].

3. All off-target homologies can have at most one bulge plus one base mismatch [8, 13]. This implies that a bulge penalty equals three base mismatches.

4. All off-target homologies can have up to two base mismatches or one indel within the seed sequence of sgRNA.

5. No off-target homology can have a DNA bulge that is of two-bases, though an off-target homology can have a RNA bulge of two-bases but with no base mismatch at the same time. No off-target homology can have a bulge of two bases inside the seed sequence.

Based on the above five assumptions, we computed the possible combinations of homologies given a sgRNA sequence. The results are summarized in Table 1. The

Table 1 Mathematical analysis of the sgRNA off-target homologies

Total combination of 20 bp	1,099,511,627,776
Mismatches in seed sequence	Number of combinations
0	1
1	36
2	594
Mismatches in non-seed sequence	Number of combinations
0	1
1	24
2	252
3	1512
4	5670
Total base mismatches	236,401
DNA bulge with 0 base mismatch	64
RNA bulge with 0 base mismatch	60
DNA bulge with 1 base mismatch	2112
RNA bulge with 1 base mismatch	1968
RNA bulge of two bases	32
Total combinations of homologies	240,637
Off-target homology probability of a 20-base DNA sequence	0.00000021886
Probability of potential PAM	0.2500
Off-target homology probability	0.00000005471

following explains how the data in Table 1 were obtained.

The number of combinations of DNA sequences with different numbers of mismatches is computed by the expression.

$$\binom{m}{n} \times 3^n,$$

where m = the length of the DNA sequence in consideration, n = number of mismatches. Thus, for the seed sequence of 12 bases, there are 1, 36 and 594 combinations respectively for zero, one and two base mismatches.

As the total base mismatches cannot exceed four, the available base mismatches for the remaining non-seed regions would be zero, one, two, three and four, and can only have a maximum of three or two base mismatches if the seed sequence has one or two base mismatches. So, the total combinations of homologies with up to four base mismatches is computed as:

$$1 \times (1 + 24 + 252 + 1512 + 5670) + 36 \\ \times (1 + 24 + 252 + 1512) + 594 \times (1 + 24 + 252) \\ = 236401$$

The computation of the number of combinations of indels deserves a detailed explanation. There are two cases, DNA bulge, i.e. there is an additional base in the DNA sequence, and RNA bulges, i.e., when there are one or two less bases in the DNA sequence. For both DNA bulge and RNA bulge, there are two sub-cases, i.e. a bulge with zero or one base mismatch. However, for RNA bulge of two bases (there are two bases less inside the aligned DNA sequence), the number of base mismatches must be zero. In addition, if the bulge is inside the seed sequence, then no base mismatch is allowed to be inside the seed sequence.

We start with the DNA bulge with zero mismatches, which means that the 20-base RNA sequence is in fact aligned with a 21-base DNA sequence and all the 20 bases of sgRNA must have an exact match to a base in the DNA sequence. In a 20 vs 20 exact alignment, there are a maximum of 20 positions in the DNA sequence to insert one additional base, and this additional base can be either one of A, C, G, T. There are two additional restrictions when considering a DNA bulge: a DNA bulge can be considered only when there are at least five base mismatches between the sgRNA and DNA sequences (20 bases vs 20 bases) and the introduction of the bulge can trade off more than the number of base mismatches that a bulge penalty equals. Thus, when introducing a bulge inside the DNA sequence, the DNA fragment left of the bulge must be at least four bases such that there are enough base mismatches to be traded off by the bulge. Therefore, there are $16 \times 4 = 64$ combinations. Via the same logic, the RNA bulge with no base mismatches will have $15 \times 4 = 60$ combinations.

When there is an indel and a mismatch, the computation becomes a bit more complicated. For DNA bulge, the bulge can be anywhere but the mismatch can only be inside the non-seed region if the bulge is already inside the seed sequence. Thus, the maximum combinations of the indel plus a base mismatch would be

$$12 \times 4 \times \binom{8}{1} \times 3 + 4 \times 4 \times \binom{20}{1} \times 3 = 2112$$

However, for RNA bulge case, the expression would be

$$11 \times 4 \times \binom{8}{1} \times 3 + 4 \times 4 \times \binom{19}{1} \times 3 = 1968$$

The last condition in consideration is the RNA bulge of two bases. Since a two-bases RNA bulge can only be inside the non-seed region, there are only two different ways to form such a RNA bulge because the introduction of such a bulge must trade off at least five base mismatches. The combinations would be

$$2 \times 4 \times 4 = 32$$

Based on data in Table 1, the probability for a 23-base single DNA region to be an off-target homology for a given sgRNA sequence is 0.00000005471. Considering the fact that there are six billion 23-base single DNA sequences, The probability for a sgRNA to have no potential off-target homology is 2.67×10^{-143}, and the expected number of off-target sites is 328.

Based on the above mathematical analysis, it seems that for a given SpCas9 sgRNA sequence, potential off-target homologies in the human genome are unavoidable.

Computational algorithm

We implemented a sgRNA design and off-target search algorithm in Java. The sgRNA design is based on the rules outlined in [6, 7] with the following exceptions: 1) sgRNA are designed only inside the first half CDS sequence; 2) all sgRNAs do not contain a run of four T or four A.

As the off-target search must be conducted through all the human chromosome sequences, the off-target search of sgRNA can be very time expensive. The high efficiency of our off-target search process comes from two critical algorithmic innovations which are explained below in detail.

The first innovation is that an indexed database based on the seed sequence variations is dynamically constructed

before any homology search work starts. Based on assumption 4, for a DNA region to be an off-target homology of a given sgRNA, it must have a good alignment with the sgRNA seed sequence such that there should be at most two mismatches or one indel. Hence, the off-target homology search starts with finding those DNA sequences that are variations of the sgRNA seed sequence. The seed sequence consists of 12 bases, so there are 4^{12} different 12-base variations in total. If we assign 0, 1, 2, 3 to A, C, G, T respectively and convert DNA sequence to a base-4 number system, then each 12-base variation can then be represented as a unique integer using the expression $\sum_{i=0}^{11} N \times 4^i$, where $N = 0, 1, 2, 3$, representing A, C, G, T respectively.

Since the package was implemented in Java whose *int* data type can only hold integers ranging from -2^{31} to 2^{31}-1 and the human genome has about three billion base pairs, i.e. six billion bases, we decided to divide the 24 chromosomes into two groups with roughly equal number of nucleotides. For each group, a two-dimensional array G_{ij} is constructed as follows: i = the integer value of each 12-base sequence, the row G[i] stores all the positions of the 12-base sequence (equivalent to integer i) in the group of chromosomes. A positive G[i][j] indicates that the position is on the sense strand while a negative G[i][j] means that the 12-base sequence is found on the anti-sense strand. Given the integer G[i][j], a conversion system matches it to a specific chromosome, a specific NT record, and a specific position inside the NT sequence. An important tip in constructing the two-dimensional array G_{ij} is that G_{ij} only stores the location information of those 12-base sequences followed by a primary PAM or a secondary PAM.

Given a 20-base sgRNA sequence, based on its 12-base seed sequence, all variations of its 12-base seed sequence are generated according to Assumption 4, which are interpreted as: 1) a variation can have at most two mismatches with this seed sequence; 2) a variation can have at most one indel when aligned against the seed sequence. The homology search algorithm then finds all the exact positions inside each NT record for all the different variations very quickly and then uses a dynamic programming algorithm to determine if there is an off-target homology at each position.

The second innovation is the efficient dynamic programming algorithm for homology determination. The dynamic programming algorithm is illustrated in Table 2.

The construction of Table 2 is explained as follows. Given a DNA sequence marked as *d* and a sgRNA sequence marked as *r*, for *d* to be an off-target homology of *r*, it must have a PAM (either primary PAM or secondary PAM) that aligns with the PAM of *r*.

Let H, L1, R1, L2, R2 = an array of integer respectively and length = 21.

Let i = the subscript of H, L1, R1, L2, and R2. Please note that i starts from 1.

Let H[21] = 0 //compute the number of base mismatches

Loop i from 20 to 1, step=1, do:

   If d[i]=r[i], then

      H[i] = H[i+1];

   else

      H[i] = H[i+1] + 1;

   End if

End Loop

For DNA bulges of 1 base or 2 bases, which are marked as L1 and L2 respectively in Table 2, the values are computed as:

      Let m = 0. Let n = 1 if for L1, otherwise n=2

      Loop i from 1 to 20 do

         If r[i] not = d[i-n] then

            m = m + 1

         End if

         L[i] = m + H[i+1]

      End Loop

For RNA bulges of 1 base or 2 bases, which are marked as R1 and R2 respectively in Table 2, the values are computed as:

      Let m = 0; n = 1 if for R1; otherwise n=2

      Loop i from 1 to (21-n-1) do

         If r[i] not = d[i+n] then

            m = m + 1

         End if

        R[i] = m + H[i+n+1]

      End Loop

The above algorithm computes the number of base mismatches only, which are the values in Table 2. For L1, L2, R1 and R2, as there is a specific bulge for each case, the total number of mismatches should add the specific bulge penalty. In our default setting, a bulge penalty equals three base mismatches (counted as two if inside the seed sequence), a RNA bulge extension penalty equals one base mismatch, and a DNA bulge extension penalty equals two base mismatches. Thus in Table 2, when L1 is computed, though it is shown that L[13] = L[14] = L[15] = L[16] = 1, they are in fact = 1 + DNA bulge penalty = 4. The result shows that by shifting the 5′ fragment (up to either the 13th, 14th, 15th, or 16th base) one base to the left, we can achieve an alignment with only one base mismatch and one DNA bulge.

The above algorithm illustrates the general condition. There are some special cases that the implementation must also consider:

- Since the seed sequence has more stringent requirements on the number of mismatches, the number of base mismatches and indels within the seed sequence should be counted and stored to determine whether or not a specific alignment should be considered as an off-target homology. In the example shown in Table 2, though the case of L1 can achieve a good alignment with only one base mismatch and one DNA bulge, d is eventually not considered a homology to r because both the DNA bulge and the base mismatch are inside the seed sequence,
- There are a total of five cases that are computed in this algorithm: H, L1, R1, L2, R2. If in one case d is found to be a homology to r, there is no need to go on to the next case.
- For cases L1, R1, L2, and R2, a shortcut can be applied. If (m + bulge penalty) become larger than the number of base mismatches allowed, there is no need to continue computing for that case because it is guaranteed that the alignment represented by this case is not a homology.

Table 2 Dynamic programming illustration. i: the subscript of the table; d: the DNA sequence; r: the sgRNA sequence; H: the number of base mismatches; L1: 1-base DNA bulge; R1: 1-base RNA bulge; L2: 2-base DNA bulge; R2: 2-base RNA bulge

i		1	2	3	4	5	6	7	8	9	10	11	12	13	14	15	16	17	18	19	20	21	
d	T	G	G	A	C	C	C	A	A	A	G	T	G	G	T	T	T	A	G	C	G	A	PAM
r		G	G	A	C	C	C	A	A	A	G	T	G	G	T	T	T	G	G	C	G	A	PAM
H		8	8	7	6	6	6	5	5	5	4	3	2	2	1	1	1	0	0	0	0	0	
L1		8	7	6	6	6	5	5	5	4	3	2	2	1	1	1	1	2	3	4	5		
R1		8	8	9	9	9	10	10	10	10	9	10	10	11	11	11	11	12	13	14			
L2		9	8	8	9	9	8	9	9	8	8	8	9	8	9	9	9	10	11	11	12		
R2		7	8	9	10	10	11	11	11	11	11	10	11	12	12	13	14	14	15				

Results and discussion

We first simulated a human genome of size three billion base pairs in which A, C, G, T are randomly distributed. With this simulated genome, we examined the off-target homologies for 1,000,000 sgRNAs randomly designed from the simulated genome and the 1,876,775 sgRNAs designed for the 19,153 human mRNA genes based on the above design rules. The off-target homology search identified 326 homologies per sgRNA in average for the group of 1,000,000 sgRNAs and 325 homologies per sgRNA in average for the group of 1,876,775 sgRNAs. Both results are fairly close to the mathematically expected 328 homologies. In fact, the mathematically expected values should be slightly larger than the computational experimental values because of two reasons. The first reason is that the mathematically calculated number of combinations for the case with one indel plus one base mismatch is the possible maximum number. The real number should be slightly smaller. The second reason can be explained by using the sequence alignment (DNA) ACCCCT/acccct (RNA) as an example. Removing any C will generate the same RNA bulge ACCCT/acccct, i.e. the computational experiment will detect one RNA bulge while the mathematical model would count four times. Overall, in agreement with our mathematical model, no sgRNA was found to be free of homologies with the simulated genome.

The computational experiment with human genome identified that only two out of the 1,876,775 sgRNAs were validated to be free of off-target homology. This confirms our mathematical analysis that theoretically, it is almost impossible for a sgRNA to have no potential off-target homologies. A total of 1,415,606,013 off-target homologies were found, indicating 754 off-target homologies per sgRNA. This number is significantly larger than the mathematical expected value. We believe that the large discrepancy was resulted from the fact that human DNA sequence is not a random composition of A, C, G, T. There are a large number of repeated sequences in human genome [19]. As we once pointed out [20], some sgRNAs with repeated sequences have an unusually large number of off-target homologies, which contributes to the large discrepancy.

It is worth to point out that of the 1,415,606,013 homologies, about 2.70% are with indels. Thus, even though the off-target homologies are mostly base mismatches, indels are a significant portion of off-target homologies and should be considered. Some sgRNA off-target search algorithms, for example, CasFinder and CRISPOR, do not detect indels, and thus miss a significant number of off-target homologies [21, 22].

The time cost to complete a whole genome sgRNA design and off-target homology examination is mostly on the homology examination. The time cost is a linear function of the number of sgRNAs. Furthermore, based on our homology examination algorithm, it is easy to understand that the time cost is also a function of the off-target homology definition. Under our default homology examination settings, the time cost to complete the whole genome design and off-target examination for the 1,876,775 sgNRAs is about 40 h. It is roughly about 77 s for every 1000 sgRNAs.

Compared with CasFinder which is built upon Bowtie, our package is much more efficient. Under a similar homology examination setting (the seed sequences allows maximum two mismatches, the 20-base sequence allows totally up to four base mismatches but no bulge, and the secondary PAM is only NAG), CasFinder took 624 h to complete the design and off-target examination of its 927,104 sgRNAs while our algorithm took about 22 h to examining 1,876,775 sgRNAs [21]. Roughly speaking, our algorithm is about 57 times faster than CasFinder.

Cas-OFFinder employed a similar strategy as our algorithm except that they first computed the variations of the 20-base guide sequence with up to certain number of mismatches [23]. With each varied sequence, they tended to find an exact match in the genome. We also compared our algorithm's efficiency with theirs under the same conditions: up to five base mismatches, no indels, and only consider the NGG and NAG PAM. Cas-Offinder's maximum speed via GPU is about 3.01 s per sgRNA sequence. However, when comparing the CPU efficiency, Cas-Offinder's maximum speed is about 60.03 s per sgRNA sequence, while ours is about 3.15 s per sgRNA sequence.

Because each sgRNA has very high probability to have off-target homologies that can result in off-target Cas9 activity, avoiding potential off-target activity is in fact the most challenging and critical factor in designing sgRNAs. In addition to its efficiency, another advantage of our algorithm is that it guarantees to find all the potential off-target homologies based on the off-target homology setting. It has been reported that a few tools are likely to miss significant number of potential homologies [22, 24]. Thus, we compared our algorithm with CRISPOR (http://crispor.tefor.net/) and Cas-OFFinder (http://www.rgenome.net/cas-offinder) which were considered to be superior in locating off-target homologies [22]. Using the EMX1 guide sequence (GAGTCCGAGC AGAAGAAGAA) as an example, Table 3 shows that our algorithm achieves as good as both Cas-OFFinder and CRISPOR.

Under exactly the same conditions, our algorithm found exactly the same off-target homologies as Cas-OFFinder and CRISPOR did. The only difference is that, by default, our algorithm searched for off-target homologies anchored with all the secondary PAMs including NAG, NCG and NGA. The web-tool of Cas-OFFinder did not search for any secondary PAM, while CRISPOR considered only a few PAMs (NAG, AGA, GGA, TGA).

The large expected number of homologies for each sgRNA has been motivating scientists to search for different solutions. A double nicking approach was then introduced to enhance genome editing specificity [11, 25]. The double nicking method is based on the Cas9 nickase mutant that can only break one single strand of DNA. To obtain a double stranded cleavage, simultaneous nicking via two individual sgRNAs each targeting a different strand is necessary [25]. The offset, the distance between the 5′ ends of the two sgRNA sequences (sgRNA pair), must be between −4 and 20 for the paired nicking to work well, and if the offset of the paired sgRNAs is less than −34 or larger than 110 bases, the paired-sgRNA-Cas9 system completely loses its efficacy [25]. Thus, a potential off-target homology for paired sgRNA nicking must have two single off-target homologies positioned in a way that their offset is between −34

and 110 bases inclusive. After 387,679 sgRNA pairs were designed for the 19,153 mRNA genes, 175,712 sgRNA pairs were found to be free of off-target homologies, covering 14,665 mRNA genes. This confirms that the double nicking method is much more reliable than the original SpCas9-sgRNA system in avoiding off-target homologies, a finding reported before [16, 25].

Conclusions

A novel and efficient sgRNA homology search algorithm was introduced in this article. Via this algorithm, genome wide sgRNA design and off-target analysis were conducted and the results confirmed the mathematical analysis that for a sgRNA sequence, it is almost impossible to escape potential off-target homologies. Future innovations on the CRISPR Cas9 gene editing technology need to focus on how to eliminate the Cas9 off-target activity.

Acknowledgements
Not applicable.

Funding
The publication costs were funded by Hong Zhou's institutional award from University of Saint Joseph.

About this supplement
This article has been published as part of *BMC Genomics* Volume 18 Supplement 9, 2017: Selected articles from the IEEE BIBM International Conference on Bioinformatics & Biomedicine (BIBM) 2016: genomics. The full contents of the supplement are available online at https://bmcgenomics.biomedcentral.com/articles/supplements/volume-18-supplement-9.

Authors' contributions
All authors participated in the analysis of the data, interpretation of the results, and review of the paper. MZ, DL, and HZ performed the mathematical analysis, EL and JM reviewed the mathematical analysis. HZ and HW designed the algorithm, HZ implemented the algorithm, HZ and XZ generated the data. HZ, DL and MZ drafted the paper, JM and EL revised the paper. All authors read and approved the final version of the manuscript.

Competing interests
The authors declare that they have no competing interests.

Author details
[1]Department of Mathematical Science, University of Saint Joseph, 1678 Asylum Avenue, West Hartford, CT 06117, USA. [2]Hall High School, 975 N Main Street, West Hartford, CT 06117, USA. [3]Susan L. Cullman Laboratory for Cancer Research, Department of Chemical Biology and Centre for Cancer Prevention Research, Ernest Mario School of Pharmacy, Rutgers, The State University of New Jersey, 164 Frelinghuysen Road, Piscataway, NJ 08854, USA. [4]PBSG, LLC, P. O. Box 771, Braddock Heights, MD 21714, USA.

Table 3 Comparison between CRISPOR, Cas-OFFinder and the proposed algorithm on off-target homology search for EMX1 sgRNA guide sequence

Number of base mismatches	Number of off-target homologies identified				
	0	1	2	3	4
CRISPOR	0	0	6	38	296
Cas-OFFinder	0	0	1	18	273
Our Algorithm (with Secondary PAM)	0	0	6	87	1227
Our Algorithm (without secondary PAM)	0	0	1	18	273

References

1. Jinek M, Chylinski K, Fonfara I, Hauer M, Doudna JA, Charpentier E. A programmable dual-RNA-guided DNA endonuclease in adaptive bacterial immunity. Science. 2012;337:816–21.
2. Mali P, Yang L, Esvelt KM, Aach J, Guell M, DiCarlo JE, Norville JE, Church GM. RNA-guided human genome engineering via Cas9. Science. 2013;339: 823–6.
3. Hsu PD, Lander ES, Zheng F. Development and applications of CRISPR-Cas9 for genome engineering. Cell. 2014;157:1262–78.
4. Wright AV, Nuñez JK, Doudna JA. Biology and applications of CRISPR systems: harnessing nature's toolbox for genome engineering. Cell. 2016; 164:29–44.
5. Travis J. Making the cut CRISPR genome-editing technology shows its power. Science. 2015;350:1456–7.
6. Doench JG, Hartenian E, Graham DB, Tothova Z, Hegde M, Smith I, Sullender M, Ebert BL, Xavier RJ, Root DE. Rational design of highly active sgRNAs for CRISPR-Cas9-mediated gene inactivation. Nat Biotechnol. 2014; 32:1262–7.
7. Doench G, Fusi N, Sullender M, Hegde M, Vaimberg EW, Donovan KF, Smith I, Tothova Z, Wilen C, Orchard R, Virgin HW, Listgarten J, Root DE. Optimized sgRNA design to maximize activity and minimize off-target effects of CRISPR-Cas9. Nat Biotechnol. 2016;34:184–91.
8. Cradick TJ, Fine EJ, Antico CJ, Bao G. CRISPR/Cas9 systems targeting β-globin and CCR5 genes have substantial off-target activity. Nucleic Acids Res. 2013;41:9584–92.
9. Fu Y, Foden JA, Khayter C, Maeder ML, Reyon D, Joung JK, Sander JD. High-frequency off-target mutagenesis induced by CRISPR-Cas nucleases in human cells. *Nat Biotechnol*. 2013;31:822–6.
10. Hsu PD, Scott DA, Weinstein JA, Ran FA, Konermann S, Agarwala V, Li Y, Fine EJ, Wu X, Shalem O, Cradick TJ, Marraffini LA, Bao G, Zhang F. DNA targeting specificity of RNA-guided Cas9 nucleases. Nat Biotechnol. 2013;31: 827–32.
11. Mali P, Aach J, Stranges PB, Esvelt KM, Moosburner M, Kosuri S, Yang L, Church GM. Cas9 transcriptional activators for target specificity screening and paired nickases for cooperative genome engineering. Nat Biotechnol. 2013;31:833–8.
12. Pattanayak V, Lin S, Guilinger JP, Ma E, Doudna JA, Liu DR. High-throughput profiling of off-target DNA cleavage reveals RNA-programmed Cas9 nuclease specificity. Nat Biotechnol. 2013;31:839–43.
13. Lin Y, Cradick TJ, Brown MT, Deshmukh H, Ranjan P, Sarode N, Wile BM, Vertino PM, Stewart FJ, Bao G. CRISPR/Cas9 systems have off-target activity with insertions or deletions between target DNA and guide RNA sequences. Nucleic Acids Res. 2014;42:7473–85.
14. Wu X, Scott DA, Kriz AJ, Chiu AC, Hsu PD, Dadon DB, Cheng AW, Trevino AE, Konermann S, Chen S, Jaenisch R, Zhang F, Sharp PA. Genome-wide binding of CRISPR endonuclease Cas9 in mammalian cells. Nat Biotechnol. 2014;32:670–6.
15. Kuscu C, Arslan S, Singh R, Thorpe J, Adli M. Genome-wide analysis reveals characteristics of off-target sites bound by the Cas9 endonuclease. Nat Biotechnol. 2014;32:677–83.
16. Cho SW, Kim S, Kim Y, Kweon J, Kim HS, Bae S, Kim JS. Analysis of off-target effects of CRISPR/Cas-derived RNA-guided endonucleases and nickases. Genome Res. 2014;24:132–41.
17. Friedland AE, Tzur YB, Esvelt KM, Colaiacovo MP, Church GM, Calarco JA. Heritable genome editing in C. Elegans via a CRISPR-Cas9 system. Nat Methods. 2013;10:741–3.
18. Li JF, Norville JE, Aach J, McCormack M, Zhang D, Bush J, Church GM, Sheen J. Multiplex and homologous recombination-mediated genome editing in Arabidopsis and Nicotiana Benthamiana using guide RNA and Cas9. Nat Biotechnol. 2013;31:688–91.
19. http://www.repeatmasker.org/, Accessed 23 Feb 2017.
20. Zhou M, Li D, Huan X, Manthey J, Lioutikova E, Zhou H. Mathematical and computational analysis of CRISPR Cas9 sgRNA off-target homologies. In: Proceedings of the IEEE International Conference on Bioinformatics and Biomedicine: 15-18 December 2016. China: Shenzhen. p. 449–54.
21. Aach J, Mali P, Church GM: CasFinder: Flexible algorithm for identifying specific Cas9 targets in genomes. *bioRxiv* 2014, doi:https://doi.org/10.1101/005074.
22. Haeussler M, Schonig K, Eckert H, Eschstruth A, Mianne J, Renaud J, Schneider-Maunoury S, Shkumatava A, Teboul L, Kent J, Joly J, Concordet J. Evaluation of off-target and on-target scoring algorithms and integration into the guide RNA selection tool CRISPOR. Genome Biol. 2016;17:148.
23. Bae S, Park J, Kim JS. Cas-OFFinder: a fast and versatile algorithm that searches potential off-target sites of Cas9 RNA-guided endonuclease. Bioinformatics. 2014;30:1743–5.
24. Tsai SQ, Zheng Z, Nguyen NT, Liebers M, Topkar VV, Thapar V, Wyvekens N, Khayter C, Iafrate AJ, Le LP, Aryee MJ, Joung JK. Guide-seq enables genome-wide profiling of off-target cleavage by CRISPR-Cas nucleases. Nat Biotechnol. 2015;33:187–97.
25. Ran FA, Hsu PD, Lin CY, Gootenberg JS, Konermann S, Trevino AE, Scott DA, Inoue A, Matoba S, Zhang Y, Zhang F. Double nicking by RNA-guided CRISPR Cas9 for enhanced genome editing specificity. Cell. 2013;154:1380–9.

Permissions

List of Contributors

Noriyuki Fuku, Hirofumi Zempo, Hisashi Naito and Nobuyoshi Hirose
Graduate School of Health and Sports Science, Juntendo University, Chiba, Japan

Roberto Díaz-Peña
Hospital Universitari Institut Pere Mata, IISPV, URV. CIBERSAM, Reus, Spain
Facultad de Ciencias de la Salud, Universidad Autónoma de Chile,Talca, Chile

Yasumichi Arai and Yukiko Abe
Center for Supercentenarian Medical Research, Keio University School of Medicine, Tokyo, Japan

Haruka Murakami and Motohiko Miyachi
Department of Physical Activity Research; National Institutes of Biomedical Innovation, Health and Nutrition, Tokyo, Japan

Carlos Spuch
Neurology Group, Galicia Sur Health Research Institute (IIS Galicia Sur), Centro de investigación biomédica en red del área de salud mental (CIBERSAM), Vigo, Spain

José A. Serra-Rexach
Centro de investigación biomédica en Envejecimiento y Fragilidad (CIBERFES), Madrid, Spain

Enzo Emanuele
E Science, Robbio, (PV), Italy

Alejandro Lucia
European University and Research Institute i+12, Madrid, Spain

James Lara, Mahder Teka and Yury Khudyakov
Division of Viral Hepatitis, National Center for HIV, Hepatitis, TB and STD Prevention, Centers for Disease Control and Prevention, Atlanta, GA 30333, USA

Huaping Liu
Department of Bioinformatics, Key Laboratory of Ministry of Education for Gastrointestinal Cancer, School of Basic Medical Sciences, Fujian Medical University, Fuzhou 350122, China

Department of Systems Biology, College of Bioinformatics Science and Technology,Harbin Medical University, Harbin 150086, China

Yawei Li, Jun He, Qingzhou Guan, Rou Chen, Haidan Yan, Weicheng Zheng, Hao Cai, You Guo and Xianlong Wang
Department of Bioinformatics, Key Laboratory of Ministry of Education for Gastrointestinal Cancer, School of Basic Medical Sciences, Fujian Medical University, Fuzhou 350122, China

Kai Song
Department of Systems Biology, College of Bioinformatics Science and Technology, Harbin Medical University, Harbin 150086, China

Zheng Guo
Department of Bioinformatics, Key Laboratory of Ministry of Education for Gastrointestinal Cancer, School of Basic Medical Sciences, Fujian Medical University, Fuzhou 350122, China
Fujian Key Laboratory of Tumor Microbiology, Fujian Medical University, Fuzhou 350122, China
Department of Systems Biology, College of Bioinformatics Science and Technology, Harbin Medical University, Harbin 150086, China
Key Laboratory of Medical bioinformatics, Fujian Province, China

Zhiwen Chen, Hushuai Nie, Haili Pei, Shuangshuang Li and Jinping Hua
Laboratory of Cotton Genetics, Genomics and Breeding /Key Laboratory of Crop Heterosis and Utilization of Ministry of Education/Beijing Key Laboratory of Crop Genetic Improvement, College of Agronomy and Biotechnology, China Agricultural University, Beijing 100193, China

Yumei Wang
Institute of Cash Crops, Hubei Academy of Agricultural Sciences, Wuhan, Hubei 430064, China

Lida Zhang
Department of Plant Science, School of Agriculture and Biology, Shanghai Jiao Tong University, Shanghai 200240, China

Rubén Álvarez-Álvarez, Yolanda Martínez-Burgo and Paloma Liras
Microbiology Section, Faculty of Biological and Environmental Sciences, University of León, León, Spain

Antonio Rodríguez-García
Microbiology Section, Faculty of Biological and Environmental Sciences, University of León, León, Spain
Institute of Biotechnology of León, INBIOTEC, León, Spain

Emma Whittington, Kirill Borziak and Steve Dorus
Center for Reproductive Evolution, Department of Biology, Syracuse University, Syracuse, NY, USA

Desiree Forsythe
Science Education and Society, University of Rhode Island, Kingston, RI, USA

Timothy L. Karr
Ecology and Evolutionary Biology, Kansas University, Lawrence, KS, USA

James R. Walters
Department of Genomics and Genetic

Fang Du
College of Horticulture, Shanxi Agricultural University, Taigu 030801, China
Department of Horticulture, College of Agriculture & Biotechnology, Zhejiang University, Hangzhou 310058, China

Junmiao Fan and Ting Wang
College of Horticulture, Shanxi Agricultural University, Taigu 030801, China

Yun Wu, Zhongshan Gao and Yiping Xia
Department of Horticulture, College of Agriculture & Biotechnology, Zhejiang University, Hangzhou 310058, China

Donald Grierson
Department of Horticulture, College of Agriculture & Biotechnology, Zhejiang University, Hangzhou 310058, China
Plant & Crop Sciences Division, School of Biosciences, University of Nottingham, Sutton Bonington Campus, Loughborough LE12 5RD, UK

Glenna J. Kramer and Justin R. Nodwell
Department of Biochemistry, University of Toronto, MaRS Centre, West
Tower, 661 University Avenue, Toronto, ON M5G 1M1, Canada

Hiroki Takahashi
Medical Mycology Research Center, Chiba University, 1-8-1 Inohana, Chuo-ku, Chiba 260-8673, Japan
Molecular Chirality Research Center, Chiba University, 1-33 Yayoi-cho, Inage-ku, Chiba 263-8522, Japan

Yoko Kusuya, Daisuke Hagiwara, Azusa Takahashi-Nakaguchi, Kanae Sakai and Tohru Gonoi
Medical Mycology Research Center, Chiba University, 1-8-1 Inohana, Chuo-ku, Chiba 260-8673, Japan

Chao-Nan Fu
Key Laboratory for Plant Diversity and Biogeography of East Asia, Kunming Institute of Botany, Chinese Academy of Sciences, Kunming 650201, China
University of Chinese Academy of Sciences, Beijing 100049, China

Hong-Tao Li, Ting Zhang, Peng-Fei Ma and Jing Yang
Germplasm Bank of Wild Species in Southwest China, Kunming Institute of Botany, Chinese Academy of Sciences, Kunming 650201, China

Richard Milne
Institute of Molecular Plant Sciences, University of Edinburgh, King's Buildings, Edinburgh, Scotland EH9 JH, UK

De-Zhu Li
Key Laboratory for Plant Diversity and Biogeography of East Asia, Kunming Institute of Botany, Chinese Academy of Sciences, Kunming 650201, China
University of Chinese Academy of Sciences, Beijing 100049, China
Germplasm Bank of Wild Species in Southwest China, Kunming Institute of Botany, Chinese Academy of Sciences, Kunming 650201, China

Lian-Ming Gao
Key Laboratory for Plant Diversity and Biogeography of East Asia, Kunming Institute of Botany, Chinese Academy of Sciences, Kunming 650201, China

Shaoxuan Li, Xiling Fu, Wei Xiao, Ling Li, Ming Chen, Mingyue Sun, Dongmei Li and Dongsheng Gao
College of Horticulture Science and Engineering, Shandong Agricultural University, Tai'an 271018, People's Republic of China
State Key Laboratory of Crop Biology, Shandong Agricultural University, Tai'an 271018, People's Republic of China

Zhanru Shao
Key Laboratory of Experimental Marine Biology, Institute of Oceanology, Chinese Academy of Sciences, Qingdao 266071, People's Republic of China
Laboratory for Marine Biology and Biotechnology, Qingdao National Laboratory for Marine Science and Technology, Qingdao 266237, People's Republic of China

Chao Tong
Key Laboratory of Adaptation and Evolution of Plateau Biota, Qinghai Key Laboratory of Animal Ecological Genomics, Laboratory of Plateau Fish Evolutionary and Functional Genomics, Northwest Institute of Plateau Biology, Chinese Academy of Sciences, Xining 810001, China
University of Chinese Academy of Sciences, Beijing 100049, China
Department of Biology, University of Pennsylvania, Philadelphia, PA 19104-6018, USA

Fei Tian and Kai Zhao
Key Laboratory of Adaptation and Evolution of Plateau Biota, Qinghai Key Laboratory of Animal Ecological Genomics, Laboratory of Plateau Fish Evolutionary and Functional Genomics, Northwest Institute of Plateau Biology, Chinese Academy of Sciences, Xining 810001, China

Kyle Duyck, Limei Ma and Ariel Paulson
Stowers Institute for Medical Research, 1000 East 50th Street, Kansas City, MO 64110, USA

Vasha DuTell
Stowers Institute for Medical Research, 1000 East 50th Street, Kansas City, MO 64110, USA
Redwood Center for Theoretical Neuroscience, University of California, 567 Evans Hall, Berkeley 94720, USA

C. Ron Yu
Stowers Institute for Medical Research, 1000 East 50th Street, Kansas City, MO 64110, USA
Department of Anatomy and Cell Biology, University of Kansas Medical Center, 3901 Rainbow Boulevard, Kansas City, KS 66160, USA

Ernur Saka and Eric C. Rouchka
Department of Computer Engineering and Computer Science, University of Louisville, Louisville, KY, USA.

Benjamin J. Harrison
Department of Anatomical Sciences and Neurobiology, School of Medicine University of Louisville, Louisville, KY, USA.
Department of Biological Sciences, University of New England, Biddeford, ME, USA

Kirk West
Department of Biochemistry and Molecular Biology University of Arkansas for Medical Science, Little Rock, AR, USA.

Jeffrey C. Petruska
Department of Anatomical Sciences and Neurobiology, School of Medicine University of Louisville, Louisville, KY, USA

Chandran Nithin and Ranjit Prasad Bahadur
Computational Structural Biology Lab, Department of Biotechnology, Indian Institute of Technology Kharagpur, Kharagpur, India

Amal Thomas
Computational Structural Biology Lab, Department of Biotechnology, Indian Institute of Technology Kharagpur, Kharagpur, India

Jolly Basak
Department of Biotechnology, Visva-Bharati, Santiniketan, India

Linlin Xing, Xiaoyan Liu and Chunyu Wang
School of Computer Science and Technology, Harbin Institute of Technology, Harbin, China

Maozu Guo
School of Electrical and Information Engineering, Beijing University of Civil Engineering and Architecture, Beijing, China

Lei Wang and Yin Zhang
Institute of Health Service and Medical Information, Academy of Military Medical Sciences, Beijing, China

Hong Zhou, Joseph Manthey and Ekaterina Lioutikova
Department of Mathematical Science, University of Saint Joseph, 1678 Asylum Avenue, West Hartford, CT 06117, USA

Michael Zhou and Daisy Li
Hall High School, 975 N Main Street, West Hartford, CT 06117, USA

Hong Wang
Susan L. Cullman Laboratory for Cancer Research, Department of Chemical Biology and Centre for Cancer Prevention Research, Ernest Mario School of Pharmacy, Rutgers, The State University of New Jersey, 164 Frelinghuysen Road, Piscataway, NJ 08854, USA

Xiao Zeng
PBSG, LLC, P. O. Box 771, Braddock Heights, MD 21714, USA

Index

www.ingramcontent.com/pod-product-compliance
Lightning Source LLC
Chambersburg PA
CBHW082037190326
41458CB00010B/3395